Rice Production: Emerging Trends and Technology

Rice Production: Emerging Trends and Technology

Edited by Radley Harwood

SYRAWOOD
PUBLISHING HOUSE

New York

Published by Syrawood Publishing House,
750 Third Avenue, 9th Floor,
New York, NY 10017, USA
www.syrawoodpublishinghouse.com

Rice Production: Emerging Trends and Technology
Edited by Radley Harwood

© 2019 Syrawood Publishing House

International Standard Book Number: 978-1-68286-673-3 (Hardback)

Cataloging-in-Publication Data

Rice production : emerging trends and technology / edited by Radley Harwood.
    p. cm.
Includes bibliographical references and index.
ISBN 978-1-68286-673-3
1. Rice farming. 2. Rice--Planting. 3. Rice. 4. Agricultural innovations. I. Harwood, Radley.
SB191.R5 R53 2019
633.18--dc23

TABLE OF CONTENTS

PREFACE

This book has been an outcome of determined endeavour from a group of educationists in the field. The primary objective was to involve a broad spectrum of professionals from diverse cultural background involved in the field for developing new researches. The book not only targets students but also scholars pursuing higher research for further enhancement of the theoretical and practical applications of the subject.

Rice is one of the most widely grown crops in the world. A greater part of Asia, Africa and parts of Europe and Latin America depend on rice as a staple diet. It is also important from an economic perspective as it is considered a nutritional and affordable food in low and middle-income countries in the world. Progress in this field has been achieved by innovations in biotechnology, genetics, GPS, drone technology, and image sensing which are used to map and analyze farm lands, soil quality, agricultural practices, irrigation, drought management practices, among many others. This book discusses modern rice growing methodologies and practices. It further elucidates new techniques and their applications in rice production. Students and researchers who want to stay abreast of the latest innovations and studies in this field will greatly benefit from this book.

It was an honour to edit such a profound book and also a challenging task to compile and examine all the relevant data for accuracy and originality. I wish to acknowledge the efforts of the contributors for submitting such brilliant and diverse chapters in the field and for endlessly working for the completion of the book. Last, but not the least; I thank my family for being a constant source of support in all my research endeavours.

Editor

Domestication and Long-Distance Dissemination of Rice: A Revised Version

Hiroshi Ikehashi*

Ex. Prof. Kyoto University, Fujisawa city, Japan

Abstract

Historically, in 1930s two sub-species, *indica* and *japonica*, were proposed on the basis of sterility in F1 hybrids between them. Subsequently, the two types were classified by associations of a number of genetically independent traits. The characteristic associations of traits have been explained by a series of hybrid sterility or reproductive barriers, in which a duplicate recessive gene model was proposed to cause varietal differentiation. In 1990s, most of hybrid sterility genes were proved to be caused by an interaction of allelic genes at each locus. In a well-known case of hybrid sterility, the *indica*, *japonica*, and wide-compatibility type which gives fertile hybrids in its cross to *indica* and *japonica* type, are proved to contain an allele, S5-i, S5-j and S5-n, respectively at a locus on Chromosome 6. Those gametes having S5-j allele is found to be partially aborted in the hybrid genotypes of S5-i/ S5-j while no gamete abortion occurred in S5-i/S5-n and S5-j/S5-n genotype. Since then, the S5-n has been used in hybrid rice breeding to obtain fertile and vigorous hybrids between *indica* and *japonica* types. For the last decade the action of S5-n is found to be due to a deletion in its gene sequence, while functional enzymes are produced by other alleles. Thus, the earlier assumed role of hybrid sterility genes as to lead to varietal differentiation are no more supported. Today, the characteristic association of traits found in each of varietal groups is better explained by founder effects. Recently , genetic analyses have been advanced for the contrasting traits between *indica* and *japonica* group. For an instance, genetic bases for long slender grain of *indica* and short wide grain of *japonica* are analyzed, and the short wide type is found to be caused by a deletion in its gene sequence. Thus, the short wide grain is considered to be mutational origin during domestication. On the other hand, a large number of native cultivars of rice were surveyed with enzyme polymorphism in 1980s, and later with molecular markers. As the results, profound genetic diversity is found in the genetic background of rice and wild rice. As the results, some scientist proposed the idea of multiple independent domestications of rice. However, before reaching to such a conclusion, at least two factors, i.e., long-distance-dissemination of some genotypes and possibilities of introgression by local wild rice to primitive cultivars need to be examined. Taking the two factors as well as historical events into consideration, the author considered that perennial *japonica* cultivars were first domesticated in Changjiang river basin, and disseminated to East India through Assam or along Bengal Bay, meanwhile their genetic background were transformed under introgressions of local wild rice which resulted to form the *indica* types. Thus–formed *indica* types seem to be disseminated to Southeast Asia, for an instance to Champa, under the expansion of Hinduism. Later some of the genotypes were introduced into China. Recently, this explanation has been verified by extensive analyses of gene sequence for wild and cultivated rice.

Keywords: Hybrid sterility; Reproductive barriers; Varietal differentiation; Genetic diversity; Geographical distribution; Founder effect; Transgression of wild rice; Domestication

Introduction

The center of domestication of rice has been assumed to be in tropical or subtropical regions. *Indica* type (Hsien) has been considered to be an ancestral type from which *japonica* type (Keng) and *javanica* types or tropical *japonica* types were differentiated [1]. But the earlier view on the origin of cultivated rice has to be changed by a number of reasons. First, archeological findings of the oldest rice cultivation provided evidence that rice must have been domesticated in the mid and lower Changjiang river basin. Second, wild rice populations have been discovered in the same region by extensive surveys in 1980s [2]. Third, a large number of local cultivars have been surveyed by enzyme polymorphism since 1980s, and recently by molecular markers. As the results a profound genetic diversity has been found among different groups. The new results suggest the process of domestication might have been much complex than that conceived earlier.

Today a re-construction of earlier concepts seems to be necessary to understand a consistent picture of distribution and domestication of cultivated rice. Initially, the author published an extensive review in other paper [3]. Since then, there are some important research results. Therefore, while avoiding details of the early version, here above mentioned three points are to be detailed in the light of recent progress.

Earlier Concepts for *Indica-Japonica* Differentiation in Rice

It has long been recognized that there are two types of rice,

'Hsien' in the South and 'Keng' in the temperate or northern region in China. The two types were first proposed as two sub-species, namely *indica* and *japonica*, by Kato et al. [4], who found varying degrees of spikelet sterility in F1 hybrids between the two types. By the time Kato distinguished the two types on the basis of hybrid sterility, a general understanding had been formulated that the sterility of inter-specific hybrids is caused by failure of chromosome paring at the reductive cell division in F1 hybrids due to structural differences between parental chromosomes. Following that idea some scientist interpreted the level of hybrid sterility as a criterion to see the degree of differentiation at the chromosomal level, and conceived the hybrid sterility as a measure of taxonomic distances.

After the initial taxonomical works, the *indica* and *japonica* groups were classified by associations of a number of characteristics within each group [5]. For an instance, long grain and less sticky rice after cooking are commonly found in *indica* types, while short grain and sticky rice are in *japonica* types.

***Corresponding author:** Hiroshi Ikehashi, Ex. Prof. Kyoto University, Kataseyama 3-10-6, Fujisawa city, Japan, 251-0033
E-mail: hiro-i@feel.ocn.ne.jp

Terminology for *Javanica* or Tropical *Japonica* type.

In addition to *indica* and *japonica* types, another term 'javanica type' or tropical *japonica* type has been introduced. The term of *javanica* was first proposed by Morinaga [6] on the basis of an extensive survey of Asian rice cultivars in Japan by Matsuo [7] who first classified them into three groups, A, B and C, of which a distinct varietal group B was considered to have originated in Java while A and C were considered to be *japonica* and *indica* type, respectively. Oka and Chang [8] further classified the *japonica* into two subdivisions, namely, tropical *japonica* type and temperate *japonica* type, and these terms have been popularly used.

Explanation of the *indica-japonica* differentiation

Oka proposed an idea that the observed characteristic associations of various traits in *indica* and *japonica* type may have arisen by a mechanism of reproductive barrier which reveals partial hybrid sterility [9]. He emphasized the importance of partial sterility in F1 hybrids between *indica-japonica* type, and attributed its genetic basis to a set of duplicate gametophytic sterility genes. Here, the detail of his explanation is not repeated, as it was given elsewhere [9,11], but his conclusion can be cited as follows. Oka considered there are many pairs of such 'duplicate recessive lethal genes' in the rice genome (Figure 1). Thus, the characteristic associations of genetic traits in *indica* or *japonica* type were explained on the basis of genetic mechanism for reproductive barriers. Further, the differentiation of rice into *indica* and *japonica* groups was explained on the basis of such reproductive barriers, which are interpreted as an inner genetic mechanism to lead the differentiation.

Despite the lack of firm experimental evidence Oka's genetic model [11] has long been accepted, because it explained the hybrid sterility and characteristic association of traits among different groups of rice as well as their differentiations. Oka himself has repeated his idea until he passed away. He summarized the idea as follows: 'In hybrids, genes tend to be associated in a certain manner across independent loci', and that 'this can be partly elucidated by the presence of many

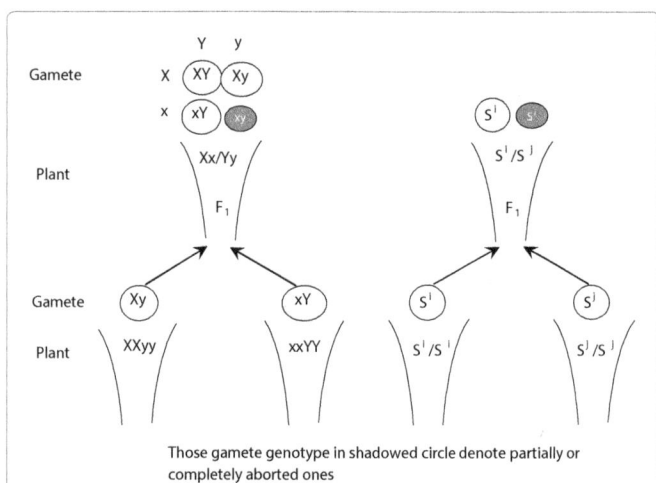

Those gamete genotype in shadowed circle denote partially or completely aborted ones

Figure 1: Genetic Model for Hybrid Sterility. Two locus model (left) and one locus model (right). In the two locus model, two isolated genotypes contain the gametophytic lethal gene genotype of *XXyy* and *xxYY*, respectively. The genotype of hybrid between them should be *XxYy*, from which four gamete genotypes will be produced after the meiosis, i.e., *XY*, *Xy*, *xY* and *xy*. Gametes with the two recessive genes (*xy*) will be aborted on the F₁ hybrid before fertilization, and lead to partial sterility of panicle or pollen in anther. In the one locus model, those gametes having *S-j* gene are partially aborted in the hybrid genotypes *S-i/ S-j*.

sets of duplicate or complementary genes for gametophytic and sporophytic sterilities, and seems to suggest an internal mechanism of genetic differentiation, which could be a complementary system of adaptive gene blocks' [12].

A New Understanding of Hybrid Sterility Genes, and their Origin and Function

The present author did have a chance to test Oka's assumption in early 1980s. After identifying some varieties which showed 'no hybrid sterility in their crosses to *indica* and *japonica* testers, the author named this character as 'Wide Compatibility (WC)' for convenience, and analyzed its genetic nature.

Hybrid sterility by allelic interaction at a locus

It was found that *indica*, *japonica* and the WC type contains single allele, respectively, S5-i, S5-j and S5-n at a locus on the short arm of Chromosome 6. And those female gametes having S5-j gene are partially aborted in the hybrid genotypes S5-i/ S5-j as shown in the right side of Figure 1, while no gamete abortion occurred in S5-i/S5-n and S5-j/S5-n genotype [13,14].

After testing of about one thousand varieties in China, the hybrid sterility in Chinese cultivars was found to be caused mostly by the allelic interaction at S5 locus. A few varieties showed hybrid sterility in their crosses to Wide Compatibility Varieties (WCV) [15]. In a wide range of varietal testing a few varieties from Indian Subcontinent also exhibited hybrid sterility when crossed to WCVs. Then, further genetic analyses of hybrid sterility gene loci (HSGLi) were conducted following the identification of S5 locus. As the results, some more loci following the same genetic model were identified on different chromosomes in other hybrids [16]. Studies to identify HSGLi have been continued until now [17].

Application of wide-compatibility gene to hybrid rice breeding.

As soon as the simple monogenic nature of hybrid sterility was understood, it was applied to hybrid rice breeding to enhance the level of hybrid vigor. Since early 1990s the S5-n has been incorporated to various rice cultivars to obtain fertile hybrids in hybrid rice breeding, because the hybrid sterility was overcome by the incorporation of S5-n allele into *indica* or *japonica* type. In practice, it has been found that the hybrids between *indica* type and *javanica* type are more productive in many areas than those between *indica* and *japonica* types.

Initially, the hybrid sterility alleles were confirmed by female gamete abortion in embryo sac, as it is easily detectable by partial sterility in panicles, but the same genetic mechanism for male game abortion has been described. When *indica -japonica* hybrids were first released in China, some of the hybrids were extremely sensitive to cold temperature due to a high level of pollen sterility, which was found to be caused by the hybrid sterility, as it is for female gametes. Accordingly, a set of neutral alleles for such a hybrid sterility in male gametes were studied and applied to obtain a sound pool of pollen in inter-subspecific hybrids [18,19].

Molecular analyses of hybrid sterility genes

The role of hybrid sterility genes at molecular levels seems to be diverse. They constitute a part of genes for normal gamete formation, and are detected by a partial gamete abortion in *indica-japonica* hybrid due to an interaction between different types of protein. The S5 region has been mapped [20] and covers up to five open reading frames (ORF1 to ORF5). Transformation studies of ORF3 to ORF5 from an *indica* variety into a *japonica* variety showed reduced fertility, due to embryo-sac abortion, for transformants harboring *indica* ORF5, whereas the fertility of transformants of ORF3 and ORF4 was not affected. The *indica* and *japonica* alleles of ORF5, which encodes an aspartic protease, differ by two nucleotides, whereas the wide compatibility allele has a large deletion in the N terminus of the predicted protein, causing subcellular mislocalization of the protein.

It is noteworthy that a case of mutational change of a hybrid sterility allele was found in an experimental line, 02428 from China which possesses the S5-n allele. The parents for 02428, Pangxiegu and Jibangdao, were proved to possess S5-j. The neutral allele S5-n in 02428 was considered to be induced from S5-j by irradiation of 60Co to the parents [21].

Initially the author considered that the Wide Compatibility Genotype (WC) is an ancestral type and the *indica* or *japonica* type is produced by defects in their gene sequence. It was assumed that such a neutral allele as S5-n should be original allele having been conserved in an open-pollinating population while other alleles like S5-i and S5-j might have been lost in an open pollinating population due to its disadvantage of gamete abortion. But, since the WC is produced by a deletion in the gene sequence, the author's initial assumption cannot be supported.

After molecular analyses of some other hybrid sterility genes, it was found that the mechanism to reveal gamete abortion is much diverse and involves complex interactions of proteins which are related to a pathway to programed cell death.

Even after the analyses of hybrid sterility genes at molecular level, a group of scientists are attempting to explain the differentiation between *indica* and *japonica* rice by the action of hybrid sterility genes as a major source of genetic diversity in the rice gene pool. The author here would not detail their exhaustive studies. As discussed later in this paper the origin of *indica-japonica* differentiation can be explained by other way of consideration.

Founder effect for the association of traits.

As explained above the hybrid sterility in rice has long been assumed to be major reproductive barriers, which have caused isolation of varietal groups and led the crop into differentiation. Oka, as explained above, ascribed the association of characters in respective varietal group to the role of duplicate recessive genes. There are still a few idea to relate such reproductive barriers to the varietal differentiation, but the characteristic associations of traits in variety groups can be easily understood by selection and propagation of a few source genotypes or a founder to a wide area or distant regions. The domestication of rice is essentially propagations of such genotypes as found better than others, and multiplication of them by farmers via seed or vegetative stocks. If a set of traits are possessed by a pool of ancestor plants, the set of traits will be largely conserved by its progeny plants retaining the characteristic combination of the traits (Figure 2). Thus, the founder effect seems to be a very important mechanism to form a varietal group in domestication or dissemination. There is no room for any sterility gene to contribute to the association of characters.

Distribution of Rice Genotypes Examined by Enzyme Polymorphism and Molecular Markers

So far the relatively clear separation of *indica* type and *japonica* type has been observed in mainland China and its vicinity. By surveying isozyme variation for a large number of cultivated varieties Glaszmann

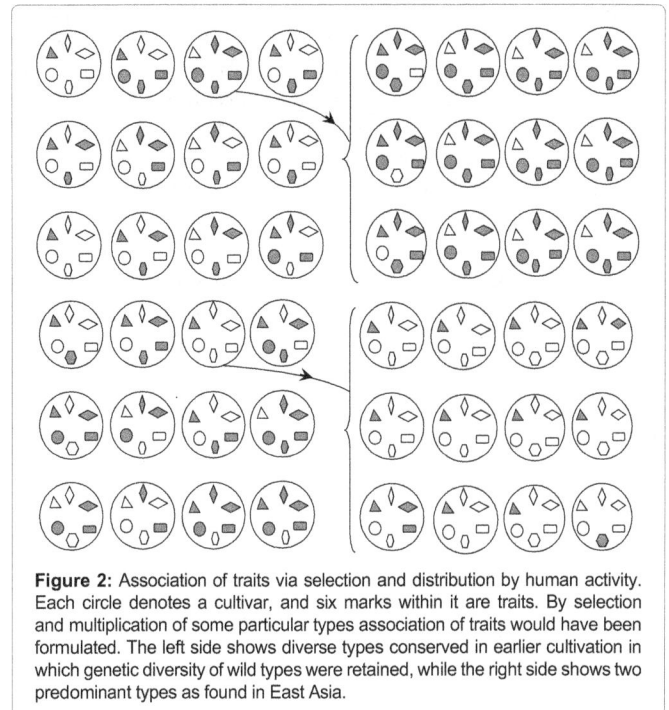

Figure 2: Association of traits via selection and distribution by human activity. Each circle denotes a cultivar, and six marks within it are traits. By selection and multiplication of some particular types association of traits would have been formulated. The left side shows diverse types conserved in earlier cultivation in which genetic diversity of wild types were retained, while the right side shows two predominant types as found in East Asia.

[22] has shown that there are predominantly *indica* type and *japonica* type in the East Asia while there are a set of intermediate types from Burma to the Indian subcontinent. Having observed the contrasting pictures of distribution of rice genotypes, i.e., the region with two major types and the region from Myanmar to Indian subcontinent with various intermediate types, he indicated that there are two alternative gene pools, one in the East Asia and another in Indian subcontinent. It is noteworthy that the similar tendency has been observed in the picture of variation for the hybrid sterility loci.

Indica-Japonica differentiation found with the use of molecular markers

It is interesting to see whether above-mentioned pictures of differentiation of rice are shown or not with molecular markers since the late of 1990s. One approach to describe genetic diversity of varietal group is to compare differential sequences in the vicinity of known genes. Londo et al. [23] studied three gene regions of 203 cultivars of O. sativa and 129 populations of O. rufipogon. They classified such haplotypes as A for wild rice in Thailand and India, B and C for *japonica*s and *javanica*s, of which B seemed to closely correspond to *javanica*s. The haplotypes D and E are found in *indica* rice and in some wild rice in india and indochina, of which haplotype D are found in india and Malaysia suggesting the domestication of Aus varieties in India and Malaysia. They concluded that the two major haplotypes of cultivated rice i.e., *japonica* and *indica* were domesticated independently from wild rice gene pools from southern China and India/Indochina, respectively. In a similar approach Zhu and Ge [24] analyzed introns of four nuclear single copy genes on different chromosomes for 37 accessions representing two cultivated and six wild species. Two subspecies of *Oryza sativa* ssp. *indica* and ssp. *japonica* formed two separate monophyletic groups, suggestive of their polyphyletic origin. It is reasonable that the pictures obtained earlier by Glaszmann with isozymes and those by molecular markers seem to be similar, because the materials in the test were derived from the similar source.

As an another approach to measure degrees of divergence, LTR(long terminal repeat) transposable elements have been used to see the divergence of varietal groups, because at the time of insertion the two LTRs of a given element were identical in the sequence but would have differed due to accumulated mutations. The extent of divergence may be proportional to the time elapsed since the insertion. Ma and Bennetzen [25] surveyed the divergence of genomes of *indica* and *japonica* types. The genome sizes of both *indica* and *japonica* have substantially increased since their divergence from a common ancestor mainly because of amplification of LTR-transposons. The sequences of genes were observed to have a very high rate of divergence. By comparing the genomic divergence with that observed for *Oryza galaberrima* as a reference genome, the divergence of *Indica* and *Japonica* is considered to date back long before the domestication of these two types. Vitte et al. [26] compared 110 LTR retrotansposons in the published DNA genome sequence of Nipponbare and 93-11 (an *Indica* type), and found that the two types diverged from one another at least 200,000years ago, older than the date of domestication. They also applied the insertion polymorphism to a wide range of traditional rice varieties of both *indica* and *japonica* types, and found that they arose from two independent domestication events in Asia.

Genetic diversity of wild rice as the source for heterogeneity of cultivated rice

Even within a few sample populations of wild rice, it is shown that there are diversified 'allelic differences' far beyond the range of *Indica* or *Japonica* type. Genetic diversity of 16 vegetative strains of wild rice (*O. rufipogon*) from three regions of Myanmar was evaluated by a genome-wide survey with SSR (single sequence repeat) markers [27]. Allelic diversity among the wild rice genome was evaluated by primer pairs each of which was assigned to 74 loci over 12 chromosomes of the genome. The loci in the wild rice genome revealed a large number of specific alleles in much wider ranges than those detected in six cultivars of O. sativa. Around 50-60 percent of alleles are specific to wild rice, while those identical to *indica* or *japonica* alleles, or common to both the cultivated and wild rice are only 10-15 percent of the total allelic diversity (Figure 3). It is noteworthy that the vegetatively propagated clones of wild rice contain such diversity.

The observation of diversity of wild rice population may lead us to an idea that only a part of their genetic diversity have been taken in the

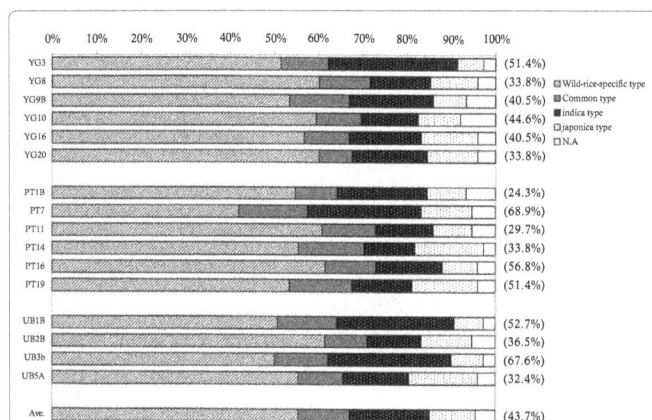

course of domestication of rice and that the domestication itself may not be the reason for the diversification of rice.

Domestication of rice in the light of archaeological evidence

The genetic divergence of *indica* and *japonica* types has been supported by isozymes, RFLP, transposon insertions. All of these studies showed a clear genetic differentiation between the two varietal groups, and seem to lead to a set of independent domestication of the two types. The diversity and distribution of existing native cultivars are often cited as the evidence for multiple domestications. But, that would be correct only if there were no long-distance-disseminations of genotypes by human activity and no introgression of local wild rice into primitive cultivars. To re-examine the ideas of single or double domestication hypothesis for rice we may have to see some historical and cultural aspects.

Two key findings in China

Since 1970s archaeological evidences have indicated that rice must have been domesticated in the mid or lower Changjiang river basin. Following the discoveries, by an extensive survey [2], a wild rice population is found at a shallow swamp in the same region at northern latitude of 35°in Jiangxi province. Another wild rice population is found in "t'u ling" county in the south west of Hunan province, China. In the two sites perennial populations of rice are regenerating via vegetative stems every summer after freezing winter cold.

Genetics of some contrasting traits in rice

Perennial versus annual habit of growth: The cultivated rice in the temperate zone is potentially perennial. It is seeded in spring and harvested in autumn, but keeps an ability to regenerate its stem and flower again in next season when climatic conditions allow it. In this sense the cultivated rice in temperate zones is direct descendants from perennial wild rice in Changjiang river basin. The perennial habit has been inscribed through the process of domestication, as many groups of primitive rice must have been propagated in the time of domestication by dividing and transplanting of sprouting stems in early summer in shallow marsh, which was nothing but the origin of transplanting cultivation of rice [28]. In that process, early types with bigger panicles and weak seed dormancy might have been selected and converted into 'cultivated types'. Contrastingly, rice genotypes in Indian subcontinent and improved *Indica* cultivars are predominantly annual. According to a QTL analysis of *indica-japonica* hybrids, the annual type or those with earlier senescence are genetically recessive [29]. Therefore, the annual type may have arisen from the perennial type through a set of recessive mutations.

Short versus long grain and the mutation origin of short type: The long slender (LS) versus short wide(SW) grain have long been cited as one of most contrasting characters between *indica* and *japonica* type. The author has searched for as many reports as possible to see any description of wild rice with SW grain, but never encountered such a report. Then, I came to consider that the SW grain might have been originated from mutations after domestication. In our earlier study [30] the quantitative locus (QTL) for the contrasting trait was analyzed by comparing two back-crossed populations, i.e., IR36/NK2//NK2 and IR36/NK2//IR36, in which NK is a *japonica* type with SW grain and IR36 is an *indica* type with LS grain (Figure 4A and 4B). In the result, the most effective QTL was confirmed only in the population of IR36/NK2//NK2 (Figure 4), and the difference of grain width was not detected in IR36/NK2//IR36. The result indicated that SW grain

Figure 3: Evaluation of genetic diversity of 16 strains of wild rice (*Oryza rufipogon* Griff.)collected in three regions, Myanmar using using simple sequence repeats (SSRs).

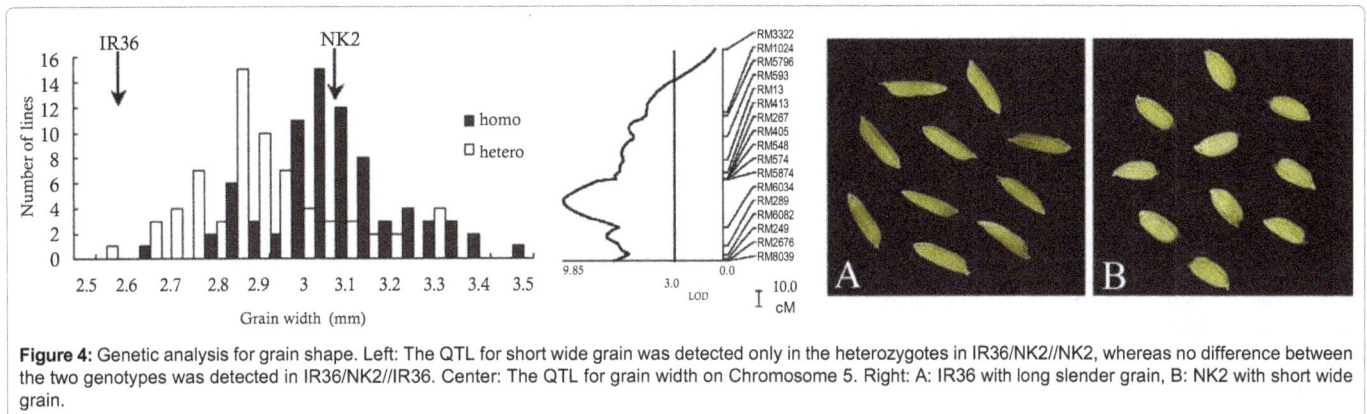

Figure 4: Genetic analysis for grain shape. Left: The QTL for short wide grain was detected only in the heterozygotes in IR36/NK2//NK2, whereas no difference between the two genotypes was detected in IR36/NK2//IR36. Center: The QTL for grain width on Chromosome 5. Right: A: IR36 with long slender grain, B: NK2 with short wide grain.

of *japonica* type is genetically recessive and revealed in the recessive homozygotes (Figure 4 (left)). Later a group of scientists analyzed the gene sequence of this QTL, and found that the SW type is caused by a deletion in the functional nucleotide polymorphisms [31]. The change of grain shape is related to a level of yield increase [31]. They considered that grain shape might have been selected as a favorable trait by ancient humans during rice domestication. But the selection toward SW grain type may not have been easy for ancient farmers, because so far all the grain samples excavated from archeological study in China are more or less a mixture of LS and SW type as indicated later in this review.

Differentiated farming culture in India from East Asia

To discuss the possibility of independent domestication of rice in India, first, we have to consider the role and characteristics of rice cultivation in India and East Asia. In East Asia we are able to see much diversified and elaborated use of rice products including sticky or glutinous rice, rice wine, and etc., while rice is less important and its use is often similar to other cereals in India. If rice had been originally domesticated in India, there should have been unique or original usage of rice and its byproducts. When we see a poor rice-based cultural complex in india, rice may have been brought there from Assam or Yunnan province of China.

Second, it should be considered that the present style of farming in India was introduced by the people who had migrated to India with cattle, cattle-drawn carts and upland cultivation of cereal crops. For them perhaps there was no need to domesticate any indigenous wild plants like wild rice in marshes.

Rice in tuber crop-based farming in the south-east Asia

As the plains in the Indian subcontinent are today dominated by the cattle-drawn farming, it is difficult to see any primitive cultivation of cereal crops before their migration to India. Only through the life styles of minority groups in hilly areas to which cattle-drawn carts did not reach, it is possible to see some sorts of traditional cultivation that were inherited by native people and common from Shan states in Burma to the western end of Assam highland. There are dominant remains of traditional tuber crops in the eastern parts of India, like banana, sugar cane and turmeric, which had been domesticated in Southeast Asia under a root-crop-based agricultural system. As discussed by Sauer [32], rice might have been also domesticated through vegetative propagation in the same system and been a component in such an earliest agriculture, whereas the present cereal agriculture in India could have been brought by a series of migration from elsewhere not earlier than BC 2000 or so.

Thus, rice must have been introduced into India by minority peoples from initially domesticated areas through Assam highland or the coastal zone along Bengal Bay. The independent domestication of rice in Indian subcontinent seems to be unlikely. Whereas, the very ancient remains of rice cultivation are found in the mid and lower basin of Changjiang river, where rice was primarily domesticated.

How did the Cultivated Rice Acquired their Genetic Diversity

There are a large number of traits which are characteristic to so-called *indica* and *japonica* type, out of which some can be attributed to peoples' preference or to an agronomic importance. The *indica-japonica* difference is not only relating to practically important traits but also covers the entire genome, as detected by molecular markers. How was the difference in the entire genetic background introduced or retained into cultivated rice ?

To answer this question, we have to think of interference by wild rice populations to cultivated rice in its course of dissemination. In that course perhaps major qualities essential to cultivation must have been conserved, but wild rice in the way of dissemination of cultivated types must have contributed to re-building or modification of genetic background of cultivated types through occasional hybridizations or introgressive pollination. Even today, it is not rare that wild rice populations as weed overwhelm the cultivated rice in Assam region of India (personal communication in Assam).

Implication of nuclear-substituted types in Assam

It is interesting that a substantial number of cultivated rice in Assam region are found to keep cytoplasma of *japonica* type while keeping a set of nuclear genes of *indica* type. According to Kaneda et al. [33] out of 51 native cultivars from Assam twelve showed nuclear genes of *Indica* type under the cytoplasm of *Japonica* type. The fact that a substantial number of rice cultivars in Assam region contained *japonica* type cytoplasm while showing a genetic background of *indica* type may imply a process of introgressive hybridization, which can be explained as follows.

The *japonica* type of rice has perennial habit, which could have been vegetatively propagated by transplanting in primitive cultivation in Changjiang river basin in China [3]. That type may have been introduced as a component of tuber crop-based-farming from its original region into Indian subcontinent through Assam highland as explained before. The *Japonica* type may have been pollinated by wild rice around farms and thus-formed seed was dropped around the mother plant. The progeny of such plants may have been selected to restore a set of characters for cultivated rice for some generations, but repeatedly been pollinated by surrounding wild rice plants which contain a different genetic background that is today recognized as the background of *indica* type. Such pollination by wild rice could have been repeated with a long interval during which a set of wild rice-affected traits would be selected out, thus, the cultivated rice must have been converted to *indica* likes in its genetic background while maintaining cytoplasm of *Japonica* rice together with basic characters as cultivated rice.

Introgression by wild rice as a differential genetic background of cultivated types

The nuclear-substituted types may be an extreme case of introgressive transformation, but suggest a frequent introgressive pollination by surrounding wild types which contain a vast range of genetic diversity. Such an introgressive pollinations may explain a reasonable origin of cultivated rice with the diversity of *indica* type. As the cultivated rice was hardly isolated from wild rice except for the area where no wild rice existed, it is rather unusual to conceive that the cultivated rice was expanded into various parts of Asia without any modification by local wild rice populations. The initially domesticated types in China could have been reformed by such an extensive transgression by local wild rice populations in its dissemination through Assam highland or along Bengal Bay to East India, where the transformed rice may have constituted the secondary center for divergence under the influence of traditional Indian way of cereal cultivation. To confirm this hypothesis further it is necessary to apply molecular markers to existing local cultivars and wild rice. So far the possibilities of transgression by wild rice do not seem to be carefully examined. Only recently, however, this aspect has been surveyed as introduced later in this review.

Distribution of *indica* and *japonica* Type under Human Activity

In the discussion of *indica-japonica* differentiation there are two aspects, i.e., one is differentiated genetic background which can be found through isozyme polymorphism or on the basis of molecular markers, and another is a series of agronomic or practically important traits like grain shape and cooking quality. The difference in terms of genetic background is shown today clearly with molecular markers, and understanding of this aspect is leading to the idea of independent origin of *indica* and *japonica* types. But the idea of the independent domestication of rice outside of the Changjiang river basin is not consistent with the historical and cultural background as discussed above. Thus, the author attempted to ascribe the genetic diversity of varietal groups to the introgressive pollination of wild rice in the course of dissemination.

Long-distance-dissemination of long grain types in China

The long grain types (hsien) with less sticky quality have been prevalent in southern region from Yangze river basin, while the short grain types (keng) in the northern regions from Huai Ha river basin. The two types were believed to have differentiated on an unknown reason and distributed in the South and North, respectively.

When we see ancient rice grains excavated in archaeological sites, it is interesting that grain samples excavated in Guangxi province in the southern border of China are round-shaped together with some intermediate type, which is dated back to 2,000 BC. Also, many grain samples from ancient remains are a mixture of long slender and short wide shaped together with some intermediate type, from the lower Changjiang river basin to Zhejiang and Hunan Province (Figure 5).

Therefore, it is rather reasonable to assume that the grain shape has been artificially selected toward a short round type from the original long type of wild rice under the preference to the round shape, which tends to give a higher milling recovery for cooking in granular shape. Those excavated from ancient archeological sites must have been domesticated in China and constituted the keng or *japonica* groups of rice.

On the other hand, the existence of long grain types with a genetic background of *indica* types in the southern region in China can reasonably be explained, if we assume that they might have been introduced from somewhere into southern region of China. Here, our attention should be called to a series of historical documents. First, in 1012 the Emperor Zhen Zong of Song Dynasty (in North) issued a decree to introduce a large quantity of seed of Champa rice seed from Fujian Province to Jiangsu, Zhejiang, Jiangxi [34]. Champa was in the mid of Vietnam where a Hindu kingdom had once been built. The background of the policy in 1012 was an urgent need to promote rice production in the granary from Changjiang river basin to the southern region, when the Yellow river basin had been threatened by a growing invasion of nomadic peoples from North and a great number of people had to migrate to the South. In 1192, it was recorded that the Champa rice was planted to 80-90% of the lower Changjiang river basin [34]. That type tends to vigorously thrive in a high water temperature than those in the temperate zone, so that presumably the Champa rice would have been more productive than local traditional ones. The prevalence of long grain rice in southern China in Song Dynasty is also noticeable in Japan to which long grain rice had been introduced sometime since the 13-14th century, where traditionally round grain rice had been prevalent.

It is also important to understand that the Champa was one of the kingdoms established under cultural and political influence of Hinduism from East India which could have been a secondary center of rice diversification as explained above. In India a number of agronomic characters seem to be altered. For instance, the shape of rice grain had not been converted into short or round under the preference to cooking by flour. The annual type of rice might have been preferred in an advanced upland cultivation with cattle-drawn plows, in which annual types do not hinder land preparation with surviving stocks. The typical *indica* type of rice as detected in Glaszman's work in East Asia, is composed of a dominant type, which may have been selectively introduced to Southeast Asia out of many intermediate types in the secondary center in India and provided a founder effect.

Another case of long-distance-dissemination

The importance of human activity for varietal characteristic as indicated above can also explain the extension of tropical *japonica* or *javanica* type into hilly areas of Southeast Asia through Indochina peninsular to Java and Philippine archipelago. This type of rice which is characterized by traits of upland rice like as thick stems with deep root might have been extended to the South by rice-cultivating tribes in hilly areas from China before the *indica* type was widely introduced into this region (Figure 5). If both *indica* and *japonica* types had been mixed in the southern China in ancient times, as often understood by archeologists, both types should have been brought to southern regions from the center of domestication. The fact that those similar to *japonica* type had been extended to the South implies that those which were domesticated earlier in the Changjiang river basin were *japonica* type and that the *indica* type would have been introduced later as discussed above. The entire picture of development and distribution of rice genotypes throughout Asia can thus be illustrated by Figure 6.

Figure 5: Carbonated grain samples excavated from one of the earliest archeological sites in t'ien lo shan ' near a famous archeological site, Hemudu, Zhejiang Province, China. The grain samples contain various types of grain including long slender and short wide. The excavated site seemed to be earlier than the era of Hemudu site (ca. BC 5000). Photo was taken by the author via the courtesy of Dr. Zheng Yunfei, Zhejian Provincial Institute of Cultural Relics and Archeology.

Verification of the hybrid origin of *indica* type

The hybrid origin of *indica* type of rice was first proposed as mentioned above by the author. Recently, it is most interesting that such an explanation is verified by an extensive survey of DNA sequence of cultivated and wild rice by Xuehui Huang et al. [35]. They generated genome sequences from 446 geographically diverse accessions of the wild rice species Oryza rufipogon and from 1,083 cultivated *indica* and *japonica* varieties to construct a comprehensive map of rice genome variation. The most persuading approach in their method is the identification of 55 gene sequences for such traits as advantageous in cultivation and assumed to have occurred during domestication. These gene sequences were proved to be commonly derived from a group of *Oryza sativa japonica* rice and conserved among all the cultivated rice. Then, they concluded that the first domesticated is from a specific population of *O. rufipogon* around the middle area of the Pearl River in southern China, and that *Oryza sativa indica* rice was subsequently developed from crosses between *japonica* rice and local wild rice as the initial cultivars spread into South East and South Asia. Their conclusion verifies the author's earlier view shown in Figure 6, although whether or not the initial domestication was from a population in the Pearl River basin seems to be a question, because most of ancient wild rice had been wiped out in China, and is not available for such analyses except for a few.

Conclusion

Two sub-species, *indica* and *japonica*, which are characterized by associations of a number of genetically independent traits, were long assumed to be differentiated through reproductive barriers or hybrid sterility. Since 1980s, the hybrid sterility has been found to be caused mostly by an independent allelic interaction at some loci. At the same time at each locus for the hybrid sterility, a neutral allele which does not cause hybrid sterility in *indica-japonica* hybrids has been found and applied to produce *indica-japonica* hybrid rice. On the other hand, the distributions of rice genotypes have been surveyed with enzyme polymorphism in 1980s and recently with molecular markers. In the light of recent studies, first, the action of each set of hybrid sterility genes are diverse at a molecular level, and not readily associated to the varietal differentiation. The characteristic association of traits in each group of rice can reasonably be explained by founder effects, as some limited types can be widely propagated by human activities.

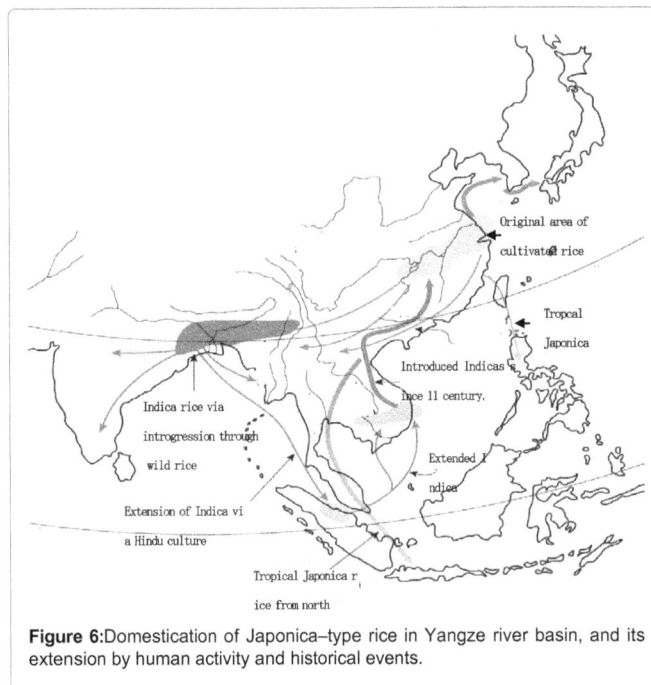

Figure 6: Domestication of Japonica–type rice in Yangze river basin, and its extension by human activity and historical events.

Second, the outstanding contrast of grain shape, i.e., long slender and short wide grain has been genetically analyzed. Then the contrast was found to be caused by a deletion in the original gene sequence for long grain in the course of domestication. Third, considering historical events, the author considered that perennial *japonica* cultivars which is close to wild rice in terms of perennial habit first domesticated in Changjiang river basin were disseminated to East India through Assam or along Bengal Bay. There they were transformed under occasional introgressions of local wild rice and formed a secondary center from which some genotypes seem to be disseminated to colonies under the influence of Hinduism in Southeast Asia. Later some of the genotypes were introduced into China. This is the answer to the question of *indica-japonica* differentiation. Recently, this earlier view is verified by an extensive survey of DNA sequence of wild and cultivated rice of the world.

References

1. Chang TT (1976) The origin, evolution, dissemination and diversification of Asian and African rice. Euphytica. 25: 425-441.

2. Wu MS (1990) Collected papers on wild rice resources. Beijing: China Science and Technology Press.

3. Ikehashi H (2009). Why are there Indica type and Japonica type in rice?— History of the studies and a view for origin of two types.Rice Science 16: 1-13.

4. Kato S, Kosaka H, Maruyama Y and Takiguchi Y (1930) On the affinity of cultivated varieties of rice plants, *Oryza sativa* L. Jour. Dept. Agr. Kyushu Imp. University 2: 241-276.

5. Oka HI(1958) Intervarietal variation and classification of cultivated rice. Indian J. Genet. Pl. Breed 18: 79-89.

6. Morinaga T (1954) Classification of rice varieties on the basis of affinity. In: Report for 5th meeting of the IRC's working party on rice breeding, Tokyo, Ministry of Agr. and Forest, Japan: 1-19.

7. Matsuo T (1952) Genecological studies on cultivated rice. Bull. D, Natl. Inst. Agr. Sci., Japan.1-111

8. OKa HI, Chang WT (1962) Rice varieties intermediate between wild and cultivated forms and the origin of the Japonica type. Bot Bull Acad Sinica 109-131.

9. Oka HI (1964) Consideration on the genetic basis of intervarietal sterility in

Oryza sativa. In: Rice Genetics and Cytogenetics. Amsterdom-London-New York: Elsevier Pub. Co.1964. 158-174.

10. Ikehashi H (2007) The Origin of Flooded Rice Cultivation. Rice Science 14: 161-171.

11. Oka H (1974) Analysis of genes controlling f(1) sterility in rice by the use of isogenic lines. Genetics 77: 521-534.

12. Oka HI (1988) Origin of cultivated rice. Amsterdom-Oxford-New York-Tokyo: Japan Scientific Societies Press, Elsevier Pub. Co.

13. Ikehashi H. Araki H (1986) Genetics of F_1 sterility in remote crosses of rice. *In* "Rice Genetics", Manila, Philippines: IRRI : 119-130.

14. Ikehashi H (1991) Genetics of hybrid sterility in wide hybridization in rice. *In*: Biotechnology in Agriculture and Forestry, Vol. 14. Rice" Y.P.S. Bajaj (ed.), Berlin Heidelberg: Springer-Verlag, 113-127.

15. Wan JM, Ikehashi H (1995) Identification of new locus S-16 causing hybrid sterility in native rice varieties (*Oryza sativa* L.) from Tai-hu Lake region and Yunnan Province, China. Breeding Science, 45: 461-470.

16. Wan J, Yamaguchi Y, Kato H, Ikehashi H (1996) Two new loci for hybrid sterility in cultivated rice (Oryza sativa L.) Theor Appl Genet 92: 183-190.

17. Li D, Chen L, Jiang L, Zhu S, Zhao Z, et al. (2007) Fine mapping of S32(t), a new gene causing hybrid embryo sac sterility in a Chinese landrace rice (Oryza sativa L.)Theor Appl Genet 114: 515-524.

18. Lu CG, Zou JS, Ikehashi H (2004) Developing rice lines possessing neutral alleles at sterility loci to improve the width of compatibility. Plant Breeding, 123: 98-100.

19. Lu CG, Takabatake K, Ikehashi H (1999) Identification of segregation-distortion-neutral alleles to improve pollen fertility of *Indica –Japonica* hybrids in rice. Euphytica113: 101-107.

20. Yang J, Zhao X, Cheng K, Du H, Ouyang Y, et al. (2012) A killer-protector system regulates both hybrid sterility and segregation distortion in rice. Science 337: 1336-1340.

21. Wan JM, Ikehashi H (1996) Evidence for mutational origin of hybrid sterility genes in rice (*Oryza sativa* L.) Breeding Science 46: 169-174.

22. Glaszmann JC (1987) Isozymes and classification of Asian rice varieties. Theor Appl Genet 74: 21-30.

23. Londo JP, Chiang YC, Hung KH, Chiang TY, Schaal BA (2006) Phylogeography of Asian wild rice, Oryza rufipogon, reveals multiple independent domestications of cultivated rice, Oryza sativa. Proc Natl Acad Sci U S A 103: 9578-9583.

24. Zhu Q, Ge S (2005) Phylogenetic relationships among A-genome species of the genus Oryza revealed by intron sequences of four nuclear genes. New Phytol 167: 249–265.

25. Ma J, Bennetzen JL (2004) Rapid recent growth and divergence of rice nuclear genomes. Proc Natl Acad Sci U S A 101: 12404-12410.

26. Vitte C, Ishii T, Lamy F, Brar D, Panaud O (2004) Genomic paleontology provides evidence for two distinct origins of Asian rice (Oryza sativa L.)Mol Genet Genomics 272: 504-511.

27. Shishido R, Kikuchi M, Nomura K, Ikehashi H (2006) Evaluation of genetic diversity of wild rice (*Oryaza rufipogon* Griff) in Myanmar using simple sequence repeats (SSRs). Genetic Resource and Crop Evolution 53: 179-186

28. Ikehashi H (2007) The Origin of Flooded Rice Cultivation. Rice Science 14: 161-171.

29. Abdelkhalik AF, Shishido R, Nomura K, Ikehashi H (2005) QTL-based analysis of leaf senescence in an indica/japonica hybrid in rice (Oryza sativa L.). Theor Appl Genet 110: 1226-1235.

30. Abdelkhalik AF, Shishido R, Nomura K, Ikehashi H (2005) QTL-based analysis of heterosis for grain shape traits and seedling characteristics in an indica-japonica hybrid in rice (Oryza sativa L.) *Breeding Science* 55: 41-48.

31. Ayahiko Shomura, Takeshi Izawa, Kaworu Ebana, Takeshi Ebitani, Hiromi Kanegae, et al. (2008) Deletion in a gene associated with grain size increased yields during rice domestication, Nature Genetics.

32. Sauer CO (1952) Agricultural Origins and Dispersals. New York: The American Geographical Society. Bowman Memorial Lectures Series Two.

33. Kaneda C, Umikawa M, Rohinikumar M Singh, Nakamura C, et al. (1996) Genetic diversity and subspecies differentiation in local rice cultivars from Manipur State of India. Breeding Science 46: 159-166.

34. Amano M (1981) Studies of agricultural history in China, 2nd. edit. Tokyo: Ochanomizu-Shobou: P.105

35. Xuehui Huang,Nori Kurata,Xinghua Wei, Zi-Xuan Wang, Ahong Wang1 et al. (2012) A map of rice genome variation reveals the origin of cultivated rice.

2

Stability Analysis of Rice Root QTL-NILs and Pyramids for Root Morphology and Grain Yield

Grace Sharon Arul Selvi[1], Farhad Kahani[2] and Shailaja Hittalmani[3*]

[1]Department of Genetics and Plant Breeding, University of Agricultural Sciences, Bangalore-65, India
[2]Marker Assisted Selection Laboratory, Genetics and Plant Breeding, University of Agricultural Sciences, GKVK, Bangalore 560065, India
[3]University Head, Genetics and Plant Breeding, University of Agricultural Sciences, GKVK, Bangalore-560065, India

Abstract

Cultivation of rice in the rain fed conditions is threatened by frequent spells of water deficits and limits the productivity to a greater extent. Root system plays a major role in uptake of water and they contribute to drought tolerance in a major way. In this study, Root QTL were pyramided and evaluated under aerobic and drought conditions and the stable genotypes were identified. Two QTL and three QTL pyramid lines for roots were developed and evaluated under drought, aerobic and in different locations to study the performance. While qRT26-9 with 2 QTL pyramid performed better with respect to the root traits, qRT16-1+7 and qRT17-1+7 performed better for shoot morphology over the various growth water regimes. Among the pyramids, qRT11-7 × qRT18-1+7-17 recorded increased performance for plant height and seed yield while qRT11-7 × qRT18-1+7-32 recorded increased performance for total biomass and maximum root length. qRT24-9 × qRT11-7-32 recorded increased performance for root traits only across environments. Lines with high means and average stability were identified as suitable across growth niches, while those with low stability and high means were identified as suitable for growth under poor environments and for specific locations.

Keywords: Rice; Aerobic; Stability; Root QTLs; Environments

Introduction

Rice (*Oryza sativa* L), the second most important cereal of the world is traditionally grown under submerged anaerobic conditions. However, this cultivation is now foraying into the less traditional rain fed uplands and marginal lands with mounting pressure on land availability. This coupled with changes in the climate make cultivation in these delicate ecosystems rather intricate. Therefore, the development of genotypes that consistently perform under conditions of climate change with less moisture availability is a viable option. The constancy or preferably increase in yield potential under climate change scenario is fundamental for food sustainability in the near future, given the expected population growth projections. Cultivation in the rain fed uplands is threatened by frequent spells of water deficits being a major limiting factor directly affecting grain yields during reproductive phases. Several mechanisms that determine drought tolerance and or resistance have been outlined, of which manipulation of the root system to maintain the water status of a crop under conditions of increasing water deficits has been the choice breeding strategy for drought. Several QTLs governing root traits across populations have been identified in rice. Root studies have become very important now that there are several ways to study them [1-17].

Pyramiding of genes is conducted to develop a genotype that expresses the said genes genes appropriately, such that the phenotype is enhanced. It has been used extensively in major gene controlled rice blast, rice blight and against insect pests such as diamond back moth (Cao et. al., 2002). Pyramids enhance the phenotype effectively and can be used to analyse the effect of QTLs upon each other as they offer a common background for the QTLs to interact. Subsequently QTL pyramiding was attempted by several researchers. Consistent and quality performance of developed genotypes is always desired as it increases the longevity of the genotypes. In breeding exercises, stable and high performance of developed varieties in target growth environments or across different environments and or seasons is an important attribute. Stability of the lines is measured as a non-significant deviation from its regression coefficient and is stated with reference to its mean. Lines

with high means and average stability can be identified to suit in most environments [18-30].

Material and Methods

Plant material

A set of twenty-nine near-isogenic lines with Root QTL introgressions of IR64 (*indica*, high yielding) with QTL introgressions from Azucena (*japonica*, drought tolerant) controlling root morphology (QTL Introgressed Lines (QILs)) developed by [31] and fine mapped by [16] was used for the study. These QILs were used in a pairwise crossing programme to develop 2 and 3 QTL pyramids. The QILs, the generated pyramids along with parents: IR64, Azucena and checks: Budda and Moroberekan were evaluated in RCBD design with 2 replications over the various growth regimes (Table 1) in 2011-2012 (Tables 2-4).

Phenotypic observations

Five plants with QTL pyramids were selected at random in each entry for recording observations. The average of these five plants was used for the statistical analysis. The individual plants were observed for plant height (cm) from the base to the tip of the panicle at harvest days to 50% flowering i.e., first flowering in 50 per cent of the plants, number of tillers per plant, number of panicles per plant, panicle length (cm) from collar to the tip, number of filled grains per panicle, number

*Corresponding author: Shailaja Hittalmani, Professor and University Head, Genetics and Plant Breeding, University of Agricultural Sciences, GKVK, Bangalore 560065, India, E-mail: shailajah_maslab@rediffmail.com

Regression Coefficient	Stability	Mean yield	Remarks
b=1	Average	High	Well adopted to all environments
b=1	Average	Low	Poorly adopted to all environments
b>1	Below average	High	Specially adopted to favorable environments
b<1	Below average	High	Specially adopted to unfavorable environments

Table 1: Mean, yield and Regression Co-efficient (b) values.

Sl. No.	Genotype	QTL introgression on
1	qRT1-1	1
2	qRT2-1	1
3	qRT3-1	1
4	qRT4-2	2
5	qRT5-2	2
6	qRT6-2	2
7	qRT7-2	2
8	qRT8-2	2
9	qRT9-7	7
10	qRT10-7	7
11	qRT11-7	7
12	qRT12-7	7
13	qRT13-7	7
14	qRT14-7	7
15	qRT15-7	7
16	qRT16-1+7	1+7
17	qRT17-1+7	1+7
18	qRT18-1+7	1+7
19	qRT19-1+7	1+7
20	qRT20-1+7	1+7
21	qRT21-1+7	1+7
22	qRT22-1	1
23	qRT23-1	1
24	qRT24-9	9
25	qRT25-9	9
26	qRT26-9	9
27	qRT27-9	9
28	qRT28-9	9
29	qRT29-9	9

Table 2: List of QTL Introgressed Lines (QILs) used in the study (Vaishali,2003).

Sl. No.	Pyramids	Chro. Introgression on	No. Plants
1	qRT11-7 X qRT18-1+7	7+ (1+7)	35
2	qRT24-9 X qRT11-7	9 + 7	38
3	qRT6-2 X qRT11-7	2+ 7	32
4	qRT11-7 X qRT19-1+7	7 + (1+7)	36
5	qRT20-1+7 X qRT18-1+7	(1+7) + (1+7)	37
6	qRT11-7 X qRT6-2	7 + 2	38
7	qRT6-2 X qRT19-1+7	2+ (1+7)	32
	Total No.		248

Table 3: List of two and three QTL pyramids generated for the study.

is the regression of the mean of environmental index and deviation for regression (s²di), which is a measure of genotype -environment interaction of an unpredictable type.

The model involves the estimation of three stability parameters: mean (μ_i), regression co-efficient (b_i) and deviation from regression (S^2d_i), which are defined by the following mathematical formula.

$$Yij = \mu i + \beta_i I_j + \delta$$

Where,

Y_{ij} : mean of the ith genotype in the jth environment.

μ_i : mean of the ith genotype over environments.

β_i : regression co-efficient that measures the response of ith genotype to varying environment.

δ_{ij} : deviation from regression of the ith genotype in the jth environment and

I_j : environmental index obtained by subtracting the grand mean of the ith genotype from the mean of all genotypes in the jth environment.

Stability parameters

The mean (μ_i), the regression co-efficient (b_i) and mean square deviation from linear regression line (S^2d_i) are the three stability parameters proposed by [33] in their stability model. The three parameters are computed using the following formulae:

Mean: $\mu_i = \sum_j Y_{ij}/n$

Regression co-efficient $b_i = \dfrac{\sum_j Y_{ij} I_j}{\sum_j I^2_j}$

Deviation from regression co-efficient $\left(S^2 d_i\right) = \dfrac{\sum_j ä^2_{ij}}{n-2} - \dfrac{ä^2 e}{r}$

Where,

$\delta^2 e/r$: mean square estimate of pooled error

n : number of environments

Y_{ij} : performance of ith genotype in j[th] environment

$\sum_j ä^2_{ij}$: sum of squares of deviations from the regression line

I_j : environmental index

$I_j = \dfrac{\sum_j Y_{ij}}{v} - \dfrac{\sum_i Y_{ij}}{nv}$

of chaffy grains per panicle, panicle weight (g) total grain weight per plant (g) root length (cm) from the crown to the tip of the longest roots, root thickness (mm), root number at 15cm root depth, total biomass (g) and test weight (g) was observed.

Statistical analyses

The data that was generated was subject to a series of statistical analyses to elucidate the relative effects of the presence of various QTLs in the rice genotypes and are presented here below:

Two way Analysis of variance: The data obtained was subjected to two way analysis of variance using the method outlined by [32] for each character in order to ascertain existence of genotype x environment interaction. If interaction was found to be significant, then the data was further subjected to stability analysis.

Stability analysis: The stability model proposed by [33] was adopted to analyze the data over the studied environments. The model considers three parameters: the mean (M), the regression co-efficient (bi) which

Sl. No.	Season	Location	Conditions	Material used	Traits
1	1	MRS, Hebbal	low moisture stress	QTL-NILs	Shoot, root and yield traits
			well watered conditions	pyramids	
2	2	Farmer's field,	low moisture stress	QTL-NILs	Shoot and yield traits
		Pavagada, Tumkur	well watered conditions	pyramids	
3	3	MRS, Hebbal	low moisture stress	QTL-NILs	Shoot and root traits
			well watered conditions	pyramids	
4	3	ZARS, GKVK	low moisture stress	QTL-NILs	Shoot, root and yield traits
			well watered conditions	pyramids	
5	3	Farmer's field,	submerged	QTL-NILs	Shoot, root and yield traits
		Dodjala, Bangalore		pyramids	
6	3	Farmer's field,	aerobic, non-stress	QTL-NILs	Shoot, root and yield traits
		Shettigere, Bangalore		pyramids	

Table 4: List of field experiments conducted in this study.

Where,

n : number of environments

v : number of genotypes with $\sum_j I_j = 0$

The total variation is partitioned into genotypes, environment, environment (linear), genotype x environment (linear), pooled deviation and pooled error.

F test

a. To test the significance of differences among the genotypic means, the 'F' test followed was:

$F = MS_1 / MS_3$

Where, MS_1 : mean sum of squares of varieties

MS_3 : mean sum of squares of pooled deviation

b. To test individual from linear regression, the formula is as follows,

$F = \left(\sum_j \delta^2 / n\text{-}2 \right) / MS_e$

Where, n: number of environments

$\sum_j \delta^2_{ij}$: sum of squares of deviations from the regression line

MS_e: pooled error

c. To test the hybrids/ varieties which do not differ for their regression on the environmental index, the appropriate test was,

$$t = \frac{b_i - 1}{SE(b)} \text{ or } \frac{b_i}{SE(b)}$$

$$SE(b) = \left[\frac{\left[\sum Y^2 - \left(\sum Y \right)^2 / n \right] - \left[\sum X^2 - \left(\sum X \right)^2 / n \right]}{n\text{-}2 \left[\sum X^2 - \left(\sum X \right)^2 / n \right]} \right]$$

Where,

X : environmental index

n : number of environments

A joint consideration of the three parameters such as

1. The mean performance of the genotype over the environments (x)

2. The regression co-efficient (b)

3. The deviation from linear regression (S^2d) is used to define stability of a genotype.

The estimate of deviations from regression (S^2d) suggests that the degree of reliance that should be put to linear regression in interpretation of the data. If these values are significantly deviating from zero, the expected phenotype cannot be predicted satisfactorily. When, deviations (S^2d) are not significant, the conclusion may be drawn by the joint consideration of mean, yield and regression co-efficient (b) values as given below (Table 1).

While interpreting the results, s^2di is first looked into. A non-significant deviation from $s^2di=0$, then stability is interpreted based on bi and mean values. If bi=1, a genotype is considered to possess average stability i.e., same performance in all the growth conditions. If bi is more than unity, then the genotype is said to have less than average stability i.e., good performance under favorable environments. If bi is less than unity, then the genotype is said to possess above average stability i.e., good performance under poor environments. Thus, genotypes possessing unit regression coefficient and non-significant deviation from regression were considered ideal, widely adapted and stable genotypes.

Results

Environment- wise analysis of variance indicated a significant mean sum of squares for the QTL-NILs and the generated pyramids for most characters studied. Combined Analysis of Variance for the pyramids qRT24-9 x qRT11-7 (data not shown) indicated significant variance for genotype as well as for genotype x environment [33] model for stability analysis was applied as the genotype x environment component of variance was found significant. The performance of genotypes in different environments for five selected characters based on two-way analysis of variance and Bartlett's test are elaborated below (Table 5).

Plant height

The varying environmental indices indicated that there was a significant difference for plant height across environments and across genotypic entries. qRT6-2 x qRT19-1+7 was the shortest (47.74cm), while qRT6-2 x qRT11-7 was the tallest (62.77 cm). The bi and s^2di were found to be non-significant for all the genotypes studied.

Seed yield per plant

As indicated by the environmental indices and the environmental means, seed yield per plant showed significant differences across

Genotypes	Source	df	PHT	NOT	NOP	PWT	PL	DFL	NS	NC	TWT	SWT/PLT	TBM	MRL	RTN	RTT
QTL-NILS	Genotype	28	1094.50**	36.30**	10.85**	65.67**	10.19**	23.71**	390.17**	100.23**	2.44**	34.01**	414.09**	319.86**	1123.27**	0.01**
	Env+(Genotype x Envt)	261	55123.9	2222.91	2195.92	72521.79	9984.15	1149.08	103713.8	18870.1	163.87	8091.21	347401	47868.35	144351.2	40.06
	Envt (linear)	1	37864.8	1676.49	1811.52	64602.31	9786.47	1148.46	96768.12	17994.6	144.85	4407.73	296222	33922	64009.25	38.98
	Genotype x Envt (linear)	28	44	0.99	3.55**	132.42**	2.21**	0.001	102.02**	16.35**	0.13**	21.40**	392.98**	373.76**	263.53**	0.01**
	Pooled Deviation	232	60.04	1.95	1.03	13.95	0.48	0.002	13.94	1.34	0.06	11.36	146.23	7.52	272.38	0.003
	Pooled Error	280	6.93	1.13	0.6	3.34	0.41	0.002	6	2.34	0.03	10.72	49.08	1.4	19.92	0.001
qRT11-7 x qRT18-1+7	Genotype	40	876.55**	24.49**	9.77**	87.17**	8.99**	13.75**	383.28**	82.08**	0.97**	31.35**	321.83**	345.57**	1982.65**	0.03**
	Env+(Genotype x Envt)	369	226.88	8.65	6.91	75.99	15.72	3.95	350.09	51	1.07	39.64	37	132.48	659.26	0.15
	Envt (linear)	1	52978.2	1937.42	1514.66	13773.78	4628.64	1256.84	88796.11	12487.9	360.68	8906.99	690332	32944.4	112931.1	51.19
	Genotype x Envt (linear)	40	113.47**	1.27	3.73**	156.19**	8.02**	0.41	143.71**	25.32**	0.38**	32.88**	721.85**	348.74**	243.31**	0.02**
	Pooled Deviation	328	79.89	3.67	2.7	24.45	2.6	0.56	105.61	16.21	0.06	13.43	282.64	6.07	367.7	0.01
	Pooled Error	410	8.3	1.65	1.24	2.09	0.34	0.55	5.95	15.8	0.03	9.36	51.35	1.65	9.28	0.002
qRT24-9 x qRT11-7	Genotype	43	660.47**	25.69**	10.89**	96.77**	19.34**	20.49**	318.33**	30.55**	0.77**	36.40**	271.54**	335.19**	2294.08**	0.01*
	Env+(Genotype x Envt)	396	229	7.65	10.85	106.96	20.3	7.85	290.78	45.75	0.16	48.77	2182.51	145.14	1461.25	0.13
	Envt (linear)	1	68261.8	1641.78	3041.87	22075.17	7027.26	1113.94	88536.55	15618.3	323.9	10459.2	738492	37439.92	297064.9	44.87
	Genotype x Envt (linear)	43	67.20**	1.22	6.76**	208.98**	10.95**	3.24	138.93**	18.35**	0.35**	50.57**	811.15**	398.62**	2233.28**	0.01
	Pooled Deviation	352	55.5	3.79	2.74	32.09	1.53	5.27	58.63	4.86	0.07	18.98	258.25	8.23	527.16	0.01
	Pooled Error	440	8.18	1.99	1.45	10.73	0.43	5.19	6.02	4.58	0.03	8.35	34.77	1.85	25.58	0.002
qRT6-2 x qRT11-7	Genotype	37	719.33**	21.83**	11.95**	59.90**	14.00**	21.72**	354.92**	20.27**	479.31**	29.87**	459.23**	401.46**	1384.50**	0.02**
	Env+(Genotype x Envt)	342	259.64	7.47	8.23	146	26.41	3.11	356.3	42.61	35.49	53.28	2019.43	101.77	1319.16	0.15
	Envt (linear)	1	60865.2	1639.85	1961.91	34778.48	8008.35	998.57	94034.25	11526	142.22	11256.2	553892	12425.92	231516.1	49.73
	Genotype x Envt (linear)	37	66.16	0.96	3.34**	56.92**	11.62**	0.29**	255.65**	5.62	57.68**	46.77**	939.75**	407.11**	1885.27**	0.02**
	Pooled Deviation	304	83.83	2.9	2.39	42.92	1.95	0.18	60.4	9.33	32.44	17.23	335.47	24.06	493.04	0.005
	Pooled Error	380	7.77	1.51	0.47	2.32	0.52	0.3	5.42	2.19	0.02	9.11	16.93	1.58	21.1	0.001
qRT11-7 x qRT19-1+7	Genotype	41	698.54**	27.99**	12.16**	80.37**	10.33**	25.36**	374.19**	47.15**	1.08**	26.33**	327.13**	306.75**	1008.60**	0.01**
	Env+(Genotype x Envt)	378	239	9.22	10.4	63.23	24.61	3.14	469.09	47.08	0.92	51.03	1886.74	150.4	1102.17	0.13
	Envt (linear)	1	67363.2	2044.62	2789.46	13111.76	8617.55	1005.53	147209.6	14989.1	307.65	13241.2	632768	39825.3	262040.7	45.68
	Genotype x Envt (linear)	41	118.94**	1.18	7.16**	99.92**	2.25**	0.48**	364.72**	11.41**	0.45**	21.49**	351.51**	350.40**	1492.14**	0.01**
	Pooled Deviation	336	53.88	4.14	2.52	19.92	1.77	0.48	45.1	6.96	0.07	15.38	196.45	7.92	277.99	0.005
	Pooled Error	420	21.86	1.91	0.51	8.46	1.25	0.82	10.39	3.12	0.03	8.43	63.55	5.79	23.12	0.003

Cross	Source	df	PHT	NOT	NOP	PWT	PL	DFL	NS	NC	TWT	SWT/PLT	TBM	MRL	RTN	RTT
qRT20-1+7 x qRT18-1+7	Genotype	42	770.22**	18.57**	6.12**	34.21**	3809.05**	15.24**	426.14**	58.03**	1.29**	22.90**	621.20**	315.25**	1061.88**	0.02**
	Env+(Genotype x Envt)	387	440.29	17.78	20.34	76.19	3902.68	3.35	429.79	43.76	0.77	86.12	3459.73	66.46	2177.75	0.1
	Envt (linear)	1	128493	4624.31	5874.1	23997.43	65221.28	1229.59	139668.5	16032.7	274.78	26952.5	1141093	7716.41	627347.9	37.93
	Genotype x Envt (linear)	42	517.38**	20.07**	25.74**	68.98**	30361.98**	0.20**	235.13**	9.75**	0.18**	46.38**	2647.6	136.89**	3691.99**	0.01**
	Pooled Deviation	344	58.64	4.11	2.67	7.53	493.93	0.17	48.79	1.44	0.05	12.87	251.82**	35.63	175.52	0.003
	Pooled Error	430	11.12	1.81	0.27	2.65	3814.38	0.22	5.84	2.1	0.02	8.21	53.08	1.49	16.34	0.001
qRT11-7 x qRT6-2	Genotype	43	882.98**	19.55**	2.93**	51.19**	5.28**	42.45**	279.74**	18.38**	1.06**	26.10**	718.26**	344.57**	914.81**	0.01**
	Env+(Genotype x Envt)	396	768.12	9.43	8.22	112.61	32.31	12.71	594.65	49.9	0.77	77.82	2717.55	59.79	1374.32	0.12
	Envt (linear)	1	249272	3004.72	2765.42	34337.35	12298.66	1370.65	213710.7	18968.7	270.01	22578.5	878866	6046.2	340944.6	48.04
	Genotype x Envt (linear)	43	858.08**	2.22**	2.34**	138.73**	4.60**	5.51	335.00**	4.96**	0.14**	74.42**	2447.62**	134.52**	2721.38**	0.01**
	Pooled Deviation	352	51.15	1.8	1.1	12.19	0.84	9.73	20.93	1.65	0.08	14.32	261.46	33.65	245.07	0.003
	Pooled Error	440	10.66	1.3	0.27	2.47	0.29	9.57	5.93	1.79	0.01	8.14	33.35	1.28	23.52	0.001
qRT6-2 x qRT19-1+7	Genotype	37	1145.51**	20.58**	5.44**	41.71**	6.34**	30.56**	273.61**	42.11**	1.06**	19.34**	382.90**	398.79**	1007.56**	0.01**
	Env+(Genotype x Envt)	342	335.54	10.8	12.67	163.99	28.51	4.07	394.86	54.92	0.67	33.31	1205.99	74.74	1035.23	0.13
	Envt (linear)	1	95236.6	2721.69	3509.35	42728.02	9294.13	1314.09	104932.5	17797.4	216.81	7209.23	345259	7064.5	233538.7	44.82
	Genotype x Envt (linear)	37	164.75**	7.62**	9.02**	215.53**	5.19**	0.35**	321.00**	10.59**	0.11**	7.16	180.05	124.55**	1036.50**	0.004**
	Pooled Deviation	304	44.15	2.27	1.61	17.7	0.87	0.21	59.97	1.95	0.03	12.88	199.1	45.69	270.27	0.003
	Pooled Error	380	15.76	0.91	0.26	2.92	0.36	0.42	5.36	2.07	0.02	8.96	39.66	1.23	15.44	0.001

PHT: Plant height (cm)
NOT: Number of tillers per plant
NOP: Number of panicles per plant
PWT: Panicle weight (g)
PL: Panicle length (cm)
DFL: Days to 50% flowering
NS: Number of filled seeds per plant
NC: Number of chaffy seeds per plant

TWT: Test weight (g)
SWT/PLT: Seed yield (g) per plant
TBM: Total biomass (g)
MRL: Maximum root length (cm)
RTN: Total root number
RTT: Mean root thickness (mm)
*/**: Significance at P=0.05 and 0.01 respectively

Table 5: ANOVA for stability in QTL introgressed NILs and the generated pyramids as per Eberhart and Russel (1966).

environments. qRT11-7 x qRT6-2 recorded the highest seed yield (9.95 g), while qRT11-7 x qRT18-1+7 recorded the least seed yields (6.92 g). The bi and s²di were found to be non-significant for all the genotypes studied.

Total Biomass per plant

The varying environmental indices indicated that there were significant differences for total biomass across environments. qRT6-2 x qRT19-1+7 was the lightest (35.82 g), while qRT20-1+7 x qRT18-1+7 was the heaviest (50.92 g). The bi and s²di were found to be non-significant for all the genotypes studied.

Maximum root length

As indicated by the environmental indices and the environmental means (11.86 to 18.63), maximum root length showed significant differences across genotypes. The QTL-NILs recorded the highest mean maximum root length (18.63 cm), while qRT11-7 x qRT6-2 recorded the least root length (11.86 cm). The bi and s²di were found to be non-significant for all the genotypes studied (Table 6).

Total number of roots per plant

The varying environmental indices indicated that there were significant differences for total number of roots across environments. qRT6-2 x qRT11-7 had the highest mean number of roots (59.13), while qRT11-7 x qRT18-1+7 had the least mean number of roots (52.65). The bi and s²di were found to be non-significant for all the genotypes studied.

Discussion

Phenotype of an individual is determined by the interaction of the genotype and environment surrounding it, the effects of genotype and environment on phenotype may not always be independent. The phenotypic response to change in environment is not the same for all the genotypes. The interplay in the genetic and non-genetic effects on development is termed as "genotype environment" interaction (Comstock and Moll, 1963) and is of major consequence to the breeder in the process of evolution of improved genotypes.

In the present study, twenty nine near isogenic lines of IR64 introgressed with QTL regions on four chromosomes: 1,2,7 and 9 from Azucena, 7 pyramids generated from these lines and the checks: IR64, Azucena, Budda and Moroberekan were grown in three seasons.2011,Season I. 2012 wet and dry seasons. (Season II and Season III). During season 1, they were grown under reproductive stage low moisture stress and well watered conditions for growth, yield and root traits at MRS, Hebbal. During season 2, they were grown under reproductive stage low moisture stress and well watered conditions at Farmer's field, Pavagada for growth and yield traits. During season 3, the genotypes were grown under reproductive stage low moisture stress and well watered conditions at ZARS, GKVK, under submerged conditions at Farmer's field, Dodjala and under aerobic non-stress conditions at Farmer's field, Shettigere for growth, yield and root traits. The genotypes were also grown under vegetative stage low moisture stress and well watered conditions for growth and root traits at MRS, Hebbal.

Mean performance of the QTL-NILs and generated pyramids

The aerobic, non stress growth condition of Farmer's field, Shettigere (data not shown) during season 3 was the most conducive environment for plant height in the QTL-NILs, with a mean height of 71.85 cm. Low moisture stress condition of season 2, Farmer's field, Pavagada was the least conducive for the QTL-NILs. qRT21-1+7 was the tallest genotype across locations. All the generated pyramids recorded maximum plant height under aerobic, non-stress condition at Farmer's field, Shettigere. The tallest pyramids generated were qRT11-7 xqRT18-1+7-7, qRT24-9 x qRT11-7-12, qRT6-2 x qRT11-7-15, qRT11-7 x qRT19-1+7-21, qRT20-1+7 x qRT18-1+7-4, qRT11-7 x qRT6-2-1 and qRT6-2 x qRT19-1+7-3. Among the pyramids, qRT11-7 x qRT18-1+7 was significantly taller, while qRT6-2 x qRT19-1+7 was significantly shorter.

For grain yield in QTL-NILs, the submerged conditions provided by Farmer's Field, Dodjala during season 3 was the most conducive, while the least mean yield were recorded under low moisture stress, MRS, Hebbal during season1. A highest mean yield across environments was recorded in qRT11-7. For the pyramids, qRT11-7 x qRT18-1+7, qRT24-9 x qRT11-7, qRT11-7 x qRT19-1+7, qRT11-7 x qRT6-7 and qRT6-2 x qRT19-1+7, the most conducive environment was the aerobic non-stress condition during season 3 at Farmer's field Shettigere while qRT6-2 x qRT11-7 and qRT20-1+7 x qRT18-1+7 recorded highest mean yields under submerged conditions, season 3, Farmer's field, Dodjala. The highest yielding pyramids were qRT11-7 x qRT18-1+7-15, qRT24-9 x qRT11-7-1, qRT6-2 x qRT11-7-15, qRT11-7 x qRT19-1+7-16, qRT20-1+7 x qRT18-1+7-17, qRT11-7 x qRT6-2-1 and qRT6-2 x qRT19-9.

The highest mean total biomass was recorded by the QTL-NILs under well watered condition, season 2, Farmer's field, Pavagada, while the least biomass was recorded under season 1, low moisture stress at MRS, Hebbal. Highest biomass across location was recorded by qRT18-1+7. For the pyramids qRT11-7 x qRT18-1+7, qRT6-2 x qRT11-7, qRT11-7 x qRT19-1+7 and qRT6-2 x qRT19, the most conducive environment for total biomass production was season 2, Farmer's field, Pavagada, while for qRT24-9 x qRT11-7, qRT20-1+7 x qRT18-1+7 and qRT11-7 x qRT6-2 yielded highest biomass under aerobic, non-stress conditions during season 3 at Farmers field, Shettigere. Highest biomass were produced by qRT11-7 x qRT18-1+7-4, qRT24-9 x qRT11-7-13, qRT6-2 xqRT11-7-25, qRT11-7 x qRT19-1+7-20, qRT20-1+7 x qRT18-1+7-8, qRT11-7 x qRT6-2-38 and qRT6-2 x qRT19-30.

Mean maximum root length was recorded by the QTL-NILs under low moisture stress, season 1, MRS, Hebbal, with the maximum root length being recorded in qRT26-9. For the pyramids qRT11-7 x qRT18-1+7, qRT24-9 x qRT11-7, qRT6-2 x qRT11-7, qRT11-7 x qRT19-1+7, and qRT6-2 x qRT19, the most conducive environment for longer root production was low moisture stress condition, season 1, MRS, Hebbal, while qRT20-1+7 x qRT18-1+7 and qRT11-7 x qRT6-2 recorded longest roots under well watered conditions, ZARS, GKVK during season 3. Longest roots were produced by qRT11-7 x qRT18-1+7-27, qRT24-9 x qRT11-7-32, qRT6-2 xqRT11-7-28, qRT11-7 x qRT19-1+7-10, qRT20-1+7 x qRT18-1+7-33, qRT11-7 x qRT6-2-12 and qRT6-2 x qRT19-24.

Highest number of roots were produced by the QTL-NILs under well watered conditions, season 3, ZARS, GKVK, with the highest number of roots being produced by qRT17-1+7. For the pyramid qRT11-7 x qRT18-1+7 the most conducive environment for number of root production was well watered conditions, season 3, ZARS, GKVK, while for qRT24-9 x qRT11-7, qRT11-7 x qRT19-1+7, qRT11-7 x qRT6-2 and qRT6-2 x qRT19, the most conducive environment for number of root production was season 3, aerobic, non-stress condition, Shettigere and for qRT6-2 x qRT11-7 and qRT20-1+7 x qRT18-1+7 the most conducive environment was submerged condition, season 3, Farmer's field, Dodjala. Highest number of roots were produced by qRT11-7 x qRT18-1+7-14, qRT24-9 x qRT11-7-1, qRT6-2 xqRT11-7-31, qRT11-7 x qRT19-1+7-6, qRT20-1+7 x qRT18-1+7-8, qRT11-7 x qRT6-2-13 and qRT6-2 x qRT19-4.

The aerobic non-stress condition therefore is the most conducive environment to grow the present genotypes. Similar results were obtained by [3,34]. This was opined to be due to aeration of roots leading to efficient utilization of resources [35-38].

Genotype x environment interaction

Prior to stability analysis, Bartlett test was done. Based on this test, five characters were selected. Homogeneity in the error variances allowed pooled analysis.

The mean sum of squares due to genotypes as well as environments was found to be significant in the two way analysis. Significant GXE interaction was obtained both by two-way analysis and the [32] model. The analysis of variance for stability indicated significant differences among the QTL-NILs as well as between and within the generated pyramids for all the characters. The significant environment (linear) variance indicated considerable additive environmental variance. Variance due to GX E interaction was found to be significant for all the characters indicating differential response of the genotypes in different environments. G X E (linear) was significant for all the characters indicating a contribution of linear portion of GE interaction. The more pronounced linearity of characters indicated that variation among the genotypes could be largely explained by the differences [3,39-41].

Stability parameters

Five characters were selected on the basis of homogeneity of error variances and after significance of G X E interactions. Identification of genotypes that perform stably over a range of growth environs would therefore be necessary. Of the many models proposed to this effect, the Eberhart and Russel model was used in the present study. Taller plants were preferred as these could lend to fodder yield in a mixed cropping system. Increase in height coupled with increase in total biomass content could result with increase in number of tillers per plant and flowering mattered for the escapes [42]. Maximum root length as a mechanism of drought tolerance ensures higher crop yields under stress situations. Based on the five characters taken together, among the QTL-NILs, qRT24-9 was found to be best in performance across locations, moisture regimes and seasons. While qRT26-9 performed better with respect to the root traits, qRT16-1+7 and qRT17-1+7 performed better for shoot morphology over the various growth regimes. Among the pyramids, qRT11-7 x qRT18-1+7-17 recorded increased performance for plant height and seed yield while qRT11-7 x qRT18-1+7-32 recorded increased performance for total biomass and maximum root length. qRT24-9 x qRT11-7-32 recorded increase in performance for root traits only across environments. qRT6-2 x qRT11-7-15 recorded increase in performance for plant height and seed yield across environments qRT20-1+7 x qRT18-1+7-15 was the best in performance considering seed yield, total biomass and number of roots per plant. qRT6-2 xqRT19-1+7-30 recorded best performance for seed yield per plant and total biomass. Since all these genotypes recorded non-significant deviation of the regression coefficient (bi) from 1 and s^2di approaching zero, we can conclude that these genotypes have average stability across locations [3,38].

Conclusion

The pyramids of root QTL are very relevant as root morphological parameters are controlled by quantitative genes and these do not act independently. When they are moved into new background the effect and the stability of the QTL in environment play a major role. Both environment specific and pyramids suitable for wider range of environments are useful. Hence root QTL pyramiding is useful for developing genotypes for using in water saving technologies like aerobic cultivation or in case where dry spells prevail and roots help to tide over and minimize the economic loss to the rice growing farmers.

References

1. Dey MM, Upadhaya HK (1996) Yield loss due to drought, cold and submergence in Asia. In: Rice Research in Asia, Progress and Priorities (Eds: Evenson, R.E., Herdt, R.W. and Hossain, M.) Oxford University Press, NC, 231-242.

2. Krishna TV, Shailaja Hittalmani (2009) Genetic assessment of root morphology under well water and low moisture stress condition at Reproductive stage. Bull Biol Sci 7: 179-188.

3. Naresh Babu N, Hittalmani S, Shivakumar N, Nandini C (2011) Effect of drought on yield potential and drought susceptibility index of promising aerobic rice (Oryza sativa L) genotypes. Electronic Journal of Plant Breeding 2: 295-302.

4. Keshava Murthy BC, Arvind Kumar, Shailaja Hittalmani (2011) Response of rice (Oryza sativa L) genotypes under aerobic conditions. Electronic J Pl Breed 2:194-199.

5. Lafitte R, Blum A, Atlin G (2003) Using secondary traits to help identify drought-tolerant genotypes. In: Fischer KS, Lafitte R, Fukai S, Atlin G, Hardy B (Ed) Breeding rice for drought- prone environments 37-48.

6. Yoshida S, Hasegawa S (1982) The rice root system: Its development and function In: Drought resistance in cereals crops with emphasis on rice, IRRI, Philippines 53-68.

7. Ekanayake IJ, De Datta, Steponkus (1989) Spikelet sterility and flowering response of rice to water stress at anthesis. Ann Bot 63: 257-264.

8. O'Toole JC, Bland WL (1987) Genotypic variation in crop plant root systems. Adv Agron 41: 91-145.

9. Thanh NO, Zheng HG, Dong NV, Trinh LN, Ali ML, et al. (1999) Genetic variation in root morphology and microsatellite DNA loci in upland rice (Oryza sativa L) from Vietnam Euphytica 105: 43-51.

10. Venuprasad R, Bool ME, Quiatchon L, Atlin G (2012) A QTL for rice grain yield in aerobic environments with large effects in three genetic backgrounds. Theor Appl Genet 124: 323-332.

11. Murthy KBC, Kumar A, Hittalmani S (2011) Response of rice genotypes under aerobic conditions, Electronic J of Plant breed 2: 194-96.

12. Price AH, Tomos AD (1997) Genetic dissection of root growth in rice (Oryza sativa L) II. Mapping quantitative trait loci using molecular markers Theor Appl Genet 95: 143-152.

13. Yadav R, Courtois B, Huang N, McLaren G (1997) Mapping genes controlling root morphology and root distribution in a doubled haploid population of rice. Theor App Genet 94: 619-632.

14. Courtois B, Mclaren G, Singh PK, Yadav R, Shen L (2000) Mapping QTLs associated with drought avoidance in upland rice. Mol Breed 6: 56-66.

15. Kamoshita A, Zang J, Siopongco J, Salarimg S, Nguyen HT, et al. (2002) Effect of Phenotyping environment on identification of quantitative trait loci for rice root morphology under anaerobic condition. Crop Sci 42: 255-265.

16. Vaishali MG, Hanamareddy B, Mane S, Gireesha TM, Hittalmani S (2003) Graphical genotyping using DNA markers andevaluation of QTL-NILs of IR64 for root morphological traits disease resistance and yield traits in rice. Plant and Animal Genome XI Conference, San Diego, USA, 335.

17. Lartaud M, Perin C, Courtois B, Thomas E, Henry S, et al (2014) PHIV-Root Cell, a supervised image analysis tool for rice root anatomical parameter quantification. Frontiers in Plant Science 5:790.

18. Ahmadi N, Audebert A, Bennett M, Bishopp A, Costa de Oliveira A (2014) The roots of future rice harvests. Rice 7:29.

19. Hittalmani S, Parco A, Mew TV, Zeigler RS, Huang N (2000) Fine mapping and

DNA marker- assisted pyramiding of the three major genes for blast resistance in rice. Theor Appl Genet 100: 1121-1128.

20. Singh S, Sidhu JS, Huang N, Vikal Y, Li Z, et al. (2001) Pyramiding three bacterial blight resistance genes (*xa5, xa13* and *Xa21*) using marker assisted selection into rice cultivar PR106. Theor Appl Genet 102: 1011-1015.

21. Steele KA, Price AH, Witcombe JR (2006) Marker assisted selection to introgress rice QTLs controlling root traits into an Indian upland variety. Theor Appl Genet 112: 208-221.

22. Wang P, Xing Y, Li Z, Yu S (2012) Improving rice yield and quality by QTL pyramiding. Mol Breed 29: 903-913.

23. Ahmed MI, Vijayakumar CHM, Viraktamath BC, Ramesha MS, Singh S (1998) Yield stability in rice hybrids. *IRRN* 23: 12.

24. Hegde S, Vidyachandra B (1998) Yield stability analysis of rice hybrids *IRRN* 23: 14.

25. Lohithaswa HC, Bhushana HO, Basavarajaiah D, Prasanna HC, Kulkarni RS (1999) Stability analysis of rice (*Oryza sativa* L) hybrids. Karnataka J Agri Sci 12: 48-52.

26. Hittalmani S, Huang N, Courtois B, Venuprasad R, Zhuang JY, et al. (2003) Identification of QTL for growth and grain yield related traits in rice across nine locations of Asia. Theoretical and Appl Genet 107: 679-690.

27. Shanmuganathan M (2005) Phenotypic stability by AMMI analysis for single plant yield in rice hybrids. *Intl J Agril Sci* 1: 50-52.

28. Latif T, Khan MA, Khan MG, Mahmood S, Butt MA (2007) Evaluation of rice genotypes for stability in paddy yield. Pakistan J Agric Res 20: 7-10.

29. Ao HJ, Wang SH, Zou YB, Peng SB, Tang QY, et al. (2008) Study on yield stability and dry matter characteristics of super hybrid rice. Scientia-Agricultura-Sinica 41: 1927-1936.

30. Mosavi AA, Jelodhar NB, Kazemitabar K (2013) Environmental responses and stability analysis for grain yield of some rice genotypes. World Appl Sci J 21: 105-108.

31. Shen L, Courtois B, Mcnally KL, Robin S, Li Z (2001) Evaluation of near- isogenic lines of rice introgressed with QTLs for root depth through marker-aided selection. Theor Appl Genet 103: 75-83.

32. Sunderarajan N, Nagarau S, Venkataramu MN, Jaganath MK (1972) In: Designs and analysis of field experiments, UAS, Bangalore, Karnataka.

33. Eberhart SF, Russell WA (1966) Stability parameters for comparing varieties. Crop Sci 6: 36-40.

34. Kanbar A, Manjunatha K, Hittalmani S (2004) Genetics of root morphology and related characters in doubled haploid mapping populations of rice (*Oryza sativa* L) Indian JGenet 64: 58.

35. Bouman BAM, Twong TP (2001) Field water management to save water and increase productivity in irrigated rice. Agric Water Management 49: 11-30.

36. Yang XG, Wang HI, Wang ZZ, Junfang CB, Boumann BAM (2002) Yield of aerobic rice (Han Dao) under different water regimes in North China. In: Water-wise rice production. IRRI, Los Banos, Philippines, 155-164.

37. Tao HG, Brueck HDK, Kreye C, Lin S, Sattel MB (2002) Growth and yield formation of rice (*Oryza sativa* L). In: The water facing ground cover rice production system, Institute of Plant Nutrition.

38. Atlin GN, Laza M, Amante M, Laffitte HR (2004) Agronomic performance of tropical aerobic rice varieties. In: Fourth Intl. Crop Sci. Congress, Philippines.

39. Shanmuganathan M, Ibrahim SM, Jeshima KY (2004) Stability analysis for yield in rice hybrids by AMMI model. *Pl. Archives* 4: 307-310.

40. Deshpande VN, Waghmode BD, Rewale AP, Vanave PB (2002) Stability performance of different rice hybrids at different locations in Maharashtra State. Crop Improv 29: 203-207.

41. Kumar BMD, Shadakshari YG (2008) Genotype x environment interaction and stability analysis for grain yield and its components in BKB local rice mutants. *Env And Eco* 26: 1667-1669.

42. Gómez-Ariza J, Brambilla V, Shrestha R, Galbiati F, Pappolla A (2015) Loss of floral repressor function adapts rice to higher latitudes in Europe. J Exp Bot 66: 2027-2039.

3

Genotype × Environment Interaction and Stability Estimate for Grain Yield of Upland Rice Genotypes in Nigeria

Maji AT[1], Bashir M[2], Odoba A[1], Gbanguba AU[1] and Audu SD[1]*

[1]National Cereals Research Institute Badeggi, Niger state. Nigeria
[2]National Biotechnology Development Agency, Lugbe, Abuja, Nigeria

Abstract

Genotype × environmental interaction and stability estimate were investigated on grain yield of 30 upland rice varieties at Sabon Daga, Amakama, Yandev and Uyo in 2003. The experiments were laid out in a randomised complete block design with three replications. AMMI Anova for grain yield revealed no significant different among genotypes (P<0.01), but there is significant difference on environments and the interaction. The significant different on the interaction indicates that, the genotypes respond differently across the different environments. The partitioning of GGE through GGE biplot analysis showed that principal component1 and principal component 2 accounted for 62.21% and 28.57% of GGE sum of squares respectively, explaining a total of 90.78% variation. AMMI 2 biplot revealed that, genotype ART16-9-3-15-3-B-1-1 (8) gave the highest mean yield of 2925 kg/ha with high main additive effect better than the check varieties. Hence, the genotype would be considered more adapted to wide environments than the rest of genotypes. Environments, such as Sabon Daga and Amakama could be regarded as a more stable site for high yielding rice varieties compare to the other locations.

Keywords: Genotype; Stability; ANOVA

Introduction

Rice (*Oryza sativa* L.) is the second most important cereals crop, grown in more than 144 million farm worldwide, most certainly than any other crop on a harvested area of about 162 million ha [1] .The author also reported that, global rice production rise more than tripled between 1961 and 2010, with a compound growth rate of 2.24% per year, most of the increase in rice production was due to higher yields, which increase at annual average rate of 1.74%, compared with an annual average growth rate of 0.49% for area harvested. He further stated that, per capital consumption of rice continues to grow fast particularly in most sub-Saharan Africa, where high population growth with changing consumer preferences is causing rapid expansion in rice consumption. In countries such as Kenya, Niger, Nigeria and Tanzania people are moving away from maize and cassava to rice as their income rises. Along with strong population growth, the rapid rise in per capita consumption also contributed to such rapid growth in rice demand.

In Nigeria, rice is a leading staple crop cultivated in virtually all the agro-ecological zones of country, from the mangrove and swamps environment of the coastal areas, to the dry zones of the Sahel in the North [2]. On the other side, the demand for rice has been soaring over years, since mid 1970's rice consumption in Nigeria has risen tremendously growing by 10.3% per annum, as a result of accelerating population growth rate, increasing per capita consumption, rapid urbanization, increase income levels, and associated changes in family occupational structures [2-4]. GRISP [1] reported that, Nigeria is blessed with three major rice production environments and their coverage is rainfed lowland (69.0%), irrigated lowland (2.7%) and rainfed upland (28.3%). More than 90% of Nigeria's rice is produced by resource poor small-scale farmers, while the remaining 10% is produced by cooperate/commercial farmers.

Upland rice is grown in rainfed, naturally well drained soils without surface water accumulation, normally without pyretic water supply, and normally not bunded. In the upland environment, rice cultivation is challenged by drought, low adoption of improved varieties, soil acidity and general soil infertility, poor weed control, limited capital investments, labor shortages and low mechanization, resulting in low yield range from 1.0 to 1.7 t/ha compared with a potential of 2.0-4.0 t/ha. Most upland rice is grown on small subsistence farms with few purchased inputs and most production is for family consumption. Therefore developing high yielding upland varieties combine with tolerant to biotic and abiotic stress will contribute substantially to poverty alleviation, especially, for resource constrained households and can increase household food security.

Numerous statistical methods have been developed for the analysis of Genotype by Environment Interactions (GEI) and phenotypic stability [5-8]. Regression technique has been widely used [9,10] due to its ease and the fact that its information on adaptive response is easily applicable to locations. The Principal Component Analysis (PCA) method that shows the mean squares of the principal components axes [11] has also been used. [12] Zobel*et al.* compared the traditional statistical analysis such as Analysis of Variance (ANOVA), Principal Component Analysis (PCA) and Linear Regression with AMMI analyses, and showed that the traditional analyses were not always effective in analyzing the multi-environment trial data structure. The ANOVA is an additive model that describes main effects effectively and determines if GE interaction is a significant source of variation, but it does not provide insight into the patterns of genotypes or environments that give rise to the interaction. The PCA is a multiplicative model that contains no sources of variation for additive G or E main effects and does not analyze the interactions effectively. The linear regression method uses environmental means, which are frequently a poor estimate of environments, such that the fitted lines in most cases

*Corresponding author: Audu SD, National Cereals Research Institute Badeggi, Niger state, Nigeria, E-mail: Saudatu28@gmail.com

account for a small fraction of the total GE and could be misleading [13-15].

Additive main effects and multiplicative interaction (AMMI) has been proved to be a suitable method for depicting adaptive responses [15-17]. AMMI analysis has been reported to have significantly improved the probability of successful selection [17] and has been used to analyse GxE interaction with greater precision in many crops [13,15,18]. The model combines the conventional analysis of variance for genotype and environment main effects with principal components analysis to decompose the GEI into several Interaction Principal Component Axes (IPCA). With the biplot facility from AMMI analysis, both genotypes and environments are plotted together on the same scatter plot and inferences about their interaction can be made.

This study, reports the use of AMMI model to analyse yield data of thirty genotypes of upland rice evaluate in four locations. The objectives is (1) to determine the nature and magnitude of G × E interaction effect on grain yield in diverse environment (2) to determine environment where upland rice genotypes would be adapted and produce economically competitive yield.

Materials and Methods

Thirty upland rice varieties selected from breeding task force upland mega environmental trial (MET) of 2012 are composed as preliminary yield trial (PET) in National Cereals Research Institute, Badeggi rice breeding unit, evaluated during 2013 cropping season at four locations as shown in Table 1.The experiment was conducted in a randomised complete block design in three replication, The plot size was 4 m × 3 m square with 20 cm inter and intra row spacing. Fertilizer application was 40 kg N, 40 Kg P_2O_5 and 40 Kg K_2O at transplanting, while additional 40 kg N per ha was used as top dressing at vegetative and panicle initiation in equal split. Weed control was by chemical at 21 days after transplanting (DAT) using a formulation of Propanil and 2-4-D (Orizo Plus[R]), and was followed by hand weeding at 43 days after transplanting. Grain yield was recorded after harvest at 14% moisture content and was subjected to analysis of variance (ANOVA) using Crop Stat statistical package. In order to determine the effect of genotype × environment interaction on rice grain yield, the data was further subjected to an additive main effect and multiplicative interaction (AMMI), GGE-biplot and Boxplot analysis using Breeding View (BV) statistical package.

Results and Discussion

AMMI analysis of variance

The fit of an additive model to the rice grain yield data are presented in Table 2. It showed that, there is no significant difference in genotypes main effect. However, significant differences (P<0.01) exist among environments and genotypes × environment (G × E) interaction, PCA1 and PCA2 main effects. The environments are characterised by the average performance of the genotypes at a particular environment and the results indicates that, the environments differ significantly. Marcos et al. [19] reported that, environmental difference is not a major concern, but the differences that exist between the genotypes. No significant genotype main effect indicates that genotypes are not different in their mean performance across environment. Although genotypic and environmental scores are deemed to represent genetic and environmental qualities, they come from a mathematical procedure, a principal components analysis on the GEI [12,20] that maximizes the variation explained by the products of the genotypic and environmental scores. The first two PCA explains most of the variation,

in grain yield. This is reflected in Table 2, which shows the results from the AMMI model to the grain yield data. In the AMMI model, GEI is explained by two axes (principal component 1, PCA1, and principal component 2, PCA2) that are highly significant respectively, both with an associated (**P**<0.001). Thus the interaction of the 30 genotypes across four environments was best predictable by the first two principal components.

Box Plot is a convenient way of graphically depicting group of numerical data through their qualities. It displays varieties in samples of a statistical population without making any assumptions of the underlying statistical distribution [21]. The spacing between the different parts of the box indicates the degree of dispersion (spread) of the data and allows visually estimate of inter-quartile mean, median and mode. Result in Figure 1 is showing the distribution pattern of grain yield of 30 rice genotypes across four environments. The result revealed that, Sabon daga has the highest mean grain yield of 3692 kg/ha (Table 3) with large variance followed by Amakama with mean yield of 2940 kg/ha, while Yandev and Uyo discriminate less between genotypes with mean of 1719 and 1846 kg/ha, respectively. This is reflected in

Location	Longitude	Latitude	State	Agro-Ecological Zones
Sabon Daga	09⁰.73'N	06⁰.52' E	Niger	Southern Guinea Savannah
Amakama	05⁰.29' N	07⁰. 33' E	Abia	Rain Forest zone
Uyo	04⁰. 50' N	07⁰. 56' E	Akwa Ibom	Rain Forest zone
Yandev	08⁰.47' N	07⁰.22' E	Benue	Southern Guinea Savannah

Table 1: Geographic description of coordinates of the trial location in 2013 cropping season.

Source	d.f.	s.s.	m.s.	v.r.	F pr
Genotypes	29	3120724	107611	1.20	0.2529
Environments	3	79273389	26424463	295.33	<0.001
Interactions	87	7784210	89474	3.37	<0.001
IPCA 1	31	4842595	156213	5.88	<0.001
IPCA 2	29	2224187	76696	2.89	0.0035
Residuals	27	717428	26571		

Table 2: AMMI analysis of variance for grain yield of 30 rice genotype across 4 environments.

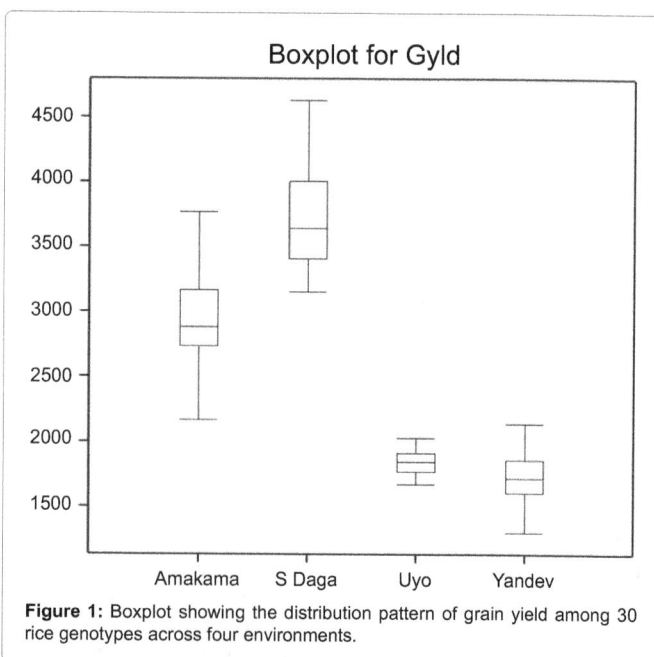

Figure 1: Boxplot showing the distribution pattern of grain yield among 30 rice genotypes across four environments.

Location	Range	Lower quartile	Upper quartile	Std. d	Mean	%cv	s. sq
S Daga	1477	3405	4004	398.4	3692	10.79	4603301
Amakama	1600	2733	3167	406.3	2940	13.82	4787551
Uyo	358	1768	1911	98.7	1846	5.35	282616
Yandev	839	1606	1861	206.1	1719	11.99	1231466

Table 3: Showing the statistical distribution of environmental performance.

the smaller variance Przystalski [22] reported that, the genetic variance tends to be larger in better environments than in poorer environments.

A desirable property of the AMMI model is that, the genotypic and environmental scores can be used to construct powerful graphical representations called biplots [19] that help to interpret the GEI, the biplot showing both genotypes and environments in the same plot. The author further stated that, biplots facilitate the exploration of relationships between genotypes and/or environments. Genotypes that are more similar to each other are closer to each other in the plot than genotypes that are less similar. The same is true for environments. Genotypes/environments that are alike tend to cluster together. Result in Figure 2 indicates that, S Daga location has the highest mean yield 3692 kg/ha, while ART12-1L6P7-8-1-B-1 (2) is the genotype with the highest mean yield. The result also shows that, there is no correlation between Amakama and Yandev/ Uyo locations. The projection of ART12-1L6P7-8-1-B-1 (2) and ART16-9-3-15-3-B-1-1 (8) on to S Daga axis reflects the higher mean yield performance of the genotypes. Similarly in Amakama genotype ART3-9L9P3-1-B-2 (22) and ART2-6L6P6-1-B-1(10) performed best in the location, while genotype ART12-1L6P7-8-1-B-1 (2) and ART16-9-3-15-3-B-1-1 (8) has positive interaction with S Daga. It is also predicted that, genotype ART3-3L12P9-1-1-B (15), ART3-7L9P8-3-B-B-2(20) and ART3-6L3P9-B-B-2 (16) has negative GEI values in S Daga because their projections were towards the negative direction of S Daga arrow. Also genotype FARO55 (23), ART16-22-1-1-2-B-1-1 (7) and WAB706-27-K5-KB-2 (28) have negative interaction with Amakama location. Generally, there was a poor yield performance in Yandev and Uyo locations as shown in Figure 2.

AMMI 2 biplot display

In the AMMI 2 biplot, (Figure 2) the environmental scores (locations) are joined to the origin by side lines. Sites with short vectors do not exert strong interactive forces (Uyo and Yandev). While those that long vectors exert strong interaction (S.Daga and Amakama). Weikai Yan reported that, a short vector indicates a location in which there is a small range of genotype performance.

The vertical Y axis is showing the best one dimension measure of the GE effect for each genotype. Thus, genotypes close to the X axis have a small GE effect, while those far away the X axis in either the positive or negative directions has a large GE effect. Figure 2 shows that, genotype ART10-1L12P2-1-B-1(1) and ART16-16-5-23-1-B-1-1 (6) has a small GE effect, which is considered stable and less influenced by the environments.

Weikai Yan reported that, If the angle between two genotype vectors is less than 90 degrees, then the genotypes are positively correlated, tending to do well, or badly, in the same environment. But if the angle between the vectors of two genotypes is greater than 90 degrees, then they tend to perform differently over the trial environments. If the angle between two genotype vectors is 90 degrees, their performance is independent, of each other. Figure 2 shows that, ART16-9-3-15-3-B-1-1 (8), ART3-9L9P3-1-B-2 (22) and ART3-6L3P9-B-B-3 (17) are

positively correlated. However, there is negative correlation between ART16-9-3-15-3-B-1-1 (8), ART10-1L12P2-1-B-1 (1) and ART16-16-5-23-1-B-1-1(6). Also, there is no correlation between ART16-9-3-15-3-B-1-1 (8) and ART3-12L11P2-B-B-1 (11) ART16-12-22-4-1-B-1-1 (5), ART3-8L6P6-5-B-2(21) and FARO58 (24). The ideal genotype is the genotype with high performance combined with good stability.

GGE biplot also allows the partitioning of environment into

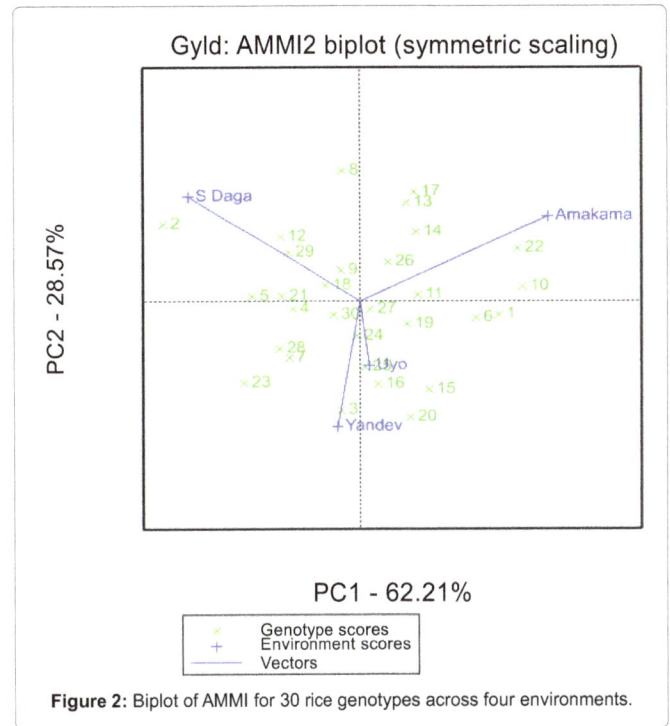

Figure 2: Biplot of AMMI for 30 rice genotypes across four environments.

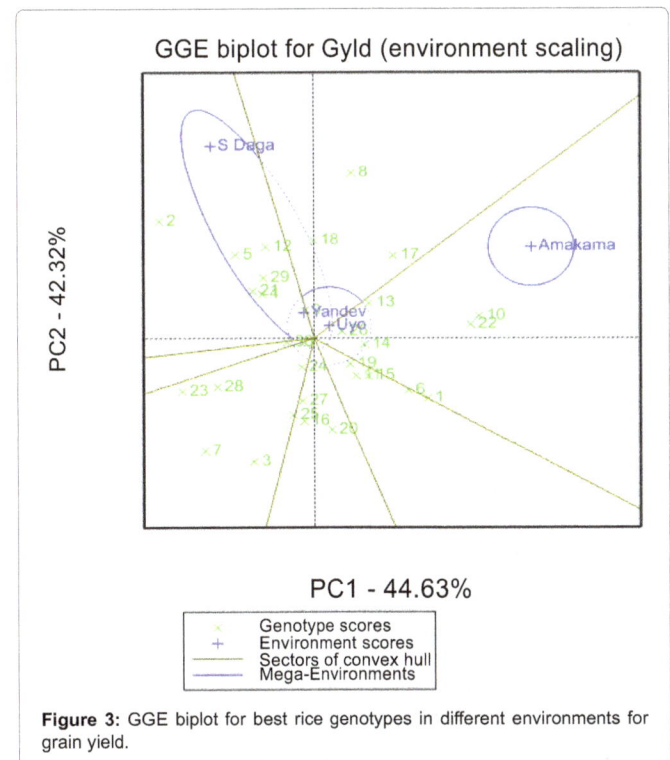

Figure 3: GGE biplot for best rice genotypes in different environments for grain yield.

groups. In this study, three environmental groups are identified as shown in the Figure 3. S Daga and Amakama in the upper part are two different environments, while Yandev and Uyo close to each other at the origin form one similar environment. The partitioning of GGE through GGE biplot analysis of grain yield showed that, PC1 and PC2 accounted for 44.63% and 42.32% of GGE sum of squares respectively, explaining a total of 86.95% variation. GGE biplot shows the cosine of the angle between two environment vectors is proportional to the correlation between those two environments that is an angle of less than 90 degrees. The environments are positively correlated [21]. The result in Figure 3 shows a negative correlation between S Daga and Amakama indicating that different genotypes performed differently across the two environments. The distance between S Daga and Amakama in the GGE biplot is related to the independence of the genotype performance in the two environments, while the closeness of Yandev and Uyo location signifies that genotypes response patterns are similar in yield performance. Therefore to save resources, it is better to select only one location out of this group for further trials, Yandev location could be selected in group 3 as it has the longest vector (Tables 4 and 5).

Conclusion

AMMI statistical model is a tool in selecting the most suitable and

S/no.	Genotype	Location Mean Grain Yield (Kg/ha)				Mean Grain yield (Kg/ha)
		Sabon Daga	Amakama	Uyo	Yandev	
1	ART10-1L12P2-1-B-1	3150	3333	1870	1611	2491
2	ART12-1L6P7-8-1-B-1	4627	2500	1743	1739	2652
3	ART15-4-14-63-2-B-1	3214	2367	1810	1678	2267
4	ART16-12-17-3-4-B-1-1	4004	2800	1849	1936	2647
5	ART16-12-22-4-1-B-1-1	4248	2767	1861	2028	2726
6	ART16-16-5-23-1-B-1-1	3225	3267	1810	1681	2496
7	ART16-22-1-1-2-B-1-1	3417	2167	1794	1369	2187
8	ART16-9-3-15-3-B-1-1	4429	3500	2016	1756	2925
9	ART16-9-4-17-3-B-1	3856	2967	1771	1708	2576
10	ART2-6L6P6-1-B-1	3405	3767	1854	1939	2741
11	ART3-12L11P2-B-B-1	3428	3067	1712	1611	2455
12	ART3-12L2P1-B-B-1	4255	2933	1804	1778	2693
13	ART3-1L6P5-1-B-1	3783	3300	1714	1450	2562
14	ART3-2L4P5-1-B-1	3599	3167	1910	1300	2494
15	ART3-3L12P9-1-1-B	3334	3100	1918	2083	2609
16	ART3-6L3P9-B-B-2	3291	2700	1767	1817	2394
17	ART3-6L3P9-B-B-3	3954	3500	2028	1494	2744
18	ART3-7L3P3-B-B-2	4132	3167	1843	2139	2820
19	ART3-7L9P8-1-B-B-2	3468	3067	1671	1861	2517
20	ART3-7L9P8-3-B-B-2	3166	2800	1931	1917	2454
21	ART3-8L6P6-5-B-2	4048	2767	1790	1894	2625
22	ART3-9L9P3-1-B-2	3423	3700	1992	1583	2675
23	FARO55(NERICA1)	3712	2200	1740	1833	2371
24	FARO58(NERICA7)	3576	2800	1986	1689	2513
25	FARO59(NERICA8)	3354	2667	1722	1756	2375
26	NERICA11	3680	3100	1768	1597	2536
27	NERICA18	3446	2733	1911	1394	2371
28	WAB706-27-K5-KB-2	3679	2367	1903	1606	2389
29	WAB788-16-1-1-2-HB	4121	2833	1993	1606	2638
30	WAB891-SG12	3732	2800	1887	1728	2537
	LSD @ 0.05%	NS	NS	NS	NS	NS
	% CV	19.7	24.3	7.4	10.2	14.3

Table 4: Mean grain yield of upland rice varieties across four location in Nigeria.

s/no.	Trt/no.	Genotype	Sensitivity	Mean	Mean square deviation
1	3	ART15-4-14-63-2-B-1	0.6666	2453	7633
2	13	ART3-1L6P5-1-B-1	0.7323	2267	22924
3	18	ART3-7L3P3-B-B-2	0.7335	2609	49011
4	28	WAB706-27-K5-KB-2	0.7814	2394	5782
5	22	ART3-9L9P3-1-B-2	0.8351	2374	5527
6	5	ART16-12-22-4-1-B-1-1	0.8481	2491	201906
7	7	ART16-22-1-1-2-B-1-1	0.8652	2496	141668
8	9	ART16-9-4-17-3-B-1	0.8931	2187	114581
9	29	WAB788-16-1-1-2-HB	0.8947	2371	199216
10	16	ART3-6L3P9-B-B-2	0.9020	2513	10073
11	10	ART2-6L6P6-1-B-1	0.9223	2517	52360
12	30	WAB891-SG12	0.9226	2741	337543
13	23	FARO55(NERICA1)	0.9326	2388	107737
14	4	ART16-12-17-3-4-B-1-1	0.9480	2371	38952
15	15	ART3-3L12P9-1-1-B	0.9724	2454	43037
16	25	FARO59(NERICA8)	0.9836	2537	8713
17	11	ART3-12L11P2-B-B-1	0.9957	2675	324508
18	1	ART10-1L12P2-1-B-1	1.0467	2647	55744
19	14	ART3-2L4P5-1-B-1	1.0767	2536	17852
20	26	NERICA11	1.0896	2625	68563

Table 5: Twenty environmentally sensitive rice varieties in grain yield across four upland rice growing environments in 2013 using AMMI analysis.

stable high yielding crop genotype for specific as well as for diverse environments. In the present study, AMMI model has shown that the largest proportion of the total variation in rice grain yield in the genotypes is attributed to environments. Most of the genotypes showed environment specificity. The mean grain yield value of genotypes averaged over environments indicated that ART16-9-3-15-3-B-1-1 (8) had the highest mean grain yield 2925 kg/ha. Genotypes ART10-1L12P2-1-B-1(1) and ART16-16-5-23-1-B-1-1 (6) has a small GE effect, which is considered as stable and less influenced by the environment.

References

1. GRISP (Global Rice Science Partnership) (2013). Rice almanac, 4th edition. Los Banos (Philippines) International Rice Research Institute: 283.

2. Akande SO (2002) An Overview of the Nigerian Rice Economy, NISER, Ibadan.

3. Akpokodje GF, Lançon O, Erenstein (2001) Nigeria's Rice Economy: State of the Art. Project Report - The Nigerian Rice Economy in A Competitive World: Constraints, Opportunities And Strategic Choices. Bouake: WARDA: 11-55.

4. UNEP (United Nations Environment Programme) (2005). Annual report; Division of Early Warning and Assessment (DEWA) Nairobi, 00100, Kenya

5. Hill,J (1975) Genotype- environment interaction- A challenge for plant breeding. J. Agri Science, Cam 85: 477-493

6. Lin CS, MR Binn, LP Lefkovitch (1986) Stability an alysis: Where do we stand. Crop Sci 26: 894-900.

7. Crossa J, Gauch HGJ, Zobel RW (1990) Additive main effects and multiplicative interaction analysis of two international maize cultivar trials. Crop Science 30: 493-500

8. Flores GV, Daga A, Kalhor HR, Banerjee U (1998) Lozenge is expressed in pluripotent precursor cells and patterns multiple cell types in the Drosophila eye through the control of cell-specific transcription factors.Development 125: 3681-3687.

9. Eberhart SA, Russell WA (1966) Stability parameters for comparing varieties. Crop Sci 6: 36-40.

10. Perkins JM, Jinks JL (1968) Environmental and genotype-environmental components of variability. 3. Multiple lines and crosses.Heredity (Edinb) 23: 339-356.

11. Gauch HG, Furnas RE (1991) Statistical analysis of yield trials with MATMODEL. Agronomy Journal 83: 916-920.

12. Gauch HG (1988) Model selection and validation for yield trials with interaction. Biometrics 44: 705-15

13. Crossa J, Fox PN, Pfeiffer WH, Rajaram S, Gauch HG Jr (1991) AMMI adjustment for statistical analysis of an international wheat yield trial. Theor Appl Genet 81: 27-37

14. Zobel RW, Wright MJ, Gauch HG (1988) Statistical analysis of a yield trial. Agronomy Journal 80: 388-393.

15. Kayode A Sanni OJ, Ariyo DK, Ojo G, Gregorio, et al. (2009) Additive Main Effects and Multiplicative Interactions Analysis of Grain Yield Performances in Rice Genotypes Across Environments. Asian Journal of Plant Sciences 8: 48-53.

16. Gauch HG (1993) Matmodel version 2.0. AMMI and related analysis for two way data matrics. Micro computer power, Ithaca, New York, USA.

17. Gauch HG Jr, Zobel RW (1989) Accuracy and selection success in yield trial analyses.Theor Appl Genet 77: 473-481.

18. Gauch HG (1990) Using Interaction to Improve Yield Estimates. In: Kang MS (ed.) Genotype-by-Environment Interaction and Plant Breeding, Department of Agronomy, Louisiana State University, Baton Rouge, Louisiana, 141-150.

19. Marcos M, Jean-Marcel R, Fred A (2013) The statistical analysis of multi-environment data: modelling genotype-by-environment interaction and its genetic basis; Wageningen University, Netherland

20. Gabriel K (1978) Least squares approximation of matrices by additive and multiplicative models. J. R. Stat. Soc. B 40: 186-196

21. McDermott B, Coe R (2012). An Easy Introduction to Biplots for Multi-Environment Trials; Statistical Services Centre, University of Reading, UK and World Agroforestry Centre, Kenya

22. Przystalski M, Osman A, Thiemt EM, Rolland B,et al. (2008). Comparing the performance of cereal varieties in organic and non-organic cropping systems in different European countries. Euphytica 163: 417-433

Estimation of Gene Action and Variance Components of Some Reproductive Traits of Rice (*Oryza sativa L*) Through Line x Tester Analysis

Ariful Islam MD[1]*, Khaleque Mian MA[2], Golam Rasul[3],Khaliq QA[4] and Mannan Akanda MA[5]

[1]*Dept of GPB, EXIM Bank Agricultural University, Bangladesh*

[2,3]*Department of GPB, Bangabandhu Sheikh Mujibur Rahman Agricultural University, Bangladesh*

[4]*Department of Agronomy, Bangabandhu Sheikh Mujibur Rahman Agricultural University, Bangladesh*

[5]*Department of Plant Pathology, Bangabandhu Sheikh Mujibur Rahman Agricultural University, Bangladesh*

Abstract

Twenty one parents (five cm S lines and sixteen Restorer lines) were considered for through line x tester analysis to study gene action and combining ability effects of developedcmS and restorer lines. Experiment was carried out at the experimental farm, Department of Genetics and Plant Breeding, Bangabandhu Sheikh Mujibur Rahman Agricultural University, Bangladesh in Boro 2011-12 following RCBD design in three replications. Significant genotypic variances for yield related traits indicated that there were significant variations among the genotypes. Significant gca variances along with additive variance component for reproductive traits indicated the accessibility of additive gene action. Degree of dominance were found negative for most of traits which reveals that regression lines passing below the origin i.e., this character are responsible for over dominance. The linear non-significant relationship between female vs male indicates the reliability of the crosses to go through heterosis breeding. The contributions of lines were found significant indicating preponderance of dominant genes among the lines with tester. The significant interaction of line x tester indicated higher estimates for sca variances. Four restorer lines showed significant negative effects for days to 1st flowering, 80% flowering and days to maturity. Three pollen parents and onecmS line, showed significant positive gca effects for pollen fertility while six pollen parents showed significant positive effects for spikelet's fertility but two pollen parents showed significant positive gca effects for both panicle and stigma exertion rate. The estimated of gca effects of parents indicated that five pollen parents contributed highly significant negative effects for plant height which were responsible for dwarfing character. Fifteen crosses showed significant negative sca estimates for days to first flowering, sixteen crosses for days to 80% flowering and twenty crosses for days to maturity. Among 80 crosses fifty two crosses showed significant positive sca effects along with above average perse performances for grain yield.

Keywords: General combining ability; Specific combining ability; gca effects; sca effects; *Perse* performances

Introduction

Combining ability analysis is one of the powerful available evaluation tools to estimate the combining ability variance and effects for selecting the desirable parents and crosses for exploitation of heterosis. Combining ability variance is usually used for estimate of genetic control of a specific trait .The estimates of additive and non-additive gene action through this technique may be useful in determining the possibility of commercial exploitation of heterosis and isolation of pure line. Hybrid rice offers an opportunity to boost the yield potential of rice. It has a yield advantage of 15-20% over conventional high-yielding varieties. Hybrid rice research now concentrates on the conversion and identification of stable local cytoplasmic male sterile (CMS) lines and effective restorers from local elite lines through repeated backcrossing. To exploit maximum heterosis using male sterility system in hybrid breeding program, we must know the combining ability of different male sterile and restorer lines. The development and use of hybrid rice varieties on commercial scale utilizing cm S fertility restoration system has proved to be one of the mile stones in the history of rice improvement. General combiining ability effect is used to select the desirable parents to be used in crosses. Selection of heterotic hybrid depends on expected level of heterosis as well as the specific combining ability effect. Combining ability is a powerful tool in identifying the best combiners that may be used in crosses either to exploit heterosis or to accumulate fixable genes and obtain desirable segregates that helps to understand the genetic architecture of various characters that enable the breeder to design effective breeding plan for future up-gradation of the existing materials. Breeding strategies based on selection of hybrids require expected level of heterosis as well as the specific combining ability. In breeding high yielding varieties of crop plant, the breeders often face with the problem of selecting parents and crosses. Line x tester analysis provides information about general combining ability (gca) effects of parents and is helpful in estimating various types of gene actions. [1-3] Therefore, the present investigation was carried out to estimate combining ability effects for yield components involving cm S and restorer lines. Considering the above idea the present investigation was undertaken

1. To determine the extent of general combining ability variances in some morpho-reproductive strains for selection of suitable parents.

2. To determine the extent of specific combining ability effects and *perse* performances in some morpho-reproductive trains for selection of suitable parents.

*Corresponding author: Dr. Md. Ariful Islam, Department of Genetics and Plant Breeding, EXIM Bank Agricultural University, Bangladesh
E-mail: i.aarif@yahoo.com

Material and methods

To study combining ability effects of developedcmS and restorer lines an experiment was carried out at the experimental farm, Department of Genetics and Plant Breeding, Bangabandhu Sheikh Mujibur Rahman Agricultural University, Salna, Gazipur during Aman and Boro following RCBD design in three replications. Five knowncmS lines and sixteen developed R-lines were considered for through line x tester analysis. Analysis of variance for general and specific combining ability (gca and sca) were estimated according to line x tester method [4] Five female parents i.e., IR 58025A, BRRI 1A, GAN46A, IR 68888A and IR 62820A and sixteen male parents i.e., RG-BU08-001R, RG-BU08-002R, RG-BU08-005R, RG-BU08-006R, RG-BU08-007R, RG-BU08-013R, RG-BU08-016R, RG-BU08-018R, RG-BU08-025R, RG-BU08-034R, RG-BU08-038R, RG-BU08-046R, RG-BU08-057R, RG-BU08-063R, RG-BU08-097R and RG-BU08-105R were used in the experiment. Data were collected from 10 hills of each genotype on 10 randomly selected individual plant basis were plant height, days to 1st flowering, days to 80% flowering, days to maturity and grain yield (ton/ha). Data obtained for each character was subjected to the analysis of variance following three replicated randomized complete block design by using GENSTAT program.

Results and Discussions

Results from combined ANOVA

Plant height is considered as an important character to select a hybrid. Highly significant variances were found for male (2694.95**), female (1381.70**) and parents (529.05*) indicating the contribution of male, female and parents. The linear non-significant relationship between female vs male (56532.15 ns) indicates the reliability of the crosses to go through heterosis breeding. Significant contribution of female in hybrid (3191.60**) was estimated followed by male in hybrid (3139.24**). While parent vs hybrid (646.91**) was found highly significant. Observed significant differences between parents and hybrids. Significant gca variances along with additive variance component (12.82*) indicates the accessibility of additive gene action .Degree of dominance (0.40937) was found negative which reveals that regression lines passing below the origin i.e. this character is responsible for over dominance (Table 1). The progenies also found the

SI no	Parents/F₁s	Status	SI no	Parents/F₁s	Status
1	RG-BU 08-001 (RG-BU 08-001 R)	Male Parent (Tester)	17	IR 58025A	Female Parent (Line)
2	RG-BU 08-002 (RG-BU 08-002 R)	Male Parent (Tester)	18	GAN 46 A	Female Parent (Line)
3	RG-BU 08-005 (RG-BU 08-003 R)	Male Parent (Tester)	19	IR 62829A	Female Parent (Line)
4	RG-BU 08-006 (RG-BU 08-004 R)	Male Parent (Tester)	20	IR 68888A	Female Parent (Line)
5	RG-BU 08-007 (RG-BU 08-005 R)	Male Parent (Tester)	21	BRRI 1 A	Female Parent (Line)
6	RG-BU 08-013 (RG-BU 08-006 R)	Male Parent (Tester)			
7	RG-BU 08-016 (RG-BU 08-007 R)	Male Parent (Tester)			
8	RG-BU 08-018 (RG-BU 08-008 R)	Male Parent (Tester)			
9	RG-BU 08-025 (RG-BU 08-009 R)	Male Parent (Tester)			
10	RG-BU 08-034 (RG-BU 08-010 R)	Male Parent (Tester)			
11	RG-BU 08-038 (RG-BU 08-011 R)	Male Parent (Tester)			
12	RG-BU 08-046 (RG-BU 08-012 R)	Male Parent (Tester)			
13	RG-BU 08-057 (RG-BU 08-013 R)	Male Parent (Tester)			
14	RG-BU 08-063 (RG-BU 08-014 R)	Male Parent (Tester)			
15	RG-BU 08-097 (RG-BU 08-015 R)	Male Parent (Tester)			
16	RG-BU 08-105 (RG-BU 08-016 R)	Male Parent (Tester)			

And eighty crosses from these parents in winter 2011-12, and winter 2012-13.

SI	Crosses (F₁s)	SI	Crosses (F₁s)	SI	Crosses (F₁s)	SI	Crosses (F₁s)
1	IR 58025A × RG-BU08-001R	21	GAN 46A × RG-BU08-007R	41	BRRI 1A× RG-BU08-025R	61	IR 68888A× RG-BU08-057R
2	IR 58025A × RG-BU08-002R	22	GAN 46A × RG-BU08-013R	42	BRRI 1A× RG-BU08-034R	62	IR 68888A× RG-BU08-063R
3	IR 58025A × RG-BU08-005R	23	GAN 46A × RG-BU08-016R	43	BRRI 1A× RG-BU08-038R	63	IR 68888A× RG-BU08-097R
4	IR 58025A × RG-BU08-006R	24	GAN 46A × RG-BU08-018R	44	BRRI 1A× RG-BU08-046R	64	IR 68888A× RG-BU08-105R
5	IR 58025A × RG-BU08-007R	25	GAN 46A × RG-BU08-025R	45	BRRI 1A× RG-BU08-057R	65	IR 62829A× RG-BU08-001R
6	IR 58025A × RG-BU08-013R	26	GAN 46A × RG-BU08-034R	46	BRRI 1A× RG-BU08-063R	66	IR 62829A× RG-BU08-002R
7	IR 58025A × RG-BU08-016R	27	GAN 46A × RG-BU08-038R	47	BRRI 1A× RG-BU08-097R	67	IR 62829A× RG-BU08-005R
8	IR 58025A × RG-BU08-018R	28	GAN 46A × RG-BU08-046R	48	BRRI 1A× RG-BU08-105R	68	IR 62829A× RG-BU08-006R
9	IR 58025A × RG-BU08-025R	29	GAN 46A × RG-BU08-057R	49	IR 68888A× RG-BU08-001R	69	IR 62829A× RG-BU08-007R
10	IR 58025A × RG-BU08-034R	30	GAN 46A × RG-BU08-063R	50	IR 68888A× RG-BU08-002R	70	IR 62829A× RG-BU08-013R
11	IR 58025A × RG-BU08-038R	31	GAN 46A × RG-BU08-097R	51	IR 68888A× RG-BU08-005R	71	IR 62829A× RG-BU08-016R
12	IR 58025A × RG-BU08-046R	32	GAN 46A × RG-BU08-105R	52	IR 68888A × RG-BU08-006R	72	IR 62829A× RG-BU08-018R
13	IR 58025A × RG-BU08-057R	33	BRRI 1ª × RG-BU08-001R	53	IR 68888A × RG-BU08-007R	73	IR 62829A × RG-BU08-025R
14	IR 58025A × RG-BU08-063R	34	BRRI 1ª × RG-BU08-002R	54	IR 68888A × RG-BU08-013R	74	IR 62829A × RG-BU08-034R
15	IR 58025A × RG-BU08-097R	35	BRRI 1ª × RG-BU08-005R	55	IR 68888A × RG-BU08-016R	75	IR 62829A × RG-BU08-038R
16	IR 58025A × RG-BU08-105R	36	BRRI 1ª × RG-BU08-006R	56	IR 68888A × RG-BU08-018R	76	IR 62829A × RG-BU08-046R
17	GAN 46A × RG-BU08-001R	37	BRRI 1ª × RG-BU08-007R	57	IR 68888A × RG-BU08-025R	77	IR 62829A × RG-BU08-057R
18	GAN 46A × RG-BU08-002R	38	BRRI 1ª × RG-BU08-013R	58	IR 68888A × RG-BU08-034R	78	IR 62829A × RG-BU08-063R
19	GAN 46A × RG-BU08-005R	39	BRRI 1ª × RG-BU08-016R	59	IR 68888A × RG-BU08-038R	79	IR 62829A × RG-BU08-097R
20	GAN 46A × RG-BU08-006R	40	BRRI 1ª × RG-BU08-018R	60	IR 68888A × RG-BU08-046R	80	IR 62829A × RG-BU08-105R

Table 1: List of parental lines (five CMS and sixteen restorer lines).

similar contribution of lines was found significant (13.77**) indicating preponderance of fertile genes among the lines followed by (3.73*) with tester for plant height. The interaction of line x tester was found significant (82.50**) that indicated higher estimates for variances among the preponderance of non-additive gene action. Significant genotypic variances for days to first (310.49**) flowering, days to 80% flowering (3937.62**) and days to maturity (4501.57**) indicates that there is signicant variations among the genotypes. But significant negative variances parents of these traits; days to first (-47.52**) flowering, days to 80% flowering (-602.70**) and days to maturity (-602.70**) indicates that short duration materials are present among these parents. Like parents, significant negative variances were also recorded for female parents for these traits but significant positive variances were recorded for male parents which indicate that restorer lines are mostly long duration materials. The linear non-significant relationship between female vs male (-9704.37, -69886.36 and -82993.74) indicates the reliability of these traits to go through heterosis breeding. While parent vs hybrid (11605.32**, 93994.24** and 110554.38**) were found highly significant. [5, 6] has observed significant differences between parents and hybrids. Significant negative gca variances (-6.455*, -40.556** and -38.864**) along with negative additive variance component (-1.614*, -10.139* and -9.716*) indicates the accessibility of additive gene action towards short duration parents. Like gca variance (∂^2 gca) significant negative sca (∂^2) variances were also recorded -61.427**, -137.922**and -164.943** along with significant negative dominant variance component (-24.857**, -36.481**and-41.736**) indicates the accessibility of dominant gene action towards short duration hybrids among these 80 cross combinations. Degree of dominance (-0.2779, -0.2964 and -0.40937) were found negative which reveals that regression lines passing below the origin i.e., these characters are responsible for over dominance. Yield is considered as the most important character to select any hybrid. The analysis of variances due to genotypes showed a highly significance (6.85**) indicating a wide range of variability for grain yield highly significant variances for male (8.21**), female (2.79**) and parents (11.01**) indicating equal contribution of male, female and parents which might lead to a wide range of variability

among the offspring's. The linear non-significant relationship between female vs male (132.07 ns) indicates the reliability of the parents to go through heterosis breeding. Non-significant contribution of female in hybrid (1.78 ns) indicates there is no significant differences between the female lines for grain yield. Significant contribution of male in hybrid (8.66**) was estimated followed by parents vs hybrid (7.29**) implies that there is a significant difference between parents and hybrids. These results were supported [7,8] (Table 2).

Non-significant general combining ability variances (0.069 ns) reveals that grain yield was not controlled by the additive gene but highly significant specific combining ability variances (1.390**) were observed which reveals that this character is completely controlled by the dominant gene action. The result was completely supported by non-significant additive variance (0.017 ns) and significant dominant genetic variance (0.848*). While selection of suitable cross combination might be done on the basis of significant specific combining ability variances along with dominance variance. Observed that gca and sca variances were highly significant for yield and the yield attributing characters which indicated the importance of both additive and non-additive gene action. Degree of dominance (0.020) was found near zero which reveals that this character is responsible for complete dominance. conducted experiments on combining ability using eight lines and four testers in rice and observed non-additive gene action governing the characters. While non-significant (1.24ns) error sum squares of tester indicates no significant deviations among the testers. Reported highly significant variance for combining ability in different characters among lines, testers and line x tester interactions. In fact, five recognized cm S (Cytoplasmic male sterile) lines were used as testers and among the cm S-lines there was no significant difference for pollen fertility and spikelet fertility. The interaction of line x tester was found significant (76.26**) that indicated higher estimates for variances due to dominant gene action [9-12].

Estimation of general combining ability (GCA) effects

The general combining ability effects of the parents in the present

Source of Variation		df	Plant height (cm)	Days first flowering	Days 80 % flowering	Days to maturity	Yield (t/ha)
Replication	r-1	2	529054.36	47523.73	602697.02	689015.99	1007.38
Genotypes	g-1	100	3456.49**	310.49**	3937.62**	4501.57**	6.58**
Parents	p-1	20	529.05**	-47.52**	-602.70**	-689.02**	11.01**
females (f)	f-1	4	1381.70**	449.59**	2580.82**	3216.64**	-2.79**
males (m)	m-1	15	2694.95**	-463.70**	-3167.27**	-4521.46**	8.21**
f vs m	1	1	56532.15	-9704.37	-69886.36	-82993.74	-132.07
Hybrids (c)	c-1	79	3525.77**	-258.15**	-3947.11**	-4473.208**	6.40**
female in hybrids (fh)	f-1	4	3191.60**	546.79**	4472.00**	5324.35**	1.78 ns
male in hybrid (mh)	m-1	15	3139.24**	475.99**	3702.70**	4370.63**	8.66**
fh x mh (Line×Testers)	(f-1)(m-1)	60	4703.29**	317.74**	5304.04**	6013.42**	76.29**
Parents vs hybrids	1	1	77694.32**	11605.32**	93994.24**	110554.38**	172.36**
Error	(e-1)(r-1)	200	724.45	65.08	825.29	943.49	1.38
Total	cr-1	239	1058108.72	95047.47	1205394.05	1378031.99	2014.76
Variance components							
∂^2 gca			-51.26*	-6.455*	-40.556**	-38.864**	0.068
∂^2 sca			125.22**	-61.427**	-137.922**	-164.943**	1.390**
∂^2A			12.82*	-1.614*	-10.139*	-9.716*	-0.017 ns
∂^2 D			31.31**	-24.857**	-36.481**	-41.736**	0.848**
∂^2 gca / ∂^2sca			-0.40937	0.1051	-0.2941	-0.2356	-0.0497
∂^2A/∂^2D (Degree of dominance)			-0.2328	-0.2779	-0.2964	-0.40937	-0.0204

*p=0.05, **p=0.01 and ns=non-significant.

Table 2: Estimates of mean sum of squares for lines, testers, line x tester as well as variance components (additive, dominance and degree of dominance) in 80 F_1 hybrids and their parents.

study have brought to light the parents with high gca effects for five different traits. A total of 16 crosses were found having significant negative heterosis for plant height. Where three crosses of IR 58025A, three crosses of BRRI 1A, four crosses of GAN46A, two crosses of IR 68888A and four crosses of IR 62820A are significant negative heterosis for plant height. This result indicates that the following croosses are semi dwarf in nature. The crosses of IR58025A showed significant positive specific combining ability effects with RG-BU08-006R&RG-BU08-034R. As these crosses showed highly significant positive sca effects and above average *perse* performances, might not be selected as suitable hybrid. The crosses of GAN46A also found significant positive sca effects and above average *perse* performances with RG-BU08-007R, RG-BU08-018R & RG-BU08-105R. The crosses of IR68888A showed significant positive sca effects and above average *perse* performances with RG-BU08-002R, RG-BU08-006R, RG-BU 08-018R and RG-BU08-016R which can be considered as good specific combination for tallness. As the above crosses showed positive sca effects which could be used as above average combinations for tall stature. These results are in line with the findings of Good specific combinations for tallness were evolved from high × high, general combiner parents. Low × above average general combiner parents produced above average specific combination for tall plant height in rest of the crosses also found similar findings. Semi-dwarf plant height is needed to protect the crop from lodging. The estimated gca effects of parents indicated that the parent, RG-BU 08-001R, RG-BU 08-002R, RG-BU 08-0046R, RG-BU 08-0057R, RG-BU 08-0097R, BRRI 1A and IR62829A contributed highly significant negative effects .These facts indicated that the former parents possessed more negative alleles for the dwarf stature of the parents. *Perse* performance also supported that RG-BU 08-001R (89.03 cm), RG-BU 08-006R (96.72 cm), RG-BU 08-057R (90.39 cm), RG-BU

08-097R (81.39 cm) and BRRI 1A (90.59 cm) except RG-BU 08-002R (103.69 cm) and IR62829A (108.57cm)were dwarf in stature. Although general gca effects found significant negative but *perse* performances did not correlate with RG-BU 08-002R (103.69 cm) and IR62829A (108.57 cm) which might be due to pseudo recessive gene effect or wide environmental fluctuation. Therefore, RG-BU 08-001R, RG-BU 08-046R, RG-BU 08-057R, RG-BU 08-097R and BRRI 1A are potential parents and have highly significant gca effect in the desirable direction (negative direction) for plant height. These findings are in accordance with [13-16]. Out of sixteen restorer lines eight restorer lines showed significant negative effects for days to 1st flowering, seven showed significant negative effects for days to 80% flowering and five showed significant negative effects for days to maturity. But among these 16 restorer lines, only four showed significant negative gca effects for all these three traits. The restorer lines showing significant nigative gca effects for all three traits are RG-BU 08-005R (-7.43**, -6.847**, -7.590**), RG-BU 08-006R (-6.073**, -6.177**, -7.412**), RG-BU 08-007R (-3.743*, -5.057**, -7.564**), RG-BU 08-097R (-8.743**, -6.177**, -5.532**). Such lines could be used as male parent for development of early maturing hybrids in rice. performed line x tester analysis using five lines and four testers to study the combining ability and heterosis for yield and its contributing characters in rice and observed that IR58025A and IR62829A were good general combiners for earliness, grain yield per plant and per day productivity [17] (Table 3).

While estimating the life cycle of these fivecmS lines, it was observed that IR 62829A (-2.996**) showed significant negative effects for days to 1st flowering and IR 62829A (-3.224*) for days to 80% flowering while GAN 46A (-2.236**, -2.994**, -5.740**) showed significant negative effects for all three traits. Such estimates indicated that these female parents possessing more negative alleles for first flowering. *Perse*

Source of Variation	Plant height (cm)		Days first flowering		Days 80% flowering		Days to maturity		Yield (ton/ha)	
Lines	gi/gj	Mean	gi/gj	mean	gi/gj	Mean	gi/gj	Mean	gi/gj	Mean
IR 58025A	-0.466	89.02 ± 12.07	0.824	106.66 ± 1.12	-1.074	115.32 ± 3.06	-2.020	139.66 ± 2.45	0.074**	2.30 ± 0.04
IR 62829A	-2.797	90.59 ± 12.07	-2.996*	102.84 ± 1.12	-3.224*	113.17 ± 3.06	-0.170	141.51 ± 2.45	-0.154**	2.16 ± 0.04
GAN 46 A	6.905	93.81 ± 12.07	-2.236*	103.60 ± 1.12	-2.994*	113.40 ± 3.06	-5.740**	135.94 ± 2.45	0.016	2.32 ± 0.04
IR 68888A	0.194	107.73 ± 12.07	-0.706	105.13 ± 1.12	1.066	117.46 ± 3.06	2.660*	144.34 ± 2.45	0.346**	2.66 ± 0.04
BRRI 1 A	-3.836	108.57 ± 12.07	5.114**	110.95 ± 1.12	6.226**	112.62 ± 3.06	5.270**	146.95 ± 2.45	-0.184**	2.13 ± 0.04
SE (gi)		12.074		1.129		2.085		2.455		0.023
SE (gi - gj)		30.185		2.222		3.711		3.387		0.057
Testers										
RG-BU 08-001 R	-5.123**	89.03 ± 2.27	-2.403	103.01 ± 3.48	3.153*	117.67 ± 5.67	5.436**	146.34 ± 5.04	0.389	5.49 ± 0.784
RG-BU 08-002 R	-5.823**	103.69 ± 2.27	-4.743*	100.67 ± 3.48	-1.177	113.34 ± 5.67	1.106	142.01 ± 5.04	0.879**	5.98 ± 0.784
RG-BU 08-005 R	-1.788	99.40 ± 2.27	-7.743**	97.67 ± 3.48	-6.847**	107.67 ± 5.67	-7.590**	133.34 ± 5.04	0.439*	5.54 ± 0.784
RG-BU 08-006 R	12.536**	96.72 ± 2.27	-6.073**	99.34 ± 3.48	-6.177**	108.34 ± 5.67	-7.412**	133.37 ± 5.04	1.609*	6.71 ± 0.784
RG-BU 08-007 R	-1.394	136.52 ± 2.27	-3.743*	101.67 ± 3.48	-5.507*	109.01 ± 5.67	-7.564**	133.02 ± 5.04	0.449*	5.55 ± 0.784
RG-BU 08-013 R	-1.063	98.54 ± 2.27	4.098*	109.51 ± 3.48	1.653	116.17 ± 5.67	-1.064	139.84 ± 5.04	0.449*	5.55 ± 0.784
RG-BU 08-016 R	-0.188	99.23 ± 2.27	11.928**	117.34 ± 3.48	8.823**	123.34 ± 5.67	5.436**	146.34 ± 5.04	0.449*	5.55 ± 0.784
RG-BU 08-018 R	0.320	91.23 ± 2.27	4.598*	110.01 ± 3.48	4.153*	118.67 ± 5.67	0.106	141.01 ± 5.04	0.109	5.21 ± 0.784
RG-BU 08-025 R	0.511	86.79 ± 2.27	-3.743*	101.67 ± 3.48	-4.847*	109.67 ± 5.67	-2.564	138.34 ± 5.04	-2.141**	2.96 ± 0.784
RG-BU 08-034 R	10.682**	92.94 ± 2.27	11.258**	116.67 ± 3.48	6.493**	121.01 ± 5.67	-2.694	138.21 ± 5.04	0.909**	6.01 ± 0.784
RG-BU 08-038 R	0.199	96.11 ± 2.27	9.598**	115.01 ± 3.48	5.823**	120.34 ± 5.67	5.436**	146.34 ± 5.04	-0.431*	4.67 ± 0.784
RG-BU 08-046 R	-3.282*	96.13 ± 2.27	-9.073**	96.34 ± 3.48	-6.507**	108.01 ± 5.67	7.436**	148.34 ± 5.04	-0.341	4.76 ± 0.784
RG-BU 08-057 R	-7.027**	100.39 ± 2.27	-1.403	104.01 ± 3.48	-4.177*	110.34 ± 5.67	-7.564**	133.34 ± 5.04	-1.231**	3.87 ± 0.784
RG-BU 08-063 R	2.580*	87.41 ± 2.27	-6.073**	99.34 ± 3.48	-0.177	114.34 ± 5.67	-2.564	138.34 ± 5.04	-0.101	5.00 ± 0.784
RG-BU 08-097 R	-4.083*	81.39 ± 2.27	-8.743**	96.67 ± 3.48	-6.177**	108.34 ± 5.67	-5.532**	135.31 ± 5.04	-1.311**	3.79 ± 0.784
RG-BU 08-105 R	2.942*	97.95 ± 2.27	12.258**	117.67 ± 3.48	11.493**	126.01 ± 5.67	10.106**	151.01 ± 5.04	-0.121	4.98 ± 0.784
SE (gi)		2.278		3.483		3.067		3.581		0.426
SE (gi - gj)		4.594		5.730		5.677		5.039		0.784

*p= 0.05, **p= 0.01 and ns = Insignificant.

Table 3: GCA effects of parents for different yield and yield contributing character of rice.

performances also recorded similar results supported parents RG-BU 08-005R (97.67, 107.07, 133.44 days), RG-BU 08-006R (99.34, 108.34, 133.47 days), RG-BU 08-007R (101.67, 109.01, 133.02 days), RG-BU08-097R (96.67, 108.34, 135.31 days) and GAN 46A (103.60, 113.40, 135.94 days). These facts indicated that the above parents possessed more negative alleles for the decreasing the life cycle. Compared to BRRI than29 of? these parents matured 27 ± 2 days earlier. So, RG-BU 08-005R, RG-BU 08-006R, RG-BU 08-007R and RG-BU 08-057R might be used in the heterosis breeding. As general combining ability (gca) effects found significant negative and their *perse* performances were comparatively lower; therefore, these parents might be used as suitable parents to develop short duration hybrid variety. These findings are in accordance with Won & Yoshida 2000 Significant positive gca effects was found in pollen parents RG-BU 08-002R (0.879**), RG-BU 08-005R (0.439*), RG-BU 08-006R (1.069**), RG-BU 08-007R (0.449*), RG-BU 08-013R (0.449*), RG-BU 08-016R (0.449*), RG-BU 08-034R (0.909**) and positive general combining ability effects ofcmS parents IR 58025A (0.074**) and IR 68888A (0.346**). *Perse* performances revealed that among 21 parents (16 pollen parents and 5cmS parents) seven pollen parents RG-BU 08-002R (5.98 t/ha), RG-BU 08-005R (5.54 t/ha), RG-BU 08-006R (6.71 t/ha), RG-BU 08-007R (5.55 t/ha), RG-BU 08-013R (5.55 t/ha), RG-BU 08-016R (5.55 t/ha), RG-BU 08-034R (6.01 t/ha) and 2 cmS lines IR 58025A (2.40 t/ha) and IR 68888A (2.66 t/ha) were superior to others. These facts indicated that among 21 parents these nine parents possessed more positive alleles for the increase of grain yield. Observed good general combinercmS lines for grain yield along with other yield contributing characters in rice. So, among the male parents, RG-BU 08-002R, RG-BU 08-005R, RG-BU 08-006R, RG-BU 08-007R, RG-BU 08-013R, RG-BU 08-016R and RG-BU 08-034R were the best general combiner due to highly significant positive gca effects. On the other hand RG-BU 08-025R (-2.141**, 2.96 t/ha), RG-BU 08-038R (-0.431*, 4.67 t/ha), RG-BU 08-046R (-0.341*, 4.76 t/ha), RG-BU 08-057R (-1.231*, 3.87 t/ha), and RG-BU 08-097R (-1.311*, 3.79 t/ha) as well ascmS parents IR 62829A (-0.154**, 2.16 t/ha) and BRRI 1A (-0.184**, 2.13 t/ha) showed highly significant negative general combining ability effects found similar results while studying gca effects in rice [18-20].

Estimation of specific combining ability (SCA) effects

The specific combining ability effects of the trait plant height is considered as an important character to select a hybrid. The crosses of IR58025A showed significant positive specific combining ability effects with RG-BU08-006R & RG-BU08-034R. As these crosses showed highly significant positive sca effects and above average *perse* performances, might not be selected as suitable hybrid. The crosses of GAN46A also found significant positive sca effects and above average *perse* performances with RG-BU08-007R, RG-BU08-018R & RG-BU08-105R. The crosses of IR68888A showed significant positive sca effects and above average *perse* performances with RG-BU08-002R, RG-BU08-006R, RG-BU 08-018R and RG-BU08-016R which can be considered as good specific combination for tallness. As the above crosses showed positive sca effects which could be used as above average combinations for tall stature. These results are in line with the findings of. Good specific combinations for tallness were evolved from high x high, general combiner parents. Low × above average general combiner parents produced above average specific combination for tall plant height in the rest of the crosses also found similar findings [13-15]. Out of 80 test crosses fifteen crosses showed significant negative sca estimates for days to first flowering and sixteen crosses showed significant negative sca estimates for days to 80% flowering. Out of 80 crosses twenty crosses showed significant negative sca estimates for days to maturity, where seven with IR 58025A, two with GAN46A, six with IR 62829A, two with IR 68888A and three with BRRI 1A. In all the cases it was observed that maximum number of crosses were found showing significant negative sca estimates with IR 58025A. The F_1s crosses of IR 58025A with seven restorer lines showed significant negative sca estimates for days to first flowering, 6 F_1s for days to 80% flowering and seven for days to maturity (Tables 4,5).

Line	Plant height									
	IR 58025A		GAN 46A		IR 62829A		IR 68888A		BRRI 1A	
Testers	Sij effect	mean	Sij effect	mean	Sij effect	mean	Sij effect	mean	Sij effect	Mean
RG-BU 08-001 R	-4.98	98.79 ± 7.26	-4.28	99.49 ± 7.78	-9.95**	93.82 ± 7.85	-5.93*	92.84 ± 7.39	-3.45	107.32 ± 7.25
RG-BU 08-002 R	-5.50*	98.27 ± 7.26	-3.91	99.86 ± 7.78	-8.86**	94.91 ± 7.85	4.44	108.22 ± 7.39	-15.27**	88.50 ± 7.25
RG-BU 08-005 R	-2.74	101.03 ± 7.26	-7.50*	96.27 ± 7.78	6.60*	110.38 ± 7.85	-1.01	102.76 ± 7.39	-4.28	99.49 ± 7.25
RG-BU 08-006 R	10.71**	114.49 ± 7.26	1.59	105.37 ± 7.78	46.40**	150.18 ± 7.85	7.88**	111.66 ± 7.39	-3.91	99.86 ± 7.25
RG-BU 08-007 R	-0.34	103.43 ± 7.26	4.87	108.65 ± 7.78	-2.31	101.46 ± 7.85	-1.67	102.10 ± 7.39	-7.50*	96.27 ± 7.25
RG-BU 08-013 R	-3.49	100.28 ± 7.26	-15.64**	88.13 ± 7.78	12.34**	116.12 ± 7.85	-0.11	103.66 ± 7.39	1.59	105.37 ± 7.25
RG-BU 08-016 R	-0.74	103.03 ± 7.26	-3.54	100.23 ± 7.78	-3.47	100.30 ± 7.85	1.95	105.73 ± 7.39	4.87	108.65 ± 7.25
RG-BU 08-018 R	1.06	104.84 ± 7.26	4.76	108.54 ± 7.78	8.05**	111.83 ± 7.85	3.36	107.14 ± 7.39	-15.64**	88.13 ± 7.25
RG-BU 08-025 R	0.10	103.88 ± 7.26	-1.91	101.86 ± 7.78	5.37*	109.15 ± 7.85	2.53	106.31 ± 7.39	-3.54	100.23 ± 7.25
RG-BU 08-034 R	8.60**	112.38 ± 7.26	-1.75	102.02 ± 7.78	45.17**	148.95 ± 7.85	-3.38	100.39 ± 7.39	4.76	108.54 ± 7.25
RG-BU 08-038 R	-0.20	103.57 ± 7.26	-2.21	101.56 ± 7.78	9.04*	112.82 ± 7.85	-3.71	100.06 ± 7.39	-1.91	101.86 ± 7.25
RG-BU 08-046 R	-3.93	99.84 ± 7.26	-7.22	96.55 ± 7.78	-5.74*	98.03 ± 7.85	2.25	106.03 ± 7.39	-1.75	102.02 ± 7.25
RG-BU 08-057 R	-6.56*	97.21 ± 7.26	-4.25	99.52 ± 7.78	-17.54**	86.23 ± 7.85	-4.55	99.22 ± 7.39	-2.21	101.56 ± 7.25
RG-BU 08-063 R	1.71	105.49 ± 7.26	-2.62	101.15 ± 7.78	17.52**	121.30 ± 7.85	3.51	107.29 ± 7.39	-7.22*	96.55 ± 7.25
RG-BU 08-097 R	-5.38*	99.39 ± 7.26	-5.88*	97.89 ± 7.78	-5.66**	98.44 ± 7.85	-6.55*	97.22 ± 7.39	-4.25	99.52 ± 7.25
RG-BU 08-105 R	3.24	107.02 ± 7.26	4.78	108.56 ± 7.78	7.19*	110.97 ± 7.85	2.10	105.88 ± 7.39	-2.62	101.15 ± 7.25
Mean		103.31		100.98		110.68		103.97		99.94
SE (sij)	3.113									
SEd (Sij-Sik)	5.153									
SEd (Sij-Skj)	7.561									
t = Sij/SEI (Sij) at error df=239										

*p=0.05, **p=0.01 and ⁿˢ=non-significant.

Table 4: SCA effects (*Sij*) vis-à-vis *per-se* mean performance of hybrids for plant height (cm) in 80 F₁ hybrids.

Line	Days to first flowering									
	IR 58025A		GAN 46A		IR 62829A		IR 68888A		BRRI 1A	
Testers	Sij effect	mean	Sij effect	mean	Sij effect	mean	Sij effect	mean	Sij effect	Mean
RG-BU 08-001 R	-4.214*	105.90 ± 2.63	4.450*	114.56 ± 3.63	0.12	114.56 ± 3.63	-5.54*	110.23 ± 3.63	4.450*	104.56 ± 3.63
RG-BU 08-002 R	-1.550	108.56 ± 2.63	0.340	110.45 ± 3.63	-0.88	110.45 ± 3.63	7.12*	109.23 ± 3.63	0.340	117.23 ± 3.63
RG-BU 08-005 R	-6.547*	103.56 ± 2.63	-0.880	109.23 ± 3.63	-1.88	109.23 ± 3.63	-13.21**	107.23 ± 3.63	-0.880	119.23 ± 3.63
RG-BU 08-006 R	-8.650*	101.46 ± 3.63	0.570	110.68 ± 3.63	-4.88*	102.23 ± 3.63	-13.21**	107.23 ± 3.63	-17.880**	96.90 ± 3.63
RG-BU 08-007 R	2.340	112.45 ± 3.63	-5.880*	104.23 ± 3.63	-1.88	111.23 ± 3.63	3.12	108.23 ± 3.63	1.120	113.23 ± 3.63
RG-BU 08-013 R	-7.547*	102.56 ± 2.63	-3.214	106.90 ± 3.63	3.12	107.90 ± 3.63	3.12	113.23 ± 3.63	-2.214	113.23 ± 3.63
RG-BU 08-016 R	-9.547*	100.56 ± 2.63	-9.547*	100.56 ± 3.63	2.12	110.23 ± 3.63	6.13*	112.23 ± 3.63	0.120	116.23 ± 3.63
RG-BU 08-018 R	-7.880*	102.23 ± 2.63	-7.880*	102.23 ± 3.63	-1.88	113.56 ± 3.63	8.78*	108.23 ± 3.63	3.453	118.23 ± 3.63
RG-BU 08-025 R	-5.547*	104.56 ± 2.63	-5.547*	104.56 ± 3.63	0.12	120.23 ± 3.63	2.78	110.23 ± 3.63	10.120**	112.90 ± 3.63
RG-BU 08-034 R	-3.214	106.90 ± 2.63	-3.214	106.90 ± 3.63	1.12	113.23 ± 3.63	-1.21	111.23 ± 3.63	3.120	108.90 ± 3.63
RG-BU 08-038 R	-0.660	109.45 ± 2.63	-0.660	109.45 ± 3.63	2.12	113.23 ± 3.63	-2.54	112.23 ± 3.63	3.120	107.56 ± 3.63
RG-BU 08-046 R	-0.214	109.90 ± 2.63	-0.214	109.90 ± 3.63	0.12	116.23 ± 3.63	0.12	110.23 ± 3.63	6.120*	110.23 ± 3.63
RG-BU 08-057 R	3.120	113.23 ± 2.63	3.120	113.23 ± 3.63	-0.88	118.23 ± 3.63	-0.88	109.23 ± 3.63	8.120*	109.23 ± 3.63
RG-BU 08-063 R	2.453	112.56 ± 2.63	2.453	112.56 ± 3.63	5.12*	112.90 ± 3.63	-0.88	115.23 ± 3.63	2.786	109.23 ± 3.63
RG-BU 08-097 R	5.120*	115.23 ± 2.63	5.120*	115.23 ± 3.63	3.12	108.90 ± 3.63	5.19*	113.23 ± 3.63	-1.214	115.23 ± 3.63
RG-BU 08-105 R	1.453	111.56 ± 2.63	1.453	111.56 ± 3.63	2.12	107.56 ± 3.63	3.12	112.23 ± 3.63	-2.547	113.23 ± 3.63
Mean		107.54		108.89		111.87		110.67		111.58
SE (sij)		4.126								
SEd (Sij-Sik)		11.692								
SEd (Sij-Skj)		13.205								
t = Sij/SEl (Sij) at error df = 239										

*p=0.05, **p=0.01 and ⁿˢ=non-significant.

Table 5: SCA effects (*Sij*) vis-à-vis *per-se* mean performance of hybrids for days to first flowering in 80 F_1 hybrids.

Crosses of IR 58025A showed significant negative relationship for days to 1st flowering and 80% flowering with RG-BU08-005R (-6.547*, -6.827*, -6.407* and 103.56, 110.42, 137.15 days), RG-BU08-006R (-8.65*, -8.950**, -9.23** and 101.46, 108.30, 134.33 days), RG-BU08-013R (-7.54*, -7.71*, -7.99* and 102.56, 109.43, 134.56 days), RG-BU08-016R (-9.54*, -9.17*, -9.35* and 100.56, 108.07, 134.20 days), RG-BU08-018R (-7.88*, -7.56*, -7.54* and 102.23, 109.69, 135.92 days) and RG-BU08-025R (-5.54*, -5.23*,-5.01 and 104.56, 112.01, 138.54 days). As these combinations showed significant negative sca effects that could be used as above average specific combinations for earlier flowering [10] in rice found significant negative sca values in days to 1st flowering, 80% flowering and maturity. From this table earliness considering both sca effects and *perse* performances for the characters days to 1st flowering and 80% flowering crosses of IR 58025A with RG-BU08-005R, RG-BU08-006R, RG-BU08-016R, RG-BU08-018R and RG-BU08-025R might be recommended. So, among these 80 cross combinations the following crosses might be selected for earliness. These results are in line with the findings [21].

Ten crosses of IR 58025A, seven crosses of GAN46A, fourteen crosses of IR 62829A, nine crosses of IR 68888A and ten crosses of BRRI 1A showed significant positive sca effects along with mean values. Ten crosses of IR 58025A with the restorers showed significant positive specific combining ability effects along with above average *perse* performances of the crosses were found in RG-BU 08-001R (0.881*, 4.97 t/ha), RG-BU 08-002R (2.67*, 6.71 t/ha), RG-BU 08-006R (2.024**, 6.11 t/ha), RG-BU 08-013R (2.852**, 6.94 t/ha), RG-BU 08-018R (1.269**, 5.36 t/ha), RG-BU 08-046R (1.425**, 5.52 t/ha), RG-BU 08-063R (1.513**, 5.60 t/ha), RG-BU 08-063R (2.95**, 6.73 t/ha) and RG-BU 08-105R (1.774**, 5.86 t/ha). Seven Crosses of GAN46A with the restorers also showed highly significant positive sca effects and above average *perse* performances for grain yield were found in RG-BU 08-001R (0.578*, 4.67 t/ha), RG-BU 08-002R (3.50*, 7.59 t/ha), RG-BU

08-007R (3.426**, 7.33 t/ha), RG-BU 08-018R (0.722*, 4.86 t/ha), RG-BU 08-063R (1.544**, 5.63 t/ha) and RG-BU 08-097R (2.452**, 6.54 t/ha).

Fourteen crosses of IR62829A resulted highly significant positive specific combining ability effects. These were found in the crosses with RG-BU08-001R (1.16**, 5.26 t/ha), RG-BU 08-002R (2.06*, 6.02 t/ha), RG-BU08-005R (2.350**, 6.44 t/ha), RG-BU08-013R (1.138**, 5.23 t/ha), RG-BU08-016R (0.960*, 5.05 t/ha), RG-BU08-025R (2.239**, 6.33 t/ha), RG-BU08-034R (1.056**, 5.15 t/ha), RG-BU08-057R (2.885**, 6.97 t/ha), RG-BU08-063R (1.459**, 5.63 t/ha) and RG-BU08-097R (1.83**, 6.26 t/ha) (Tables 6,7). Nine crosses of IR68888A showed significant positive sca effects and above average *perse* performance were found in the crosses with RG-BU 08-001R (0.98**, 5.10 t/ha), RG-BU 08-006R (2.66*, 5.93 t/ha), RG-BU 08-007R (1.62**, 5.71 t/ha), RG-BU 08-018R (3.09**, 7.18 t/ha), RG-BU 08-057R (0.84**, 4.93 t/ha), RG-BU 08-063R (0.81**, 4.88 t/ha) and RG-BU 08-097R (1.16**, 5.26 t/ha). Ten crosses of BRRI1A showed significant positive sca effects and above average *perse* performance were found in the crosses with RG-BU 08-001R (1.972**, 6.06 t/ha), RG-BU 08-002R (1.97*, 6.06 t/ha), RG-BU 08-005R (1.570**, 5.66 t/ha), RG-BU 08-006R (1.47**, 5.62 t/ha), RG-BU 08-007R (2.21**, 6.10 t/ha), RG-BU 08-016R (3.491**, 7.48 t/ha), RG-BU 08-025R (3.278**, 6.37 t/ha), RG-BU 08-063R (1.572**, 5.66 t/ha) and RG-BU 08-097R (2.21**, 6.18 t/ha) [10,11] observed non-additive gene action governing the characters. Banumathy and Thiyagarajan 2005 also found similar results while studying sca variances of rice. The crosses of five R-lines, RG-BU08-001R, RG-BU08-002R, RG-BU08-006R, RG-BU08-007R and RG-BU08-097R were found resulting significant positive sca effects and above average *perse* performances with all five cm S lines for grain yield. RG-BU08-002R and RG-BU08-097R were found resulting significant positive sca effects with all yield contributing characters like pollen fertility, spikelets fertility, panicle exertion rate, stigma exertion rate, effective tillers per plant, primary branches per panicle and secondary primary branches

Line / Testers	Days to 80% flowering									
	IR 58025A		GAN 46A		IR 62829A		IR 68888A		BRRI 1A	
	Sij effect	mean	Sij effect	mean	Sij effect	mean	Sij effect	mean	Sij effect	Mean
RG-BU 08-001 R	-3.844	113.41 ± 3.20	4.090	121.34 ± 3.20	4.770*	122.02 ± 3.20	-0.150	117.10 ± 3.20	-5.8278	111.42 ± 3.20
RG-BU 08-002 R	-1.070	116.18 ± 3.20	0.070	117.32 ± 3.20	0.650	117.90 ± 3.20	-0.510	116.74 ± 3.20	6.820*	124.07 ± 3.20
RG-BU 08-005 R	-6.827*	110.42 ± 3.20	-0.510	116.74 ± 3.20	-1.150	116.10 ± 3.20	-1.510	115.74 ± 3.20	8.760**	126.01 ± 3.20
RG-BU 08-006 R	-8.95**	108.30 ± 3.20	0.890	118.14 ± 3.20	-8.15**	109.10 ± 3.20	-4.400*	114.85 ± 3.20	-13.48**	103.77 ± 3.20
RG-BU 08-007 R	1.980	119.23 ± 3.20	-5.570*	111.68 ± 3.20	0.850	118.10 ± 3.20	-2.160	115.09 ± 3.20	3.490	120.74 ± 3.20
RG-BU 08-013 R	-7.817*	109.43 ± 3.20	-3.484	113.77 ± 3.20	-1.844	115.41 ± 3.20	2.820	120.07 ± 3.20	3.440	120.69 ± 3.20
RG-BU 08-016 R	-9.17**	108.07 ± 3.20	-9.817**	107.43 ± 3.20	0.600	117.85 ± 3.20	1.760	119.01 ± 3.20	6.430*	123.68 ± 3.20
RG-BU 08-018 R	-7.56**	109.69 ± 3.20	-8.150**	109.10 ± 3.20	3.173	120.42 ± 3.20	-2.150	115.10 ± 3.20	7.850**	125.10 ± 3.20
RG-BU 08-025 R	-5.237*	112.01 ± 3.20	-5.177*	112.07 ± 3.20	9.820**	127.07 ± 3.20	0.490	117.74 ± 3.20	2.516	119.77 ± 3.20
RG-BU 08-034 R	-3.484	113.77 ± 3.20	-2.844	114.41 ± 3.20	2.760	120.01 ± 3.20	1.440	118.69 ± 3.20	-1.484	115.77 ± 3.20
RG-BU 08-038 R	-0.930	116.32 ± 3.20	-0.180	117.07 ± 3.20	2.850	120.10 ± 3.20	2.430	119.68 ± 3.20	-2.177	115.07 ± 3.20
RG-BU 08-046 R	-0.484	116.77 ± 3.20	-0.494	116.76 ± 3.20	6.490*	123.74 ± 3.20	-0.150	117.10 ± 3.20	0.490	117.74 ± 3.20
RG-BU 08-057 R	3.490	120.74 ± 3.20	2.820	120.07 ± 3.20	8.440**	125.69 ± 3.20	-1.150	116.10 ± 3.20	-0.400	116.85 ± .20
RG-BU 08-063 R	2.933	120.18 ± 3.20	2.093	119.34 ± 3.20	3.096	120.35 ± 3.20	4.850*	122.10 ± 3.20	-1.160	116.09 ± 3.20
RG-BU 08-097 R	4.840*	122.09 ± 3.20	4.850*	122.10 ± 3.20	-1.484	115.77 ± 3.20	3.490	120.74 ± 3.20	4.820*	122.07 ± 3.20
RG-BU 08-105 R	1.153	118.40 ± 3.20	1.823	119.07 ± 3.20	-2.817	114.43 ± 3.20	2.600	119.85 ± 3.20	2.760	120.01 ± 3.20
Mean		114.69		116.03		119.00		117.86		118.67
SE (sij)		4.317								
SEd (Sij-Sik)		7.366								
SEd (Sij-Skj)		3.205								
t = Sij/SEI (Sij) at error df=239										

*p=0.05, **p=0.01 and ns=non-significant.

Table 6: SCA effects (Sij) vis-à-vis per-se mean performance of hybrids for days to 80% flowering in 80 F_1 hybrids.

ine / Testers	Days to maturity									
	IR 58025A		GAN 46A		IR 62829A		IR 68888A		BRRI 1A	
	Sij effect	mean	Sij effect	mean	Sij effect	mean	Sij effect	mean	Sij effect	Mean
RG-BU 08-001 R	-3.92	139.64 ± 6.87	-5.29*	137.47 ± 6.87	-4.91*	140.85 ± 6.87	-4.43*	139.15 ± 6.87	-6.07*	137.55 ± 6.87
RG-BU 08-002 R	-1.25	142.31 ± 6.87	1.17	143.45 ± 6.87	-0.11	144.73 ± 6.87	-0.59	142.97 ± 6.87	6.64*	150.20 ± 6.87
RG-BU 08-005 R	-6.40*	137.15 ± 6.87	-1.13	142.97 ± 6.87	-0.59	142.43 ± 6.87	-1.59	141.97 ± 6.87	8.68**	152.24 ± 6.87
RG-BU 08-006 R	-9.23**	134.33 ± 6.87	-8.23**	144.97 ± 6.87	1.41	135.33 ± 6.87	-1.88	141.68 ± 6.87	-13.26**	130.30 ± 6.87
RG-BU 08-007 R	1.70	145.26 ± 6.87	0.67	137.91 ± 6.87	-5.65*	144.23 ± 6.87	-1.64	141.92 ± 6.87	3.21	146.77 ± 6.87
RG-BU 08-013 R	-7.99**	135.56 ± 6.87	-2.02	139.90 ± 6.87	-3.66	141.54 ± 6.87	2.84	146.40 ± 6.87	3.36	146.92 ± 6.87
RG-BU 08-016 R	-9.35**	134.20 ± 6.87	0.52	134.16 ± 6.87	-9.39**	144.08 ± 6.87	1.68	145.24 ± 6.87	6.35*	149.91 ± 6.87
RG-BU 08-018 R	-7.64**	135.92 ± 6.87	3.69	135.13 ± 6.87	-8.43**	147.25 ± 6.87	-2.33	141.23 ± 6.87	8.37**	151.93 ± 6.87
RG-BU 08-025 R	-5.01*	138.54 ± 6.87	9.64**	138.10 ± 6.87	-5.45*	153.20 ± 6.87	0.31	143.87 ± 6.87	3.03	146.60 ± 6.87
RG-BU 08-034 R	-3.76	139.80 ± 6.87	3.18	140.54 ± 6.87	-3.02	146.74 ± 6.87	1.36	144.92 ± 6.87	-1.46	142.10 ± 6.87
RG-BU 08-038 R	-1.01	142.55 ± 6.87	2.57	143.20 ± 6.87	-0.36	146.13 ± 6.87	2.95	146.51 ± 6.87	-2.25	141.30 ± 6.87
RG-BU 08-046 R	-0.56	143.00 ± 6.87	6.21*	142.99 ± 6.87	-0.57	149.77 ± 6.87	-0.23	143.33 ± 6.87	0.31	143.87 ± 6.87
RG-BU 08-057 R	4.01	147.57 ± 6.87	8.26**	146.60 ± 6.87	3.04	151.82 ± 6.87	-1.33	142.23 ± 6.87	-0.58	142.98 ± 6.87
RG-BU 08-063 R	3.45	147.01 ± 6.87	2.91	145.37 ± 6.87	1.81	146.48 ± 6.87	5.27*	148.83 ± 6.87	-1.24	142.32 ± 6.87
RG-BU 08-097 R	-4.86*	138.42 ± 6.87	-7.56	132.33 ± 6.87	-4.77*	140.07 ± 6.87	-7.21**	136.77 ± 6.87	-5.34*	140.90 ± 6.87
RG-BU 08-105 R	1.07	144.63 ± 6.87	-2.59	145.30 ± 6.87	1.74	140.96 ± 6.87	2.32	145.88 ± 6.87	2.68	146.24 ± 6.87
Mean		140.99		142.27		145.35		144.18		145.01
SE (sij)		4.013								
SEd (Sij-Sik)		6.870								
SEd (Sij-Skj)		7.442								
t=Sij/SEI (Sij) at error df=239										

*p=0.05, **p=0.01 and ns=non-significant.

Table 7: SCA effects (Sij) vis-à-vis per-se mean performance of hybrids for days to maturity in 80 F_1 hybrids.

per panicle with all five cm S lines. Increased sca effect in yield might be due to significant positive sca values in pollen fertility, spikelet's fertility, panicle exertion rate, stigma exertion rate, effective tillers per plant, primary branches per panicle, secondary primary branches per panicle and significant negative sca values in days to 1st flowering, 80% flowering and maturity. In rice found similar results in sca effects of several cross combinations [3] found high specific combinations of crosses of rice from high × high general combiner parents (Table 8).

Conclusion

Significant genotypic variances for yield related traits indicated that there were significant variations among the genotypes. Significant

Line / Testers	Grain yield									
	IR 58025A		GAN 46 A		IR 62829A		IR 68888A		BRRI 1 A	
	Sij effect	mean	Sij effect	mean	Sij effect	mean	Sij effect	mean	Sij effect	mean
RG-BU 08-001 R	0.88*	4.97 ± 0.35	0.57*	4.67 ± 0.21	1.16**	5.26 ± 0.22	0.98**	5.10 ± .35	1.97**	6.06 ± 0.35
RG-BU 08-002 R	2.67*	6.71 ± 0.35	3.50**	7.59 ± 0.35	2.06**	6.02 ± 0.35	1.24*	5.73 ± 0.35	1.23**	5.86 ± 0.35
RG-BU 08-005 R	-1.22**	2.87 ± 0.35	-1.36**	2.73 ± 0.21	2.35**	6.44 ± 0.35	-2.91**	1.18 ± 0.21	1.57**	5.66 ± 0.35
RG-BU 08-006 R	2.02**	6.11 ± 0.35	1.17**	5.91 ± 0.21	0.38*	4.48 ± 0.21	2.66**	5.93 ± 0.35	1.47**	5.62 ± 0.35
RG-BU 08-007 R	1.55**	5.53 ± 0.35	3.23**	7.33 ± 0.35	1.54*	5.64 ± 0.22	1.62**	5.71 ± 0.35	2.21**	6.10 ± 0.35
RG-BU 08-013 R	2.85**	6.94 ± 0.35	-0.93*	3.16 ± 0.21	1.13**	5.23 ± 0.21	-1.02**	3.06 ± 0.21	-0.55*	3.53 ± 0.21
RG-BU 08-016 R	-1.69**	2.39 ± 0.35	-0.35	3.74 ± 0.21	0.96*	5.05 ± 0.21	-0.70*	3.39 ± 0.22	3.39**	7.48 ± 0.35
RG-BU 08-018 R	1.26**	5.36 ± 0.35	0.77*	4.86 ± 0.21	0.44*	4.53 ± 0.21	3.09**	7.18 ± 0.35	-0.63*	3.46 ± 0.21
RG-BU 08-025 R	-0.54*	3.54 ± 0.35	0.22	4.31 ± 0.21	2.23**	6.33 ± 0.35	-2.18**	1.91 ± 0.21	3.27**	7.37 ± 0.35
RG-BU 08-034 R	0.30	4.39 ± 0.35	0.31	4.40 ± 0.21	1.05**	5.15 ± 0.21	0.33	4.43 ± 0.22	-0.91*	3.17 ± 0.21
RG-BU 08-038 R	-0.71*	3.37 ± 0.35	-0.08	4.01 ± 0.21	2.64**	6.74 ± 0.35	-1.54**	2.54 ± 0.25	-2.75**	1.33 ± 0.21
RG-BU 08-046 R	1.42**	5.52 ± 0.35	-0.77*	3.32 ± 0.21	-0.47*	3.61 ± 0.21	-0.70*	3.38 ± 0.21	-0.22	3.86 ± 0.21
RG-BU 08-057 R	-2.81**	1.27 ± 0.35	-2.26**	1.83 ± 0.21	2.88**	6.97 ± 0.35	0.84*	4.93 ± 0.21	0.89*	4.98 ± 0.21
RG-BU 08-063 R	1.51**	5.60 ± 0.35	1.54**	5.63 ± 0.21	1.45**	2.63 ± 0.21	0.81*	4.88 ± 0.21	1.57**	5.66 ± 0.21
RG-BU 08-097 R	2.95**	6.73 ± 0.35	2.45**	6.34 ± 0.35	1.83**	6.26 ± 0.35	1.16**	5.26 ± 0.35	2.21**	6.18 ± 0.35
RG-BU 08-105 R	1.77**	5.86 ± 0.35	0.13	4.23 ± 0.21	-0.68*	3.41 ± 0.22	-0.05	4.04 ± 0.35	0.01	4.10 ± 0.21
Mean	4.08		3.89		4.67		3.60		4.20	
SE (sij)	0.354									
SEd (Sij-Sik)	0.977									
SEd (Sij-Skj)	1.104									
t = Sij/SEl (Sij) at error df = 239										

*p= 0.05, **p=0.01 and [ns] =Insignificant.

Table 8: SCA effects (Sij) vis-à-vis per-se mean performances of hybrids for grain yield (ton/ha) in 80 F1 hybrids.

gca variances along with additive variance component for reproductive traits indicated the accessibility of additive gene action. Degree of dominance were found negative for most of traits which reveals that regression lines passing below the origin i.e., this character are responsible for over dominance. The linear non-significant relationship between female vs male indicates the reliability of the crosses to go through heterosis breeding. The contributions of lines were found significant, indicating preponderance of dominant genes among the lines with tester. The significant interaction of line x tester indicated higher estimates for sca variances. Four restorer lines showed significant negative effects for days to 1st flowering, 80% flowering and days to maturity. Three pollen parents and one cm S line, showed significant positive gca effects for pollen fertility while six pollen parents showed significant positive effects for spikelets fertility but two pollen parents showed significant positive gca effects for both panicle and stigma exertion rate. The estimated gca effects of parents indicated that five pollen parents contributed highly significant negative effects for plant height which were responsible for dwarfing character. Fifteen crosses showed significant negative sca estimates for days to first flowering, sixteen crosses for days to 80%flowering and twenty crosses for days to maturity. Among 80 crosses fifty two crosses showed significant positive sca effects along with above average perse performances for grain yield.

References

1. Islam MA (2009) Synchronization and stability analysis of hybrid seed production of rice in different environment. A Master Degree (MS) thesis of GPB Dept. BSMRAU, Salna, Gazipur.

2. Virmani SS, Young JB, Moon HP, Kumar I, Finn JC (2000) Increasing Rice Yields through Exploitation of Heterosis. IRRI. Los Baños, Laguna, Philippines.

3. Chen YJ, Ding XH, Zhang GQ, Lu YG (2002) Studies on heterosis of F1 hybrids in candidate Japonica lines in rice (Oryza sativa L.) J of South China Agril Univ 23: 1-4.

4. Kempthorne EA (1957) Biometrical Genetics, Combining ability through Line × Tester Method. Ed.,3. Chapman and Hall, London.

5. Soni DK, Arvind K, Lakeswar S (2005) Study of heterosis by utilizing cytoplasmic-genetic male sterility system in rice (Oryza sativa L). Plant Archives 5: 617-621.

6. Tang, DC, Huang SK, Duan YG,Wang YH (2002) Studies on relationships of flowering time and pollination time with outcrossing rate of male sterile lines in hybrid rice seed production. Hybrid-Rice; 19: 50-54.

7. Agrawal KB (2003) Heterosis in rice. Annals of Agricultural Research 24: 375-378.

8. Faiz FA, Sabar M, Awan TH, Ijaz M, Manzoor Z (2006) Heterosis and combining ability analysis in Basmati rice hybrids. J Animal and Pl Sci 16:56-59.

9. Kumar S, Senadhira TD, Chandrappa JK (2007) combining ability analysis for grain yield and other associated traits in rice. Oryza 44: 108-114.

10. Hossain A, Mujtaba NSH, Khoyumthem FJ (2005).The isolation and identification of volatile components from basmati rice (Oryza sativa L). Flavor Science and Technology Proc. 5th Weurman Flavour Res. Symp. Wiley: New York.

11. Venkatesan ND, Maurya DM,Verma GP, Vishwakarma SR (2007) Heterosis for yield components in rice hybrids (Oryza sativa). IJ Agril Sci 99: 1120-1122.

12. Shanthi P, Yadav DV, Singh AK, Yadav G, Singh J (2003) (Oryza sativa L). Res Corps 2: 390-392.

13. Roy AK, Mandal EF (2001) Development of aromatic cytosource for hybrid rice production PhD Dissertatio n, Bangabandhu Sheikh Muzibur Rahman Agricultural University, Gazipur-1706, Bangladesh 166.

14. Singh RJ, Kumar AK (2004) Evaluation of CMS lines for various floral traits influence outcrossing in rice. International-Rice-Research-Note 28: 24-26.

15. Xiao GY, Yuan LP,Tang L (2003) Studies on heterosis of Indica/Javanica and Japonica/Javanica hybrids rice. Acta Agronomica Sinica 29:169-174.

16. Su XJ, Chen CH (2006) Selection and utilization of a new rice restorer line Gui 1025 with small and high quality grains. Hybrid-Rice 21: 21-23.

17. Rao AM, Ramesh S,Kulkarni RS, Savithramma DL, Madhusudhan K (2006) Heterosis and combining ability in rice. Crop Improvement 23: 53-56.

18. Dorosti H, Ali AJ, Nematzadeh G, Ghodsi H, Alinia F (2006) IRRI, the first hybrid rice in Iran. International Rice Research Notes 31: 31-32.

19. Banumsathy S, Thiyagarajan K (2005) Heterosis of rice hybrids for yield and its yield components. Crop Res Hisar 25: 287-293.

20. Khoyumthem P, Sharma PR, SinghNB, Singh MRK (2005) Heterosis for grain yield and its component characters in rice (*Oryza sativa* L). Environment and Ecology 23: 687-691.

21. Singh RJ, Maurya A (1999) Evaluation of CMS lines for various floral traits influence outcrossing in rice. *International-Rice-Research-Notes* 28: 24-26.

22. Biju S, Manonmani S, Mohanasundaram K (2006) Studies on heterosis for yield and related characters in rice hybrids. *Plant Archives* 6:549-551.

23. Salgotra RK, Gupta BB, Praveen Singh (2009) Combining ability studies for yield and yield components in Basmati rice. *Oryza* 46: 12-16.

24. Zhang J, Chen GR, Huang DJ, Liu KH, Tan XL (2002) Genetic relationship of stigma exterior between maintainer lines and sterile lines for Dian type japonica hybrid rice. *Journal-of-Yunnan-Agricultural-University* 20: 459-461, 477.

5

Phenotypic Diversity Studies on Selected Kenyan and Tanzanian Rice (*Oryza sativa* L) Genotypes Based on Grain and Kernel Traits

Mawia Musyoki A[1]*, **Wambua Kioko F**[1], **Agyirifo Daniel**[1], **Nyamai Wavinya D**[1], **Matheri Felix**[1], **Langat Chemtai R**[1], **Njagi Mwenda S**[1], **Arika Arika W**[1], **Gaichu Muthee D**[1], **Ngari Ngithi L**[1], **Ngugi Piero M**[1] and **Karau Muriira G**[2]

[1]Department of Biochemistry and Biotechnology, School of Pure and Applied Sciences, Kenyatta University, P.O. Box 43844-00100, Nairobi, Kenya
[2]Molecular Biology Laboratory, Kenya Bureau of Standards, Nairobi, Kenya

Abstract

Phenotypic characterization of rice (*Oryza sativa L.*) provides useful information regarding preservation of diversity and selection of parental genotypes with superior traits in plant breeding program. The main objective of the present study was to characterize 13 selected rice genotypes from Kenya and Tanzania based on 7 grain and kernel traits. Minitab 15.0 software was used to analyze data. The 7 traits showed highly significant differences among the improved and local landraces. A dendrogram was constructed from data set of mean values of grain and kernel traits and showed two super clusters; I and II. Principal component analysis revealed that all the seven quantitative traits significantly influenced the variation in these genotypes.

Keywords: *Oryza sativa L*; Dendrogram; Principal component analysis

Introduction

Rice (*Oryza spices*) is a monocotyledonous plant belonging to the family Granineae and subfamily Oryzoidea. It is ranked second to wheat among the most cultivated cereals in the world. Due to its importance as a food crop, rice is being planted on approximately 11% of the Earth's cultivated land area. During crop improvement strategies, selection on breeding lines depends on a given set of criteria found suitable to a particular environment and for specific application. This process has led to development of morphologically related genotypes. Phenotypic similarity poses threats of epidemic of pests and diseases. To address this problem, phenotypic characterization is important in breeding program to avoid this inherent danger of phenotypic uniformity. In addition, landraces offers valuable genetic materials that can be utilized in future crop development and improvement programs. High yielding varieties which are the back bone of green revolution have led to erosion of landraces and wild varieties of rice[5]. Importance of landraces can never be ignored in agriculture system. This is because improvement in existing varieties depends upon desirable genes which are possibly present in landraces and wild varieties only. Therefore, characterization of phenotypic diversity on existing landraces of rice reveals important traits of interest that can be utilized in rice improvement programs. A number of research studies on phenotypic diversity assessment of various rice varieties around the world based on grain and kernel traits have been carried out. However, phenotypic diversity studies on rice genotypes from Kenya and Tanzania based on grain and kernel traits has not yet been studied before. Therefore, the objectives of this study were to determine the phenotypic diversity on selected rice (*Oryza sativa L*) genotypes from Kenya and Tanzania along other 2 genotypes from Philippine based on 7 grain and kernel traits and to identify the traits that contribute to the total variation among the rice genotypes studied. Information generated from phenotyping these genotypes can be utilized in rice breeding programs [1-6].

Materials and Methods

Plant materials

A total of 13 rice genotypes comprising of local landraces and improved rice genotypes were collected from Mwea Irrigation Agricultural Development Centre (MIAD) in Mwea, Kenya and Kilimanjaro Agricultural Research Institute in Moshi, Tanzania. The name, country of origin and category of the rice genotype chosen for the study are given in (Table 1).

Measurement of grains and kernel traits

The following seven (7) phenotypic traits were measured in this study; grain length (GL), grain breadth (GB), grain length/breadth (G-L/B), kernel length (KL), kernel breadth (KB), kernel length/breadth (K-L/B) and 100 grain weight (100 GW). For each of the rice genotype, 10 grains were randomly selected and their measurement taken using a digital vernier caliper. Weight of 100 rice grains from each genotype was determined using an electronic weighing balance (Mettler toledo), and average weight ± standard error mean recorded.

Data management and analysis

To determine phenotypic relatedness based on grain and kernel traits of the 13 rice genotypes, the data was analyzed statistically for the difference in means for the 7 grain and kernel traits measurement through ANOVA followed by Tukey's post hoc to separate means. Cluster analysis yielded a dendrogram that was used to examine the phenotypic relatedness among the 13 rice genotypes. To assess the underlying source of variation in morphology based on the 7 grain and kernel traits among the 13 rice genotypes, Principal component analysis (PCA) was carried out. All analyses were done using software Minitab 15.0 (State College Pennsylvania-USA).

*Corresponding author: Mawia Musyoki A, Department of Biochemistry and Biotechnology, School of Pure and Applied Sciences, Kenyatta University, Nairobi, Kenya, E-mail: amosyokis@gmail.com

Variety	Origin	Category of rice
R 2793	Kenya	Improved variety
BS 217	Kenya	Improved variety
BS 370	Kenya	Improved variety
BW 196	Kenya	Improved variety
ITA 310	Kenya	Improved variety
SARO 5	Tanzania	Improved variety
IR 64	Philippine	Improved variety
KILOBERO	Tanzania	Local land race
RED AFAA	Tanzania	Local land race
KAHOGO	Tanzania	Local land race
SUPA	Tanzania	Local land race
IR 54	Philippine	Improved variety
WAHIWAHI	Tanzania	Local land race

Table 1: Names, origin and category of the rice genotypes used in this study.

Results

Measurement of grain and kernel traits

The measurement of the grain and kernel traits for the 13 rice genotypes and their mean values are shown in Table 2. From the table, it can be seen that *Supa*, a local landrace rice genotype from Tanzania had the highest value of the grain length (10.48) while *ITA 310*, an improved rice genotype from Kenya had the lowest value of the grain length (9.00). In terms of grain breadth, *Supa*, a local landrace rice genotype from Tanzania had the highest value of the grain breadth (2.14) while *ITA 310*, an improved rice genotype from Kenya had the lowest value of grain breadth (1.79). Furthermore, it was observed that *Wahiwahi*, a local landrace rice genotype from Tanzania reported the highest value of grain length/breadth ratio (5.26) whereas *IR 64*, an improved rice genotype from Philippine had the lowest value of grain length/breadth ratio (4.60). Moreover, Supa, a local landrace rice genotype from Tanzania had the highest value of the kernel length (7.78) while *ITA 310*, an improved genotype from Kenya recorded the lowest value of kernel length (6.54). In terms of kernel breadth, *Supa*, a local landrace rice genotype from Tanzania had the highest value of the kernel breadth (1.90) while *ITA 310*, an improved genotype from Kenya recorded the lowest value of kernel breadth (1.60). In terms of kernel length/breadth ratio, *Kilombero*, a local landrace rice genotype from Tanzania had the highest value of the kernel length/breadth ratio (4.45) while *BW 196*, an improved genotype from Kenya had the lowest

value of the kernel length/breadth ratio (3.68). In terms of grain weight, *Supa*, a local landrace rice genotype from Tanzania had the highest value of the grain weight (2.91) while *ITA 310*, an improved genotype from Kenya recorded the lowest value of grain weight (1.67) (Table 2).

Cluster analysis

A dendrogram was constructed from data set of mean values of the 7 grain and kernel traits and showed two super clusters; I and II as shown in (Fig 1). Super cluster I comprised of *Kilombero*, *Wahiwahi* and *Supa* genotypes both indigenous rice genotypes from Tanzania. On the other hand, super cluster II comprised of 10 rice genotypes; 5 improved rice genotypes from Kenya, 2 improved rice genotypes from Philippine, 1 improved rice genotype from Tanzania and 2 indigenous rice genotypes from Tanzania. Super cluster II was more diverse and could further be divided into 2 sub groups.

Principal component analysis (PCA)

The PCA was performed for all the 7 grain and kernel traits among the 13 rice genotypes as indicated in table. Out of seven, two principal components exhibited more than one Eigen value and showed about 91.3% variability among the traits studied. The PC1 had 53.8%, PC2 showed 37.5% and PC3 exhibited 5.4% variability among the genotypes for the traits under study. Principal component one (PC1), principal component two (PC2) and principal component three (PC3) had Eigen values of 3.77, 2.62, 0.38 and 0.14 respectively. The first PC was highly and positively correlated to grain length, grain breadth, kernel length and 100 grain weight. However, it was negatively correlated to grain length/breadth ratio. In the second PC, grain length/breadth ratio, kernel length and kernel length/breadth ratio were highly and positively correlated traits. On the other hand, it was highly and negatively correlated to grain breadth and kernel breadth. The third PC exhibited high positive correlation between kernel length and kernel length/breadth ratio. However, it was highly and negatively correlated to grain length and grain length/breadth ratio as shown in fig. Based on the 7 grain and kernel traits, the 13 rice genotypes were clustered into four groups as shown in figure 1. The first quadrant was made up of Kilombero and Wahiwahi. The second quadrant grouped Saro 5, BS 217, ITA 310 and BS 370 together. R 2793, IR 64 and BW 196 were grouped into third quadrant. The forth quadrant consisted of Red Afaa, Kahogo, IR 54 and Supa genotypes (Table 3 and Figure 2).

Discussion

Diversity analyses of germplasm collections of several crop species

Genotype	Traits						
	G.L	G.B	G L/B	K.L	K.B	K L/B	100 GW
R 2793 (A)	9.49 ±0.18bcd	1.96 ±0.05bc	4.86± 0.11ab	6.87±0.08bc	1.74±0.03bcd	3.95±0.03bcde	2.46±0.17abc
BS 217 (B)	9.34±0.25cd	1.87 ±0.02bcd	5.00 ±0.11ab	6.82±0.14bc	1.66±0.02cd	4.11 ±0.04abcd	2.70±0.12ab
BS 370 (C)	9.03±0.14d	1.80 ±0.02cd	5.01± 0.08ab	7.02±0.10bc	1.65 ±0.02cd	4.27 ±0.08ab	1.96 ±0.19cde
BW 196 (D)	9.52±0.08bcd	1.99 ±0.04ab	4.79 ±0.09ab	6.72±0.06c	1.83± 0.02ab	3.68 ±0.05e	2.18 ±0.18bcde
ITA 310 (E)	9.00 ±0.06d	1.79 ±0.02d	5.03 ±0.06ab	6.54±0.08c	1.60±0.04d	4.10 ±0.08abcd	1.67 ±0.09e
Saro 5 (F)	9.75 ±0.15abcd	1.93 ±0.04bcd	5.08 ±0.12ab	6.95±0.16bc	1.68 ±0.04bcd	4.16±0.11abc	2.61±0.13ab
IR 64 (G)	9.02 ±0.14d	1.96 ±0.02b	4.60 ±0.08b	6.56±0.09c	1.76±0.02abc	3.72 ±0.05de	2.27±0.04bcd
Kilobero (H)	9.97 ±0.28abc	2.01± 0.03ab	4.96 ±0.09ab	7.37±0.19ab	1.68 ±0.05cd	4.45±0.16a	1.84±0.08de
Red Afaa (I)	9.43 ±0.15bcd	2.02 ±0.05ab	4.70 ±0.12b	6.62±0.05c	1.79 ±0.05abc	3.73 ±0.13de	2.84±0.04a
Kahogo (J)	9.49± 0.17bcd	2.01± 0.03ab	4.72± 0.10b	6.88±0.10bc	1.79±0.02abc	3.85 ±0.06cde	2.52 ±0.09ab
Supa (K)	10.48 ±0.26ab	2.14± 0.02a	4.82 ±0.12ab	7.78±0.19a	1.90±0.01a	4.09 ±0.10abcd	2.91±0.02a
IR 54 (L)	9.75 ±0.11abcd	2.02± 0.03ab	4.85 ±0.10ab	7.02±0.05bc	1.77 ±0.03abc	3.98 ±0.04bcde	2.43 ±0.08abc
Wahiwahi (M)	10.44 ±0.28a	1.99± 0.03ab	5.26 ±0.13a	7.30±0.12ab	1.74 ±0.04bcd	4.20 ±0.07abc	2.83 ±0.06a

Table 2: Analysis of variance of the 7 grain and kernel traits among the 13 rice genotypes.

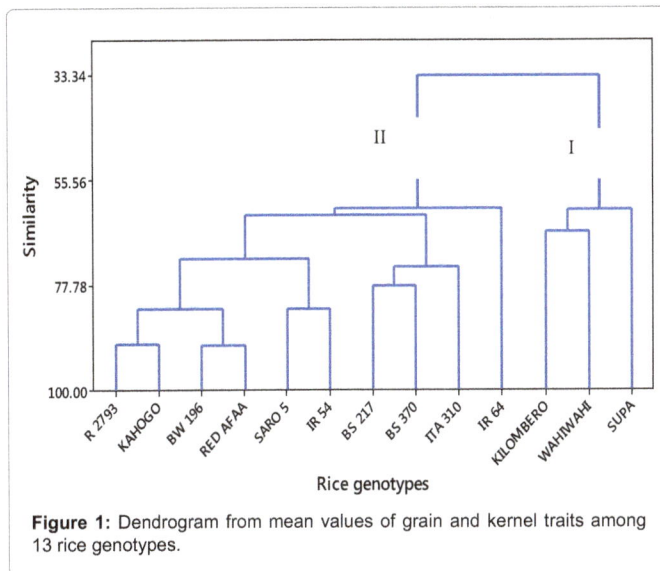

Figure 1: Dendrogram from mean values of grain and kernel traits among 13 rice genotypes.

	PC 1	PC 2	PC 3
Eigen value	3.77	2.62	0.38
% Total Variance	53.8	37.5	5.4
% cumulative	53.8	91.3	96.8
Traits	**Eigen vectors**		
Grain Length (cm)	0.457	0.234	0.352
Grain Breadth (cm)	0.473	-0.214	0.101
Grain Length/Breadth	-0.034	0.566	-0.613
Kernel Length (cm)	0.408	0.312	0.474
Kernel Breadth (cm)	0.395	-0.368	-0.072
Kernel Length/Breadth	0.023	0.587	0.482
100 Grain weight (g)	0.494	0.036	-0.169

Table 3: Eigen vectors, Eigen values, total variance and cumulative variance for 13 genotypes based on 7 grain and kernel traits.

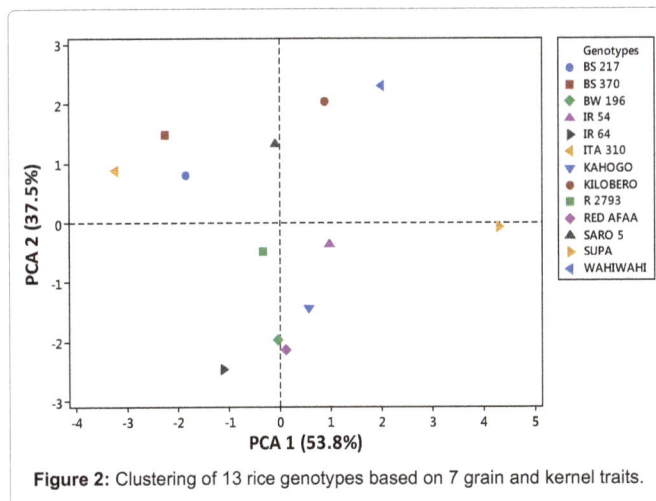

Figure 2: Clustering of 13 rice genotypes based on 7 grain and kernel traits.

have revealed considerable variability for a wide range of traits. The wide range of diversity in phenotypic traits have proved a useful tool in classification of plants and the information obtained could be of high interest to plant breeders in development of plant species with desirable agronomic and nutritional qualities [7,8]. From the ANOVA table, it was evident that the means of the measured traits varied significantly

among the 13 rice genotypes included in this study. The differences in mean values of grain length across the 13 rice genotypes studied could be as a result of different expression levels of GS3 gene that confers grain length in rice. In addition, the differences in mean values of grain breadth and grain weight across the 13 rice genotypes studied could be as a result of different expression levels of GW2 gene within the rice genome. Reduced expression of GW2 gene increases grain breadth, resulting in enhanced grain weight, whereas over expression decreases grain breadth and weight. Moreover, the differences in mean values of kernel length across the 13 rice genotypes studied could be as a result of different expression levels of *Kne* gene that confers kernel length in rice. These results were consistent with findings of. Genotypes that were clustered together indicated that they shared similar traits. For example, three improved rice genotypes namely *BS 370, BS 217* and *ITA 310* both from Kenya were grouped together and this showed that they have similar grain and kernel traits. Genetic improvement of these genotypes could have resulted in sharing of similar genes that influenced their phenotypic characteristics. On the other hand, genotype *R 2793* and *BW 196* both improved genotypes from Kenya formed separate clusters implying that they are phenotypically distinct from each other as far as grain and kernel traits are concerned [9-12].

Principal component analysis measures the importance and contribution of each component to the total variance while each coefficient of proper vectors indicates the degree of contribution of every original variable with which each principal component is associated. In PC1, the traits that accounted for most of the 53.8% observed variability among 13 genotypes included grain length, grain breadth, kernel length, kernel breadth and 100 grain weight. Therefore, this implied that these traits contributed most to the diversity captured in the first principal component. Grain length/breadth ratio, kernel length and kernel length/breadth ratio were the most important traits that highly contributed to variation patterns among the genotypes studied in PC2. Moreover, in PC3 only 4 traits contributed highly to variation patterns among the genotypes studied. They included: grain length, grain length/breadth ratio, kernel length and kernel length/breadth ratio. Kernel length contributed to variation across all the three principal components indicating that is one of the most important agronomic traits in rice. These results agree with the findings of. The phenotypic diversity witnessed among the rice genotypes studied could be used as possible parents for rice improvement program [13,14].

Conclusion

Phenotyping of germplasm materials is an important undertaking in genetic resource conservation to ensure efficient conservation management as well as its effective utilization especially in breeding programs. In this study, 13 rice genotypes were characterized to assess their phenotypic diversity and showed considerable variability based on the 7 grain and kernel traits. There was a moderate significant difference in mean values of the measured traits. Out of the 7 traits measured, *Supa* had the highest number of 5 traits with highest mean values, that is, grain length, grain breadth, kernel length, kernel breadth and grain weight while *ITA 310* had the highest number of 5 traits with lowest mean values, that is, grain length, grain breadth, kernel length, kernel breadth and grain weight. Cluster analysis based on the 7 grain and kernel traits clustered the 13 rice genotypes into 2 major groups. Cluster I was less diverse since it consisted of only 3 genotypes while cluster II was more diverse and consisted of 10 genotypes. Such groupings are useful to breeders in identifying possible genotypes that may be used as parents in breeding for any of the phenotypic traits that were studied. The PCA revealed that all the traits contributed significantly to variation patterns among the genotypes studied. Above all, the information generated will reduce the overall time required by rice breeders to screen large populations for potential breeding stock.

References

1. Vaughan DA, Morishima H, Kadowaki K (2003) Diversity in the *Oryza* genus. *Current opinion in plant biology*, 6: 139-146.

2. Walia US, Walia MK (2007) Scope of direct seeded rice in India. In *Proc. Biennial Conf. ISWS on New and Emerging Issues in Weed Science*, 2-3.

3. Khush GS (2005) What it will take to feed 5.0 billion rice consumers in 2030. *Plant molecular biology*, 59: 1-6.

4. Patra BC (2000) Collection and characterization of rice genetic resources from Keonjhar district of Orissa. *Oryza*, 34: 324-326

5. Fowler C, Mooney P (1990) *Shattering: Food, Politics and Loss of genetic diversity*. University of Arizona Press, Tucson.

6. Holden J, Peacock J, Williams T (1993) *Gens, Crops and the Environment*. Cambridge University Press, Cambridge.

7. Yang RC, Jana S, Clarke JM (1991) phenotypic diversity and associations of some potentially drought-responsive characters in durum wheat. *Crop Science*, 31: 1484-1491.

8. Maduakor HO, Lal R (1989) Root system top-growth and yield of cassava (*Manihot esculenta*) as affected by plant population in an arid areas. *Root Crops*, 15: 115-122.

9. Takano-Kai N, Jiang H, Kubo T, Sweeney M, Matsumoto T, et al. (2009) Evolutionary history of GS3, a gene conferring grain length in rice. *Genetics* 182: 1323-1334.

10. Xian JS, Wei H, Min S, Mei Z, Zhu HXL (2007) A QTL for rice grain width and weight encodes a previously unknown RING-type E3 ubiquitin ligase. *Nature genetics*, 39: 623-630.

11. Ahn SN, Bollich CN, McClung AM, Tanksley SD (1993) Restriction Fragment Length Polymorphism analysis of genomic regions associated with cooked-kernel elongation in rice. *Theoretical and Applied Genetics* 87: 27-32.

12. Basabdatta D, Samik S, Mrityunjay G, Tapas KG (2012) Assessment of diversity amongst a set of aromatic rice genotypes from India. *International Journal of Biodiversity and Conservation*, 4: 206-218.

13. Sanni KA, Fawole I, Ogunbayo A, Tia D, Somado EA, et al. (2012) Multivariate analysis of diversity of landrace rice germplasm. *Crop Science* 52: 494-504.

14. Takeda K (1990) Inheritance of grain size and its implications for rice breeding. *In: Proceedings of the Second International Rice Genetics Symposium*. IRRI, Manila, 181-189.

6

Effect of Rice Varieties on Digestive Enzymes Some Components in Intermediary Metabolism of *Chilo suppressalis* Walker (Lepidoptera: Crambidae)

Hedieh Jafari[1], Gadir Nouri-Ganbalani[1], Bahram Naseri[1] and Arash Zibaee[2]*

[1]Department of Plant Protection, College of Agriculture science, University of Mohaghegh Ardebili, Ardebil, Iran
[2]Department of Plant Protection, Faculty of Agricultural Sciences, University of Guilan, Rasht, Iran

Abstract

In this study, the effects of four rice varieties including Gohar, Khazar, Kazemi and Hashemi were studied on digestive enzymatic activities and some components involved in intermediary metabolism of *Chilo suppressalis* Walker, the most important pest of rice in Iran. Activities of general protease, elastase and chymotrypsin showed statistical differences among rice varieties but no significant differences were found in case of other proteases. Feeding of larvae on the rice varieties showed the lowest activity of glucosidases and alpha-amylase in the larvae fed on Kazemi and Gohar, respectively. No statistical differences were found in lipase activity as activities of ALT, ACP and ALP. The highest activities of AST and LDH were found in the larvae fed on Hashemi. These results may helpful to have a background regarding planted varieties in the infested rice fields to *C. suppressalis*.

Keywords: *Chilo suppresslis*; Digestive enzyme; Rice variety; Intermediary metabolism

Introduction

Rice striped stem borer, *Chilo suppressalis* Walker (Lepidoptera: Crambidae) is a serious pest of rice in Iran [1], south-east Asia, India, China, and southern Europe. The pest causes whitehead and dead heart during reproductive and vegetative stages of rice, respectively [2]. Cultural methods, light traps, pheromone traps, biological control and chemical methods are using to decrease pest damages in Iran.

Digestion in insects depends on activities of several enzymes including amylases, glycosidases, lipases and proteases. A-Amylases (EC 3.2.1.1) are the glucano hydrolases that catalyze hydrolysis of α-D(1,4)-gluncan linkage in glycogen, starch and other carbohydrates [3]. β-Glucosidases (EC 3.2.1.20) and β-glucosidase (EC 3.2.1.21) hydrolyze α-D (1,4)-glucose linkage such as p-nitrophenyl-α/β-D-glycoside in oligo and disaccharides [4]. TAG-lipases (triacylglycerol-acyl-hydrolase) (EC 3.1.1.3) hydrolyze the outer ester bonds of triacylglycerides as storage lipids in insects and plants [5]. Proteases are divided into exo- and endo-peptidases. Exo-peptidase remove amino acids from N- and C-terminals of amino acids known as amino-(EC 3.4.11.2) and carboxypeptidase (EC3.4.17) [5]. Endo-peptidases or proteinases are divided into several classes based on their active sites such as serine (EC 3.4.21), cysteine (EC 3.4.22) and aspartic proteases [5]. Serine proteases divide into three classes by their substrate specificity as trypsin (EC 3.4.21.4), chymotrypsin (EC 3.4.21.1) and elastase (EC 3.4.21.36). They have catalytic triad of serine, histidine and aspartic acid which are the common digestive proteases in lepidopteran larvae [5].

After digestion and absorption of dietary food, insects store them as macromolecules protein, glycogen and triglyceride in their fat bodies. These stored molecules are utilizes during biological processes or activities such as flight, reproduction, tissue repair and etc. several cycles and enzymes are involved to obtain energy from dietary storages in fat bodies. These metabolic and digestive processes could be altered by feeding on various food sources or presence in environmental conditions. In the infested areas by *C. suppressalis*, various varieties of rice cultivated that definitely affect digestion and metabolic rate of the larvae. Moreover, determination of these differences are helpful to find the resistant variety since it can be recommended to have the lowest damage. So, objectives of the current study were to find effects of four rice varieties on digestive enzymatic activities and some enzymes involved in intermediary metabolism of *C. suppressalis* larvae.

Material and Methods

Insect rearing

Eggs of *C. suppressalis* were collected from rice fields of Mazandaran province, Northern Iran. The larvae were reared in plastic boxes at 25 ± 2°C, 65 ± 5 of relative humidity and 16 L: 8D of photoperiod. The stems of Gohar, Khazar, Kazemi and Hashemi varieties were separately provided for larvae and 4th larval instars were used for biochemical analyses.

Sample preparation

Fourth larval instars (10 individuals) were randomly selected and dissected under a stereoscopic microscope in ice cold saline buffer (NaCl, l0 mM). Larval midguts were seperated, unwanted tissues removed and homogenized with a handling glass homogenizer on ice. Samples were transfer to ependorf tubes contain 1 ml of distilled water and centrifuged at 13000 rpm at 5°C for 20 min. The supernatants were transferred in new ependorf tubes and kept at -20°C. For intermediary metabolism assays, total body of larvae homogenized and centrifuged as described above.

*Corresponding author: Arash Zibaee, Assistant of Insect Physiology, Department of Plant Protection, Faculty of Agricultural Sciences, University of Guilan, Rasht, Iran 41635-1314
E-mail: arash.zibaee@gmx.com; arash.zibaee@guilan.ac.ir

Proteolytic assay

Azocaseain 2% was used for general proteolytic assay based on Elpidina et al. [6]. Substrate (40µl) and phosphate buffer (pH: 7.1, 0.02 M) were incubated prior to addition of enzyme extract (40 µl). The reaction was continued for 120 min at 30°C before adding Trichloroacetic Acid (TCA 30%, 100 µl) as stopper. Precipitations were achieved by centrifugation of 13000 rpm for 5 min at 4°C. Finally, equal volume of NaOH was added and absorbance was read at 450 nm.

Specific proteolytic activity

Trypsin, Chymotrypsin and Elastase activities were assayed based on Oppert et al. [7] using βA$_p$Na (Nabenzoyl-L-arginine-p-nitroanilide), SAAPPpNA (N-succinyl-alanine-alanine-proline-phenylalanine-p-nitroanilide, SAAApNA (N-succinyl-alaninealanine-alanine-p-nitroanilide) as substrates. Forthy µl of substrate added to 80 µl of Tris-HCl (pH 8, 20 mM) buffer and 10 µl of Enzymes. Then, reaction incubated in 30°C for 10 min prior to adding TCA 30%. The absorbance was measured in 405.

Exopeptidase assay

Hippuryl L-arginine and hippuryl L-phenylalanine were used to find aminopeptidase and carboxypeptidase activities in the midgut of C. suppressalis. The reaction mixture consisted of 80 µL of phosphate buffer (pH 10.0), 10 µL of each mentioned substrate and 5 µL of enzyme solution. The reaction mixture was incubated at 30°C for 5-60 min before adding 30% of TCA to terminate the reaction. The absorbance was then read at 340 nm.

Glucosidase activity

α- and β -Glucosidases were assayed based on Ferreira et al. [8] method. The activities were determined by (pNαG) p-nitro phenyl-α-D-glycopyranoside (5 mM) and (pN$_β$G) p-nitro phenyl-β-D-glycopyranoside (5 mM) as substrates, respectively. Phosphate buffer (0.02 M, pH: 7.1) (35 µl) and enzyme (5 µl) were mixed and incubated for 10 min. NaOH (1 N, 50 µl) were added and absorbance was read at 405.

α-Amylase activity

Based on Bernfeld [9], the raction mixture contained starch as substrate (30 µl), phosphate buffer (50 µl) and samples (20 µl). Mixture was incubated for 30 min prior to add dinitrosalysilic acid (DNSA). Then, samples were boiled for 10 min and absorbance was read at 545 nm.

Lipase assay

The reaction mixture contained p-nitro phenyl-butyrate (27 mM), phosphate buffer and enzyme extract [10]. After incubated for 10 min, NaOH was added and absorbance was read at 405 nm.

ACP and ALP assays

The method of Bessey et al. [11] were used for the assay. First, 50 µl of Tris-HCl buffer (pH 8 for ALP and pH 5 for ACP) were added to 10 µl of p-nitrophenol phosphate in addition to 5 µl of enzyme. After 10 min, absorbance was read at 405 nm.

AST and ALT assays

This assay was carried out based on Thomas et al. [12] using biochemical kit of Biochem Co, Iran. Initially, 50 µl of reagent A was added to 5 µl of reagent D. After 5 min of incubation, 10 µl of the enzymes and 5 µl of reagent E were mixed for 60 min. Finally, 50 µl of reagent C and NaOH were added and absorbance was read at 492 nm. Method for ALT was similar to AST but reagent B was used instead of reagent A.

LDH assay

King's [13] method was used to evaluate the presence of Lactate Dehydrogenase (LDH). Test tubes contained 1 µl of the buffered substrate and 0.01 µl of the sample. To standardize volumes, 0.2 µl NAD + solution was added to the test tubes of the sample group and 0.2 µl of water was added to the test tubes of the control group. Then, samples were incubated for exactly 15 min at 37°C. The reactions were then arrested by adding 1 µl of color reagent (2,4-dinitrophenyl hydrazine) to each tube, after additional 15 min. The contents were then cooled at room temperature, and 10 µl of 0.4 N NaOH was added to each tube to make the solutions strongly alkaline. At exactly 60 s after the addition of alkaline to each tube, the intensity of color was measured at 454 nm.

Protein assay

The method of Lowry et al. [14] was used to assay amount of total protein in samples.

Statistical analysis

Data was compared by one-way analysis of ANOVA followed by Tukey's student test. The data was considered to be significantly different within the treatments at probability less than 0.05.

Results and Discussion

Proteins have critical roles physiological systems of insects via involvement as enzymes, tissue repair, metabolic demand, reproduction and etc. Any disruption in utilizing dietary protein will alter insect success in both physiological and ecological aspects [15]. In the current study, the highest activity of general and specific proteases were observed in the larvae fed on Hashemi variety except for carboxypeptidas that the highest activity were found in case of Gohar and Kazemi (Table 1). Although activities of the enzymes were not statistically different in other varieties but it is imperative to indicate that the highest activities of Elastase and aminopeptidase were obtained in the larvae fed on both Hashemi and Khazar varieties (Table 1). Zibaee reported six types of proteases namely trypsin-like, chymotrypsin-like and elastase for proteinases, and amino and carboxypeptidases for exopeptidases. These two results indicate that larvae of C. suppressalis significantly depend on dietary proteins for its growth and development. So, it is expected that larvae fed on Hashemi variety will have well ecological success in comparison with others.

Table 2 shows activity profiles of lipase, α-glucosidase, β-glucosidase and α-amylase in the larvae fed on various rice varieties. The highest activity of lipase were observed in the larvae fed on Hashemi and no significant differences were found among other treatments (Table 1). The highest activities of two glucosidases were found in the larvae fed on Hashemi and Gohar varieties but the highest activity of α-amylase were obtained in case of larvae fed on Khazar and Kazemi varieties (Table 2). Dietary lipids mainly triglyceride is the main energetic macromolecule for insects because those release more energy than other molecules [16]. So, the higher activity of lipase in the larvae fed on Hashemi variety may indicate the higher amount of dietary lipid in the variety to be obtain by the target pest. In case of carbohydrases, two conclusions could be described. It seems that varieties Khazar and Kazemi contain more content of starch in comparison with others because the higher activity of α-amylase was found in the relevant

Host (Variety)	*General protease	Carboxy-Peptidase	Amino-Peptidase	Elastase	Chymotrypsin	Trypsin
Hashemi	0.0008a ± 0.08	0.009b ± 0.044	0.06a ± 0.21	0.001a ± 0.05	0.001a ± 0.031	0.009a ± 0.036
Gohar	0.016b ± 0.02	0.019a ± 0.103	0.022b ± 0.15	0.017ab ± 0.03	0.0003b ± 0.013	0.006b ± 0.012
Kazemi	0.006b ± 0.009	0.023a ± 0.09	0.015b ± 0.17	0.004b ± 0.004	0.003ab ± 0.028	0.005b ± 0.012
Khazar	0.002b ± 0.03	0.011b ± 0.05	0.023a ± 0.19	0.002a ± 0.07	0.006b ± 0.021	0.005b ± 0.019

Table 1: Activities of Enzymes on varieties of Hosts.

Host(Variety)	Lipase	β*-Glucosidase	*α-Glucosidase	α-Amylase
Hashemi	0.017a ± 0.380	0.001a ± 0.244	0.000a ± 0.181	0.014ab ± 0.058
Gohar	0.005b ± 0.02	0.016a ± 0.288	0.006a ± 0.178	0.0008b ± 0.029
Kazemi	0.002b ± 0.02	0.007b ± 0.175	0.008b ± 0.111	0.011a ± 0.096
Khazar	0.003b ± 0.028	0.069a ± 0.255	0.085b ± 0.148	0.009a ± 0.086

Statistical differences are shown by different letter in Tukey's test; (mean ± SE, n=3).
* (μmol/ min/ mg protein)

Table 2: Effect of various rice varieties on non-protease enzymes of *C. suppressalis* (U/mg protein).

Host (variety)	AST	ALT	LDH	ACP	ALP
Hashemi	0.38 ± 0.01a	0.31 ± 0.0008a	0.19 ± 0.02a	0.01 ± 0.002b	0.24 ± 0.03b
Gohar	0.13 ± 0.03b	0.09 ± 0.008c	0.03 ± 0.017b	0.23 ± 0.015a	1.27 ± 0.05a
Kazemi	0.21 ± 0.02b	0.21 ± 0.09b	0.06 ± 0.003b	0.008 ± 0.004c	0.95 ± 0.28b
Khazar	0.13 ± 0.01b	0.14 ± 0.02bc	0.1 ± 0.015ab	0.01 ± 0.003b	0.48 ± 0.005b

Statistical differences are shown by different letter in Tukey's test; (mean ± SE, n=3).
*(mg/ ml)

Table 3: Effect of various rice varieties on intermediate metabolism enzymes of *C. suppressalis* (U/ mg protein).

larvae. Although, the higher activity of glucosidases may indicate efficiency of carbohydrate digestion in insects, but it could show higher amounts of plant secondary metabolites. So, any conclusion might be considered regarding to these phenomena.

Intermediary metabolism in insects is a multi-complex processes in which several energetic pathways are recruit to supply required energy for biological processes. Several enzymatic and non-enzymatic are involved in the process. In the current study, activities of some enzymes including ALT, AST, LDH, ACP and ALP changed in the larvae fed on various host plants. ALT (EC 2.6.1.2) and AST (EC 2.6.1.1) are the two enzymes in transamination reaction in animals for amino acid recruitment. On the other hands, ALT catalyzes the two parts of the alanine cycle in proline metabolism but AST facilitates the conversion of aspartate and α-ketoglutarate to oxaloacetate and glutamate [17]. It has been obtained that the larvae fed on Hashemi variety had the highest activities of these two enzymes (Table 3). The result corresponds with higher activity of proteases in the larvae fed on the variety since the obtained amino acids are processed to be stored or used in the given processes. LDH (1.1.1.27) catalyzes the conversion of pyruvate to lactate or NADH to NAD+ *vice versa* (King, 1965). The higher activity of LDH in the fed larvae on Hashemi variety could be attributed to energy allocation via glycolysis cycle. ACP (EC 3.1.3.2) and ALP (EC 3.1.3.1) hydrolyze phosphate groups from several molecules including nucleotides, proteins, and alkaloids in alkaline and acidic conditions through dephosphorylation [18]. In the current study, the highest activities of these enzymes have been observed in the larvae fed on Gohar variety (Table 3). This could be due to efficiency of digestion and transportation of nutrients in the midgut as well as hemolymph significantly aff ect activities of these enzymes.

In conclusion, variety Hashemi seems to be the most suitable variety for damages of *C. suppressalis*. This is a conclusion based on higher activities of the enzymes in protein digestion or transamination. In contrast, varieties Khazar and Kazemi are not suitable for development of the pest. So, those could be recommended as the appropriate varieties to be planted in the infested areas. This conclusion might be supported by ecological variable such as life table to have a better background in this case.

References

1. Zibaee A (2012) Proteolytic profile in the larval midgut of Chilo suppressalis Walker (Lepidoptera: Crambidae). Entomol Res 42: 142-150.

2. Dale D (1994) Insect pests of the rice plant their biology and ecology, in Biology and Management of Rice Insects, (Heinrichs EA ed.). IRRI, Los-Banos, the Philippines: 363-485.

3. Franco OL, Redgen DJ, Melo FR, Grossi-de-Sa MF (2002) Plant alpha amilase inhibitors and their interaction with insect alpha amylases. Structure, function and potential for crop protection. Euro J Biochem 269: 397-412.

4. Zibaee A, Bandani AR, Ramzi S (2009) Characterization of α and β-glucosidases in midgut and salivary glands of Chilo suppressalis Walker (Lepidoptera: Pyralidae), rice striped stem borer. Comp Ren Biol 332: 633-641.

5. Terra WR, Ferreira C (2005) Biochemistry of digestion. In: Gilbert LI, Iatrou K (eds.) Comprehensive Molecular Insect Science. Vol. 3. Elsevier: 171-224.

6. Elpidina EN, Vinokurov KS, Gromenko VA, Rudenskaya YA, Dunaevsky YE, et al. (2001) Compartmentalization of proteinases and amylases in Nauphoeta cinerea midgut. Insect Biochem Physiol 48: 206-216.

7. Oppert B, Kramer KJ, Mc Gaughey WH (1997) Rapid microplate asay of proteinase mixtures. Bio Technol 23: 70-72.

8. Ferriera C, Riberio AF, Garcia ES, Terra WR (1988) Digestive enzymes trapped between and associate with the double plasma memberanes of Rhodnius prolixus posterior midgut cells. Insect Biochem 18: 521-530.

9. Bernfeld P (1955) Amylases, α and β. Meth Enzymol 1: 149-158.

10. Tsujita T, Ninomiya H, Okuda H (1989) p-Nitrophenyl butyrate hydrolyzing activity of hormone -sensitive lipase form bovine adipose tissue. J Lipid Res 30: 997-1004.

11. Bessey OA, Lowry OH, Brock MJ (1946) A method for the rapid determination of alkaline phosphatase with five cubic millimeters of serum. J Biochem Chem 164: 321-329.

12. Thomas L (1998) Clinical Laboratory Diagnostic, first edithion. TH Books Verlasgesellschaft, Frankfurt: 89-94.

13. King J (1965) The dehydrogenases or oxidoreductases. Lactate dehydrogenase, in: D.Van Nostrand. Practical Clinical Enzymology, London: 83-93.

14. Lowry OH, Rosebrough NJ, Farr AL, Randall RJ (1951) Protein measurement with the Folin phenol reagent. J Biol Chem 193: 265-275.

15. Gatehouse AMR, Norton E, Davidson GM, Babbe SM, Newell CA, et al. (1999) Digestive proteolytic activity in larvae of tomato moth, Lacenobia oleracea; effects of plant protease inhibitors in vitro and in vivo. J Insect Physiol 45: 545-558.

16. Klowden MJ (2007) Physiological Systems in Insects, Academic Press: 697.

17. Nation JL (2008) Insect Physiology and Biochemistry, 2nd edn. CRC Press, London.

18. Zibaee A, Zibaee I, Sendi JJ (2011) A juvenile Hormone analogue, Pyriproxifen, affects some biochemical components in the hemolymph and fat bodies of Eurygaster integriceps Puton, (Hemiptera: Scutelleridae). Pestic Biochem Physiol 100: 289-298.

7

Phenotypic Characterization on Selected Kenyan and Tanzanian Rice (*Oryza sativa L*) Populations Based on Grain Morphological Traits

Wambua F Kioko[1]*, Musyoki A Mawia[1], Ngugi M Piero[1], Karau G Muriira[2], Nyamai D Wavinya[1], Lagat R Chemutai[1], Matheri Felix[1], Arika W Makori[1] and Njagi S Mwenda[1]

[1]Department of Biochemistry and Biotechnology, School of Pure and Applied Sciences, Kenyatta University, P.O. Box 43844-00100, Nairobi, Kenya
[2]Molecular Biology Laboratory, Kenya Bureau of standards, P.O. Box 54974-00200, Nairobi, Kenya

Abstract

Phenotypic characterization of rice varieties is a good approach for assessing genetic and phenotypic variability among varieties and is key in grading of rice varieties. The objective of this study was to determine the major determinants of phenotypic diversity and the strength of segregation among aromatic and non-aromatic rice (*Oryza sativa* L) populations collected from Kenya and Tanzania. Multivariate analyses including principal component analysis (PCA) and cluster analysis were carried out assess the overall patterns of morphological variation. Using Principal component analysis, it was found that grain length, kernel length, grain weight and kernel length/breadth ratio are major drivers of the huge phenotypic diversity observed. Cluster analysis was found to effectively distinguish the majority of aromatic from non-aromatic varieties based on the grain quality traits evaluated where two distinct clusters were formed. The results obtained from this study demonstrated that phenotypic trait measurement can be relied upon in diversity studies among diverse and closely related genotypes. We conclude that this research which forms first part of rice grading gives an insight in to the general patterns of phenotypic diversity and finds out the most important distinguishing characters. This will be validated with subsequent molecular analysis.

Keywords: Rice; Phenotypic diversity; Cluster analysis

Introduction

Rice (*Oryza sativa* L) is a member of the grass family (Gramineae) belonging to the genus *Oryza*. The genus *Oryza* includes 23 wild species and 2 cultivated species. of the two cultivated species, African rice (*Oryza glaberrima*) is highly grown in West Africa whereas the Asian rice (*Oryza sativa* L) has spread over time and is grown in all continents in the world. Being able to grow in a wide spectrum of climates and conditions, rice is a staple food for one third of the world's population [1].

Rice (*Oryza sativa* L) is regarded as one of the major cereal crops with high agronomic and nutritional importance. The current global production of rice is about 738.1 million metric tonnes per year. This constitutes more than a quarter of all cereal grains. of these, Asia accounts for the largest production totaling to about 584 million tones, whereas Africa produces approximately 21.9 million tones. In Kenya, rice is the third most important staple food after maize and wheat. The local production is estimated at between 45,000 to 80,000 tones whereas its consumption is about 300,000 tones. This huge production - consumption gap is met through imports. About 80% of the rice grown in Kenya is from irrigation schemes in Mwea, Ahero, Bunyala, West Kano and Yala swamp. The remaining 20% is produced under rain fed conditions [2].

Landraces are the local or traditional varieties of a domesticated plant species which have developed over time through adaptation to their natural environment [3]. The demand for productive and homogeneous crops has led to development of a small number of standard, high yielding varieties. This has consequently resulted to tremendous loss of heterogeneous traditional cultivars through genetic erosion. Landraces preserve much of this lost diversity and are known to harbor great genetic potential for breeding new crop varieties that can cope with environmental and demographic changes. There are more than 400,000 rice varieties worldwide but the major categories include; indica, japonica, basmati and glutinous. These varieties differ in their grain qualities which include: milling quality, grain shape, cooking quality, nutritional quality and aroma. These traits are crucial determinants of cooked rice grain quality [4].

Kenya is home to many varieties of rice varieties and land races. These varieties were developed through selection based on agronomic traits. This resulted in a wide spectrum of varieties that are highly valued both in domestic and foreign markets. In Kenya, rice consumers prefer the aromatic rice, which is high in quality, and hence price. Unscrupulous traders often blend this fragrant rice which has good cooking quality traits with low quality non-fragrant rice to make more profit from their trade. Various methods routinely used to evaluate and grade rice varieties are inconsistent and have failed to address these concerns due to low sensitivity, time consumption and large sample volume requirement. Therefore, phenotypic characterization based on grain morphological traits that describe the uniqueness of a variety is imperative in morphological distinction between aromatic and nonaromatic rice varieties [5].

Materials and Methods

Plant material

A total of 500 g of thirteen different rice varieties were collected from

*Corresponding author: Wambua F Kioko, Department of Biochemistry and Biotechnology, School of Pure and Applied Sciences, Kenyatta University, P.O. Box 43844-00100, Nairobi, Kenya
E-mail: festuswambua101@gmail.com, festus.w@students.ku.ac.ke

Mwea Irrigation Agricultural Development (MIAD) and Kilimanjaro Agricultural Training Center (KATC). The names and attributes of the rice varieties and the names of the corresponding sources are detailed in Table 1. The rice seeds were stored in Molecular Biology laboratory at Kenya Bureau of Standards, Nairobi, Kenya (Table 1).

Determination of phenotypic diversity

A total of seven traits were measured in this study. They included; grain length (GL), grain breadth (GB), grain length/breadth (G-L/B), grain weight (GW), kernel length (KL), kernel breadth (KB), kernel length/breadth (K-L/B). 100 randomly selected raw rice grains and kernels from each rice variety were measured for their length and breadth traits using a digital vernier caliper. The measurements were repeated 10 times in each and thus an average of 10 replicates was recorded. The grain weight of 100 randomly counted rice kernels from each variety was determined using a weighing balance (METTLER TOLEDO) and an average recorded[5]. The grain and kernel length/breadth ratio (measure of slenderness) for each variety was obtained by dividing length/breadth.

Data Analysis

The phenotypic data was analysed using Analysis of Variance (ANOVA) followed by Tukey's post hoc statistical tools as implemented in Minitab 17 software package (State College, Pennsylvania). A dendrogram was obtained from the mean values of the seven traits across all the test varieties with the help of Minitab 15 software package. Principal Component Analysis (PCA) was carried out to investigate the overall pattern of phenotypic diversity and the individual trait contributions to observed phenotypic diversity [6].

Results

Seven grain and kernel trait measurements were found to vary across the 13 studied rice varieties as shown in Table 2 of all the traits, the highest variation was observed in grain weight where most of the rice varieties significantly differed (P<0.05). *Supa* rice variety showed the highest grain weight followed by *IR 2793* and *IR 54* whereas *BS 370*, *ITA 310* and *BS 217* showed the lowest grain weight mean values. It was observed that short and bold grains were heavier compared to long and slender grains. High grain length coupled with grain breadth was associated with high weight values for *Supa* and most of improved rice varieties. However, *Kahogo*, *Saro 5*, *Kilombero* and *Red Afaa* had no significant variation in grain weight (P>0.05; Table 2). Moderate variation was observed in kernel length where dimensions ranged

from 6.052 mm to 7.586 mm. Based on this trait, *Supa* and *Wahiwahi* which showed the highest kernel length mean values were significantly different from the rest of the test varieties (P<0.05; Table 2). The lowest kernel length mean values were identified in *ITA 310* and *Red Afaa* and the two varieties significantly differed from other varieties (P<0.05; Table 2). All the varieties that had high and low values for grain length also showed high and low values for kernel length (Table 2).

Low variation was observed in grain and kernel breadth traits where grain breadth dimensions across the rice varieties ranged from 1.846 to 2.055 mm. The highest grain breadth mean values were observed in *Supa*, followed closely by *Red Afaa* and *IR 54* and they significantly differed from the rest of the varieties (P<0.05); Table 2). The lowest grain breadth mean values were observed in *BS 370*, *BS 217* and *ITA 310* respectively and based on this trait, they were significantly different from other test varieties (P<0.05; Table 2).

On the other hand, kernel breadth dimensions ranged from 1.64 mm to 1.87 mm where the highest mean values were observed in *Red Afaa*, *Supa* and *Kilombero*. The three rice varieties had almost similar kernel breadth dimensions but differed significantly when compared to the rest of the varieties in this study (P<0.05; Table 2). The lowest kernel breadth mean values for were identified in *BS 217*, *BS 370* and *ITA 310*. These results indicated that there was an association between grain and kernel breadth traits since similar varieties showed consistency in high and low kernel breadth values.

Grain length measurements ranged from 8.999 mm to 10.666 mm. *Wahiwahi* had the longest grain size followed by *Supa* and *Kilombero*. Unlike other traits, the three rice varieties that had the longest grain sizes were significantly different from each other (P<0.05; Table 2). On the other hand, *ITA 310* and *IR 64* had the shortest grain sizes and were significantly different from other rice varieties. It was observed that aromatic landraces had the longest grains among the test varieties and shared a common source, Tanzania. On the other hand, non-aromatic improved varieties were found to have the shortest grains and shared a common origin, as shown by the IR codes which indicates are improved varieties from Philippine.

Grain length/breadth ratio was calculated and the highest mean values were observed in *Wahiwahi*, *BS 217* and *Saro 5* varieties. The lowest values were observed in *Red Afaa*, *IR 54 and IR 2793*. Combination of the two traits depicted *IR 54* and *IR 2793* as short and bold grains. Kernel length/breadth ratio which is the measure of slenderness mean values ranged from 3.45 mm to 4.34 mm. The highest mean values for this trait were observed in *BS 217*, *BS 370* and *Wahiwahi* where *BS 217*, an improved aromatic variety from Kenya, was the most slender kernel and significantly differed from the rest of the rice varieties (P<0.05; Table 2). On the other hand, *Red Afaa*, *IR 64* and *IR 2793* had the lowest mean values for kernel length/breadth ratio.

Red Afaa, *IR 64* and *IR 2793* with KL/B ratio of less than 3.80 was categorized as short grain varieties whereas *Kahogo*, *IR 54*, *ITA 310*, and *BW 196* rice varieties having KL/B ratio of less than 4.0 were considered as medium grain varieties. *BS 217*, *BS 370*, *Kilombero*, *Saro 5*, *Wahiwahi* and *Supa* showed a KL/B ratio greater than 4.0 and were categorized as long grain varieties. From these results, it was inferred that basmati varieties had long and slender rice kernels followed by *Supa*, *Wahiwahi*, *Kilombero* and *Saro 5* which had medium grain sizes. *Red Afaa*, *IR 64* and *IR 2793* varieties had short and bold kernels.

High variability was revealed by analysis of variance of the seven traits across all the varieties are shown in Table 2. The ANOVA table

Sr.no	Genotype	Source	Attribute
1	IR 2793	Kenya	Improved variety
2	BS 217	Kenya	Improved variety
3	BS 370	Kenya	Improved variety
4	BW 196	Kenya	Improved variety
5	ITA 310	Kenya	Improved variety
6	Red Afaa	Tanzania	Landrace
7	IR 54	Tanzania	Improved variety
8	Kilombero	Tanzania	Landrace
9	IR 64	Tanzania	Improved variety
10	Kahogo	Tanzania	Landrace
11	Saro 5	Tanzania	Improved variety
12	Wahiwahi	Tanzania	Landrace
13	Supa	Tanzania	Landrace

Table 1: Profiles of rice varieties used in the study.

VARIETIES TRAITS							
	GL (mm)	GB (mm)	GL/B	KL (mm)	KB (mm)	KL/B	GW (g)
IR 2793 (1)	9.199 ± 0.37 [ef]	1.994 ± 0.10 [ab]	4.625 ± 0.31 [cde]	6.619 ± 0.30 [cd]	1.762 ± 0.05 [abc]	3.759 ± 0.19 [de]	28.9 ± 0.01 [b]
BS 217 (2)	9.543 ± 0.50 [bcdef]	1.85 ± 0.08 [b]	5.159 ± 0.18 [ab]	7.112 ± 0.39 [abc]	1.641 ± 0.08 [c]	4.336 ± 0.18 [a]	23.5 ± 0.02 [e]
BS 370 (3)	9.225 ± 0.38 [def]	1.843 ± 0.06 [b]	5.005 ± 0.10 [abc]	6.931 ± 0.27 [cd]	1.659 ± 0.05 [bc]	4.177 ± 0.09 [ab]	18.2 ± 0.01 [g]
BW 196 (4)	9.302 ± 0.36 [cdef]	2.011 ± 0.13 [ab]	4.640 ± 0.30 [cde]	6.625 ± 0.29 [cd]	1.749 ± 0.1 [abc] 6	3.808 ± 0.26 [cd]	26.2 ± 0.01 [d]
ITA 310 (5)	8.999 ± 0.31 [f]	1.846 ± 0.06 [b]	4.877 ± 0.15 [abcde]	6.522 ± 0.35 [d]	1.643 ± 0.08 [c]	3.971 ± 0.15 [bcd]	20.6 ± 0.01 [f]
Red Afaa (6)	9.138 ± 0.22 [ef]	2.028 ± 0.16 [a]	4.531 ± 0.37 [e]	6.435 ± 0.29 [d]	1.868 ± 0.13 [a]	3.452 ± 0.16 [e]	27.2 ± 0.01 [bcd]
IR 54 (7)	9.929 ± 0.35 [bcd]	2.049 ± 0.10 [a]	4.852 ± 0.20 [bcde]	7.139 ± 0.35 [abc]	1.788 ± 0.11 [abc]	3.997 ± 0.15 [bcd]	28.5 ± 0.01 [bc]
KILOMB(8)	10.02 ± 0.60 [abc]	1.991 ± 0.17 [ab]	5.057 ± 0.41 [abc]	7.501 ± 0.32 [ab]	1.814 ± 0.06 [a]	4.169 ± 0.22 [ab]	27.3 ± 0.01 [bcd]
IR 64 (9)	9.072 ± 0.63 [f]	1.989 ± 0.07 [ab]	4.565 ± 0.33 [de]	6.600 ± 0.50 [cd]	1.786 ± 0.06 [abc]	3.697 ± 0.27 [de]	26.5 ± 0.01 [cd]
Kahogo (10)	9.952 ± 0.47 [abc]	2.012 ± 0.12 [ab]	4.961 ± 0.35 [abcde]	6.981 ± 0.27 [bcd]	1.783 ± 0.09 [abc]	3.924 ± 0.24 [bcd]	27.5 ± 0.01 [bcd]
Saro 5 (11)	9.855 ± 0.42 [bcde]	1.939 ± 0.12 [ab]	5.096 ± 0.32 [ab]	7.078 ± 0.31 [abc]	1.744 ± 0.14 [abc]	4.075 ± 0.26 [abc]	27.5 ± 0.01 [bcd]
Wahiwahi(12)	10.666 ± 0.65 [a]	2.017 ± 0.16 [ab]	5.302 ± 0.29 [a]	7.540 ± 0.44 [a]	1.804 ± 0.08 [ab]	4.130 ± 0.15 [ab]	28.4 ± 0.02 [bc]
Supa (13)	10.243 ± 0.66 [ab]	2.055 ± 0.11 [a]	4.989 ± 0.29 [abcd]	7.586 ± 0.54 [a]	1.849 ± 0.11 [a]	4.108 ± 0.25 [abc]	32.7 ± 0.03 [a]

The values are mean ± SEM of ten independent determinations at 5% level of significance. Data was analysed using Analysis Of Variance (ANOVA) followed by Tukey's post hoc test. In this table, means that do not share a superscript are significantly different (P>0.05).

Table 2: Analysis Of Variance (ANOVA) of seven grain and kernel traits of the 13 studied rice genotypes.

TRAITS	PRINCIPAL COMPONENT ANALYSIS						
	PC1	PC2	PC3	PC4	PC5	PC6	PC7
Eigen values	0.519	0.0647	0.0265	0.0026	0.00025	2E-05	0
Proportion of variance	84.6	10.6	4.3	0.4	0	0	0
Cumulative % variance	84.6	95.2	99.5	99.9	100	100	100
Eigen vectors							
GL	0.73	-0.456	0.408	-0.235		0.093	0.153
GB	0.029	-0.281	-0.113	-0.247	0.048	-0.517	-0.76
GL/B	0.294	0.099	0.489	0.568	-0.143	-0.177	-0.3
KL	0.562	0.455	-0.742	0.271	0.029	0.188	-0.12
KB	0.027	-0.261	-0.15	0.356	-0.03	-0.729	0.499
KL/B	0.25	0.645	-0.064	-0.591	-0.016	-0.352	0.211
GW	0.323	-0.112	-0.064	-0.105	-0.985	0.04	-0

Table 3: Eigen values and percent of variation for 7 principal component axes in 13 rice varieties.

clearly showed that the means of the characters measured varied significantly across all the varieties.

Principal component analysis

The principal component analysis (PCA) was carried out to investigate the morphological traits that played a key role in phenotypic diversity among the rice varieties. It provided the Eigen values and percent of variation for seven principal component axes across 13 rice varieties as shown in Table 3. It was found that the first three principal components jointly accounted for 99.5% of the total variation among all the studied varieties. Combination of the first and the second principal components accounted for 95.2% of the total variation among the seven component axes of the total rice varieties (Table 3).

Principal component 1 (PC1) had 84.6% of the total variation where all the traits; grain length, grain breadth, grain length/breadth ratio, kernel length, kernel length/breadth ratio and grain weight contributed positively. Of all, three traits; grain length, kernel length and grain weight had a notably major contribution to PC1. In the case of Principal Component 2 (PC2), three traits; grain length/breadth ratio, kernel length/breadth ratio and kernel length contributed positively and accounted for 10.6% of the total morphological variability.

On the other hand, grain length, grain breadth, kernel breadth and grain weight traits were negatively associated with PC2. Figure 1

shows a plot for the first two vectors of PCA. A total of 4.30% variation was could be explained by the third principle component and was greatly negatively characterized by grain breadth, kernel length, kernel breadth, kernel length/breadth ratio and grain weight. However, it was found that two traits; grain length and grain length/breadth ratio contributed positively to Principal Component 3 (PC3).

The first two principal components efficiently separated most of the improved varieties from landraces with varieties possessing long grains clustering close together as shown in the scatter plot, Figure 1. Basmati varieties clustered in a separate group distantly from the other varieties and this correspond well with their slender grains. The GL, KL, GW and KL/B were found to be the major contributors of PC1 and PC2 (Figure 1).

Cluster analysis of rice varieties based on morphological traits

Cluster analysis grouped the 13 rice varieties into two distinct major clusters I and II with a similarity index of 1.25 thereby revealing presence of high diversity as shown in Figure 2. Cluster I was the largest with 8 rice varieties whereas cluster II had only 5. Cluster I was further subdivided into three other sub clusters CIA, CIB and CIC where *Wahiwahi*, a landrace, formed its own sub cluster, CIA. Sub cluster CIB contained two other smaller groups I and II.

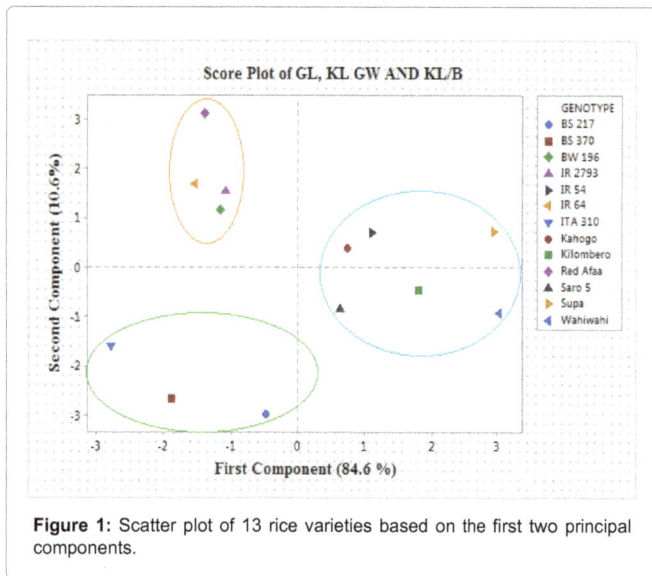

Figure 1: Scatter plot of 13 rice varieties based on the first two principal components.

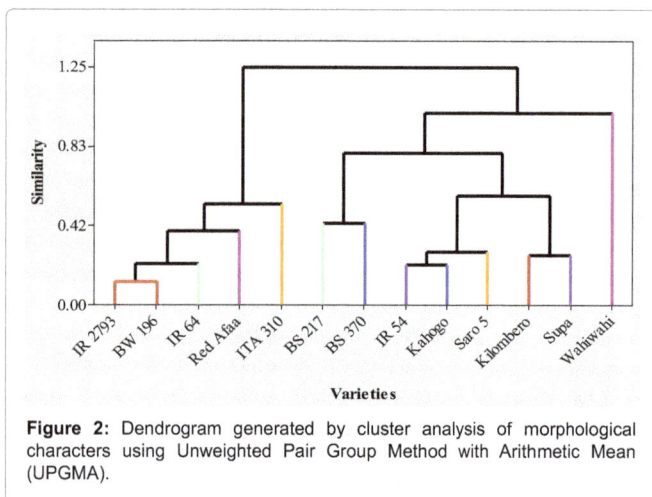

Figure 2: Dendrogram generated by cluster analysis of morphological characters using Unweighted Pair Group Method with Arithmetic Mean (UPGMA).

Among these two groups, *Supa*, an improved aromatic variety, clustered close together with *Kilombero*, a semi aromatic variety in group i. In group ii, *Saro 5*, an improved aromatic variety clustered together with two other varieties from the same origin. In sub cluster CIC, improved aromatic Basmati genotypes clustered together in with a similarity coefficient of 0.43.

Cluster II contained four improved rice varieties and only one land race from both source countries where *ITA 310* showed parentage to the rest of the varieties on the pedigree. Two improved varieties from Kenya in this cluster, *IR2793* and *BW 196* were the most similar with a similarity coefficient of 0.21. The relationship among the 13 rice varieties was revealed by the dendrogram as shown in (Figure 2).

Discussion

Analysis of variance of the grain and kernel traits measurements indicated a wide spectrum of variation among all the rice varieties for all the characters. The means of the characters measured varied significantly across all the varieties and this demonstrated that based on these traits, the varieties were distinct from each other. The highest variability was observed in grain weight ranging from 18.2 g to 32.7 g. The maximum grain weight was recorded in *Supa* with 32.7 g followed

by *IR 2793* with 28.9 g and then *IR 54* with 28.5 g. The lowest grain weight was observed among basmati varieties, *BS 370* and *BS 217* with 18.2 g and 23.5 g, respectively. These results concur with the report of who reported immense variation in grain weight among Bangladesh rice varieties. Low grain weight among BS varieties can be explained by the fact that these varieties possess longer and slender grains. This study observation relates to the report of [6] who found that most non aromatic landrace varieties had large and heavier grains. On the other hand, majority of aromatic varieties had small and lighter grains using a set of Indian aromatic and non-aromatic rice varieties.

On the other hand, varieties showed moderate variation in kernel length ranging from 6.052 mm for *ITA 310* to 7.586 mm for *Supa* with landrace varieties having the highest grain and kernel length. The same varieties also showed very high values for grain and kernel breadth ranging from 1.804 mm for *Kilombero* to 1.849 mm for *Supa*. Maximum grain length coupled with grain breadth gave these rice varieties highest grain weight above the other varieties in this study. These results agree with the findings of [7] who reported moderate variability for kernel length among Indian rice varieties. However, these results contrast with the report of who did not find any association between grain/kernel length and grain/kernel breadth.

Low variation was observed in grain and kernel breadth. The results clearly indicated that *BS 217* and *BS 370*, both of which are aromatic varieties had a small grain breadth that contributes to [8] heir slender appearance. *Red Afaa, IR 54 and IR 2793* had the lowest values of grain length/breadth ratio. Combination of these two traits depicted *IR 54* and *IR 2793* as short and bold grains. *Supa*, an improved aromatic variety and *Kilombero*, an aromatic landrace showed the highest grain length and breadth mean values and this corresponded well with their long and bold rice grains. High grain lengths coupled with high grain breadths contributed round appearance of these varieties. Generally, most non-aromatic varieties had low variability in GB and KB traits hence appeared as short and bold. These results agree with earlier findings of who reported the existence of long bold aromatic varieties [6].

In the case of grain and kernel length/breadth ratio (a measure of slenderness) the highest mean values were observed in *BS 217* with 4.336 followed by *BS 370* with 4.177 and *Kilombero* with 4.169. The rice varieties with a KL/B ratio of >4.0 were categorized as slender and long grains [9]. Most slender rice varieties are desirable and preferred by consumers [10]. It was observed that most of aromatic landraces and improved varieties belonged to this category with KL/B ratios ranging from 4.075 for *Saro 5* to 4.336 *BS 217*. Generally, Basmati varieties showed the highest kernel L/B values. Similar observations were made by who reported Basmati varieties as being slender and light using Indian rice varieties. This type of rice is long and slender in shape and possess most desirable grain and cooking quality traits. *Wahiwahi* and *Kilombero* were the most slender landrace varieties with a KL/B ratio of 4.180 and 4.139 respectively. On the other hand, non-aromatic varieties had the lowest KB/L values of <4.0-hence had bold appearance.

The varieties with high grain and kernel length/breadth ratios may be utilized as sources of these traits in breeding for long grain varieties. Immense variation in grain and kernel L/B ratios has also been reported by [11]. Grain shape and size are important traits that determine the market value of rice. These traits are highly considered by breeders in developing new varieties for commercial release [12,13].

Principal Component Analysis (PCA) provided an insight of the contribution of each of the trait towards divergence among the characteristics of the rice varieties. Principal component analysis grouped the rice varieties into three clusters indicating the presence of considerable phenotypic diversity among the varieties. The first two principal components were utilized for principal component analysis because they expressed better total variability (95.2%) of the plant material. The most predominant traits that contributed to the observed phenotypic diversity were; grain length, kernel length, grain weight, and kernel length/breadth ratio. This analysis validates their use as the main discriminating traits among the test varieties. The other traits were found to have a minimal contribution to variability. These results are consistent with the report of [10] who found major contributions of grain length, kernel length, grain weight and grain length/breadth ratio to phenotypic diversity in a set of Indian rice. In contrast [7] reported that grain breadth, grain length and grain weight were the major contributors of phenotypic diversity. This discrepancy could be due to use of different set of rice varieties. The major contribution of grain length, kernel length, grain weight, and kernel length/breadth ratio could be perhaps due to the fact that they are the most important agronomic traits subjected to selection by farmers and breeders over time [13,15].

Cluster analysis provided a good opportunity to identify and group the rice varieties into distinct categories with respect to similarity levels based on the phenotypic traits. The dendrogram showed two major clusters, I and II. Cluster I contained eight rice varieties from both source countries. All the aromatic rice varieties, both improved and landraces were grouped in cluster I. This indicated the extent in which most aromatic varieties share phenotypic grain quality traits. On the other hand, cluster II contained five rice varieties from both source countries where all were non aromatic varieties, representing both improved and landrace varieties. Cluster analysis indicated that there was no association between the observed pattern of variation and the geographical origin of the rice varieties. Similar observations were made by [6] where cluster analysis effectively distinguished the majority of aromatic from non-aromatic landraces. However, these results disagree with the findings of who [11] reported an association between the pattern of phenotypic variation with geographical origin of the varieties.

Cluster analysis grouped the basmati varieties, *BS 217* and *BS 370* together and close to other aromatic varieties from different sources and this presented aroma as a grain quality trait that distinguished rice varieties into different categories. This also suggests a possibility that the widely preferred basmati type of rice may have evolved through natural mutation from non-basmati genotypes. These results are consisted with the findings of who observed clustering together of the basmati group based on morphological traits using aromatic and non-aromatic rice varieties from India [6,14,15].

Only one improved aromatic variety, *Saro 5,* clustered with an improved non-aromatic variety *IR 54,* and *Kahogo,* a landrace semi aromatic variety, which is rational since they are collections from Kilimanjaro Agricultural Research Training College and possibly have common ancestors. These varieties had similar phenotypic traits such as grain length, kernel breadth and grain weight but differed in grain breadth, grain length/breadth ratio and kernel length/breadth ratio.

These differences contribute to the slender nature of *Saro 5* and distinguish it from other varieties with almost similar size. Other long grain non-aromatic varieties fell in one sub-cluster corresponding well with their grain characters. These varieties, although have varying cooking and eating qualities, lack the desirable basmati traits. Generally, cluster analysis gave an insight into the diversity of the rice varieties as

shown in the dendrogram in Figure 2. Variety groups were primarily associated with phenotypic differences among them and with variety type. These results agree with earlier report of [16], who also found that in cluster analysis, varieties grouped together with greater phenotypic similarity but the groups did not essentially include varieties from the same origin[17,18].

Conclusion

From this study, it can be concluded that phenotypic analysis of the 13 studied rice varieties revealed an enormous diversity across all the varieties for all the traits evaluated. The most distinct phenotypic traits among the studied Kenyan and Tanzanian rice varieties were GL, KL, GW and KL/B ratio. This analysis recommends their use as the main discriminating traits among the test varieties and this can be validated by follow up molecular analysis. Based on these traits, at least three landrace rice varieties; *Kilombero*, *Wahiwahi* and *Kahogo* were found to possess good grain quality traits. These promising landraces should be conserved as reservoir of beneficial gene pool for improvement of grain quality traits in rice varieties. From this study it is recommended that agro- morphological traits can be employed as a common approach for assessing genetic and phenotypic variability among varieties. Hybridization of different distantly related rice varieties from different clusters should be carried out to obtain segregants with high degree of hybrid vigor for the traits studied.

References

1. Chakravarthi BK, Naravaneni R (2006) SSR marker based DNA fingerprinting and diversity study in rice (*Oryza sativa* L) African Journal of Biotechnology 5: 684-688.

2. Ouma-Onyango A (2014) Promotion of Rice Production: A Likely Step to Making Kenya Food Secure. An Assessment of Current Production and Potential. Developing Country Studies 4: 26-31.

3. Jones H, Lister DL, Bower MA, Leigh FJ, Smith LM, et al. (2008) Approaches and constraints of using existing landrace and extant plant material to understand agricultural spread in prehistory. Plant Genetic Resources: Characterization and Utilization 6: 98-112.

4. Shahidullah SM, Hanafi MM, Ashrafuzzaman M, Ismail MR, Khair A (2009) Genetic diversity in grain quality and nutrition of aromatic rices. African Journal of Biotechnology 8: 1238-1246.

5. Esquinas-Alacazar J (2005) Protecting crop genetic diversity for food security: political, ethical and technical challenges. Nature Revision on Genetics 6: 946-953.

6. Vanniarajan C, Vinod KK, Pereira A (2012) Molecular evaluation of genetic diversity and association studies in rice (*Oryza sativa* L). Journal of Genetics 91: 9-19.

7. Varnamkhasti MG, Mobli H, Jafari A, Keyhani AR, Soltanabadi MH, et al. (2008) Some physical properties of rough rice (Oryza Sativa L) grain. Journal of Cereal Science 47: 496-501.

8. Ray A, Deb D, Ray R, Chattopadhayay B (2013) Phenotypic characters of rice landraces reveal independent lineages of short-grain aromatic indica rice. Journal for plant sciences 5: 258-264.

9. Hien NL, Sarhadi WA, Oikawa Y, Hirata Y (2007) Genetic diversity of morphological responses and the relationships among Asia aromatic rice (*Oryza sativa* L) cultivars. *Tropics* 16: 343-355.

10. Khush GS (1987) Rice breeding: past, present and future. Journal of Genetics 66: 195-216.

11. Parikh M, Rastogi NK, Sarawgi AK (2013) Variability in grain quality traits of aromatic rice (Oryza sativa L) Bangladesh. Journal of Agricultural Research 37: 551-558.

12. Fitzgerald MA, McCouch SR, Hall RD (2009) Not just a grain of rice: the quest for quality. Trends in plant science 14: 133-139.

13. Yadav RB, Khatkar BS, Yadav BS (2007) Morphological, physicochemical and

cooking properties of some Indian rice (*Oryza sativa* L) cultivars. Journal of Agricultural Technology 3: 203-210.

14. Cruz ND, Khush GS (2000) Rice grain quality evaluation procedures. Aromatic Rices 15-28.

15. Pusadee T, Jamjod S, Chiang YC, Rerkasem B, Schaal BA (2009) Genetic structure and isolation by distance in a landrace of Thai rice. Proceedings of the National Academy of Sciences 106: 13880-13885.

16. Joshi RK, Behera L (2007) Identification and differentiation of indigenous non-Basmati aromatic rice genotypes of India using microsatellite markers. African Journal of Biotechnology 6: 348-354.

17. Kaur S, Panesar PS, Bera MB (2011) Studies on evaluation of grain quality attributes of some basmati and non-basmati rice cultivars. Journal of Food Quality 34: 435-441.

18. Sarawgi AK, Parikh M, Sharma B (2012) Agro-morphological and quality characterization of Dubraj group from aromatic rice germplasm of Chhattisgarh and Madhya Pradesh. Vegetos-An International Journal of Plant Research 25: 387-394.

8

Optimizing Sowing Methods Through Estimates of Interactions between Sowing Date and Genotypes in Some Morpho-Physiological Traits in Rice (*Oryza sativa L*)

Fatema-Tuj-Johora[1]*, Jalal Uddin Ahmed[2], Morshed[3] M, Mian MAK[4] and Ariful Islam M[5]

[1]Department of Crop Botany, EXIM Bank Agricultural University, Bangladesh

[2]Dept of Crop Botany, BSMRAU, Gazipur1706, Bangladesh

[3]Senior Scientist, IRRI, Philippines

[4]Department of Genetics and Plant Breeding, BSMRAU, Gazipur1706, Bangladesh

[5]Department of GPB, EXIM Bank Agricultural University, Bangladesh

Abstract

For optimizing some morpho-physiological traits in two different sowing methods of boro rice genotypes at varying seeding time, a field experiment was carried out by IRRI at the research farm of Bangabandhu Sheikh Mujibur Rahman Agricultural University, Gazipur; Bangladesh during 2010-12. A decreasing trend of plant height was observed at sowing dates after 8th December and the decreasing trend continued until January 07. The smaller plant in delayed seeding indicate the sensitivity of late seeding low temperature on the dry matter production of the genotype, unlike other genotypes which expressed no significant reduction in plant height due to late seeding. Significant responses of environments were found either in between the varieties but in between the sowing dates. Maximum number of effective tillers was observed in both direct seeded and transplanted condition in December 18. The tiller number increased with delayed seeding until December 18. But tillers numbers were found to decline afterwards. PSBRC82 took maximum days to flowering sowing at December 08th in transplanting condition but first flowering in direct seeding condition which is statistically different. And the earliest flowering was observed in BRRI dhan28 sowing at January 07. In direct seeding condition than in transplanting.OM2718 gave the maximum sterility when it was practiced in transplanting method than the direct seeded method at January 07. The increase of filled grains reflects clearly that the number of filled grains exhibited higher in direct seeded methods than transplanting method in all genotypes interaction with sowing date of December18th. Such difference in number of filled grains might be due to undisturbed root establishment in direct seeding method than transplanted Method. PSBRC82 sowing at December 18 resulted the highest grain yield in direct seeding methods followed by in BRRI dhan28 in transplanting.

Keywords: Optimizing; Direct seeding; Transplanting; Sowing dates; Genotypes; Effective tillers and harvest index

Introduction

Rice (*Oryza sativa* L.) is the world's second most important food crop after wheat. It is the most extensively cultivated crop and viewed as staple food in Bangladesh. More than 90% of the people depend on rice for their diets. Among the major rice growing countries of the world, Bangladesh ranks third in rice cultivated area and fourth in production. Before 1970, 95% of the total rice in the world was being produced by the method of direct seeding. The modernization in rice cultivation started with the advent of IRRI (International Rice Research Institute). Afterword's, transplanting was found to be the most popular method of rice cultivation. But in the extreme conditions due to meet up the need to feed the hunger we had to go through direct seeding method. The transplanting method requires a large amount of labor. In addition, under the changing socioeconomic environment, workers are not available or reluctant to undertake operations like nursery transplanting. These situations further escalate labor costs. Alternate methods of establishing crops, especially rice, that require less labor and water without sacrificing productivity are needed. Direct-seeded rice can achieve equivalent yields to transplanted rice across a range of rice cultivars in common use. Although no statistical significant differences in yield so a higher cost: benefit ratio for direct-seeded rice (1.59) than for transplanted rice (0.96). Net labor savings with direct-seeded rice compared with transplanting averaged 27 days/ha [4]. Direct seeding methods encompasses a set of principles, each of them

fairly simple, but working synergistically with the others in order to achieve higher grain yield. In Bangladesh, direct seeded rice produced about 2-12% higher grain yield than transplanting. These situations further escalate labor costs. In boro season poor seedling growth is a main problem in transplanting method compared to direct seeding. Transplanting of aged seedlings caused considerable reduction in yield and yield components, since grain yield and its components depended mainly on seedling age at the time of transplanting beside the other factors controlling growth and yield [1-7].

Objective:

To evaluate the morpho-physiological traits of rice in direct seeding method as compared to transplanting method.

*Corresponding author: Fatema-Tuj-Johora, Lecturer, Department of Crop Botany, EXIM Bank Agricultural University, Bangladesh
E-mail: fatemamoona@ymail.com

Materials and Methods

The experiment was carried out at the research farm of Bangabandhu Sheikh Mujibur Rahman Agricultural University (BSMRAU), Salna, Gazipur under the department of Crop Botany during Boro (Winter) seasons from November 2009 to May 2010. The experimental site is located at the centre of Madhupur Tract at 24°09ʹ North latitude and 90°26ʹ East longitude with elevation of 8.4 meter from sea level. The Soil of the experimental site was clay loam of shallow red brown in texture belonging to Salna series. The seeds were collected from genetic resources center (GRC) of International Rice Research Institute (IRRI) and Bangladesh Rice Research Institute (BRRI). Different sowing dates were, Sowing date 01, December 08/2009: Sowing date 02, December 18/2009: Sowing date 03, December 28/2009 and Sowing date 04, January 07/2010. Seeds were sowed directly for direct seeding and soaked separately for 48 hr in clothes bag for the method of transplanting. Soaked seeds were picked out from water and wrapped with straw and gunny bag to increase the temperature for facilitating germination. Fertilizers were applied according to manual of BRRI; N @ 110 kg/ha, P_2O_5 90 kg/ha, K_2O 75 kg/ha, and Gypsum 60 kg/ha. Total Nitrogen was applied in four installments at just before seeding, at 15 DAT (Days after transplanting), 30 DAT and 45 DAT recommended by BRRI. Other fertilizers were applied during the final land preparation. Land was well plowed, well puddled and leveled for the seedbed. About 250 g seeds were soaked for 24 hr in transplanting methods. Soaked seeds were kept under straw for 3 days. Sprouted seeds of each variety were sown. The channels (15 cm deep and 50 cm wide) were used for irrigation and drainage. A foliar spray of Furadan 5 G was applied to protect the seedlings against insect pests. Healthy seedlings of 40 days old were transplanted on each plot of the experimental field. In each plot for transplanting 20cm × 20cm (line to line and seedling to seedlings) spacing and for direct seeded condition 3-4 seeds were sown per pit as a distance of 20 cm from line to line and 15 cm from plant to plant as recommended. In direct seeded condition 3-4 seeds were sown per pit as a distance of 20 cm from line to line and 15 cm from plant to plant as recommended. A bundle of seedlings were kept at the side of each plot for gap filling and gap filling was done at seven days of transplanting to replace the dead ones. Hand weeding was done to keep the experimental field free from weed infestation throughout the crop growth phases. The field was irrigated properly and 5-10 cm water depth was maintained up to heading in transplanted method. A good drainage system was also maintained specially after flowering. The excess water was drained out leaving the soil only at saturated condition. Data were collected on seedling mortality, days to panicle initiation, total number of effective tillers per panicle and biological yield. Three factor Split-split plot design where the used methods of sowing (main plot), sowing dates (sub plot) and genotypes were in sub-sub plot. The whole experimental area was divided into different blocks and sub-blocks and sub-sub-blocks, representing three replications. Four rice varieties were assigned randomly to each unit plot of 5.8 m × 2.6 m dimension. All the collected data of the present study were statistically analyzed using MSTAT-C Program. The statistical analysis for various characters under investigation were done and the analysis of variance for each of the characters was performed by F test and mean values were separated by lsd.

Genotypes used

Four different genotypes were used in the study .The seeds were collected from genetic resources center (GRC) of the International Rice Research Institute (IRRI) and Bangladesh Rice Research Institute (BRRI) (Table 1).

Sl no.	Germplasms Identity	Source of planting materials
1	OM 2718	International Rice Research Institute (IRRI)
2	PSBR 82	International Rice Research Institute (IRRI)
3	BRRI dhan28	Bangladesh Rice Research Institute (BRRI)
4	BRRI dhan36	Bangladesh Rice Research Institute (BRRI)

Table 1: Source of planting materials and their identity.

Result and Discussion

Plant height (cm)

Varietal mean represents PSBRC-82 as the tallest (122.71 cm) among the genotypes whereas the dwarfing natures of BRRI dhan36 (106.20 cm). Statistically significant differences were recorded from genotypes to genotypes but neither for sowing dates nor sowing methods. In combined interaction it is clearly observed that the tallest plant was found in PSBRC82 sowing on 8 December in direct seeding methods and the smallest plant was found in BRRI dhan36 sowed on 7th January 2010 in transplanted condition. A decreasing trend of plant height was observed on sowing dates after 8th. December and the decreasing trend continued until 7th January. The smaller plant in delayed seeding indicate the sensitivity of late seeding low temperature on the dry matter production of the genotype unlike other genotypes which expressed no significant reduction in plant height due to late seeding. In transplanting method comparative low plant height was observed which might be due to low soil temperature than the direct sowing condition. In direct seeded condition the field was dry whereas in transplanting condition an adequate amount of moisture was maintained within the field (Table 2).

Number of effective tillers per hill

In respect of direct seeded boro rice, the highest number of effective tillers per hill was found in PSBRC to be 82 sowing at December 18 (20.247) and the lowest number of effective tillers per hill was found in OM 2718 sowing in January 07 (5.40).In transplanted boro rice, the highest number of effective tiller was found in PSBRC 82 sowing at December 18 (19.60). And the lowest number of effective tiller per plant (5.73) was found in OM 2718 sowing at January 07. The results indicated that direct seeded boro rice is superior over transplanted rice in respect of effective tiller per hill. However significant responses for environments were found either in between the varieties but not in between the sowing dates. Maximum number of effective tillers was observed in both direct seeded and transplanted condition executed on December 18. The tiller number increased with delayed seeding until 18 December but tillers numbers were found to decline afterwards. Although transplanting is the common rice growing practice in Bangladesh but for obtaining maximum effective tillers on December 18 in PSBRC 82 direct seeded method might be practiced (Table 3).

Days to first flowering

In combined interaction PSBRC 82 took maximum days to flowering (124.667 days) sowing on December 08 in transplanting condition but first flowering in direct seeding condition which is statistically different (116.826 days). The earliest flowering (90.39 days) was observed in BRRI dhan28 sowing at 7 January in direct seeding condition than (93.335 days) in transplanting. It is clear that in both the cases direct seeding is earlier than transplanting. Similar results were observed in respect of panicle initiation, so it might be concluded that panicle initiation is positively correlated with days to first flowering i.e., the earlier the panicle initiation the earlier the flowering. It indicated that a genotype may give better yield whether it is earlier or late. Days

Genotypes	Plant height (cm)								Mean
	Direct seeding				Transplanting				
	8th Dec/09	18th Dec/09	28th Dec/09	7th Jan/10	8th Dec/09	18th Dec/09	28th Dec/09	7th Jan/10	
BRRI Dhan 28	117.732	115.095	111.992	108.541	113.021	110.482	107.514	104.202	111.07
BRRI Dhan 36	112.571	110.044	107.073	103.784	108.073	105.643	102.797	99.635	106.20
PSBRC-82	130.063	127.153	123.736	119.922	124.862	122.064	118.785	115.136	122.71
OM-2718	125.035	122.237	118.946	115.283	120.034	117.341	114.185	110.676	117.96
Mean	121.355	118.636	115.436	111.883	116.497	113.883	110.813	107.409	11.67
SE	1.998	1.954	1.902	1.844	1.918	1.876	1.826	1.771	
Lsd	7.53								
CV %	12.843								

SE = Standard Error

Lsd = Least significant difference (Genotype × Methods of Sowing × Date of sowing)

CV % = Co-efficient of variation

Table 2: Plant height (cm) of four different genotypes in four different sowing dates in both direct seeding and transplanting conditions.

Genotypes	Number of effective tillers per hill								Mean
	Direct seeding				Transplanting				
	8th Dec/09	18th Dec/09	28th Dec/09	7th Jan/10	8th Dec/09	18th Dec/09	28th Dec/09	7th Jan/10	
BRRI Dhan 28	14.759	18.357	12.936	10.427	14.112	17.630	12.299	9.790	13.289
BRRI Dhan 36	11.348	14.465	6.850	8.702	10.711	13.818	9.153	8.065	10.389
PSBRC-82	15.582	20.247	17.542	11.358	14.935	19.600	13.955	10.711	15.491
OM-2718	8.379	10.006	9.575	5.400	9.692	14.259	8.928	5.733	8.996
Mean	12.517	14.769	11.726	8.972	12.363	16.327	11.084	8.575	12.041
SE	1.656	2.101	2.304	1.312	1.276	1.384	1.228	1.095	
Lsd	2.62								
CV %	7.993								

SE = Standard Error

Lsd = Least significant difference (Genotype × Methods of Sowing × Date of sowing)

CV % = Co-efficient of variation

Table 3: Number of effective tillers per hill of four different genotypes at different sowing dates in both direct seeding & transplanting conditions.

to flowering were reduced in case of direct seeding than transplanting methods in all genotypes. This can be used to interpret that in the transplanting method seedling establishment, duration might cause much variation (Table 4).

Days to maturity

Among the genotypes, PSBRC 82 took the highest days for maturation followed by for OM 2718 whereas the lowest days for maturation was shown by BRRI dhan28.In respect of sowing dates, sowing on 8th December took the highest days for maturation whereas the lowest number of dates taken for maturation was in sowing on January 7th.In direct seeded boro rice the highest days to maturity was taken by PSBRC 82 sowing at 8th December while the lowest days to maturity was observed in BRRI dhan28 sowing on the 7th.of January. In combined interaction the interaction highest days for maturation (157.04 days) was observed in PSBRC 82 sowing at December8 in transplanting and 153.63 days in direct seeding condition. The earliest flowering (124.53) was observed in BRRI dhan28 sowing on 7 January in direct seeding condition than 130.32 in transplanting. It is clear that in both the cases direct seeding the flowering is earlier than transplanting. Similar results were observed with respect of panicle initiation, with regard to days to first flowering and days to 100% flowering (Table 5).

Spikelet's sterility percent

The inverse relationship existing between spikelet's sterility percent and yield was observed. Variety, OM 2718 gave the maximum sterility (43.550) when it was practiced in transplanting method than the direct seeded method (40.701) on January 07th. From the temperature graph

it was clearly observed that during the 1st week of January it was below 10 °C what is very detrimental. During the germination to seedling stage the severe cold injury affects the number of tillers per plant as well as spikelet's fertility. Whereas favorable temperature in 1st and 2nd week of December decrease spikelet's sterility as well as increased effective tillers per plant. Infact the Lsd values signifies that there is no statistical differences in BRRI dhan28 although germinated from 8th December to 7th January either in direct seeding or transplanting. Physiological homeostasis (maintenance of relative stable condition in varying environments) is much higher in BRRI dhan28 than the other three varieties. Thus, BRRI dhan28 has the ability to adopt to a varying range of temperatures. Although higher fertility (low sterility percent) was found in BRRI dhan36 but its low effective tillers per plant invaueds its suitability. Among the genotypes the maximum fluctuation in spikelet's sterility percent was found in OM 2718 (25.026-43.550 %) whereas the minimum fluctuation in BRRI dhan28 (20.942-29.321 %). From this table it might be concluded that OM 2718 has the higher sensitivity while BRRI dhan28 has the minimum sensitivity (Table 6).

Filled grains (number) per panicle

An important observation with respect to filled grains per plant indicated that in all the interactions between genotypes and sowing dates, direct seeded condition exhibited superior performances than transplanted condition. Higher filled grains per plant is a major criteria of increasing grain yield i.e., conversion of photosynthates (source) to sink. The highest number of filled grains per plant (196.323) was found in PSBRC 82 sowed on 18th.December, by direct seeding methods while statistically, similar result (192.432) was also found in transplanting

Genotypes	Days to first flowering								Mean
	Direct seeding				Transplanting				
	8th Dec/09	18th Dec/09	28th Dec/09	7th Jan/10	8th Dec/09	18th Dec/09	28th Dec/09	7th Jan/10	
BRRI Dhan 28	109.966	101.685	94.531	90.395	112.906	104.625	97.471	93.335	100.614
BRRI Dhan 36	111.926	103.645	96.491	92.355	119.766	111.485	104.331	100.195	105.024
PSBRC-82	116.826	108.545	101.391	97.255	124.666	116.385	109.231	105.095	112.924
OM-2718	114.866	106.585	99.431	95.295	122.706	114.425	107.271	103.135	107.964
Mean	113.396	105.115	97.961	93.825	120.011	111.730	104.576	100.440	105.882
SE	1.523	1.543	1.052	1.623	2.453	2.098	2.008	2.097	
Lsd	5.356								
CV %	4.821								

SE = Standard Error
Lsd = Least significant difference (Genotype × Methods of Sowing × Date of sowing)
CV %= Co-efficient of variation

Table 4: Days to first flowering of four different genotypes at different sowing dates interaction with both direct seeding & transplanting methods.

Genotypes	Days to maturity								Mean
	Direct seeding				Transplanting				
	8th Dec/09	18th Dec/09	28th Dec/09	7th Jan/10	8th Dec/09	18th Dec/09	28th Dec/09	7th Jan/10	
BRRI Dhan 28	140.34	131.67	127.74	124.53	148.44	139.61	132.56	130.32	134.15
BRRI Dhan 36	143.05	138.33	134.38	127.23	145.27	141.46	137.34	136.11	140.40
PSBRC-82	153.63	148.96	144.02	138.85	157.04	153.21	148.18	142.88	148.97
OM-2718	146.96	139.27	134.33	128.18	149.24	147.37	143.35	138.08	141.85
Mean	146.99	139.56	134.62	129.70	151.25	144.91	140.36	136.35	141.334
SE	1.960	1.838	1.928	1.625	1.276	1.562	1.764	1.154	
Lsd	5.35								
CV %	4.071								

SE = Standard Error
Lsd = Least significant difference (Genotype × Methods of Sowing × Date of sowing)
CV % = Co-efficient of variation

Table 5: Days to maturity of four different genotypes at different sowing date's interaction with both direct seeding & transplanting method.

Genotypes	Spikelet's sterility percent								Mean
	Direct seeding				Transplanting				
	8th Dec/09	18th Dec/09	28th Dec/09	7th Jan/10	8th Dec/09	18th Dec/09	28th Dec/09	7th Jan/10	
BRRI Dhan 28	22.184	20.942	27.218	27.403	23.737	22.408	29.123	29.321	25.292
BRRI Dhan 36	17.926	11.921	21.767	21.301	19.180	12.756	23.290	22.792	18.867
PSBRC-82	27.645	20.079	30.962	29.682	29.580	21.485	33.130	31.760	28.040
OM-2718	27.015	25.026	36.394	40.701	28.906	26.778	38.942	43.550	33.414
Mean	23.692	19.492	29.085	29.772	25.351	20.857	31.121	31.856	26.403
SE	1.960	1.838	1.928	1.625	1.276	1.562	1.764	1.154	
Lsd	5.35								
CV %	4.071								

SE = Standard Error
Lsd = Least significant difference (Genotype × Methods of Sowing × Date of sowing)
CV % = Co-efficient of variation

Table 6: Spikelet's sterility percent of four different genotypes at different sowing date's interaction with both direct seeding & transplanting method.

methods. Number of filled grain was increased up to a level(state the level) (December 18) and after attaining this critical level it decreased up to last sowing dates (January 07). The highest filled grains might be due to higher number of effective tillers per panicle, lower spikelet sterility (higher spikelet fertility), low seedling mortality and favorable atmospheric and adaphic temperature during that particular time. Although a moderate number of filled grains per plant (146.793) was found in BRRI dhan28 but very minimum spikelet sterility percent indicated that most grains were converted to filled grains. Therefore the increase of filled grains reflects clearly that the number of filled grains exhibited is higher in direct seeded methods than transplanting method in all genotypes interaction with sowing date of December 18. Such difference in number of filled grains might be due to undisturbed (smooth) root establishment in direct seeding method than the transplanted. Among the genotypes PSBRC 82 was found to initiate maximum number of filled grains with least difference of the process between direct seeding and transplanting method. The results thus indicate that genotypes PSBRC 82 in direct seeded condition expressed its best adaptation in respect of sowing on December 18 by minimal differences concerning seedling mortality, spikelet sterility percent, effective tillers per panicle and days to panicle initiation (Table 7).

1000 grain weight (g)

The higher genotypic mean in BRRI dhan36 reveals that among four different genotypes BRRI dhan36 is coarser to others. But from the sowing date mean it was observed that 7th. January sowing exhibits the higher 1000 grain weight in direct seeding method. In the combined interaction BRRI dhan36 that was sowed on 28th. December represented the highest 1000 grain weight in transplanting methods. Statistical very similar result (26.627 g) was found in BRRI dhan36 that was sowed on 18th December. This result exhibiting the superiority of direct seeding at early sowing than transplanting for obtaining higher yield it is better to sow the seeds on December 18 in direct seeding but December 28th. In transplanting the least significant values (2.74) indicating very strong statistical significant advantages (18 December than 28 December) of direct seeding method than transplanting. This result indicates that the BRRI dhan36 is coarser to that of other three genotypes. The lowest 1000 grain weight (18.150 g) was found in BRRI dhan28 on sowing 18th. December in transplanting condition and (18.816 g) in direct seeding condition, which signify that BRRI dhan28 was finer than that of other genotypes (Table 8).

Grain yield (ton/ha)

Among the varieties studies, the highest yield was found in PSBRC 82 followed by BRRI dhan28where as the lowest grain yield was found in BRRI dhan36. It is to be noted that among four different sowing dates, sowing on 18th. December performed best in respect of yield, and the lowest grain yield was found in sowing on 7th.

January 2010.In direct seeded condition the highest grain yield was found in PSBRC 82 which was sown at 18th. December and significant difference was found in BRRI dhan28. The lowest grain yield was found in BRRI dhan36 sown on 7thJanuaryIt implies that the variety PSBR 82 which was sowed on the 18th December gave the highest grain yield (ton/ha). And the lowest grain yield (1.69 ton/ha) was found in BRRI

dhan36sown on 7th January. But except OM 2718, the three other genotypes did not show much more variation between direct seeding and transplanting method. So if OM 2718 was selected for cultivation, than direct seeding method might be used for this genotype. In the combined interaction, the highest genotypic mean (4.548 ton/ha) in PSBRC 82 among the four varieties indicates that PSBRC 82 is more superior to that of other varieties which is statistically similar to BRRI dhan28 (4.462 ton/ha).PSBRC 82 sown on 18th. December produced the highest grain yield (5.900 ton/ha) in direct seeding methods followed by (5.841 ton/ha) in BRRI dhan28 and 5.841 ton/ha in transplanting. The highest yield might be due to low mortality percent, higher number of effective tillers per plant, low spikelet's sterility percent, higher filled grains per panicle and favorable temperature (16-18 °C) in the following interaction. The highest yield is not only due to the genotypes but also the combined action of environments and cumulative effects of all the yield and yield contributing characters. From this table it might be concluded that the interaction between genotype PSBRC 82 and BRRI dhan28 sowing at 18 December for both direct seeding & transplanting methods might be selected for higher yield potentials (Table 9).

Harvest Index (HI)

The highest harvest index was found in PSBRC 82 and the lowest harvest index was found in OM 2718 followed by in BRRI dhan36. Harvest index of boro rice was found to increase in direct seeding than transplanting method due to delay in sowing dates from 8thDecember to 7thJanuary. In direct seeding method such increase in harvest index was recorded to range from 0.392 to 0.412 in BRRI dhan28 up to December 18th and decreased up to January 07th.The higher harvest index in the genotypes PSBRC 82 and BRRI dhan28 in 18th. December

Genotypes	Filled grains (number) per panicle								Mean
	Direct seeding				Transplanting				
	8th Dec/09	18th Dec/09	28th Dec/09	7th Jan/10	8th Dec/09	18th Dec/09	28th Dec/09	7th Jan/10	
BRRI Dhan 28	131.526	146.739	105.213	101.303	127.625	138.829	116.934	96.403	118.571
BRRI Dhan 36	106.310	129.752	86.113	74.039	112.200	125.842	89.729	62.612	98.325
PSBRC-82	124.333	196.323	128.596	99.372	138.062	192.423	124.685	101.597	138.174
OM-2718	112.700	132.888	91.473	71.021	103.890	119.178	97.363	80.840	101.169
Mean	118.717	148.926	102.849	86.434	120.444	142.568	107.178	85.363	114.068
SE	2.670	5.864	3.478	3.061	2.659	6.850	3.178	3.772	
Lsd	7.35								
CV %	19.171								

SE = Standard Error
Lsd = Least significant difference (Genotype × Methods of Sowing × Date of sowing)
CV % = Co-efficient of variation

Table 7: Filled grains (number) per panicle of four different genotypes at different sowing date's interaction with both direct seeding & transplanting method.

Genotypes	1000-grain weight (g)								Mean
	Direct seeding				Transplanting				
	8th Dec/09	18th Dec/09	28th Dec/09	7th Jan/10	8th Dec/09	18th Dec/09	28th Dec/09	7th Jan/10	
BRRI Dhan 28	18.881	18.816	20.217	20.756	18.914	18.150	20.531	20.090	19.632
BRRI Dhan 36	25.117	23.804	26.627	25.000	24.451	23.138	26.293	25.960	25.049
PSBRC-82	22.413	21.746	23.059	24.686	21.746	21.080	24.353	24.020	22.888
OM-2718	23.187	21.923	22.863	24.539	22.520	21.256	24.157	23.873	23.040
Mean	22.324	21.322	22.785	24.152	21.658	20.656	23.834	23.486	114.068
SE	1.371	1.258	0.982	1.228	1.371	1.258	1.202	1.228	
Lsd	2.745								
CV %	6.982								

SE = Standard Error
Lsd = Least significant difference (Genotype × Methods of Sowing × Date of sowing)
CV % = Co-efficient of variation

Table 8: 1000-grain weight (g) of four different genotypes in four different sowing date's interaction with both direct seeding and transplanting method.

Genotypes	Grain yield								Mean
	Direct seeding				Transplanting				
	8th Dec/09	18th Dec/09	28th Dec/09	7th Jan/10	8th Dec/09	18th Dec/09	28th Dec/09	7th Jan/10	
BRRI Dhan 28	4.625	5.841	4.469	2.881	5.341	5.557	4.185	2.597	4.462
BRRI Dhan 36	4.684	4.978	3.685	1.940	4.400	4.204	4.185	1.656	3.717
PSBRC-82	5.264	5.900	4.449	2.920	4.400	5.655	4.361	2.636	4.548
OM-2718	4.782	5.233	3.587	2.038	2.538	2.754	2.323	1.754	3.126
Mean	4.589	5.488	4.047	2.445	4.170	4.543	3.763	2.161	3.913
SE	0.264	0.227	0.239	0.264	0.587	0.587	0.482	0.264	
Lsd	2.74								
CV %	6.982								

Table 9: Grain yield of four different genotypes at different sowing dates interaction with both direct seeding & transplanting method.

Genotypes	Harvest index								Mean
	Direct seeding				Transplanting				
	8th Dec/09	18th Dec/09	28th Dec/09	7th Jan/10	8th Dec/09	18th Dec/09	28th Dec/09	7th Jan/10	
BRRI Dhan 28	0.392	0.412	0.333	0.292	0.333	0.392	0.323	0.263	0.368
BRRI Dhan 36	0.333	0.363	0.323	0.363	0.343	0.323	0.353	0.314	0.339
PSBRC-82	0.333	0.431	0.314	0.221	0.294	0.294	0.314	0.219	0.337
OM-2718	0.323	0.314	0.314	0.214	0.316	0.225	0.225	0.235	0.270
Mean	0.345	0.380	0.321	0.352	0.344	0.309	0.304	0.301	0.372
SE	0.016	0.026	0.005	0.023	0.031	0.035	0.027	0.026	
Lsd	0.03								
CV %	2.331								

SE = Standard Error
Lsd = Least significant difference (Genotype × Methods of Sowing × Date of sowing)
CV % = Co-efficient of variation

Table 10: Harvest index of four different genotypes at different sowing date's interaction with both direct seeding & transplanting method.

might be due to higher accumulation of photosynthates for the favorable temperature and very minimum rainfall and higher sunshine during the active vegetative and panicle initiation stage. The higher panicle initiation was also found in this interaction followed by higher effective tillers, filled grains and biological yield. The increase of harvest index was significant between 8 and 18 and 28 Decemberbut non-significant between 28th December and 7th January in most genotypes. Therefore the increase of harvest index reflects it clearly that harvest index occurs higher in direct seeded methods than transplanting method in all genotypes of boro rice. Such difference in increase of harvest index might be due to the time taken to root establishment of newly transplanted seedlings which was absent in direct seeding method. The lowest harvest index (0.214) was found in OM 2718 at sown on 7th January in direct seeding method. From this table it might be concluded that, PSBRC 82 and BRRI dhan28 in 18 December might be selected for a better contribution in yield (Table 10).

Conclusion

The dwarf plant in delayed seeding indicate the sensitivity of late seeding low temperature on the dry matter production of the genotype unlike other genotypes which expressed no significant reduction in plant height due to late seeding. Significant responses of environments were found either in between the varieties but in between the sowing dates. Maximum number of effective tillers was observed in both direct seeded and transplanted condition in December 18. The tiller number increased with delayed seeding until 18 December but tillers numbers were found to decline afterwards interaction PSBRC 82 took maximum days to flowering sowing at December 08 in transplanting condition but first flowering in direct seeding condition which is statistically different. And the earliest flowering (90.39 days) was observed in BRRI dhan28 sowing at 7 January in direct seeding condition than (93.335 days) in

transplanting. PSBRC 82 took the highest days for maturation followed by for OM 2718 whereas the lowest day for maturation was taken by BRRI dhan28. In respect of sowing dates, sowing at 8 December took the highest days for maturation whereas the lowest number taken for maturation was in sowing at January 7. In direct seeded boro rice the highest days to maturity was taken in PSBRC 82 sowing at 8 December while the lowest days to maturity was observed in BRRI dhan28 sowing at 7 January. OM 2718 gave the maximum sterility (43.550) when it was practiced in transplanting method than the direct seeded method (40.701) at January 07. During the germination to seedling stage the severe cold injury affect spikelet's fertility. The increase of filled grains reflects it clearly that the number of filled grains exhibited higher in direct seeded methods than transplanting method in all genotypes interaction with sowing date of December 18. Such difference in number of filled grains might be due to undisturbed (smooth) root establishment in direct seeding method than transplanted. Among the genotypes PSBRC 82 was found to initiate maximum number of filled grains with least difference of the process between direct seeding and transplanting method. The lowest 1000 grain weight (18.150 g) was found in BRRI dhan28 at sowing 18 December in transplanting condition and (18.816 g) in direct seeding condition, which signify that BRRI dhan28 was finer than that of other genotypes. PSBRC 82 sowing at 18 December resulted the highest grain yield (5.900 ton/ha) in direct seeding methods followed by (5.841 ton/ha) in BRRI dhan28 and 5.841 ton/ha in transplanting.

References

1. Wang OP, Srivastava HK (2002) Studies on the nature of relationship between grain size, spikelet number, grain yield and spikelet filling in late duration varieties of rice.(Oryza Sativa L) Plant and Soil .60: I23-130.

2. Hill AN, Liang MZ, Xiao HH, Chen LB (2002) Correlation analysis between

pollen fertility and panicle exsertion in P(T) GMS rice with eui gene. *Acta-Agronomica-Sinica* 32: 1311-1315.

3. Islam MA, Johora FT, Sarker UK, Mian MAK (2012) Adaptation of Chinese CMS lines interaction with seedling age and row ratio on hybrid seed production of Rice (*Oryza sativa L*) Bangladesh Journal of Agronomy. 25:178-183.

4. Ahmed GJU et al. (2006) Effect of different weed management systems on weed control, plant growth and grain yield of lowland rice 84–93. In: *Weed Science, Agricultural Sustainability and GMOs*, Vol.1, 19th Asian–Pacific Weed Science Society Conference, 17–21 March 2003, Manila, the Philippines.

5. Islam MA, Mian MAK, Rasul G, Khaliq QA, Bashar MK (2013) Estimation of Specific Combining Ability (GCA) Effects and *Per-se* performances in Some Yield Related Traits of Rice (*Oryza sativa* L) Echo-Friendly Agriculture Journal. 7:143-145.

6. Hossain M, Deb UK (2003) Liberalization of Rice Sector: Can Bangladesh withstands Regional Competition? Poster paper presented at PETRRA Communication Fair 2003 held at Hotel Sheraton, Dhaka on Aug. 10-11.

7. Islam MA, Sarker U, Mian MAK (2009) Synchronization and Stability Analysis of Hybrid Seed Production of Rice (*Oryza sativa L*) M. S.Thesis. Department of Genetics and Plant Breeding. Bangabandhu Sheikh Muzibur Agricultural University, Gazipur 1706, Bangladesh.

9

Root QTL Pyramiding through Marker Assisted Selection for Enhanced Grain Yield Under Low Moisture Stress in Rice (*Oryza sativa L*)

Grace Sharon Arul Selvi, Shailaja Hittalmani* and Uday G

Department of Genetics and Plant Breeding, University of Agricultural Sciences, Bangalore-65, India

Abstract

Drought is the foremost limiting factor affecting rice production especially in the critical reproductive growth phase. Several mechanisms that resist/ tolerate or avoid drought stress have been elucidated and documented in many crops. In this study, Quantitative trait loci pyramiding (QTL) pyramiding for root morphological traits and an evaluation of the effects of QTLs in combination. Four QTLs present on chromosomes 1, 2, 7 and 9 in a near-isogenic line (NIL) population derived by backcrossing 4 doubled haploid lines of IR64/Azucena cross into IR64 were used to develop pyramids. These NILs were crossed pair-wise to arrive at combinations of two-QTL and three-QTL pyramids. Among the generated QTL pyramids, qRT6-2 x qRT11-7 and qRT6-2 x qRT19-1+7 were found to have higher yields than parent IR64 under moisture stress situation. QTLs present on different chromosomes showed maximum antagonistic effects in comparison with the different QTLs present on the same chromosome. The pyramids also show a relative increase of the NILs in terms of height, number of tillers and number of panicles, however, pyramids not show an increase in the dry matter accumulation which implies that the partitioning of the biomass is directed towards the below ground parts, leading to less sturdy plants that perform averagely under moisture stress situations.

Keywords: Rice; Pyramids; Root morphology; Drought; Stress situation

Introduction

Rice (*Oryza sativa L*), the staple cereal of the world's population is heavily dependent on availability of copious amounts of water for optimum productivity. With the diminishing water supplies in the rain fed regions that occupy predominant rice growing belts, frequent and intermittent drought that coincide with the critical reproductive growth periods often affect rice production. Severity of drought stress has led to a high yield decline in rice in recent years in drought-affected areas. This therefore necessitates the development of genotypes that survive on reduced water supply and which can adapt and cope with the climate changes [1-6].

Several physio-morphological characters contribute towards drought resistance or tolerance in plants, among which, root traits are known to impart resistance under conditions of increasing water deficits with substantial yields [5] several quantitative trait loci that affect root morphology have been identified in rice across mapping populations [7].

Pyramiding of genes controlling a trait in order to enhance the trait expression and also impart stability, is a widely followed approach in disease resistance breeding, where a crop is plagued by a spectrum of races/pathovars of a pathogen. The individual genes pyramided offer specific resistance against a race/pathovars. A combination thereof will therefore, ensure a wider disease resistance spectrum. Pyramiding is made more efficient through the use of molecular markers that are tightly linked to the gene of interest as is evident in the studies of [8] for blast resistance, for bacterial blight, for basmati quality traits, for fungal pathogen resistance and for heading date in rice [9-12].

Drought being a major abiotic factor affecting rice production especially in the critical reproductive growth phase. Novel approach, such as QTL pyramiding through marker assisted selection for root morphological traits and evaluation of the effects of root QTLs in combination is attempted to develop new drought tolerant rice genotypes.

Material and Methods

Plant material

One hundred and thirty five doubled haploid lines (DHLs) were derived from an IR64/ Azucena cross by Guiderdoni et al. in 1992 at IRRI. [13] identified the four target Quantitative trait loci pyramided (QTL) regions between the markers RZ19, RG690, RZ730 and RZ801 on chromosome # 1, RM29, RG171, RG157 and RZ318 on chromosome # 2, RM234, CDO418, RZ978, CDO38, and RM248 on chromosome # 7 and RZ228 and RZ12 on chromosome # 9. The markers were used by [1] to develop Near-Isogenic Lines (NILs) of IR64 with Azucena alleles at the target loci through backcross breeding. The selections were made strictly based on marker genotypes with the exception of RZ228 and RZ12 on chromosome # 9, which were replaced by RM201 and RM242 PCR-SSR markers after the selection in BC_2 generation [7] identified 25 new rice microsatellite markers to the QTL regions spanning 51.8 cm on chromosome 1, 88.7 cm on chromosome 2, 19.7cm on chromosome 7 and 15.1 cm on chromosome 9. These NILs were crossed pair-wise to develop two and three QTL combination pyramids and were evaluated in well-watered and low moisture stress used in this study (Table 1 and Figure 1).

Validation of pyramids

The generated pyramids were validated for the presence of the

---

*Corresponding author: Shailaja Hittalmani, Professor and University Head, Genetics and Plant Breeding, University of Agricultural Sciences, GKVK, Bangalore-560065, India, E-mail: shailajah_maslab@rediffmail.com

Sl. No.	Pyramids	Chro. Introgression on	QTL No	No. Plants
1	qRT11-7 X qRT18-1+7	7+ (1+7)	3	35
2	qRT24-9 X qRT11-7	9 + 7	2	38
3	qRT6-2 X qRT11-7	2+ 7	2	32
4	qRT11-7 X qRT19-1+7	7 + (1+7)	3	36
5	qRT20-1+7 X qRT18-1+7	(1+7) + (1+7)	4	37
6	qRT11-7 X qRT6-2	7 + 2	2	38
7	qRT6-2 X qRT19-1+7	2+ (1+7)	3	32
	Total No. of plants			248

Table 1: List of two and three QTL pyramids generated for the study.

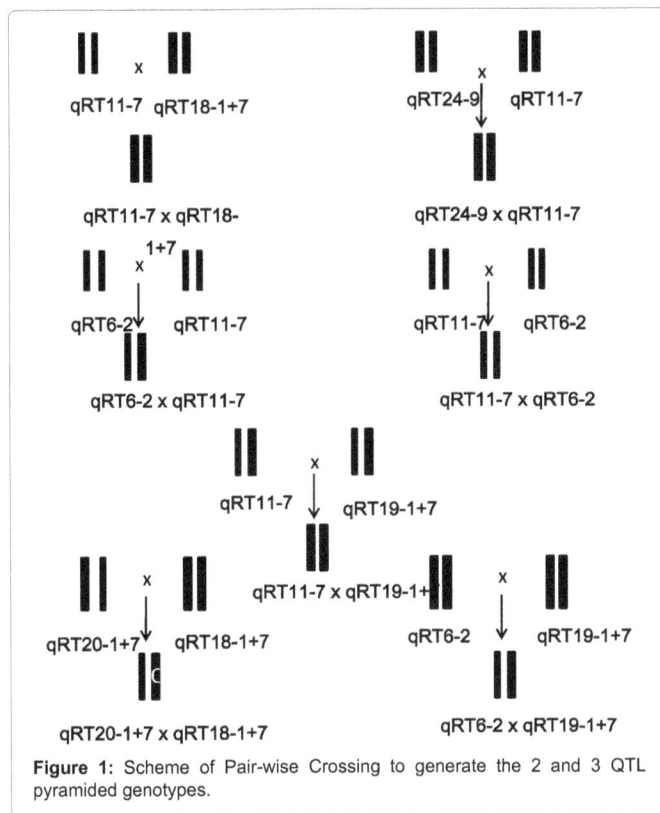

Figure 1: Scheme of Pair-wise Crossing to generate the 2 and 3 QTL pyramided genotypes.

QTLs through tightly linked and flanking SSR markers. Total DNA was extracted as per the procedure outlined by [14]. The PCR reaction mixture contained 50ng template DNA, 50 ng of each primer, 0.05 mM dNTPs, 1X PCR buffer (10mM TRIS, pH 8.4, 50 mM KCl, 1.8 mM MgCl$_2$ and 0.01 mg/ml gelatin) and 1U *Taq* DNA polymerase in a volume of 20 µl. Template DNA was initially denatured at 94°C for 5 min followed by 35 cycles of PCR amplification of 94°C denaturation for 30 seconds, primer annealing at 56°C for 1 minute, primer extension at 72°C for 2 minutes. The amplified product was electrophoretic ally resolved on a 3.5% agarose gel in 1X TBE buffer and visualized under UV light after staining with ethidium bromide. Pyramids selected strictly on the basis of presence of marker alleles were evaluated in the field (Figure 2).

Phenotypic evaluation of pyramid combination plants

The experiment was conducted during the dry season at Main Research Station, Hebbal, Bangalore, and during the wet season at Zonal Agricultural Research Station, GKVK, in a replicated randomized complete block design. The parent genotypes, the recipient IR64, donor Azucena were evaluated along with the pyramids and checks: Budda and Moroberekan under reproductive stage low moisture stress (50% field capacity between 65 to 80 DAS) and well-watered condition (100% field capacity) in 1m long, 8 inch diameter PVC pipes. The pipes were filled with sandy loam soils and the standard package of fertilizer dosage was followed. The entries were sown and upon germination, a single plant was maintained in each pipe. Regular irrigation schedule was followed until 65 days after sowing. Low moisture stress treatment was included by maintaining the pipes at 50% field capacity using controlled irrigation and a rain-out shelter was used to keep out rainwater. Well-watered treatment received the stipulated amount if irrigation so as to maintain 100% field capacity. The observations were recorded at harvest, plant height was measured from the base of the crown to the tip of the highest leaf, number of tillers, number of panicles were counted at harvest, panicle weight was recorded as an average of 5 panicles, panicle length was from an average of 5, plants measured from the collar to the tip, test weight was recorded on 100 randomly selected seeds, seed yield was recorded as the total yield produced by the plant, total biomass was the sum of both the shoot and root mass, maximum root length was recorded from the crown to the tip of the longest root, root number was recorded at 15 cm depth and root thickness was on an average of 5 roots at 15 cm depth measured with slide calipers.

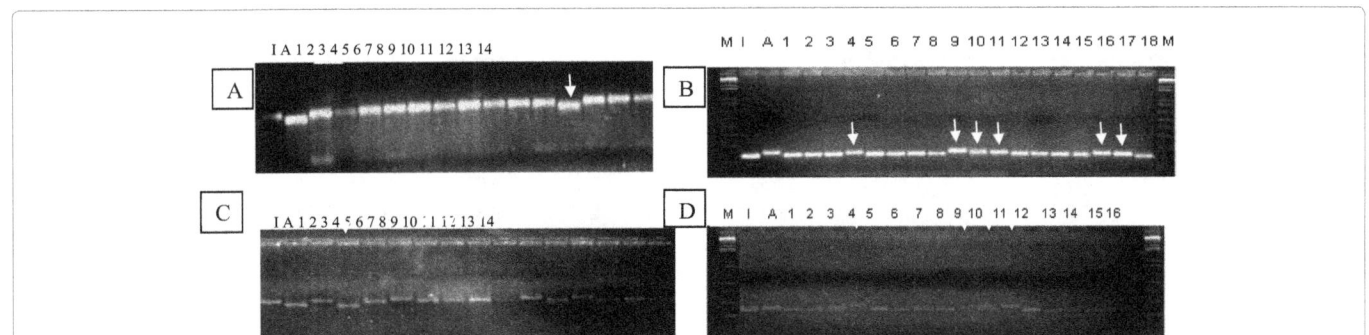

Figure 2: Gel pictures of the pyramids qRT 24-9X qRT11-7 with chromosome specific markers in the introgression regions.
A. Shows the DNA marker banding pattern of the pyramids with RM201 (158bp), a chromosome 9 specific marker.
B. Shows the marker banding pattern of the pyramids to RM 3164 (166bp), a chromosome 9 specific marker,
C. Shows the DNA marker banding pattern of the pyramids to RM234 (156bp), a chromosome 7 specific marker,
D. Shows the marker banding pattern of the pyramids to RM5380 (89bp), a chromosome 7 specific marker.
In the figures, I=IR64, the genetic background of the Near Isogenic Lines used for development of pyramids, while A=Azucena, the donor for the introgression fragments conferring root morphological trait and M=standard DNA Marker.

Statistical Analysis

The data that was generated was subject to a series of statistical analyses to elucidate the relative effects of the presence of various QTLs in the rice genotypes. The data was analyzed to individual and combined ANOVA over two moisture regimes based on RCBD for different experiments to assess the variability that is present in the various rice NILs and in the generated pyramids. The data was pooled upon testing for homogeneity of variances. The mean difference in the performance of the entries under low moisture stress in comparison with the well-watered conditions was arrived at using the formula

$$\text{Mean \% performance} = \left[(LMS-WW)/LMS \right] *100$$

The significance of the difference was tested using the t-test. All statistical analyses utilized the *SPAR 2* software analysis package.

Results and Discussion

The mean performance of the QTL-NILs, the generated pyramids and the checks are presented in Table 2. Most genotypes showed significant increases plant height over IR64 but not over Azucena and the checks (Moroberekan and Budda) under low moisture stress. Under well-watered conditions, the QTL-NILs and the pyramids are significantly shorter than the lesser parent IR64. The number of tillers and panicles are comparable with both Azucena and Budda but are significantly lower than both IR64 and Moroberekan under two

Entries		PHT	NOT	NOP	PWT	PL	TWT	SWT/PLT	TBM	MRL	RTN	RTT
QTL-NILs	LMS	46.76	7.34	4.26	5.29	12.81	1.54	4.52	25.48	16.87	49.57	0.75
	WW	61.02	9.25	7.23	8.24	14.24	2.13	7.87	47.94	14.72	59.78	0.63
	Performance	-30.51**	-26.01*	-69.54**	-55.64**	-11.2	-38.56**	-74.14**	-88.16**	12.74*	-20.60*	16.79
qRT11-7 x qRT18-1+7	LMS	50.91	7.07	4.97	5.26	13.49	1.28	4.14	34.35	15.89	37.24	0.75
	WW	55.24	8.59	7.1	7.45	14.83	2.06	6.97	62.11	12.04	59.62	0.62
	Performance	-8.5	-21.6	-43.05**	-41.71**	-9.98	-60.84**	-68.33*	-80.84**	24.19**	-60.09*	16.93
qRT24-9 x qRT11-7	LMS	45.89	7.46	4.15	6.84	13.75	1.3	4.65	29.6	17.34	36.62	0.75
	WW	55.49	9.2	6.92	8.94	15.11	2.07	7.49	53.47	14.19	66.82	0.63
	Performance	-20.92*	-23.42*	-66.53**	-30.83*	-9.83	-59.55**	-61.20**	-80.63**	18.17**	-82.46*	16.98
qRT6-2 x qRT11-7	LMS	54.95	7.66	5.3	6.3	12.95	2.14	4.61	25.95	14.01	44.47	0.79
	WW	57.31	9.13	9.1	8.36	14.19	2.96	7.75	49.72	11.82	69.56	0.62
	Performance	-4.29	-19.18*	-71.78**	-32.70**	-9.57	-38.16	-68.27**	-91.60**	15.61*	-56.42**	22.23
qRT11-7 x qRT19-1+7	LMS	43.77	7	4.35	7.07	12.38	1.38	5.17	26.47	15.9	37.7	0.73
	WW	45.99	9.28	7.91	9.39	13.64	2.09	8.18	52.52	13.24	51.78	0.58
	Performance	-5.06	-32.66*	-81.90**	-32.81*	-10.15	-52	-58.10**	-98.44**	16.71*	-37.34**	20.95
qRT20-1+7 x qRT18-1+7	LMS	45.83	6.58	3.59	6.38	12.2	1.32	5.11	26.43	12.21	34.96	0.73
	WW	47.32	9.1	7.27	9.01	13.92	2.02	7.95	49.97	10.21	53.69	0.59
	Performance	-3.27	-38.35*	-102.51**	-41.13**	-14.08	-53.70**	-55.50**	-89.09**	16.43*	-53.57**	19.32
qRT11-7 x qRT6-2	LMS	37.48	4.64	3.46	5.18	12.19	1.46	5.02	24.07	11.01	44.91	0.76
	WW	43.74	7.3	6.86	7.96	13.48	2.15	8.35	46.91	8.54	56.67	0.64
	Performance	-16.71*	-57.45*	-98.10**	-53.63*	-10.54	-46.86**	-66.31**	-94.90**	22.44*	-26.19**	15.64
qRT6-2 x qRT19-1+7	LMS	36.03	5.36	3.52	5.08	12.19	1.72	5.03	22.83	10.56	39.03	0.79
	WW	40.52	7.23	6.22	7.53	13.49	2.39	8.3	46.53	8.31	60.72	0.62
	Performance	-12.48	-35.08	-77.04**	-48.23*	-10.73	-39.23**	-65.14**	-103.79**	21.34*	-55.60**	20.54
IR64		47.07	12.5	5.33	5.28	12.94	1.38	5.45	28.98	9.81	66	0.79
Azucena	LMS	61.94	4.5	2.67	3.34	12.6	1.23	4.12	22.22	28.78	27.67	0.8
Budda		57.26	4.33	3.17	6.64	15.08	1.71	4	25.59	29.36	38.77	0.85
Moroberekan		68.4	7.17	4.83	11.33	11.97	3.03	8.95	36.24	32.16	30.9	0.76
IR64		60.54	15.67	8	9.38	15.1	2	8.8	49.38	10.13	90.5	0.66
Azucena	WW	71.5	7.33	5.5	5.96	14.29	1.79	7.35	38.7	24.94	54.5	0.68
Budda		84.2	6.5	6.33	9.38	16.16	2.26	7.28	42.86	28.46	62.43	0.67
Moroberekan		86.13	10	9.33	18.87	13.71	3.82	8.36	53.85	28.79	59.9	0.71

*: Significance at 5%. **: Significance at 1%.
LMS: low moisture stress (50% FC); WW: well watered condition (100% FC)
PHT: Plant height (cm); NOT: Number of tillers per plant
NOP: Number of panicles per plant; PWT: Panicle weight (g)
TWT: Test weight (g); SWT/PLT: Seed yield per plant (g)
MRL: Maximum root length at maturity (cm); RTN: Total root number at the crown.
TBM: Biomass accumulated at maturity (g);　RTT: Root thickness (mm).

Table 2: Mean performance of the QTL-NILs and the pyramids under low moisture stress in comparison with well watered conditions.

moisture regimes. While the panicle length and panicle weight of the pyramids is higher than Azucena and is comparable with IR64, the test weight is significantly higher than IR64 in the pyramids qRT6-2 × qRT11-7 and qRT6-2 × qRT19-1+7. The root length under low moisture stress was significantly higher than IR64 but was less than the check varieties, while under well-watered conditions; the root length was lesser than IR64. The root number was significantly lower than IR64 under two moisture regimes, while the roots were comparatively thicker than IR64 under different moisture regimes (Table 2).

The results of analysis of variance for the growth, yield and yield contributing characters and three root morphology characters were studied under well-watered (WW) and low moisture stress (LMS) condition revealed highly significant differences among the QTL-NILs (data not shown). This reveals that large amount of genetic difference is present in these QTL-introgressed lines for all other traits, thus indicating the scope for selection and to forward the best among them for further study. The mean performance of the genotypes differed for all the traits under these contrasting moisture regimes of low moisture stress (LMS) and well-watered (WW) conditions indicating an adaptive response of drought tolerance [15,16]. A significant reduction in plant height, number of tillers and panicles per plant, panicle length, panicle weight, test weight, seed yield per plant and total biomass were observed under low moisture stress condition in different growth environments.

Considering root traits, a significant increase in root length and thickness, while a significant reduction in the number of roots was reported during dry season 2004, at MRS, Hebbal. A non-significant increase in root length and thickness, with a significant increase in root number was reported during wet season 2005 at ZARS, GKVK. This difference in the performance of the genotypes is ascribable to the seasonal difference [7,17]. The pyramid qRT11-7 × qRT18-1+7 recorded significant variance for all traits except for seed yield per plant during dry season at MRS, Hebbal, while during wet season at ZARS, GKVK, significant variance was recorded only for panicle weight, panicle length, seed yield per plant, total biomass, root length and number under both the moisture regimes. Significant variance was recorded for plant height, number of panicles per plant under well-watered conditions, and for test weight and root thickness under low moisture stress conditions. Similar results were also obtained in qRT20-1+7 × qRT18-1+7, while in the pyramid qRT11-7 × qRT19-1+7, contrasting results were obtained for number of chaffy seeds per panicle, total biomass and root length. This indicates that though these crosses involve the same QTL regions being introgressed, there exists differences within the inherited fragments and hence a difference in the phenotypic expression patterns. Between the pyramids qRT6-2 × qRT11-7 and qRT11-7 × qRT6-2, similar results between both of them were obtained in both the growth environments. qRT24-9 × qRT11-7 significant variance for all traits except for panicle length and root thickness.

The mean performance of the genotypes differed for all the traits under these contrasting moisture regimes of low moisture stress (LMS) and well-watered (WW) conditions. A significant reduction in plant height, number of tillers and panicles per plant, panicle length, panicle weight, test weight, seed yield per plant, total biomass and root number, with a significant increase in root length and thickness were observed under low moisture stress condition in all the three growth environments. The initial near isogenic line population that was used for the crossing programme recorded a low number of lines that showed an increase in the root morphology. This was attributed to non-allelic interactions between the donor alleles that

may be disrupted by the recurrent parent alleles [1]. Initial mapping procedures indicated antagonistic effects between the QTLs so large so as to inhibit QTL detection [13,18]. They also stated that the QTLs present on different chromosomes showed maximum antagonistic effects in comparison with the different QTLs present on the same chromosome. The pyramids also show a relative increase of the NILs in terms of height, number of tillers and number of panicles, however, did not show an increase in the dry matter accumulation which implies that the partitioning of the biomass is directed towards the below ground parts, leading to less sturdy plants that perform averagely under stress situations. The increases in the root morphology performances in the present study could be attributed to the disruption of such deleterious interaction effects through recombination.

Conclusion

In the present study, it was demonstrated that it is possible to combine QTLs in a common genetic background using co-dominant markers that reside in the QTL region. The combination of QTLs for root morphology had both additive as well as interactive effect epistatic effect on grain yield. The combinations of more than one QTL present in a single genotype that performed superior in both stress and well-watered condition is being tested on large scale and in natural water deficit situation.

Acknowledgement

The authors gratefully acknowledge the funding from the Rockefeller Foundation, USA, for carrying out this research work to the senior author SH.

References

1. Shen L, Courtois B, Mcnally KL, Robin S, Li Z (2001) Evaluation of near-isogenic lines of rice introgressed with QTLs for root depth through marker-aided selection. Theor Appl Genet 103: 75-83.

2. Widawsky DA, O'Toole JC (1990) Prioritizing the Rice Biotechnology Research Agenda for Eastern India. New York: The Rockefeller Foundation.

3. Lafitte R, Blum A, Atlin G (2003) Using secondary traits to help identify drought-tolerant genotypes. In: Fischer KS, Lafitte R, Fukai S, Atlin G, Hardy B Breeding rice for drought- prone environments 37-48.

4. Naresh Babu N, Hittalmani S, Shivakumar N, Nandini C (2011) Effect of drought on yield potential and drought susceptibility index of promising aerobic rice (Oryza sativa L) genotypes. Electronic J Pl Breed 2: 295-302.

5. Keshava Murthy BC, Arvind Kumar, Shailaja Hittalmani (2011) Response of rice (Oryza sativa L) genotypes under aerobic conditions. Electronic J Pl Breed 2: 194-199.

6. Arvind Kumar, Shalabh Dixit T, Ram RB, Yadaw KK, Mandal NP (2014) Breeding high-yielding drought-tolerant rice: genetic variations and conventional and molecular approaches. J Exp Bot 60: 6265-6278.

7. Vaishali MG, Hanamareddy B, Mane S, Gireesha TM, Hittalmani S (2003) Graphical genotyping using DNA markers andevaluation of QTL-NILs of IR64 for root morphological traits, disease resistance and yield traits in rice. Plant and Animal Genome XI Conference, San Diego, USA, 335.

8. Hittalmani S, Parco A, Mew TV, Zeigler RS, Huang N (2000) Fine mapping and DNA marker- assisted pyramiding of the three major genes for blast resistance in rice. Theor Appl Genet 100: 1121-1128.

9. Singh S, Sidhu JS, Huang N, Vikal Y, Li Z, et al. (2001) Pyramiding three bacterial blight resistance genes (xa5, xa13 and Xa21) using marker assisted selection into rice cultivar PR106. Theor Appl Genet 102: 1011-1015.

10. Joseph M, Gopalakrishnan S, Sharma RK, Singh VP, Singh AK (2004) Combining bacterial blight resistance and Basmati quality characteristics by phenotypic and molecular marker-assisted selection in rice. Mol Breed 100: 1-11.

11. Maruthasalam S, Kalpana K, Kumar KK, Loganathan M, Poovannan K, et al. (2007) Pyramiding transgenic resistance in elite indica rice cultivars against sheath blight and bacterial blight. Pl. Cell. Rep 26: 791-804.

12. Wang P, Xing Y, Li Z, Yu S (2012) Improving rice yield and quality by QTL pyramiding. Mol Breed 29: 903-913.

13. Yadav R, Courtois B, Huang N, McLaren G (1997) Mapping genes controlling root morphology and root distribution in a doubled haploid population of rice. Theor Appl Genet 94: 619-632.

14. Cao D, Oard JH (1997) Pedigree and RAPD-based DNA analysis of commercial U.S. Rice cultivars. Crop Sci 37: 1630-1635.

15. O' Toole JC (1982) Adaptation of rice to drought prone environments In: Drought resistance in crops with emphasis on rice. IRRI, Philippines 195-213.

16. Shivapriya M (2000) Phenotypic evaluation for agronomically useful traits, blast desease, DNA fingerprinting and diversity studies in local rices (Oryza sativa L) of Karnataka. M. Sc (Agri) thesis, Univ. Agril Sci, Bangalore, 139.

17. Mahesh HB (2007) DNA marker assisted validation and evaluation of root QTL pyramids under drought for root morphology and yield parameters in rice (Oryza sativa L) M. Sc Thesis, Univ Agric Sci, Bangalore, 120.

18. Champoux MC, Wang G, Sarkarung S, Mackill DJ, O'Toole JC (1995) Locating genes associated with root morphology and drought avoidance in rice via linkage to molecular markers. Theor Appl Genet 90: 969-981.

10

Development of Component Lines (CMS, Maintainer and Restorer lines) and their Maintenance Using Diversed Cytosources of Rice

Ariful Islam[1]*, Mian MAK[2], Rasul G[2], Bashar K[3] and Fatema-Tuj-Johora[4]

[1]*Department of GPB, EXIM Bank Agricultural University, Bangladesh*
[2]*Department of Genetics and Plant Breeding, BSMRAU, Gazipur1706, Bangladesh*
[3]*International Potato Research Centre (CIP), Bangladesh*
[4]*Department of Crop Botany, EXIM Bank Agricultural University, Bangladesh*

Abstract

The practice of hybridization has greatly contributed to the increase in crop productivity. A major component that exploits heterosis in crops is the cytoplasmic male sterility (CMS)/nucleus-controlled fertility restoration (*Rf*) system. The development and use of hybrid rice varieties on commercial scale utilizing male sterility and fertility restoration system has proved to be one of the mile stones in the history of rice improvement. Pollen sterility status of 148 exotic rice germplasm was assessed at flowering stage. Sixteen genotypes showed 100% pollen sterility status which was considered as completely male sterile lines (A-line). Sixteen genotypes were also identified as completely fertile due to 80% and above pollen and spikelet fertility. For identification of proper maintainer lines, the identified 16 CMS lines chance crossed with established known maintainer lines *viz.* IR 58025B, IR 62829B, GAN46B, IR 68888B and BRRI1B. Based on pollen male sterility status of the F_1s lines it was indicated that 10 out of 16 were maintained by IR 58025B line, 8 CMS lines were maintained by IR 62829B, three CMS were maintained by IR 68888B and one CMS line was maintained by GAN46B and BRRI1B. Restoration potentiality of identified 16 suspected restorer genotypes were assessed throgh judgement of pollen and spikelet fertility of F_1s developed through crossing with five standard CMS lines. Based on 80% and above pollen and spikelet fertility of F_1s, seven suspected restorers were identified as restoerer against IR 58025A, two against GAN 46A, five against IR 62829A and two against IR 68888A.

Keywords: Pollen; Spikelet; Fertility; Sterility; Fertility restoration; CMS line; Maintainer line; Synthesized

Introduction

The development and use of hybrid rice varieties on commercial scale utilizing male sterility and fertility restoration system has proved to be one of the mile stones in the history of rice improvement. The hybrid rice technology now in operation, aims at yield increment through higher exploitable heterosis levels [1]. In hybrid rice technology most usually two sterility systems *viz.*, CMS and EGMS are used for commercial seed production. In three line system of hybrid rice variety development system, three lines, A, B and R are required. A line is the cytoplasm-genetic male sterile line where the male sterility is jointly controlled by recessive nuclear gene and sterile cytoplasm. B-line is isogenic line of A-line, only difference in male sterility and fertility. R-line possesses fertility restoration gene [2]. A commercial A-line is characterized by the absence of pollen grains or rudimentary pollens, argonomically superiority, stable sterility, wide regeneration spectrum, abortive anther and highly synchronized [3]. B line is the maintainer line characterized by normal anthers, functional pollens and seed setting on selfing. While normal anthers, functional pollens, abundant pollen producing capacity, strong restoring ability, good combining ability, high out crossing rate, and genetically diverse from CMS line [4,5] are the main characteristics of R-line. It is 30 years since the first commercial release of hybrid rice. Plant cytoplasmic male sterility (CMS), a maternally inherited trait that prevents plants from producing functional pollen, has been identified in many higher plants, including rice, cotton, maize, and sorghum. CMS restorer systems have been widely exploited to produce hybrids that outperform their inbred parents in yield, biomass, or other traits. CMS is usually attributed to an unusual chimeric gene in the mitochondrial genome. In many cases, a nuclear-encoded fertility restorer gene (*Rf*) can restore fertility of the cytoplasmic male-sterile plants. Therefore, the CMS/*Rf* system is an ideal model for dissecting the interaction between mitochondrial and

nuclear genomes. A variety of mechanisms of fertility restoration by the *Rf* genes have been reported for different CMS systems. T-urf13, a mitochondrial gene encoding a 13 kDa protein, has been detected only in maize carrying T male-sterile cytoplasm. The first restorer allele cloned, the maize *Rf2* gene, does not affect the expression of urf13 and encodes aldehyde dehydrogenase (ALDH), which is located in the mitochondrial matrix in a homotetrameric. Hybrid rice has spread such that now it commands about 50% of the total rice area in China but only 7% in Bangladesh [4]. New male sterile cytoplasm sources and inter-subspecies crosses have contributed to the development of super rice breeding. However, sustainable improvements of hybrid rice yield potential, grain quality, and tolerance to biotic and abiotic stresses continue to be a great challenge. Exploitation of new germplasm has always played a critical role in rice breeding, and this will continue [6]. Component lines development is considered as the backbone of any sustainable hybrid rice program. Successful and long lasting hybrid rice program depends on development of diversified component lines utilizing CMS source, local and exotic genetic resources. In this country rice hybrid programs mostly depend on the component lines developed by IRRI and China [7]. There is a shortage of research works in component line development of hybrid rice activities performing by different organizations of Bangladesh [8].

Corresponding author: Dr. Ariful Islam M, Department of GPB, EXIM Bank Agricultural University, Bangladesh, E-mail: i.aarif@yahoo.com

As a part of that program the present investigation was initiated with the following objectives:

Objectives:

1. Identification of male sterile (CMS) and fertile (B) lines from the exotic rice germplasm.

2. Identification of Restorer (R-lines) based on pollen and spike let fertility test.

Materials Used

A series of experiments were conducted from Winter (Boro) 2011 to Summer (Aman) 2013 in the experiment field of Department of Genetics and Plant Breeding, Bangabandhu Sheikh Mujubur Rahman Agricultural University, Gazipur Bangladesh using the following materials (germplasm) (Tables 1 and 2).

Methods of pollen fertility test

Pollen fertility test was done by Potassium Iodide solution (KI). At flowering stage young spikelets were collected early in the morning (7.30-8.50 AM) from the field and kept in the jar for opening the spikelets after about 2 hours. One or two anthers were kept on a glass slide and smashed with KI solution and covered with a cover slip and then observed under a compound Microscope. Records on fertile and sterile pollens from three microscopic focuses were noted. Round well developed stained pollens were considered as viable and non-stained irregular shaped pollens were counted as sterile pollens and then

Genotypes	Genotypes	Genotypes	Genotypes	Genotypes	Genotypes
RG-BU-08-001	RG-BU-08-026	RG-BU-08-051	RG-BU-08-076	RG-BU-08-101	RG-BU-08-126
RG-BU-08-002	RG-BU-08-027	RG-BU-08-052	RG-BU-08-077	RG-BU-08-102	RG-BU-08-127
RG-BU-08-003	RG-BU-08-028	RG-BU-08-053	RG-BU-08-078	RG-BU-08-103	RG-BU-08-128
RG-BU-08-004	RG-BU-08-029	RG-BU-08-054	RG-BU-08-079	RG-BU-08-104	RG-BU-08-129
RG-BU-08-005	RG-BU-08-030	RG-BU-08-055	RG-BU-08-080	RG-BU-08-105	RG-BU-08-130
RG-BU-08-006	RG-BU-08-031	RG-BU-08-056	RG-BU-08-081	RG-BU-08-106	RG-BU-08-131
RG-BU-08-007	RG-BU-08-032	RG-BU-08-057	RG-BU-08-082	RG-BU-08-107	RG-BU-08-132
RG-BU-08-008	RG-BU-08-033	RG-BU-08-058	RG-BU-08-083	RG-BU-08-108	RG-BU-08-133
RG-BU-08-009	RG-BU-08-034	RG-BU-08-059	RG-BU-08-084	RG-BU-08-109	RG-BU-08-134
RG-BU-08-010	RG-BU-08-035	RG-BU-08-060	RG-BU-08-085	RG-BU-08-110	RG-BU-08-135
RG-BU-08-011	RG-BU-08-036	RG-BU-08-061	RG-BU-08-086	RG-BU-08-111	RG-BU-08-136
RG-BU-08-012	RG-BU-08-037	RG-BU-08-062	RG-BU-08-087	RG-BU-08-112	RG-BU-08-137
RG-BU-08-013	RG-BU-08-038	RG-BU-08-063	RG-BU-08-088	RG-BU-08-113	RG-BU-08-138
RG-BU-08-014	RG-BU-08-039	RG-BU-08-064	RG-BU-08-089	RG-BU-08-114	RG-BU-08-139
RG-BU-08-015	RG-BU-08-040	RG-BU-08-065	RG-BU-08-090	RG-BU-08-115	RG-BU-08-140
RG-BU-08-016	RG-BU-08-041	RG-BU-08-066	RG-BU-08-091	RG-BU-08-116	RG-BU-08-141
RG-BU-08-017	RG-BU-08-042	RG-BU-08-067	RG-BU-08-092	RG-BU-08-117	RG-BU-08-142
RG-BU-08-018	RG-BU-08-043	RG-BU-08-068	RG-BU-08-093	RG-BU-08-118	RG-BU-08-143
RG-BU-08-019	RG-BU-08-044	RG-BU-08-069	RG-BU-08-094	RG-BU-08-119	RG-BU-08-144
RG-BU-08-020	RG-BU-08-045	RG-BU-08-070	RG-BU-08-095	RG-BU-08-120	RG-BU-08-145
RG-BU-08-021	RG-BU-08-046	RG-BU-08-071	RG-BU-08-096	RG-BU-08-121	RG-BU-08-146
RG-BU-08-022	RG-BU-08-047	RG-BU-08-072	RG-BU-08-097	RG-BU-08-122	RG-BU-08-147
RG-BU-08-023	RG-BU-08-048	RG-BU-08-073	RG-BU-08-098	RG-BU-08-123	RG-BU-08-148
RG-BU-08-024	RG-BU-08-049	RG-BU-08-074	RG-BU-08-099	RG-BU-08-124	BRRI Dhan28
RG-BU-08-025	RG-BU-08-050	RG-BU-08-075	RG-BU-08-100	RG-BU-08-125	BRRI Dhan29

Table 1: 148 Exotic rice genotypes and two checks BRRI dhan28 and BRRI dhan29.

Suspected CMS lines	16 suspected R lines	5 known CMS lines	5 known B lines
1. RG BU 08-053 A	1. RG-BU 08-001 R	1. GAN46A	1. GAN46B
2. RG BU 08-058 A	2. RG-BU 08-002 R	2. BRRI1A	2. BRRI1B
3. RG BU 08-061 A	3. RG-BU 08-005 R	3. IR58025A	3. IR58025B
4. RG BU 08-066 A	4. RG-BU 08-006 R	4. IR62820 A	4. IR62820 B
5. RG BU 08-069 A	5. RG-BU 08-007 R	5. IR68888A	5. IR68888B
6. RG BU 08-084 A	6. RG-BU 08-013 R		
7. RG BU 08-086 A	7. RG-BU 08-016 R		
8. RG BU 08-087 A	8. RG-BU 08-018 R		
9. RG BU 08-107 A	9. RG-BU 08-025 R		
10. RG BU 08-125 A	10. RG-BU 08-034 R		
11. RG BU 08-126 A	11. RG-BU 08-038 R		
12. RG BU 08-129 A	12. RG-BU 08-046 R		
13. RG BU 08-132 A	13. RG-BU 08-057 R		
14. RG BU 08-136 A	14. RG-BU 08-063 R		
15. RG BU 08-137 A	15. RG-BU 08-097 R		
16. RG BU 08-141 A	16. RG-BU 08-105 R		

Table 2: Sixteen suspected CMS lines, 16 suspected restorer lines, 5 established known CMS lines and 5 established known maintainer lines.

converted into percentage. At maturity stage filled spikelets per panicle were counted in each genotype and then converted into percentage.

Data Recorded on

➢ Pollen fertility

➢ Spikelet fertility

Identification of A and R lines

The plants having 100% pollens sterility were considered as male sterile (CMS) and the plants having 80% and above pollen fertility and spikelet fertility were initially suspected as restorer line.

Identification of B-lines and maintenance of CMS lines

The identified CMS lines from Chinese germplasm were chance crossed with the established maintainer lines, GAN 46B, BRRI 1B, *viz.*, IR 58025B, IR 62820B and IR 68888B as the corresponding maintainer lines were not known to us. All the 16 CMS lines were crossed with each of the above five maintainer lines. F1 seeds from each cross were harvested separately and grown in next season along with their respective maintainer lines. At flowering stage pollen fertility status of each plant of each cross was recorded. If all the plants of a cross showed 100% pollen sterility then the corresponding male plant was considered as maintainer line.

Performance of restorer lines against established CMS lines

The identified sixteen restorer lines were crossed with five CMS lines viz. GAN46A, BRRI1A, IR58025A, IR62820A and IR68888A. The F1s were grown in the next rice growing season and performance based on pollen and spikelet fertility was analyzed. The F1(s) having above 80% pollen and spikelet fertility indicated that the male parent was an effective restorer.

Results and Discussion

Identification of A and R lines based on Pollen sterility and fertility status

Results of pollen sterility and fertility are shown in Tables 3 and 4.

Among the 148 exotic rice genotypes different levels of pollen sterility and fertility were observed. Such variation in pollen fertility indicated the existence genetic variation in respect of these reproductive traits among the genotypes. Among one hundred and forty eight exotic rice germplasm sixteen genotypes (RG-BU 08-053, RG-BU 08-058, RG-BU 08-061, RG-BU 08-066, RG-BU 08-069, RG-BU 08-084, RG-BU 08-086, RG-BU 08-087, RG-BU 08-107, RG-BU 08-125, RG-BU 08-126, RG-BU 08-129, RG-BU 08-132, RG-BU 08-136, RG-BU 08-137 and RG-BU 08-141) showed 100% pollen sterility status which was considered as completely male sterile lines or A line. The amount is not so less but 10.81% of the total germplasm. Ten genotypes were found sterile having pollen fertility 0-9%. Miyagawa and Nakamura [9] classified 85 rice cultivars based on the regional differences in varietal characteristics and found some elite male fertile lines as well as their counterpart maintainer lines. Chetia et al. [10] evaluated five cytoplasmic male sterile (CMS) rice lines (PMS 2A, PMS 3A, PMS 10A, IR 58025A and IR 62829A) and observed that PMS 2A, PMS 10A and IR 62829A were recorded as complete pollen and spikelet sterility. Sun et al. [11] determined the genetic effects of male sterile cytoplasm on major characters of rice hybrids. Ramesha et al. [12] identified three new and diversified CMS sources and many CMS lines possessing sporophytic type of male sterility with a very high frequency of typically abortive pollen. The new CMS lines were compared with other CMS lines belonging to wild abortive. Besides stable sterility, the new CMS lines had very high panicle exertion rate (92-96%) and good stigma exertion (48-65%) ability. Twenty nine genotypes were recorded partially sterile which is 19.59% of total. But most of the genotypes were categorized in partially fertile based on pollen fertility test. About fifty one genotypes were found partially fertile which 36.46% of total. Such type of genotypes may open the scope of development of restorer lines of hybrid program. And sixteen genotypes (RG-BU 08-001, RG-BU 08-002, RG-BU 08-005, RG-BU 08-006, RG-BU 08-007, RG-BU 08-013, RG-BU 08-016, RG-BU 08-018, RG-BU 08-025, RG-BU 08-034, RG-BU 08-038, RG-BU 08-046, RG-BU 08-057, RG-BU 08-063, RG-BU 08-097 and RG-BU 08-105) were identified as completely fertile as these genotypes had above 80% pollen and spikelet fertility which is 10.81% of the total genotypes. Research results indicated that the

Sl no	Fertility Status	Symbol	Number	Genotypes
1	Completely Sterile	CS	16 (10.81%)	RG-BU 08-053, RG-BU 08-058, RG-BU 08-061, RG-BU 08-066, RG-BU 08-069, RG-BU 08-084, RG-BU 08-086, RG-BU 08-087, RG-BU 08-107, RG-BU 08-125, RG-BU 08-126, RG-BU 08-129, RG-BU 08-132, RG-BU 08-136, RG-BU 08-137 and RG-BU 08-141
2	Sterile	S	10 (6.76%)	RG-BU 08-045, RG-BU 08-049, RG-BU 08-055, RG-BU 08-068, RG-BU 08-072, RG-BU 08-082, RG-BU 08-095, RG-BU 08-099, RG-BU -08-101 and RG-BU 08-122
3	Partially Sterile	PS	29 (19.59%)	RG-BU 08-003, RG-BU 08-015, RG-BU 08-017, RG-BU 08-020, RG-BU 08-030, RG-BU 08-033, RG-BU 08-042, RG-BU 08-043, RG-BU 08-044, RG-BU 08-047, RG-BU 08-056, RG-BU 08-065, RG-BU 08-070, RG-BU 08-077, RG-BU 08-080, RG-BU 08-083, RG-BU 08-092, RG-BU 08-093, RG-BU 08-094, RG-BU 08-106, RG-BU 08-107, RG-BU 08-110, RG-BU -08-115, RG-BU 08-119, RG-BU 08-124, RG-BU 08-134, RG-BU 08-146, RG-BU 08-147 and RG-BU 08-148
4	Partially fertile	PF	51 (34.46%)	RG-BU 08-004, RG-BU 08-009, RG-BU 08-010, RG-BU 08-012, RG-BU 08-014, RG-BU 08-021, RG-BU 08-023, RG-BU 08-026, RG-BU 08-028, RG-BU 08-029, RG-BU 08-031, RG-BU 08-034, RG-BU 08-037, RG-BU 08-039, RG-BU 08-041, RG-BU 08-046, RG-BU 08-048, RG-BU 08-059, RG-BU 08-060, RG-BU 08-062, RG-BU 08-064, RG-BU 08-071, RG-BU 08-073, RG-BU 08-075, RG-BU 08-076, RG-BU 08-078, RG-BU 08-079, RG-BU 08-085, RG-BU 08-088, RG-BU 08-091, RG-BU 08- 096, RG-BU 08-098, RG-BU 08-101, RG-BU 08-102, RG-BU 08-104, RG-BU 08-105, RG-BU 08-109, RG-BU 08-112, RG-BU 08-113, RG-BU 08-114, RG-BU 08-118, RG-BU 08-121, RG-BU 08-123, RG-BU 08-128, RG-BU 08-131, RG-BU 08-133, RG-BU 08-135, RG-BU 08-140, RG-BU 08-141, RG-BU 08-143 and RG-BU 08-145,
5	Fertile	F	26 (17.57%)	RG-BU 08-008, RG-BU 08-011, RG-BU 08-019, RG-BU 08-024, RG-BU 08-033, RG-BU 08-035, RG-BU 08-040, RG-BU 08-042, RG-BU 08-043, RG-BU 08-050, RG-BU 08-051, RG-BU 08-052, RG-BU 08-054, RG-BU 08-056, RG-BU 08-074, RG-BU 08-085, RG-BU 08-088, RG-BU 08-090, RG-BU 08-096, RG-BU 08-100, RG-BU 08-103, RG-BU 08-117, RG-BU 08-128, RG-BU 08-131, RG-BU -08-139 and RG-BU 08-146
6	Highly/fully Fertile	FF	16 (10.81%)	RG-BU 08-001, RG-BU 08-002, RG-BU 08-005, RG-BU 08-006, RG-BU 08-007, RG-BU 08-013, RG-BU 08-016, RG-BU 08-018, RG-BU 08-025, RG-BU 08-034, RG-BU 08-038, RG-BU 08-046, RG-BU 08-057, RG-BU 08-063, RG-BU 08-097 and RG-BU 08-105

Table 3: Classification of 148 (Chinese) exotic rice germplasm based on pollen fertility status.

Pollen fertility status	Fertility Percent
1. FS: Fully Sterile/Completely sterile	0%
2. HS: Sterile	1-9%
3. PS: Partially Sterile	10-29%
4. HF: Partially Fertile	30-69%
5. F: Fertile	70-79%
6. FF: Fully Fertile	80 % and above

Table 4: Classification of 148 (Chinese) exotic rice germplasm based on pollen fertility status.

frequency of restorer lines in Chinese rice germplasm was found very high as compared to local rice genotypes. Thus the above mentioned genotypes having restorer genes may be utilized as a good reservoir of restorer genes for development of efficient restorer lines. Abeysekera et al. [13] studied 53 cytoplasmic male sterile (CMS) lines of rice and observed that pollen fertility and spikelet fertility had significant effects on the out crossing rate (Tables 3 and 4).

Restorability of suspected restorer lines

Sixteen suspected restorer lines were used to test their restoration ability with four established CMS lines, IR 58025A, GAN 46A, IR 62820A and IR 68888A. Results of restoration ability based on pollen and spikelet fertility of F1s between CMS and restorer lines are shown in Tables 5 and 6. Out of sixteen F_1 ten crosses (IR 58025A× RG-BU 08-001R, IR 58025A× RG-BU 08-006R, IR 58025A× RG-BU 08-013R, IR 58025A× RG-BU 08-016R, IR 58025A× RG-BU 08-018R, IR 58025A× RG-BU 08-025R, IR 58025A× RG-BU 08-034R, IR 58025A× RG-BU 08-046R, IR 58025A× RG-BU 08-063R and IR 58025A× RG-BU 08-0105R) showed above 80% pollen fertility but seven crosses showed 80% above both pollen and spikelet's fertility (IR 58025A× RG-BU 08-001R, IR 58025A× RG-BU 08-006R, IR 58025A×RG-BU 08-018R, IR 58025A×RG-BU 08-034R, IR 58025A×RG-BU 08-046R, IR 58025A×RG-BU 08-063R and IR 58025A×RG-BU 08-0105R) which indicated that seven suspected restorers had desirable restoration

ability with the above CMS line. When these sixteen suspected restorers were crossed with GAN 46A six F_1s (GAN46A×RG-BU 08-007R, GAN46A×RG-BU08-025R, GAN46A×RG-BU 08-034R, GAN46A×RG-BU 08-038R, GAN46A×RG-BU 08-097R and GAN46A× RG-BU 08-105R) showed above 80% pollen fertility but only two crosses (GAN46A×RG-BU 08-007R and GAN46A×RG-BU08-025R) showed 80% above both pollen and spikelet's fertility. Eight crosses of restorer lines with IR 62820A showed above 80% pollen fertility (IR 62820A×RG-BU 08-001R, IR 62820A×RG-BU 08-005R, IR 62820A×RG-BU 08-006R, IR 62820A×RG-BU 08-007R, IR 62820A×RG-BU 08-013R, IR 62820A×RG-BU 08-025R, IR 62820A×RG-BU 08-038R and IR 62820A×RG-BU 08-057R) where as five F1s were found having 80% and above both pollen and spikelet's fertility i.e., IR 62820A×RG-BU 08-005R, IR 62820A×RG-BU 08-013R, IR 62820A×RG-BU 08-025R, IR 62820A×RG-BU 08-038R and IR 62820A×RG-BU 08-057R. Five crosses of restorer lines with IR 68888A showed above 80% pollen fertility (IR 62820A×RG-BU 08-001R, IR 62820A×RG-BU 08-002R, IR 62820A×RG-BU 08-007R, IR 62820A×RG-BU 08-097R and IR 62820A×RG-BU 08-105R) where as only two F1s were found having 80% and above both pollen and spikelet's fertility i.e., IR 62820A×RG-BU 08-002R and IR 62820A×RG-BU 08-007R. Li et al. [14] identified R899 as an early-maturing restorer line. Its characteristics include desirable agronomic traits, strong restoring ability, strong resistance to rice blast, fine grain quality and high hybrid seed yield. The hybrid combinations displayed high and stable grain yield, fine grain quality, suitable growth period and wide adaptability. Five out of sixteen F_1 crosses with IR 62820A showed above 80% pollen and spikelet fertility. Such results indicated that only five lines have desirable restoration ability in hybrids with IR 62820A. In case of IR 68888A only two restorers were found effective [15-17]. Further study on the performance of F_1 hybrids obtained from crosses between IR 62820A and IR 68888A and the selected five and two restorer lines may done for identification of desirable rice hybrid.

Test crosses	IR 58025B		BRRI 1B		GAN 46B		IR 62820B		IR 68888B	
	Pollen Fertility (PF)	Spikelet's fertility (SF)	Pollen Fertility (PF)	Spikelet's fertility (SF)	Pollen Fertility (PF)	Spikelet's fertility (SF)	Pollen Fertility (PF)	Spikelet's fertility (SF)	Pollen Fertility (PF)	Spikelet's fertility (SF)
RG BU 08-053 A	0.00 ± 0.00	0.00 ± 0.00	0.00 ± 0.00	0.00 ± 0.00	0.00 ± 0.00	0.00 ± 0.00	3.00 ± 0.32	12.40 ± 0.34	13.00 ± 0.32	29.12 ± 0.37
RG BU 08-058 A	7.00 ± 0.14	15.68 ± 0.17	6.00 ± 0.27	12.44 ± 0.27	6.00 ± 0.27	12.44 ± 0.32	9.00 ± 0.27	20.16 ± 0.32	11.00 ± 0.32	24.64 ± 0.32
RG BU 08-061 A	8.00 ± 0.13	17.92 ± 0.18	0.00 ± 0.00	0.00 ± 0.00	0.00 ± 0.00	0.00 ± 0.00	8.00 ± 0.27	17.92 ± 0.32	13.00 ± 0.32	29.12 ± 0.31
RG BU 08-066 A	11.00 ± 0.15	24.64 ± 1.19	0.00 ± 0.00	0.00 ± 0.00	0.00 ± 0.00	0.00 ± 0.00	0.00 ± 0.00	0.00 ± 0.00	0.00 ± 0.00	0.00 ± 0.00
RG BU 08-069 A	12.00 ± 0.19	26.88 ± 4.15	6.00 ± 0.24	13.44 ± 0.27	6.00 ± 0.27	13.44 ± 0.27	9.00 ± 0.27	20.16 ± 0.32	5.00 ± 0.27	11.20 ± 0.32
RG BU 08-084 A	0.00 ± 0.00	0.00 ± 0.00	8.00 ± 0.27	17.92 ± 0.32	8.00 ± 0.23	17.92 ± 0.32	0.00 ± 0.00	0.00 ± 0.00	6.00 ± 0.27	13.44 ± 0.32
RG BU 08-086 A	0.00 ± 0.00	0.00 ± 0.00	9.00 ± 0.26	20.16 ± 0.32	9.00 ± 0.22	20.16 ± 0.32	23.00 ± 0.19	21.52 ± 0.32	9.00 ± 0.27	20.16 ± 0.32
RG BU 08-087 A	0.00 ± 0.00	0.00 ± 0.00	12.00 ± 0.32	26.88 ± 2.12	12.00 ± 0.32	26.88 ± 0.32	0.00 ± 0.00	0.00 ± 0.00	9.00 ± 0.27	20.16 ± 0.32
RG BU 08-107 A	0.00 ± 0.00	0.00 ± 0.00	13.00 ± 0.32	29.12 ± 2.12	13.00 ± 0.31	29.12 ± 2.19	0.00 ± 0.00	0.00 ± 0.00	8.00 ± 0.27	17.92 ± 0.32
RG BU 08-125 A	0.00 ± 0.00	0.00 ± 0.00	8.00 ± 0.24	17.92 ± 0.32	8.00 ± 0.27	17.92 ± 0.32	9.00 ± 0.27	20.16 ± 0.32	12.00 ± 0.32	26.88 ± 0.32
RG BU 08-126 A	9.00 ± 0.15	20.16 ± 2.19	9.00 ± 0.27	20.16 ± 0.32	9.00 ± 0.21	20.16 ± 0.32	8.00 ± 0.27	17.92 ± 0.32	1.22 ± 0.27	3.54 ± 0.27
RG BU 08-129 A	0.00 ± 0.00	0.00 ± 0.00	7.00 ± 0.22	15.68 ± 0.27	7.00 ± 0.27	15.68 ± 0.32	0.00 ± 0.00	0.00 ± 0.00	3.00 ± 0.27	6.72 ± 0.27
RG BU 08-132 A	8.00 ± 0.16	17.92 ± 0.87	13.00 ± 0.32	29.12 ± 2.10	13.00 ± 0.32	29.12 ± 2.19	7.00 ± 0.27	15.68 ± 0.32	14.00 ± 0.32	31.36 ± 0.39
RG BU 08-136 A	0.00 ± 0.00	0.00 ± 0.00	14.00 ± 0.32	31.36 ± 2.10	14.00 ± 0.32	35.72 ± 2.19	0.00 ± 0.00	0.00 ± 0.00	11.00 ± 0.32	24.64 ± 0.32
RG BU 08-137 A	0.00 ± 0.00	0.00 ± 0.00	18.00 ± 0.38	20.32 ± 2.12	18.00 ± 0.32	20.32 ± 0.32	0.00 ± 0.00	0.00 ± 0.00	11.00 ± 0.32	24.64 ± 0.32
RG BU 08-141 A	0.00 ± 0.00	0.00 ± 0.00	22.00 ± 2.11	29.28 ± 2.19	22.00 ± 1.19	29.28 ± 2.19	0.00 ± 0.00	0.00 ± 0.00	10.00 ± 0.32	22.40 ± 0.32
Minimum	0.00	0.00	0.00	0.00	0.00	0.00	0.00	0.00	0.00	0.00
Maximum	12.00	26.88	23.00	31.36	23.00	35.72	23.0	22.4	13.00	31.36
Average	7.06	14.31	7.06	14.31	7.06	14.31	8.37	10.52	7.06	14.31
SD	6.05	10.90	6.05	10.90	6.05	10.90	7.87	5.24	6.05	10.90
SE	0.67	1.21	0.67	1.21	0.67	1.22	1.25	1.43	0.67	1.19
R²	-0.06	-0.04	-0.06	-0.05	-0.06	-0.04	-0.04	-0.03	-0.07	-0.04

Table 5: Maintenance of identified CMS lines using IR 58025B, BRRI 1B, GAN 46B, IR 62820B and IR 68888B.

Test crosses	IR 58025A		BRRI 1A		GAN 46A		IR 62820A		IR 68888A	
	Pollen Fertility (PF)	Spikelet's fertility (SF)	Pollen Fertility (PF)	Spikelet's fertility (SF)	Pollen Fertility (PF)	Spikelet's fertility (SF)	Pollen Fertility (PF)	Spikelet's fertility (SF)	Pollen Fertility (PF)	Spikelet's fertility (SF)
RG-BU 08-001 R	100.00 ± 0.00	87.35 ± 4.10	12.12 ± 1.19	34.77 ± 4.19	71.16 ± 4.36	56.00 ± 4.19	87.45 ± 4.19	56.99 ± 4.19	91.12 ± 4.36	43.12 ± 4.19
RG-BU 08-002 R	76.44 ± 4.19	54.45 ± 4.11	3.32 ± 2.19	13.13 ± 3.19	7.12 ± 0.32	12.12 ± 4.19	34.34 ± 4.10	43.12 ± 4.10	90.00 ± 4.36	89.01 ± 4.36
RG-BU 08-005 R	64.87 ± 4.19	47.77 ± 4.19	100.00 ± 0.00	85.11 ± 4.36	63.23 ± 4.36	45.45 ± 4.19	100.00 ± 0.00	92.65 ± 4.36	12.12 ± 1.19	34.77 ± 4.19
RG-BU 08-006 R	100.00 ± 0.00	92.53 ± 4.36	12.00 ± 4.33	84.44 ± 4.36	9.23 ± 0.32	12.34 ± 4.19	95.34 ± 4.36	34.77 ± 4.19	3.32 ± 2.19	13.13 ± 3.19
RG-BU 08-007 R	76.23 ± 4.36	34.34 ± 4.19	78.78 ± 4.36	68.94 ± 4.36	100.00 ± 0.00	87.57 ± 4.36	91.12 ± 4.36	31.77 ± 4.18	100.00 ± 0.00	85.11 ± 4.36
RG-BU 08-013 R	98.87 ± 4.36	56.44 ± 4.19	78.90 ± 4.36	81.76 ± 4.19	34.35 ± 4.19	46.49 ± 4.19	100.00 ± 0.00	83.88 ± 4.36	12.00 ± 4.33	84.44 ± 4.36
RG-BU 08-016 R	91.23 ± 4.36	56.56 ± 4.19	78.00 ± 4.36	40.01 ± 4.17	34.00 ± 4.19	48.48 ± 4.19	45.90 ± 4.19	56.56 ± 4.19	78.78 ± 4.36	68.94 ± 4.36
RG-BU 08-018 R	100.00 ± 0.00	86.34 ± 4.36	71.00 ± 4.19	82.99 ± 4.36	56.34 ± 4.19	82.25 ± 4.36	68.68 ± 4.19	71.17 ± 4.19	78.90 ± 4.36	81.76 ± 4.19
RG-BU 08-025 R	92.22 ± 4.36	67.67 ± 4.19	61.00 ± 4.19	53.77 ± 4.19	98.34 ± 4.36	82.76 ± 4.36	100.00 ± 0.00	83.99 ± 4.36	78.00 ± 4.36	40.01 ± 4.17
RG-BU 08-034 R	100.00 ± 0.00	87.09 ± 4.36	63.00 ± 4.19	81.09 ± 4.36	90.24 ± 4.36	56.00 ± 4.12	67.99 ± 4.19	42.24 ± 4.17	71.00 ± 4.19	82.99 ± 4.36
RG-BU 08-038 R	69.23 ± 4.19	34.88 ± 4.19	78.00 ± 4.19	74.26 ± 4.36	90.33 ± 4.36	32.12 ± 4.12	100.00 ± 0.00	85.22 ± 4.36	61.00 ± 4.19	53.77 ± 4.19
RG-BU 08-046 R	100.00 ± 0.00	91.78 ± 4.36	67.66 ± 4.19	46.25 ± 4.10	23.23 ± 4.10	45.45 ± 4.12	71.11 ± 4.19	56.00 ± 4.10	63.00 ± 4.19	81.09 ± 4.36
RG-BU 08-057 R	12.12 ± 0.32	23.23 ± 4.19	90.99 ± 4.36	54.76 ± 4.19	34.34 ± 4.10	42.34 ± 4.12	100.00 ± 0.00	83.32 ± 4.36	78.00 ± 4.19	74.26 ± 4.36
RG-BU 08-063 R	100.00 ± 0.00	92.65 ± 4.36	98.00 ± 4.36	40.89 ± 4.10	76.23 ± 4.36	87.57 ± 4.36	34.34 ± 4.19	76.49 ± 4.36	67.66 ± 4.19	46.25 ± 4.10
RG-BU 08-097 R	10.23 ± 0.32	23.00 ± 4.19	91.12 ± 4.36	43.12 ± 4.19	100.00 ± 0.00	76.49 ± 4.36	34.98 ± 4.19	48.48 ± 4.13	90.99 ± 4.36	54.76 ± 4.19
RG-BU 08-105 R	100.00 ± 0.00	82.76 ± 4.36	90.00 ± 4.36	89.01 ± 4.36	98.48 ± 4.36	48.48 ± 4.19	45.22 ± 4.19	56.99 ± 4.19	98.00 ± 4.36	40.89 ± 4.10
Minimum	2.672	12.123	2.673	12.127	7.172	12.123	2.672	12.123	3.324	13.127
Maximum	100.000	92.654	100.000	92.656	100.000	92.654	100.000	92.654	100.000	89.065
Average	67.847	60.803	67.841	60.806	67.847	60.803	67.847	60.803	65.846	60.806
SD	32.384	22.564	32.387	22.567	32.384	22.564	32.384	22.564	32.387	22.567
SE	3.604	2.513	3.608	2.516	3.604	2.513	3.604	2.513	3.608	2.516
R²	0.752	0.672	0.752	0.672	0.752	0.672	0.729	0.672	0.695	0.672

Table 6: Fertility Restoration of identified R-lines against IR 58025A, BRRI 1A, GAN 46A, IR 62820A and IR 68888A.

References

1. Islam MA, Mian MAK, Rasul G, Johora, Sarker UK (2010) Interaction effect between genotypes, row ratio and fertilizer dose on hybrid seed production of rice (Oryza sativa. L.). Bangladesh Society of Agronomy 4: 134-141.

2. Islam MA, Mian MA, Rasul G, Khaliq QA, Bashar MK (2014) Estimation of specific combining ability (SCA) effects and per-se performances in some reproductive traits in rice (Oryza sativa L.). Bangladesh Journal of Genetics and Plant Breeding 33: 12-16.

3. Yuan LP (1998) Genetic relationship of stigma exterior between maintainer lines and sterile lines for Dian type japonica hybrid rice. Journal-of-Yunnan-Agricultural-University 20: 459-461, 477.

4. Islam MA, Sarker UK, Mian MAK, Ahmed JU (2009) Genotype Seedling age Interaction for hybrid seed yield of rice (Oryza sativa. L.). Bangladesh Journal of Genetics and Plant Breeding 24: 23-26.

5. Mian MAK (2010) Hybrid rice breeding. Breeding for self-pollinated crops, Heterosis Breeding. BSMRAU, Summer.

6. Islam MA, Mian MAK, Rasul G, Khaliq QA, Bashar MK (2014) Estimation of general combining ability (GCA) effects and per-se performances in some reproductive traits in rice (Oryza sativa L.). Bangladesh Journal of Genetics and Plant Breeding 32: 18-13.

7. Islam MA, Johora FT, Sarker UK, Mian MAK (2012) Adaptation of Chinese CMS lines interaction with seedling age and row ratio on hybrid seed production of Rice (Oryza sativa L.). Bangladesh Journal of Agronomy 25: 178-183.

8. Islam MA, Johora FT, Hasan MR, Islam MR, Mahmud MNH (2012) Optimizing sowing date of morpho-physiologycal traits in direct seeding and transplanting rice (Oryza sativa. L.). Echo-Friendly Agriculture Journal 5: 162-167.

9. Miyagawa S, Nakamura S (1994) Regional differences in varietal characteristics of scented rice. Japan J. crop Sci 53: 494-502.

10. Chitra S, Awan E, Ijaz M, Manzoor Z (2000) Heterosis and combining ability analysis in Basmati rice hybrids. J Animal and Pl Sci 16: 56-59.

11. Sun A, Kreetapiron S, Varanyanond W, Tungtrakul P, Somboonpong S, Rattapat S (2006) Combining ability analysis for grain yield and its components in rice Karnataka J Agric Sci 16: 223-227.

12. Ramesha MS, Ahmed MI, Viraktamath BC, Vijayakumar CHM, Singh S (2008) New cytoplasmic male sterile (CMS) lines with diversified CMS sources and better out crossing traits in rice. Intl. Rice Res. Notes.23: 5.

13. Abeysekera LPR, Saini G, Sharma AK (2003) Studies on morphological traits of rice (Oryza sativa L.). New-Agriculturist 7: 79-83

14. Li ZK, Luo LJ, Mei HW, Wang DL, Shu QY, Tabien R et al., (2006) Overdominant epistatic loci are the primary genetic basis of inbreeding depression and heterosis in rice. I. Biomass and grain yield. Genetics 158 : 1737-1753.

15. Hien NL, Sarhadi WA, Oikawa Y, Hirata Yi (2007) Genetic diversity of morphological responses and the relationships among Asian rice (Oryza sativa L.) cultivars. Tropics 16: 343-355.

16. Islam MA, Mian MAK, Rasul G, Khaliq QA, Bashar MK (2013) Estimation of Specific Combining Ability (GCA) Effects and Per-se performances in Some Yield Related Traits of Rice (Oryza sativa L.). Echo-Friendly Agriculture Journal 7: 143-145.

17. Rita KT, Sarawgi A (2008) Differential fertility restoration of restorer genes to WA-cytoplasmic male sterility system in rice (Oryza sativa L.). Indian-Journal-of-Genetics-and-Plant-Breeding; 65: 207-208.

11

Performance of Rice Grown after Cassava/Legume Intercrops at Badeggi in the Southern Guinea Savanna Ecological Zone of Nigeria

Gbanguba AU[1]*, Kolo MGM[2], Odofin AJ[3] and Gana AS[2]

[1]National Cereals Research Institute, Badeggi. P. M. B. 8. Bida, Niger State. Nigeria
[2]Department of Crop Production, Federal University of Technology, Minna, Niger State, Nigeria
[3]Department of Soil Science, Federal University of Technology, Minna, Niger State, Nigeria

Abstract

An experiment was conducted on the lowland experimental field of National Cereals Research Institute, Badeggi (9°45'N, 60°7'E, ALT 70.57M) in the southern Guinea savanna zone of Nigeria in 2011, 2012 and 2013 to determine the effects of cassava/legume inter-cropping patterns on grain yield of subsequent lowland rice. The result revealed that significant higher plant height, tiller/stool and grains/panicle were obtained from with rice grown after cassava/cowpea and cassava/*Aeschynomene* compared with rice grown after cassava/soybean and cassava/*Lablab*. Significantly higher rice panicles per metre, 1000 rice grain weight and rice grain yield were also observed in the same treatments. Cassava/cowpea and cassava/*Aeschynomene* are hence recommended to be planted before lowland rice.

Keywords: Performance; Rice; Growth; Cassava; Legume

Introduction

Rice is a cereal grain belonging to the grass family of *Poaceae* and to two species namely, *Oryza sativa* and *Oryza glaberrima*. It is a major staple food in both developing and developed world and its production has been essential for many countries. More than one third of the human population rely on rice for their sustenance, making it the most important of the world's food crops [1]. The potential land area for rice production in Nigeria is between 4.6-4.9 million ha. Out of this, only about 1.7 million ha or 35 percent of the available land area is presently cropped to rice [2]. This area includes five different rice environments or ecologies [3]. The upland ecology/rain fed lowland accounts for 55 to 60 percent of the cultivated rice land. An estimated 25 percent of Nigeria's rice area is under inland valley swamp rice production. The irrigated rice ecology accounts for about 18 percent of cultivated rice land and deep water or floating rice constitutes 5 to 12 percent of the national rice production area. Tidal (mangrove) swamp ecology contributes less than 2 percent to national rice production area [3].

FAO (2010; 2011; 2012) reported that world paddy production for the 2010 season was 700.7 million tonnes719.8 million tonnes and 728.7 million tonnes for 2012 respectively. *Oryza*, (2013) reported that Nigeria paddy production for 2012 was estimated at 4.2 million tonnes. World rice requirements are predicted to increase at a rate of 1.7% per year between now and 2025 (IRRI, 1993) [4].

Crop rotation has many benefits that can influence the success of crop production enterprises. Crop rotation is an essential practice in sustainable agriculture because of its many positive effects like increasing soil fertility and reducing crop competitiveness [5]. The positive effect of long-term rotation on crop yield has been recognized and exploited for centuries. During the last few decades, however, its benefits in terms of yield seem to have been ignored by farmers [6]. It is now evident that crop rotation increases yield and promotes agricultural sustainability [7].

During the off-season, rainfed rice lands are typically followed [8]. The straw and fallow weed vegetation are subjected to grazing by livestock. In a minor fraction of the area with conducive residual soil water-holding capacity, and/or a high groundwater table, upland crops, including legumes, are grown in the post-rice season. This practice is most common where the soil texture is loamy and easy to till. In well-drained rice lands, upland crops are grown prior to rice during the dry-to-wet season transition period. Very short duration crops are advantageous to permit maturity before the soil becomes waterlogged. Mungbean is a very common grain legume in the pre-rice niche.

Throughout a two years experiment, Filizadeh et al. [9] found that rice yields in rotation with soybean were higher compared with continuous rice. Rice yields from rotation plots were 17 and 21% higher in 2002 and 2003, respectively, compared with continuous rice plots. Anders et al. [10] reported higher yield of rice in rice grown after soybean than in rice wheat rotation. Grain yields of rice which had been preceded by a legume fallow were on average 0.2 mg ha⁻¹ or about 30% greater than that preceded by a natural weedy fallow control, Becker and Johnson, [11] Lixiao et al. [12] stated that the grain yields of rice after fallow and upland crops rotation were compared in 2007. The rice yielded more after two seasons of upland crops than after two seasons of fallow. Among the three upland crops, maximum yield of rice was observed after two seasons of soybean. Toomsan et al. [13] recorded 50% higher rice yields in rice followed cover crop green mixtures than rice in bare fallow rotation. This study was therefore with the aim of determine the effects of cassava/legume inter-cropping patterns on grain yield of subsequent lowland rice.

Materials and Methods

The experiment was conducted at the lowland experimental field of National Cereals Research Institute, Badeggi (9°45'N, 60°7'E, ALT

*Corresponding author: Gbanguba AU, National Cereals Research Institute, Badeggi. P. M. B. 8. Bida, Niger State. Nigeria
E-mail: alhassangbanguba@yahoo.com

70.57 m) in the southern Guinea savannah zone of Nigeria with mean annual rainfall of 2066.3, 1163.6 and 899.7 mm distributed between April to October in 2011, 2012 and 2013 respectively with maximum and minimum temperature of 30-38°C and 14-26°C respectively. Raised beds for planting of cassava/legume intercrop were done manually. Beds were 2.5 m long, 0.5 m wide and 0.75 m high. Cassava was planted on the top sides of beds in two rows at inter and intra-row spacing of 0.5 m (ten stands per bed) and legumes were planted by the sides of the beds at inter and intra-row spacing of 0. 5 m x 0. 25 m respectively except for soybean which was drilled immediately the beds were constructed. The cassava/legume intercropping was carried out between January and August using residual moisture. Velvet bean (*Mucuna puriens*), cowpea (*Vigna unguiculata*), soybean (*Glyxine max*), hyacinth bean (*Lablab purpureus*) and Jointvetche (*Aeschynomene histrix*) legumes used as intercrops with cassava IIT 427 and there was sole cassava and natural fallow. The treatments were laid out in Randomized Complete Block Design and replicated three times. The plot size was 13.5 m x 5 m. The cassava/legume cropping lasted till August when cassava was harvested. Those plots previously cropped with cassava/legume intercrops were followed by rice.

Agronomic practices

Rice seedlings were raised in the nursery. The rice variety used was Faro 52. The nursery was done in the second week of July and the seedlings were ready for transplanting at three weeks after seeding. The beds made for cassava/legume intercrop were levelled manually and rice transplanting was done at a spacing of 20 x 20 cm at the rate of two seedlings per hill.

Urea (46%N), single superphosphate (18% $P_2O_{5)}$ and muriate of potash (60% K_2O) were applied to supply 80 kg/ha N, 40 kg/ha P_2O_5, 40 kg/ha K_2O/ha. The N was in split – applied in equal amounts at planting and panicle initiation. The fertilizers were applied by broadcasting method.

Hand weeding was done at 3 and 6 weeks after planting.

Data collected were

Soil sample, rice plant height, tiller/stool and panicle m^{-2}. Also taken were grains/panicle, weight of 1000 grain and grain yield. Rice height was done by measuring the rice plant from the ground level to the flag leaf using ruler at 3, 6, 9 and 12 weeks after transplanting. Tiller number was carried out by counting the number of rice plants grown after the initial two seedlings transplanted at 3, 6, 9 and 12 weeks after transplanting. Rice panicle number per meter by thrown one meter square quadrat inside rice plants in each plot and the panicles of rice that fell inside it were counted.

Grain number per panicle was determined by counting grains on

the sample panicles from each plot. Weight of 1000 grains was taken after counting 1000 rice grains from each plot.

Rice grain from net plot of each plot was weighed using scale. The weight was measured in kilograms per plot which was converted to kilogram per hacter and later to tones per hacter

Results

Rice plant height (cm)

The result of soil nutrient status of experimental site is presented in Table 1 and it indicates that soil pH was slightly acidic in all the intercrops. Cassava/cowpea and cassava/*Aeschynomene* plots gave higher percentage organic carbon in all study periods and were at increased with the repetition of cassava/legume/intercrops. Nitrogen obtained in cassava/cowpea and cassava/*Aeschynomene* was 83.3%, 84.2% and 86.3% higher and those of *Mucuna*, soybean and *Lablab* were 72.7-78.5%, 71.4-75% and 38.9-52.9% when compared with natural fallow plots in 2011, 2012 and 2013 respectively. Organic carbon and available P were lower in plots where rice was grown after either sole cassava or natural fallow. Exchangeable bases were affected by cassava/legume intercrop in which 16.0-36.3%, 15.9-35.9 and 26.5-37.7% Ca, 39.6-55.7%, 40.5-56.0% and 31.8-51.8 Mg, 75.0-81.8, 66.6-75.0% and 30.0-50% K were found higher when comparing with natural fallow in 2011, 2012 and 2013 respectively.

The data on rice plant height revealed that it was significantly (P<0.05) affected by pre-rice cropping of cassava/legume intercrop (Table 2). The tallest rice plant was observed after cassava/*Mucuna* and cassava/*Aeschynomemne* at 3 WAT in 2011 cropping season. The height of rice grown after cassava/*Lablab*, sole cassava and natural fallow were similar at 3 WAT. At 6 WAT rice grown after cassava/*Mucuna* produced taller rice plant which was not significantly different from those obtained after cassava/cowpea and cassava/*Aeschynomemne* intercrops. Among all rice grown after cassava/legume intercrop, rice after cassava/*Lablab* produced the shortest rice plant height at 9 WAT. Rice grown after cassava/cowpea produced taller plant at 12 WAT. Generally, rice after natural fallow had the short plants and followed by sole cassava plots. In 2012 cropping season rice after cassava/*Aeschynomene* gave significant (P<0.05) taller rice plant height at 3 WAT while the one after natural fallow was the least (Table 2). Rice grown after cassava/*Mucuna*, cassava/cowpea and cassava/*Aeschynomene* gave similar but higher plant height at 6 and 9 WAT. However, at 12 WAT rice grown after cassava/*Mucuna* and cowpea produced taller but similar plant. The height of rice plants after natural fallow was shortest throughout the growing season (Table 2).

Rice tiller number per stool

Rice tiller number per stool was significantly affected by pre-rice

Treatments	Ph (H$_2$O)			Organic carbon %			Total N (g kg^{-1})			Avail P(mg kg^{-1})			Exch. Bases (cmol kg^{-1})									CEC(cmolkg^{-1})		
													Ca			Mg			K					
	2011	2012	2013	2011	2012	2013	2011	2012	2013	2011	2012	2013	2011	2012	2013	2011	2012	2013	2011	2012	2013	2011	2012	2013
Cassava/Mucuna	5.09	5.19	5.21	2.97	3.11	3.21	0.14	0.15	0.17	30.8	32.87	34.87	5.55	5.57	5.58	3.61	3.64	3.68	0.08	0.09	0.10	9.28	9.3	9.36
Cassava/cowpea	5.12	5.22	5.26	3.15	3.17	3.27	0.18	0.19	0.22	31.7	32.75	36.75	6.10	6.12	6.22	4.47	4.57	4.59	0.11	0.12	0.13	10.6	10.8	10.9
Cassava/soybean	5.23	5.33	5.36	3.13	3.16	3.26	0.13	0.16	0.17	30.7	31.76	35.76	5.30	5.33	5.35	3.41	3.51	3.57	0.09	0.11	0.13	8.80	11.1	9.05
Cassava/Lablab	5.10	5.30	5.35	2.94	2.94	2.24	0.11	0.14	0.13	29.8	30.86	32.86	4.6	4.63	4.64	3.28	3.38	3.39	0.09	0.11	0.13	7.97	8.12	8.16
Cassava/ Aeschynomene	5.61	5.62	5.69	3.29	3.30	3.33	0.18	0.19	0.21	31.7	33.78	34.78	6.10	6.21	6.23	4.47	4.57	4.58	0.11	0.12	0.14	10.6	10.9	10.9
Sole cassava	5.03	5.00	5.06	2.21	2.22	2.32	0.03	0.05	0.07	20.8	22.8	25.8	3.95	3.90	3.96	2.78	2.88	2.89	0.03	0.04	0.08	6.77	8.71	6.93
Natural fallow	5.0	5.0	5.02	2.0	2.1	2.20	0.03	0.04	0.08	28.7	29.7	30.0	3.88	3.89	3.91	1.98	2.01	2.21	0.02	0.03	0.07	5.88	5.93	6.19

Table 1: Influence of cassava/legume intercrop on soil chemical properties in 2011-2013 cropping seasons.

cropping of cassava/legume (Table 3). Generally, rice tiller number per stool was significantly higher in the order cassava/*Aeschynomene*, cassava/cowpea and cassava/soybean pre-cultivation of rice throughout the three years of study. Rice tiller number in cassava/*Mucuna,* in 2011 and 2012 were similar to that of cassava/*Aeschynomene* while that of soybean was at par with that in cassava/*Lablab* in 2013 only. The rice tiller number per stool also increased with continuous cultivation, except in the natural fallow treatment where it was declining.

Rice panicles per meter square

There was significant (P<0.05) effect of pre-rice cropping of cassava/legume on rice panicle per meter square (Table 4). Generally, rice panicle per meter square was higher under cassava/cowpea and cassava/*Aeschynomene* but was similar throughout the years of study. This was followed by cassava/soybean intercrop while it was least in cassava/*Lablab* in the three years. The natural fallow treatment had the least number of rice panicles which was followed by sole cassava throughout the years of study. It was observed that rice panicle increased with continuous cassava/legume rotation of cassava/legume intercrop with rice but decreased in rice after natural fallow in the years of study.

Rice grains per panicle

Rice grown after cassava/cowpea gave greater number of rice grain throughout the three years of study although it did not differ statistically with what was recorded in rice grown after cassava/*Mucuna* in 2011 and cassava/*Aeschynomene* in 2012 and 2013 (Table 5). This was followed closely by rice grown after cassava/soybean intercrop. Rice after cassava/soybean and cassava/*Aeschynomene* produced similar

rice grain number per panicle in 2011. Rice following natural fallowing produced the least number of rice grains per panicle.

Weight of 1000 rice grain (g)

There was significant effect of pre-rice cropping of cassava/legume intercrop on the weight of 1000 rice grain (Table 6). In all the study periods, rice grown after cassava/cowpea and cassava/*Aeschynomene* gave the heaviest 1000 rice grain weight which was significantly greater than what was obtained in rice grown after the other treatments. This was followed by that in cassava/*Mucuna*. The rice following natural fallow had the least grain weight which was followed by sole cassava treatment.

Rice grain yield (kg/ha)

Rice grain yield was highest in the three years of study and in 2012 and 2013 in cassava/*Aeschynomene* respectively (Table 7). This was followed by rice grown after cassava/*Mucuna* intercrop in the three years of study. Rice grain yield in cassava/*Lablab* was least among the intercrops throughout the three years of study.

Discussions

Rice plant height was affected by cassava/legume intercropping. The taller rice plant observed in rice grown after cassava/legume intercrop could be due probably higher soil fertility status. The level of variation observed in rice plant height might be a function of differences in the preceeding cassava intercrop. This might be due to the fact that crops tend to establish and grow where soil nutrients tends to be improved. This result agrees with the findings of Morteza et al. [1] who observed variation in rice height, where rice was planted after different legumes.

Treatments	2011				2012				2013			
	Weeks after planting				Weeks after planting				Weeks after planting			
	3	6	9	12	3	6	9	12	3	6	9	12
Cassava Intercrop (CI)												
Cassava/*Mucuna*	16.6ᵃ	34.9ᵃ	57.5ᵃ	75.5ᵇ	20.1ᶜ	44.0ᵃᵇ	63.4ᵃˢ	84.4ᵃᵇ	21.7ᵈ	53.6ᶜ	70.3ᵇ	93.9ᵇ
Cassava/cowpea	15.8ᵇ	34.3ᵃᵇ	57.7ᵃ	77.2ᵃ	21.3ᵇ	45.0ᵃ	64.0ᵃ	86.4ᵃ	26.9ᵇ	57.0ᵇ	72.2ᵃ	98.5ᵃ
Cassava/soybean	15.6ᵇᶜ	33.7ᵇ	57.6ᵃ	71.1ᶜᵈ	19.4ᵈ	43.3ᵇ	61.7ᵇᶜ	80.2ᶜ	23.6ᶜ	46.6ᵈ	68.5ᶜ	88.1ᶜ
Cassava/lablab	15.2ᶜᵈ	31.4ᶜ	52.3ᵇ	69.7ᵈ	19.8ᵈ	40.7ᶜ	61.1ᶜ	76.2ᵈ	22.1ᵈ	45.3ᵈ	67.4ᶜ	84.4ᵈ
Cassava/*Aeschynomene*	16.5ᵃ	34.1ᵃᵇ	56.3ᵃ	72.2ᶜ	22.9ᵃ	44.0ᵃᵇ	64.0ᵃ	83.6ᵇ	27.5ᵃ	58.7ᵃ	71.4ᵃᵇ	97.9ᵃ
Sole cassava	15.1ᵈ	29.7ᵈ	50.9ᵇᶜ	67.0ᵉ	18.5ᵉ	35.1ᵈ	53.3ᵈ	69.0ᵉ	18.9ᵉ	36.6ᵈ	54.0ᵈ	70.0ᵉ
Natural fallow	14.8ᵈ	22.2ᵉ	47.5ᶜ	58.7ᶠ	15.7ᶠ	24.8ᵉ	49.5ᵉ	59.7ᶠ	15.2ᶠ	24.3ᶠ	48.7ᵉ	59.8ᶠ
Significant	*	*	*	*	*	*	*	*	*	*	*	*
SE±	0.1	0.3	1.3	0.5	0.2	0.4	0.6	0.6	0.1	0.5	0.5	0.6
CV%	3.4	3.3	8.5	2.6	3.3	3.7	3.6	2.7	3.0	4.0	2.9	2.4

Means followed by the same letter(s) within the same column/factor are not significantly different at 5% level of Probability (DMRT).

Table 2: Effects of pre-rice cropping of cassava/legume intercrop on rice plant height at 3, 6, 9 and 12 WAT in 2011-2013 cropping seasons.

Treatments	2011	2012	2013
Cassava Intercrop (CI)			
Cassava/*Mucuna*	19.0ᵃ	23.0ᵇ	27.0ᵈ
Cassava/cowpea	18.0ᵇ	23.0ᵇ	32.0ᵇ
Cassava/soybean	16.0ᶜ	22.0ᶜ	28.0ᶜ
Cassava/lablab	15.0ᵈ	21.0ᵈ	28.0ᶜ
Cassava/*Aeschynomene*	20.0ᵃ	26.0ᵃ	33.0ᵃ
Sole cassava	12.5ᵉ	14.5ᵉ	18.0ᵉ
Natural fallow	9.3ᶠ	9.0ᶠ	7.0ᶠ
Significant	*	*	*
SE ± CV%	0.2	0.3	0.2

Means followed by the same letter (s) within the same column/factor are not significantly different at 5% level of Probability (DMRT).

Table 3: Effects of pre-rice cropping of cassava/legume intercrop on rice tillers/stool at harvest in 2011-2013 cropping seasons.

Tiller number is an important yield component of rice. The variation in tiller number observed might be due to differences in legumes in the intercropping before rice.

The variation observed in rice grain per panicle might be due to the fact that legumes in intercropping system. The same was obtained by Morteza et al. [1] when rice was planted after different legumes. The higher grain number per panicle obtained in rice grown after cassava/legume intercropping than in rice after fallow might be due to low fertility statusin the fallow system and this result is in consonance with the work of Yousefnia, Tabatabae and. Hashemi. The variation observed in 1000 rice grain weight might be probably that rice grains varies in grain filled capacity which might be the function of legumes in the rotation. This result is in consonance with the work of Morteza

Treatments	2011	2012	2013
Cassava Intercrop (CI)			
Cassava/*Mucuna*	273.0a	291.0ª	307.0ᵇ
Cassava/cowpea	273.0a	297.0ª	326ª
Cassava/soybean	268.0b	276.0ᵇ	298.0ᵇ
Cassava/lablab	225.0c	250.0ᶜ	298.0ᵇ
Cassava/*Aeschynomene*	272.0a	298.1ª	327.0ª
Sole cassava	193.0d	180.0ᵈ	188.0ᶜ
Natural fallow	118.0e	116.0ᵉ	144.0ᵈ
Significant	*	*	*
SE±	1.1	4.5	4.5

Means followed by the same letter(s) within the same column/factor are not significantly different at 5% level of Probability (DMRT).

Table 4: Effects of pre-rice cropping of cassava/legume intercrop on rice panicle (m⁻²) at harvest in 2011-2013 cropping seasons.

Treatments	2011	2012	2013
Cassava Intercrop (CI)			
Cassava/*Mucuna*	126.0ᵃᵇ	128.0ᵇᶜ	129.0ᵇᶜ
Cassava/cowpea	128.0ª	131.0ᵃᵇ	133.0ª
Cassava/soybean	125.0ᵇ	128.0ᵇᶜ	130.0ᵇ
Cassava/lablab	123.0ᶜ	125.0ᶜ	127.0ᶜ
Cassava/*Aeschynomene*	125.0ᵇ	133.0ª	134.0ª
Sole cassava	112.0ᵈ	114.0ᵈ	117.0ᵈ
Natural fallow	96.0ᵉ	94.0ᵉ	93.0ᵉ
Significant	*	*	*
SE±	0.6	1.2	0.8

Means followed by the same letter (s) within the same column/factor are not significantly different at 5% level of Probability (DMRT).

Table 5: Effects of pre-rice cropping of cassava/legume intercrop on rice grains/panicle at harvest in 2011-2013 cropping seasons.

Treatments	2011	2012	2013
Cassava Intercrop (CI)			
Cassava/*Mucuna*	27.7ᵇ	29.0ᵇ	31.0ᵇ
Cassava/cowpea	30.1ª	31.0ª	33.5ª
Cassava/soybean	26.8ᶜᵈ	27.0ᶜ	30.2ᵇ
Cassava/lablab	26.8ᶜᵈ	27.0ᶜ	30.1ᵇ
Cassava/*Aeschynomene*	30.4ª	31.0ª	33.9ª
Sole cassava	26.5ᶜᵈ	26.0ᵈ	27.3ᶜ
Natural fallow	25.5ᵈ	25.0ᵈ	23.6ᵈ
Significant	*	*	*
SE±	0.3	0.3	0.4

Means followed by the same letter(s) within the same column/factor are not significantly different at 5% level of Probability (DMRT).

Table 6: Effects of pre-rice cropping of cassava/legume intercrop on 1000 rice grain weight (g) at harvest in 2011-2013 cropping seasons.

Treatments	2011	2012	2013
Cassava Intercrop (CI)			
Cassava/*Mucuna*	2733.3ᵇ	4565.6ᵇ	4684.0ᵇ
Cassava/cowpea	2933.3ª	4836.9ª	5039.6ª
Cassava/soybean	2200.0ᶜ	3963.5ᶜ	4329.5ᶜ
Cassava/lablab	2066.7ᵈ	3558.6ᵈ	3821.7ᵈ
Cassava/*Aeschynomene*	2800.0ᵇ	4718.5ª	5000.0ª
Sole cassava	1466.7ᵉ	2576.5ᵉ	2670.0ᵉ
Natural fallow	1096.7ᶠ	1042.5ᶠ	1005.0ᶠ
Significant	*	*	*
SE±	66.7	49.2	77.1

Means followed by the same letter(s) within the same column/factor are not significantly different at 5% level of Probability (DMRT).

Table 7: Effects of pre-rice cropping of cassava/legume intercrop on rice grain yield (Kg/ha) at harvest in 2011, 2012 and 2013 cropping seasons.

et al. [1] who observed highest filled grain and highest 1000 rice grain weight in rice grown after potato.

The rice grain yield obtained from rice grown after cassava/legume intercropping varies significantly compared with rice after fallow (rice-rice) which ranged from 46.9-62.2%, 70.7-78.4 and 73.7-80.0% in 2011, 2012 and 2013 respectively. This result agrees with the work of Anders et al. [9] who reported higher rice yield in rice grown after soybean than in rice -wheat rotation. The result is in tandem with the findings of Becker and Johnson [1] that grain yields of rice which had been preceded by a legume fallow were on average about 30% greater than that preceded by a natural fallow. This also is in line with the findings of Morteza et al. [1] who obtained higher rice grain yield in rice after potato than in rice after rice. The lower rice grain yield obtained in natural fallow (rice-rice) might be due to declining factors of productivity; in agreement with the findings of Hobbs and Morris [14] The result was also in line with that of Regmi, Ladha, Pasuquin, Pathak, Hobbs and Shrestha [15]. The result was comparable to the work of Duxbury, Abrol, Gupta and Bronson and [16] and Imtiaz, Hassnain, Azeem Khan, Waqa, Abdul and Mujahid [17] who observed stagnating and even yield declines in long term experiments of rice – wheat rotation system in South Asia.

Conclusions and Recommendation

This study revealed that rice yield and yield components were affected by pre-rice cropping of cassava/legume intercrop. Rice after cassava/cassava/cowpea, cassava/*Mucuna* and cassava/*Aeschynomene* produced taller rice plant height, greater number of tiller/stool, panicle per meter square and grains per panicle. Maximum 1000 rice grain weight and rice yield was also found in rice after cassava/cassava/cowpea, cassava/*Mucuna* and cassava/*Aeschynomene*. Therefore rice field should be not left fallowed. The rice field should be planted with cassava/cowpea and *Aeschynomene* to attain higher performance in subsequent rice cropping.

References

1. Morteza NY, Nicknejad H, Pirdeshti DB, Tari, Nasiri S (2008) Growth, Yield and Yield Traits of Rice Varieties in Rotation with Clover, Potato, Canola and Cabbage in North of Iran. Asian journal of plant Sciences 7: 495-499

2. Faleye T, David J, Olulani T, Dada-Joel, Y Segun A, Wakatsuki T (2012) Impact of Mechanization on Lowland Rice Production in Nigeria Journal of Agricultural Science and Technology: 114-120

3. Imolehin ED, Wada AC (2000) Meeting the rice production and consumption demands of Nigeria with improved technology, Journal International Rice Commission Newsletter 49:33-41

4. IRRI (1993) Rice Almanac, International Rice Research Institute, PO Box 933, Manila, Philippines.

5. Lieberman M, Dyck E (1993) Crop Rotation and Intercropping Strategies for Weed Management. University Of California Sustainable Agriculture Research and Education Program. Ecological Application 3: 92-122

6. Crookston RK (1984) The Rotation Effect, What Causes It To Boost Yields Crops And Soils Magazine 36: 12-14.

7. Mitchell CC, Westerman RL, Brown JR, Peck TR (1991). Overview of Long Term Agronomic Research. Agronomy Journal 83: 24-29.

8. George TJK, Ladha R, Buresh J, Garrity OP (1992). Managing native and legume-fixed nitrogen in lowland rice-based cropping systems Plant and Soil 141: 69-91.

9. Filizadeh Y, Rezazadeh A, Younessi Z (2007) Effects of Crop Rotation and Tillage Depth on Weed Competition and Yield of Rice in the Paddy Fields of Northern Iran. J. Agric. Sci. Technology9: 99-105

10. Anders MM, Winham TE, Watkins KB, Moldehaner KAK, Mcnew RW, et al. (2004) The Effect Of Rotation, Tillage, Fertility and Variety On Rice Yield and Nutrient uptake. Proceedings of the 26th Southern Conservation Tillage Conference for Sustainable Agriculture, Releigh, North Carolina, USA: 26-33.

11. Lixiao Nie, Jing Xiang, Shaobing Peng, Bas AM, Bouman, et al. (2009) Alleviating soil sickness caused by aerobic monocropping, Responses of aerobic rice to fallow, flooding and crop rotation Journal of Food, Agriculture & Environment 7: 7 2 3-7 2 7

12. Toomsan B, Cadisch G, Srichantawong M, Tongsodsaeng, Giller CK, et al. (2000) Biological N2 Fixation and residual N benefit of pre-rice grain legumes and green manures. Netherlands Journal of Agricultural Science 48: 19-29

13. Hobbs PR, Morris ML (1996) Meeting South Asia's future food requirements from rice-wheat cropping systems: Priority issues facing researchers in the post- Green Revolution era. NRG Paper 96-01. CIMMYT, Mexico.

14. Regmi AP, Ladha JK, Pasuquin EM, Pathak H, Hobbs, et al. (2002) Potassium in sustaining yields in a long-term rice-wheat experiment in the Indo-Gangetic Plains of Nepal. Biol. Fert. Soils.

15. Duxbury JM Abrol IP, Gupta RK, Bronson KF (2000) Analysis of long-term soil fertility experiments with rice-wheat rotations in South Asia. Rice-Wheat Consortium for the Indo-Gangetic Plains, New Delhi, India: 7-22.

16. Imtiaz H, Hassnain S, Azeem Khan M, Waqar A, Abdul M, Mujahid MY (2012) Productivity in Rice-Wheat Crop Rotation of punjab: An Application of Typical Farm Methodology Pakistan J. Agric. Res.

17. Hoseini SS (2003) Effect of agronomical treatments on yield and yield components of rice promising line. M.Sc. Thesis, Mazandaran University, Sari, Iran.

12

Genetic Variability, Correlation and Path Coefficient Analysis of Morphological Traits in some Extinct Local Aman Rice (*Oryza sativa* L)

Shajedur Hossain[1], Maksudul Haque MD[2]* and Jamilur Rahman[3]

[1]Supreme Seed Company Limited, Mymenshing, Bangladesh
[2]Plant Breeding Division, Bangladesh Rice Research Institute, Gazipur 1701, Bangladesh
[3]Department of Genetics & Plant Breeding, Sher-e-Bangla Agricultural University, Dhaka 1207, Bangladesh

Abstract

Thirty five local aman rice varieties were evaluated for their variability with regards to yield and yield components. Estimates of heritability and genetic advance in per cent of mean were also obtained for the above traits. In addition, studies on character associations and path coefficients were also undertaken. The highest $\sigma^2 g$ was found for number of root hair (103415.40) and the lowest magnitude of $\sigma^2 g$ was observed in number of primary branches per panicle (1.97). The highest $\sigma^2 p$ was found for number of root hair (109410.31) and the lowest magnitude of $\sigma^2 p$ was observed in number of primary branches per panicle (2.61). High GCV and PCV for number of effective tiller, root weight, number of root hair and grain yield per hill (g) indicated that selection of these traits would be effective. Correlation of Grain yield per hill was found to be highly significant and positive for number of root hair, days to flowering and plant height at both genotypic and phenotypic level and negatively significant for number of secondary branches per panicle at both level. Significant positive correlation of grain yield per hill with number of root hair, days to flowering and plant height implied that selection for these characters would lead to simultaneous improvement of grain yield in rice. Further, yield was observed to be positively associated with panicle bearing tillers and number of filled grains per panicle and these characters were noticed to exert high direct effects on grain yield per plant. High indirect effects of most of the traits were noticed mostly through panicle bearing tillers per hill indicating importance of the trait as selection criteria in crop yield improvement programs.

Keywords: Correlation, Genetic advance, Heritability, Path co-efficients, Rice (*Oryza sativa* L)

Introduction

Rice is a self-pollinated cereal crop belonging to the family Gramineae (synomym-Poaceae) under the order Cyperales and class Monocotyledon having chromosome number 2n=24 [1]. The genus *Oryza* includes a total of 25 recognized species out of which 23 are wild species and two, *Oryza sativa* and *Oryza glaberrima* are cultivated [2]. It can survive as a perennial crop and can produce a ratoon crop for up to 30 years but cultivated as annual crop and grown in tropical and temperate countries over a wide range of soil and climatic condition.

Rice and agriculture are still fundamental to the economic development of most of the Asian countries. In much of Asia, rice plays a central role in politics, society and culture, directly or indirectly employs more people than any other sector. A healthy rice industry, especially in Asia's poorer countries, is crucial to the livelihoods of rice producers and consumers alike. Farmers need to achieve good yields without harming the environment so that they can make a good living while providing the rice-eating people with a high-quality, affordable staple. Underpinning this, a strong rice research sector can help to reduce costs, improve production and ensure environmental sustainability. Indeed, rice research has been a key to productivity and livelihood.

Rice is the second largest produce cereal in the world in 158.3 million hectare area with annual production of about 685.24 million metric tons [3] and also the staple food for over one third of the world's population [4] and more than 90% to 95% of rice is produced and consumed is Asia [5]. Rice (*Oryza sativa* L) is the staple food in Bangladesh, and grown in a wide range of environments ranging from the upland areas like Chittagong Hill Tracts, Sylhet and Garo Hills, with little moisture, to situations where the water is 3-4 meter deep [6]. Bangladesh is ranked as fourth in rice production with annual production of 47.72 million metric ton in the world [3]. Bangladesh has a population density of 977/square km [7] which is the highest in the world. The land scarcity therefore, usually calls for vertical increased in yield or total production. To solve this problem, the production must be increase from less land, with less labor, less water and fewer pesticides.

Rice is the most consumed cereal grain in the world. It is staple food crop for more than half of the world's human population. Yield enhancement is the major breeding objective in rice breeding programmes and knowledge on the nature and magnitude of the genetic variation governing the inheritance of quantitative characters like yield and its components is essential for effective genetic improvement. A critical analysis of the genetic variability parameters, namely, Genotypic Coefficient of Variability (GCV), Phenotypic Coefficient of Variability (PCV), heritability and genetic advance for different traits of economic importance is a major pre-requisite for any plant breeder to work with crop improvement programs. Further, information on correlation co-efficients between grain yield and its component characters is essential for yield improvement, since grain yield in rice is a complex entity and is highly influenced by several component characters. Studies on

---

*Corresponding author: Maksudul Haque MD, Scientific officer (Golden Rice), Plant Breeding Division, Bangladesh Rice Research Institute, Gazipur 1701, Bangladesh, E-mail: maksudulhq@gmail.com

path co-efficient also provide useful information regarding the direct and indirect effects of different yield component characters on grain yield and thus aid in the identification of effective selection criteria for effective yield improvement. The present investigation was undertaken in this context to elucidate information on variability, heritability, genetic advance, character associations and path of effect in promising rice genotypes. A good knowledge of genetic resources might also help in identifying desirable genotypes for future hybridization program.

Materials and Methods

The study was conducted at the experimental farm of Sher-e-Bangla Agricultural University (SAU), Sher-e-Bangla Nagar, Dhaka-1207 during the period from July 2012 to December 2012. The experimental field was located at 90º 33.5′ E longitude and 23º 77.4′ N latitude at an altitude of 9 meter above the sea level. The soil of the experiment site was a medium high land, clay loam in texture and having pH 5.47-5.63. The land was located in Agro-ecological Zone of 'Madhupur Tract' (AEZ No. 28). The climate of the experimental site is sub-tropical characterized by heavy rainfall during April to July and sporadic during the rest of the year. The experimental plots were laid out in randomized complete block design (RCBD). The field was divided into three blocks; representing three replications. Thirty five genotypes were distributed to each plot within each block randomly. The experimental materials of the study comprised of 35 rice genotypes. Thirty five separate strips were made and sprouted seeds were sown on each strip in 14th July of 2012. The experimental plot was fertilized by applying urea, TSP, MP and Gypsum @180,100,70&60 Kg/ha, respectively. Total TSP, MP and Gypsum were applied at final land preparation. Total urea was applied in three installments, at 15 days after transplanting (DAT), 30 DAT and 50 DAT. The seeds were collected from Bangladesh Institute of Nuclear Agriculture, Mymensingh (BINA). The details of these genotypes are given in (Table 1).

Statistical Analysis of Data

Estimation of genotypic and phenotypic variance

Genotypic and Phenotypic Variances were estimated according to the formula given by [8]

Genotypic variance $(\sigma^2 g) = \dfrac{GMS - EMS}{r}$ Where,

GMS=Genotypic mean square

EMS=Error mean square

R=Number of replication

Phenotypic Variance $(\sigma^2 p) = \sigma^2 g + EMS$ Where,

$\sigma^2 g$=Genotypic variance

EMS=Error mean square

Estimation of genotypic and phenotypic co-efficient of variation

Genotypic and Phenotypic co-efficient of variation were estimated according to [9,10]

Genotypic co-efficient of variation,

$(GCV) = \dfrac{\sigma^2 g}{x} \times 100$ Where,

$\sigma^2 g$=Genotypic variance

x=Population mean

Phenotypic co-efficient of variation,

$(PCV) = \dfrac{\sigma^2 p}{x} \times 100$ Where,

$\sigma^2 p$=Phenotypic variance

x=Population mean

Estimation of heritability

Heritability was estimated in broad sense by the formula suggested by [8].

Heritability $(h^2 b) = \dfrac{\sigma^2 g}{\sigma^2 p} \times 100$ Where,

Sl. No.	Indicating Symbol	Genotypes	Source
1	G-1	KathiGoccha	BINA
2	G-2	Hamai	BINA
3	G-3	KhakShail	BINA
4	G-4	Hari	BINA
5	G-5	Tal Mugur	BINA
6	G-6	DakhShail	BINA
7	G-7	MoinaMoti	BINA
8	G-8	Nona Bokhra	BINA
9	G-9	Bogi	BINA
10	G-10	Patnai	BINA
11	G-11	LedraBinni	BINA
12	G-12	Lalanamia	BINA
13	G-13	Hogla	BINA
14	G-14	JamaiNaru	BINA
15	G-15	Jota Balam	BINA
16	G-16	KhejurChori	BINA
17	G-17	Ghunshai	BINA
18	G-18	Malagoti	BINA
19	G-19	BazraMuri	BINA
20	G-20	Nona Kochi	BINA
21	G-21	MoghaiBalam	BINA
22	G-22	Ghocca	BINA
23	G-23	Mondeshor	BINA
24	G-24	MotaAman	BINA
25	G-25	Golapi	BINA
26	G-26	BhuteShelot	BINA
27	G-27	Mowbinni	BINA
28	G-28	KaloMota	BINA
29	G-29	Ponkhiraj	BINA
30	G-30	Jolkumri	BINA
31	G-31	Lalbiroi-31	BINA
32	G-32	Karengal	BINA
33	G-33	SadaGotal	BINA
34	G-34	HoldeGotal	BINA
35	G-35	BRRI Dhan-33	BRRI

BINA: Bangladesh Institute of Nuclear Agriculture
BRRI: Bangladesh Rice Research Institute
Data were collected from 5 hills of each genotype on the following parameters: Plant height at maturity (cm), Number of effective tillers, Length of panicle (cm), Days to flowering (DAS), Number of primary branches per panicle, Number of secondary branches per panicle, Number of unfilled grain per panicle, Number of filled grain per panicle, 1000 grain weight, Root length, Number of root hair, Root weight and Grain yield per hill (g).

Table 1: List of thirty five rice genotypes along with their sources.

σ²g=Genotypic variance

σ²p=Phenotypic variance

Estimation of genetic advance

Estimation of Genetic Advance was done following formula given by [8].

Genetic Advance (GA)=h²b.K.σp   Where,

h²b=Heritability

K=Selection differential, the value of which is 2.06 at 5% selection intensity; and

σp=Phenotypic standard deviation.

Estimation of correlation coefficients

The Genotypic and Phenotypic correlation coefficients between yield and different yield contributing characters were estimated as:

$$\text{Genotypic correlation} = \frac{\text{Cov}(g)1.2}{\sqrt{\sigma^2(g)1.\sigma^2(g)2}} \quad \text{Where,}$$

Cov(g)(xy)=Genotypic covariance between the variables X and Y

σ2(g)1=Genotypic variance of the variable X1

σ2(g)2=Genotypic variance of the variable X2

Similarly,

$$\text{Phenotypic correlation} = \frac{\text{Cov}(ph)1.2}{\sqrt{\sigma^2(ph)1.\sigma^2(ph)2}} \quad \text{Where,}$$

Cov(ph)(xy)=Phenotypic covariance between the variables X and Y

σ2(ph)1=Phenotypic variance of the variable X1

σ2(ph)2=Phenotypic variance of the variable X2

Estimation of path coefficients

Path coefficient analysis was done according to the procedure employed by [11] also quoted in [10], using phenotypic correlation coefficient values. In path analysis, correlation coefficients between yield and yield contributing characters were partitioned into direct and indirect effects of yield contributing characters on grain yield per hectare.

After calculating the direct and indirect effect of the characters, residual effects (R) was calculated by using the formula given below [10].

$$P^2_{RY} = 1 - (r_{1.y}P_{1.y} + r_{2.y}P_{2.y} + \ldots\ldots\ldots\ldots\ldots + r_{12.y}P_{12.y})$$

Where,

$P^2_{RY} = R^2$ and hence residual effect, $R = (P^2_{RY})^{1/2}$

$P_{i.y}$ = Direct effect of the i th character on yield y.

$r_{i.y}$ = Correlation of the i th character with yield y.

Result and Discussion

The extent of variability for any character is very important for the improvement of a crop through breeding. The estimates of genotypic variation (σ²g), phenotypic variation (σ²p), genotypic coefficient of variation (GCV), phenotypic coefficient of variation (PCV), heritability (h²) and genetic advance (GA) for different characters have been presented in Table 2. The highest σ²g was found for number of root hair (103415.40) and the lowest magnitude of σ²g was observed in number of primary branches per panicle (1.97). The highest σ²p was found for number of root hair (109410.31) and the lowest magnitude of σ²p was observed in number of primary branches per panicle (2.61). Wide variability existed in anther length, stigma length and percent exerted stigma. The genetic variation constituted a high proportion of the total variation for these traits. Thus, selection for these characters is expected to be highly effective.

The GCV and PCV were the highest for number of effective tiller (54.39and 56.33) followed by root weight (50.97and 51.26), number of root hair (28.00 and 28.80) and grain yield per hill (g) (26.51 and 26.57). High GCV and PCV for number of effective tiller, root weight, number of root hair and grain yield per hill (g) indicated that selection of these traits would be effective. The GCV and PCV were the lowest for days to flowering (3.28and 3.43) and plant height (cm) (5.69 and 5.76). PCV were slightly higher than GCV in case of all the traits, indicating presence of environmental influence to some degrees in the phenotypic expression of the characters [12] also reported similar result (Table 2 and Figure 1).

Characters	Grand mean	(σ²g)	(σ²p)	(σ² e)	h²_b (%)	GCV (%)	PCV (%)	GA	GA (%)
PH	161.54	84.47	86.71	2.24	97.42	5.69	5.76	18.69	11.57
NET	5.04	7.50	8.05	0.54	93.25	54.39	56.33	5.45	108.20
LP	28.95	4.93	6.07	1.14	81.28	7.67	8.51	4.12	14.24
DF	108.66	12.71	13.92	1.20	91.35	3.28	3.43	7.02	6.46
NPBP	10.71	1.97	2.61	0.64	75.52	13.10	15.08	2.51	23.46
NSBP	27.25	42.99	43.96	0.97	97.79	24.06	24.33	13.36	49.02
NFGP	120.36	181.79	187.50	5.71	96.96	11.20	11.38	27.35	22.72
NUFGP	31.38	31.04	32.16	1.11	96.54	17.75	18.07	11.28	35.93
1000GW	22.91	2.19	3.08	0.89	71.00	6.46	7.67	2.57	11.21
RL	9.85	2.57	4.54	1.98	56.46	16.26	21.64	2.48	25.17
NRH	1148.66	103415.40	109410.31	5994.91	94.52	28.00	28.80	644.06	56.07
RW	57.05	845.41	855.07	9.66	98.87	50.97	51.26	59.56	104.40
GYH	23.56	39.02	39.20	0.18	99.53	26.51	26.57	12.84	54.48

PH: Plant height (cm); NET: Number of effective tiller; LP: Length of panicle; DF: Days to flowering; NPBP: Number of primary branches per panicle, NSBP: Number of secondary branches per panicle; NFGP: Number of filled grain per panicle; NUFGP: Number of unfilled grain per panicle; 1000 GW: 1000 grain weight; RL: Root length; NRH: Number of root hair; RW: Root weight, GYH: Grain yield per hill (g); **σ²g**: Genotypic variance; **σ²p**: Phenotypic variance; σ² e: Environmental variance; **h²_b**: Heritability (broad sense); GCV: Genotypic Coefficient of Variation; PCV: Phenotypic Coefficient of Variation; GA: Genetic Advance

Table 2: Estimation of genetic parameters in thirteen characters of 35 genotypes in rice.

Heritability estimates were high for all the characters studied except root length, 1000 grain weight and number of primary branches per panicle. The values were especially high for grain yield per hill (g), root weight and number of secondary branches per panicle [8] suggested that high heritability combined with high genetic advance is an indicative of additive gene action and selection based on these parameters would be more reliable. In the present investigation, high heritability estimates in conjunction with high genetic advance in present of mean were observed for number of effective tiller, root weight and number of root hair (Figure 2).

Yield is a complex product being influenced by several independent quantitative characters. Breeders always look for variation among traits to select desirable types. Some of these characters are highly associated among themselves and with seed yield. The analysis of the relationships among these characters and their associations with seed yield is essential to establish selection criteria. When more characters are involved in correlation study it becomes difficult to ascertain the characters which really contribute toward yield. The path coefficient analysis under such situations helps to determine the direct contribution of these characters and their indirect contributions via other characters. Selection for yield *per se* may not be effective unless the other yield components were having direct or indirect influence on it and are taken into consideration. When selection pressure is exercised for improvement of any character highly associated with yield, it simultaneously affects a number of other correlated traits. Genotypic and phenotypic correlation coefficients were calculated as according to [13].

Correlation coefficient

The genotypic and phenotypic correlations among yield and yield contributing characters in rice are shown in Table 3. It is evident that in majority of the cases, the genotypic correlation coefficients were higher than their phenotypic correlation coefficients indicating that of a strong inherent association between the characters studied and suppressive effect of the environment modified the phenotypic expression of these characters by reducing phenotypic correlation values. In few cases, however, phenotypic correlation coefficients were same with or higher than their corresponding genotypic correlation coefficients suggesting that both environmental and genotypic correlation in these cases act in the same direction and finally maximize their expression at phenotypic level. Accordingly, [14] reported that the genotypic correlations were greater than the phenotypic values in medium duration rice varieties (Table 3).

Character association

Correlation of grain yield per hill was found to be highly significant and positive for number of root hair, days to flowering and plant height at both genotypic and phenotypic level and negatively significant for Number of secondary branches per panicle at both level. Significant positive correlation of grain yield per hill with number of root hair, days to flowering and plant height imply that selection for these characters would lead to simultaneous improvement of grain yield in rice. Plant height recorded significant negative correlation with number of secondary branches per panicle followed by number of primary branches per panicle and number of root hair at genotypic and phenotypic level. Number of effective tiller showed significant positive correlation with number of root hair at both genotypic and phenotypic level. It is observed that length of panicle was significantly and positively associated with root length followed by root weight at both genotypic and phenotypic level.

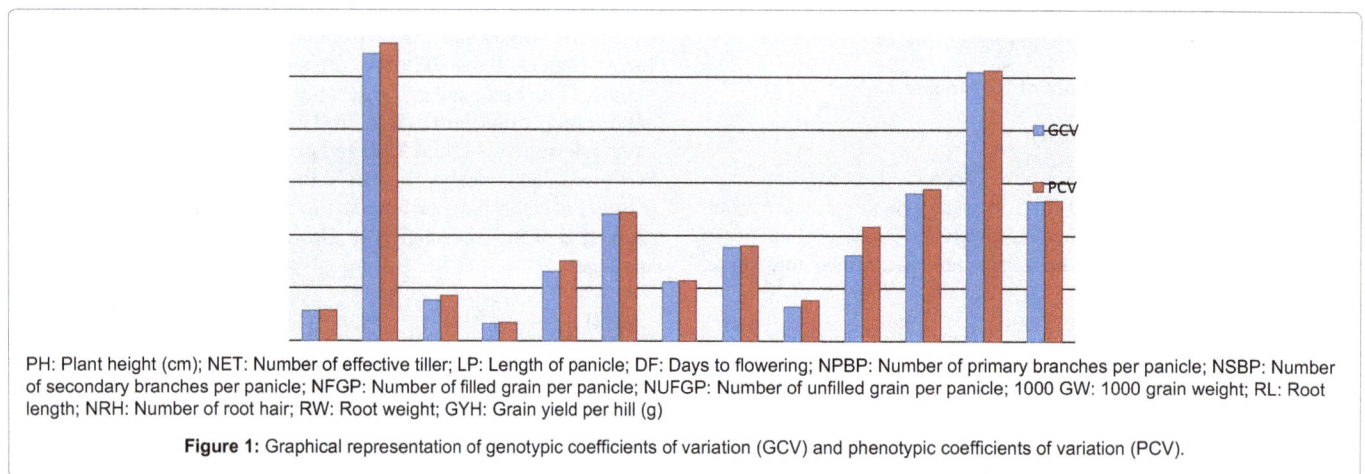

PH: Plant height (cm); NET: Number of effective tiller; LP: Length of panicle; DF: Days to flowering; NPBP: Number of primary branches per panicle; NSBP: Number of secondary branches per panicle; NFGP: Number of filled grain per panicle; NUFGP: Number of unfilled grain per panicle; 1000 GW: 1000 grain weight; RL: Root length; NRH: Number of root hair; RW: Root weight; GYH: Grain yield per hill (g)

Figure 1: Graphical representation of genotypic coefficients of variation (GCV) and phenotypic coefficients of variation (PCV).

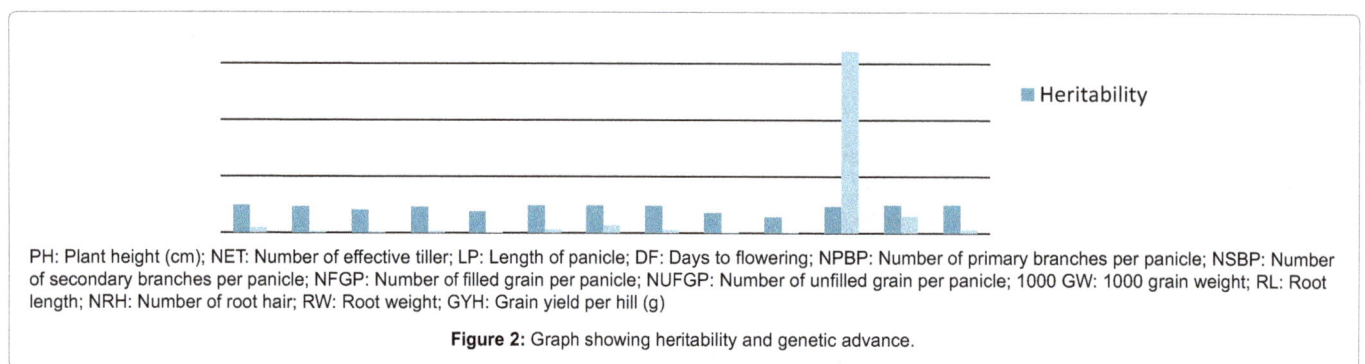

PH: Plant height (cm); NET: Number of effective tiller; LP: Length of panicle; DF: Days to flowering; NPBP: Number of primary branches per panicle; NSBP: Number of secondary branches per panicle; NFGP: Number of filled grain per panicle; NUFGP: Number of unfilled grain per panicle; 1000 GW: 1000 grain weight; RL: Root length; NRH: Number of root hair; RW: Root weight; GYH: Grain yield per hill (g)

Figure 2: Graph showing heritability and genetic advance.

Characters		NET	LP	DF	NPBP	NSBP	NFGP	NUFGP	1000 GW	RL	NRH	RW	GYH
PH	Rg	0.201	0.985**	0.648**	-0.624**	-0.996**	-0.158	-0.314	0.030	0.983**	-0.616**	0.151	0.772**
	Rp	0.192	0.882**	0.620**	-0.539	-0.970**	-0.157	-0.302	0.041	0.955**	-0.590**	0.145	0.763**
NET	Rg		-0.638**	-0.427**	-0.851**	-0.044	-0.728**	0.038	0.165	0.303	0.802**	0.309	0.161
	Rp		-0.577**	-0.382**	-0.810**	-0.041	-0.691**	0.037	0.134	0.209	0.753**	0.289	0.161
LP	Rg			0.103	0.442**	0.110	-0.323	0.235*	0.111	0.829**	0.038	0.514**	0.039
	Rp			0.102	0.325	0.098	-0.281	0.204	0.061	0.565**	0.032	0.459**	0.035
DF	Rg				0.200	-0.543**	0.663**	0.186	-0.516**	0.233	0.598**	-0.309	0.846**
	Rp				0.140	-0.503**	0.614**	0.171	-0.432**	0.157	0.528**	-0.295	0.809**
NPBP	Rg					0.543**	0.817**	0.921**	0.515**	0.426*	0.114	-0.446**	0.060
	Rp					0.470**	0.686**	0.789**	0.403*	0.308	0.106	-0.369*	0.051
NSBP	Rg						0.563**	0.770**	-0.124	-0.127	0.152	-0.214	-0.218
	Rp						0.545**	0.747**	-0.118	-0.081	0.133	-0.210	-0.215
NFGP	Rg							-0.071	0.433**	0.831**	0.182	-0.162	0.288
	Rp							-0.069	0.353*	0.612**	0.175	-0.160	0.282
NUFGP	Rg								0.531**	0.290	0.505**	0.091	0.517**
	Rp								0.442**	0.239	0.478**	0.089	0.507**
1000GW	Rg									0.963**	-0.038	-0.283	0.176
	Rp									0.630**	-0.025	-0.232	0.151
RL	Rg										0.839**	-0.125	0.645**
	Rp										0.630**	-0.088	0.479**
NRH	Rg											-0.094	0.888**
	Rp											-0.092	0.859**
RW	Rg												0.031
	Rp												0.030

**: Significant at 1%. *: Significant at 5%.
PH: Plant height (cm); NET: Number of effective tiller; LP: Length of panicle; DF: Days to flowering; NPBP: Number of primary branches per panicle; NSBP: Number of secondary branches per panicle; NFGP: Number of filled grain per panicle; NUFGP: Number of unfilled grain per panicle; 1000 GW: 1000 grain weight; RL: Root length , NRH: Number of root hair; RW: Root weight; GYH: Grain yield per hill (g); Rg: Genotypic correlation; Rp: Phenotypic correlation.

Table 3: Genotypic and phenotypic correlation coefficients among different pairs of yield and yield contributing characters for different genotypes of rice.

Characters	PH	NET	LP	DF	NPBP	NSBP	NFGP	NUFGP	1000 GW	RL	NRH	RW	GYH
PH	0.240	0.127	-0.184	0.466	-0.110	0.092	-0.027	-0.170	-0.005	0.182	0.144	0.016	0.772**
NET	0.048	0.635	0.119	-0.307	-0.150	0.004	-0.128	0.020	-0.027	0.056	-0.141	0.033	0.161
LP	0.236	-0.405	-0.186	0.074	0.078	-0.010	-0.057	0.127	-0.018	0.153	-0.008	0.055	0.039
DF	0.155	-0.271	-0.019	0.719	0.035	0.050	0.117	0.101	0.087	0.043	-0.140	-0.033	0.846**
NPBP	-0.149	-0.540	-0.082	0.143	0.177	-0.050	0.144	0.501	-0.087	0.079	-0.026	-0.048	0.060
NSBP	-0.239	-0.027	-0.020	-0.390	0.096	-0.093	0.099	0.419	0.021	-0.023	-0.035	-0.023	-0.218
NFGP	-0.037	-0.462	0.060	0.477	0.144	-0.052	0.176	-0.038	-0.073	0.154	-0.042	-0.017	0.288
NUFGP	-0.075	0.024	-0.043	0.133	0.163	-0.071	-0.012	0.544	-0.090	0.053	-0.118	0.009	0.517**
1000GW	0.007	0.104	-0.020	-0.371	0.091	0.011	0.076	0.289	-0.169	0.178	0.008	-0.030	0.176
RL	0.236	0.192	-0.154	0.167	0.075	0.011	0.146	0.157	-0.163	0.185	-0.197	-0.013	0.645**
NRH	-0.147	0.382	-0.007	0.430	0.020	-0.014	0.032	0.274	0.006	0.155	-0.235	-0.010	0.888**
RW	0.036	0.196	0.096	-0.222	-0.079	0.019	-0.028	0.049	0.048	-0.023	0.022	0.108	0.031

Residual effect: 0.137. **: Significant at 1%. *: Significant at 5%.
PH: Plant height (cm); NET: Number of effective tiller; LP: Length of panicle; DF: Days to flowering; NPBP: Number of primary branches per panicle; NSBP: Number of secondary branches per panicle; NFGP: Number of filled grain per panicle; NUFGP: Number of unfilled grain per panicle; 1000 GW: 1000 grain weight; RL: Root length; NRH: Number of root hair; RW: Root weight; GYH: Grain yield per hill (g).

Table 4: Genotypic path coefficient analysis showing direct and indirect effects of different characters on yield of 35 rice genotypes.

Path analysis

Associations of characters determined by correlation coefficient may not provide an exact picture of the relative importance of direct and indirect influence of each of yield components on grain yield per plant. As a matter of fact, in order to find out a clear picture of the interrelationship between grain yield per plant and other yield attributes, direct and indirect effects were worked out using path analysis at both genotypic and phenotypic level.

Genotypic path coefficient analysis showing direct and indirect effects of different characters and the results are presented in Table 4.

From the path coefficient analysis showed that days to flowering had maximum direct effect (0.719) on yield followed by number of effective tiller (0.635), number of unfilled grain per panicle (0.544). [13] reported that the number of branches per panicle (0.424) had the highest positive direct effect on grain yield followed by number of filled grains per panicle (0.411), and days to 50% flowering (0.07).The lowest direct effect on grain yield was exhibited by root weight(0.108) followed by number of filled grain per panicle(0.176). The highest negative indirect effects on grain yield were obtained by number of root hair (-0.235). [15] worked out on path co-efficient in 128 aromatic rice accessions for seven traits and observed greatest positive direct effect of 1000-grain

Characters	PH	NET	LP	DF	NPBP	NSBP	NFGP	NUFGP	1000 GW	RL	NRH	RW	GYH
PH	0.247	0.110	-0.461	0.329	-0.278	0.274	-0.004	-0.188	-0.020	0.519	0.206	0.029	0.763**
NET	0.047	0.575	0.302	-0.203	-0.418	0.011	-0.018	0.023	-0.067	0.113	-0.263	0.058	0.161
LP	0.218	-0.332	-0.523	0.054	0.168	-0.027	-0.007	0.127	-0.030	0.307	-0.011	0.093	0.035
DF	0.153	-0.219	-0.053	0.531	0.072	0.142	0.016	0.106	0.218	0.085	-0.184	-0.060	0.809**
NPBP	-0.133	-0.466	-0.170	0.074	0.517	-0.132	0.018	0.492	-0.203	0.167	-0.037	-0.075	0.051
NSBP	-0.239	-0.023	-0.051	-0.267	0.243	-0.282	0.014	0.466	0.059	-0.044	-0.046	-0.042	-0.215
NFGP	-0.038	-0.397	0.147	0.326	0.354	-0.154	0.027	-0.043	-0.178	0.332	-0.061	-0.032	0.282
NUFGP	-0.074	0.021	-0.106	0.090	0.407	-0.211	-0.001	0.623	-0.223	0.129	-0.167	0.018	0.507**
1000GW	0.010	0.077	-0.031	-0.229	0.208	0.033	0.009	0.275	-0.505	0.342	0.008	-0.047	0.151
RL	0.236	0.120	-0.295	0.083	0.159	0.022	0.016	0.149	-0.318	0.543	-0.220	-0.017	0.479**
NRH	-0.145	0.433	-0.016	0.280	0.054	-0.037	0.004	0.298	0.012	0.342	-0.349	-0.018	0.859**
RW	0.035	0.166	-0.240	-0.156	-0.190	0.059	-0.004	0.055	0.117	-0.047	0.032	0.203	0.030

Residual effect: 0.0055

PH: Plant height (cm); NET: Number of effective tiller; LP: Length of panicle; DF: Days to flowering; NPBP: Number of primary branches per panicle; NSBP: Number of secondary branches per panicle; NFGP: Number of filled grain per panicle; NUFGP: Number of unfilled grain per panicle; 1000 GW: 1000 grain weight; RL: Root length; NRH: Number of root hair; RW: Root weight; GYH: Grain yield per hill (g).

Table 5: Phenotypic path coefficient analysis showing direct and indirect effects of different characters on yield of rice.

weight followed by number of ear-bearing tillers per plant, number of filled grains per panicle and number of days to 50% flowering on grain yield. However, 1000-grain weight had on significant correlation with grain yield per plant due to its negative indirect on grain yield plant through number of filled grain per panicle and plant height. The results prescribed that highly significant positive correlation with positive direct effect was observed in days to flowering, number of effective tiller, number of unfilled grain per panicle, number of filled grain per panicle and root weight. The residual effect of the present study was 0.137 [16] (Table 4).

Phenotypic path coefficient analysis showing direct and indirect effects of different characters and the results are presented in Table 5. From the path coefficient analysis showed that number of unfilled grain per panicle had maximum direct effect (0.623) on yield followed by number of effective tiller (0.575), root length (0.543). The lowest direct effect on grain yield was exhibited by number of filled grain per panicle (0.027) followed by root weight (0.203). The highest negative indirect effects on grain yield were obtained by length of panicle (-0.523). The results prescribed that highly significant positive correlation with positive direct effect was observed in number of unfilled grain per panicle, number of effective tiller, root length, root weight and number of filled grain per panicle. The residual effect of the present study was 0.0055 (Table 5).

Conclusion

Thirty-four local varieties of rice along with one standard check were evaluated for thirteen yield and yield attributing traits. Results of the present investigation on variability, heritability and genetic advance indicated a scope for improvement of grain yield through selection. Further, studies on character association and path co-efficient revealed the importance of panicle bearing tillers per hill and number of filled grains per panicle as selection criteria for effective yield improvement. The study also indicated the need for balanced selection in light of negative association of panicle bearing tillers per hill with number of filled grains per panicle; and number of filled grains per panicle with 1000-grain weight in crop yield improvement programs.

References

1. Hooker JD (1979) The Flora of British India. Vol 2L Reeve Co Kent, England 25.

2. Brar DS, Khush GS (2003) Utilization of wild species of genus Oryzae in rice improvement. In: J S Nanda, Sharma SD Monograph on Genus Oryzae 283-309.

3. Anonymous (2011) Food and Agricultural Organization of the United Nations. Production: FAOSTAT.

4. Poehlman JM, Sleper DA (1995) Breeding Field Crops. Panima Publishing Corporation, New Delhi, India 278.

5. Virmani SS (1996) Hybrid Rice. Adv Agron 57: 379.

6. Alim A (1982) Bangladesh Rice 35-39.

7. BBS (Bangladesh Bureau of Statistics) (2009) Monthly Statistical Bulletin August, 2009. Bangladesh Bur. Stat Div Min Planning, Dhaka, Bangladesh P. 54.

8. Johnson HW, Robinson HF, Comstock RE (1955) Estimates of genetic and environmental variability in Soybean. Agronomy J 47: 314-318.

9. Burton GW (1952) Quantitative inheritance in grasses. Proc 6th Int Grassland, Cong 1: 277-283.

10. Singh RK002C Chaudhary BD (1985) Biometrical methods of quantitative genetic analysis. Haryana J Hort Sci 12: 151-156.

11. Deway DR, Lu KH (1959) A correlation and path coefficient analysis of components of crested wheat grass seed production. Agron J 51: 515-518.

12. Akanda MAL, Alam MS, Uddin MM (1997) Genetic variability, correlation and path analysis in maize (Zea mays L) inbreds. Bangladesh J Pl Breed Genet 10: 57-61.

13. Miller PJ, Williams JC, Robinson HF, Comstock RE (1958) Estimates of genotypic and environmental variances and covariances in upland cotton and their implications in selection. Agron J 50: 126-131.

14. Bai NR, Devika R, Regina A, Joseph CA (1992) Correlation of yield and yield components in medium duration rice cultivars. Environment and Ecology 10: 469-470.

15. Mahto RN, Yadava MS, Mohan KS (2003) Genetic variation, character association and path analysis in rainfed upland rice. Indian J Dryland Agric Res and Dev 18: 196-198.

16. Patil PV, Sarawgi AK (2005) Studies on genetic variability, aromatic rice accessions. Annals Pl Physiol 19: 92-95.

13

Complex Regulatory Networks of Flowering Time in Rice

Yang-Seok Lee and Gynheung An*

Crop Biotech Institute & Graduate School of Biotechnology, Kyung Hee University, Yongin 446-701, Korea

Abstract

Rice flowering is inhibited when days are long during Spring and early Summer. This phenomenon is mediated by several independent pathways. Several genes are preferentially expressed under such conditions, including Grain yield and heading date 7 (Ghd7), Heading date 1 (Hd1), Heading date 5 (Hd5), and Heading date 6 (Hd6). By contrast, *Oryza sativa* Constans-like 4 (OsCOL4) deters flowering regardless of day length. The AP2-like genes Supernumerary Bract (SNB) and *Oryza sativa* Indeterminate Spikelet1 (OsIDS1) constitutively inhibit flowering. However, as days become shorter in late Summer, flowering is induced. Although Hd1 protein functions as a repressor under long days, it becomes a promoter under short days (SD). Both OsMADS50 and OsDof12 induce flowering specifically under LD, while *Oryza sativa* Indeterminate 1 (OsId1) causes flowering regardless of day length. Levels of expression by the repressors decrease as plants mature. For example, Ghd7 transcripts are more abundant in young plants. Transcripts of SNB and OsIDS1 are degraded by miR172, which is induced in older plants. Most of the upstream signals are transferred to Early heading date 1 (Ehd1), an immediate upstream regulator of the florigen genes Heading data 3a (Hd3a) and Rice FT 1 (RFT1). However, some signals directly turn on the florigens that are transferred to the shoot apical meristem, where the reproductive transition occurs.

Keywords: Day length; Florigens; Flowering time; Regulatory genes

Introduction

Plant yields are maximized and the exchange of genetic information among individuals is enhanced when flowering occurs at the optimal time. This phenomenon is modulated by environmental factors such as day length as well as by endogenous cues, e.g., developmental age. Rice (*Oryza sativa*) is a facultative short-day (SD) plant that flowers earlier under SD conditions. Photoreceptors and circadian clock genes play roles in modulating the expression of genes that regulate this timing. Flowering is inhibited during the early stage of development to allow sufficient vegetative growth. Various transcription factors (TFs) modulate flowering time by several independent pathways. Chromatin remodeling factors and microRNAs also control that activity. Signals are merged to Ehd1 or directly to florigens that induce flowering when an adequate amount has accumulated. Summaries of rice research have covered photoperiodic flowering [1-4] comparative biology [5-7], and the florigen mechanism [2]. Other reviews have focused on the relationships among photoperiod, epigenetic regulation [8] (Sun et al.), and temperature [9]. Here, we review integrations among various flowering pathways.

Review

Several rice cultivars have been domesticated for cultivation in a range of growing regions. Flowering time, often defined by heading date, is one of the most important agronomic traits that can be adapted to a new environment to increase yields. Genes that influence flowering time have been identified through analyses of Quantitative Trait Loci (QTL) from cultivars growing in different geographical locations [10]. Genes that control this regulatory network have also been found by studying either naturally occurring mutants or those that are generated through T-DNA or Tos17 insertions [11-13].

Table 1 summarizes those genes and roles.

Florigens

Heading date 3a (Hd3a) protein is a florigen in rice. Generated in the leaf phloem and transferred to the shoot apical meristem (SAM), this protein induces reproductive development when a plant reaches a certain growth stage [14]. Hd3a was initially identified by QTL mapping that utilized a population derived from a cross between the photoperiod-sensitive cultivar 'Nipponbare' and the photoperiod-insensitive cultivar 'Kasalath' [15]. Through high-resolution mapping, the candidate gene was located within an approximately 20-kb region that includes four genes [16]. One of those is highly homologous to Arabidopsis Flowering Locus T (FT), which prompts flowering. A genomic DNA fragment containing the gene from 'Kasalath' causes early flowering when introduced to 'Nipponbare', indicating that the FT-homologous gene is Hd3a [17].

Among the 13 rice proteins homologous to FT, RFT1 also functions as a florigen [18,19]. RFT1 is adjacent to Hd3a, separated by 11.5 kb on Chromosome 6, suggesting that the two arose through tandem duplication [18-20]. The RFT1 amino acid sequence shares 91% identity with Hd3a [16,18,21,22] Chardon and Damerval). Transgenic plants expressing RNAi of both Hd3a and RFT1 do not flower until 300 days after sowing (DAS) [23]. Overexpression of Hd3a and RFT1 causes flowering at the callus induction stage [17,24]. These results are evidence that Hd3a and RFT1 are essential for flowering in rice [25]. Overexpression of FTL1, another FT-like gene in rice, also induces early flowering at the callus stage [22].

Hd3a is preferentially functional under SD whereas RFT1 has an LD role [19,23,25]. During short photoperiods, flowering of Hd3a RNAi plants is delayed by approximately one month, whereas timing for RFT1 RNAi plants is similar to that for the wild type, or WT [19,23,25]. By contrast, under LD conditions, RFT1 RNAi plants bolt later while the flowering time for Hd3a RNAi plants is similar to the WT [19]. Although temporal expression of both Hd3a and RFT1

***Corresponding author:** Gynheung An, Crop Biotech Institute & Graduate School of Biotechnology, Kyung Hee University, Yongin 446-701, Korea
E-mail: genean@khu.ac.kr

Gene name	TIGR locus ID	Function	Note	Elucidation method	Pathways
Hd3a	LOC_Os06g06320	Florigen	Phosphatidylethanolamine-binding protein (PEBP)	QTL/Overexpression/RNAi	
RFT1	LOC_Os06g06300	Florigen	Phosphatidylethanolamine-binding protein (PEBP)	RNAi/QTL/Overexpression	
OsMADS14	LOC_Os03g54160	Floral transition	AP1 homolog	Overexpression	
OsMADS15	LOC_Os07g01820	Floral transition	AP1 homolog	Overexpression	
Hd1/Se1	LOC_Os06g16370	SD activator/LD repressor	Constans-like (COL)	QTL/Tos17/T-DNA/ Overexpression	Hd3a
Ehd1	LOC_Os10g32600	Signal integrator	B-type response regulator	QTL/RNAi/Overexpression	Hd3a
Hd4/Ghd7/Lhd4	LOC_Os07g15770	LD repressor	Constans-like (COL)	QTL/Overexpression	Ehd1
Hd5/DTH8/Ghd8/LHD1/ Ds9	LOC_Os08g07740	SD activator/LD repressor	HAP3 subunit	QTL/Ds	Hd1,Ehd1
Hd6/CK2α	LOC_Os03g55389	LD repressor	Caseine kinase 2α	QTL	Hd1, Hd3a
Hd16/EL1/CKI	LOC_Os03g57940	LD repressor	Caseine kinase I	QTL/T-DNA	Ghd7
OsMADS50	LOC_Os03g03100	LD activator	SOC1 homolog	T-DNA/Overexpression	OsLFL1
OsMADS56	LOC_Os10g39130	LD repressor	SOC1 homolog	Overexpression	OsLFL1
OsDof12	LOC_Os03g07360	LD activator	DNA-binding with one zinc finger protein	Overexpression	Hd3a
OsMADS51	LOC_Os01g69850	SD activator	Type I MADS-box protein	T-DNA/Overexpression	OsGI,Ehd1
OsCO3	LOC_Os09g06464	SD repressor	CONSTANS-like (COL)	Overexpression	Hd3a
Ehd2/OsId1/RID1	LOC_Os10g28330	Constitutive activator	C2H2-type zinc finger protein	QTL/T-DNA/RNAi	Ehd1
Ehd4	LOC_Os03g02160	Constitutive activator	CCCH-type zinc finger protein	QTL	Ehd1
OsCOL4	LOC_Os02g39710	Constitutive repressor	CONSTANS-like (COL)	T-DNA/Overexpression	Ehd1
OsLF	LOC_Os05g46370	Constitutive repressor	Atypical HLH protein	Activation tagging by T-DNA	OsGI
SNB	LOC_Os07g13170	Constitutive repressor	AP2-like	Overexpression	miR172,Ehd1
OsIDS1	LOC_Os03g60430	Constitutive repressor	AP2-like	Overexpression	miR172,Ehd1
OsLFL1	LOC_Os01g51610	Constitutive repressor	B3 domain transcription factor	Overexpression	Ehd1
OsGI	LOC_Os01g08700	Circadian rhythm/ Flowering activator	GIGANTEA ortholog	γ-ray radiation/RNAi/T-DNA	
Hd2/OsPRR37	LOC_Os07g49460	Circadian rhythm/ SD activator/LD repressor	Pseudo response regulator	QTL/T-DNA	Hd3a
Hd3b/Hd17/Ef7/OsELF3	LOC_Os06g05060	Circadian rhythm/LD activator	ELF3 homolog	QTL/T-DNA/RNAi	Hd1,Ghd7,OsGI
OsEF3	LOC_Os01g38530	Circadian rhythm/ Flowering activator	ELF3 homolog	T-DNA	
OsPhyA	LOC_Os03g51030	SD activator/LD repressor	Phytochrome	Tos17/T-DNA	
OsPhyB	LOC_Os03g19590	Constitutive repressor	Phytochrome	γ-ray radiation/T-DNA	
OsPhyC	LOC_Os03g54084	LD repressor	Phytochrome	Tos17/T-DNA	
SE5	LOC_Os06g40080	SD activator/LD repressor	Heme oxygenase	γ-ray radiation	
OsCRY2	LOC_Os02g41550	Constitutive activator	Cryptochrome	antisense	
OsVIL1/OsVIL2	LOC_Os12g34850	Flowering activator	VIN3-like (PRC2 complex)	RNAi	OsLF
OsVIL2/LC2/OsVIL3	LOC_Os02g05840	Flowering activator	VIN3-like (PRC2 complex)	T-DNA	OsLFL1
OsEMF2b	LOC_Os04g08034	Flowering activator	Su(z)12-like (PRC2 coomplex)	T-DNA/RNAi/Overexpression	OsLFL1
OsTrx1	LOC_Os09g04890	LD activator	TrxG	T-DNA	Ghd7
Ehd3	LOC_Os08g01420	Flowering activator	Plant Homeodomain (PHD) finger motif	QTL	Ghd7
SDG724/LVP1	LOC_Os09g13740	LD activator	SET domain	Somaclonal variation by tissue culture	OsMADS50
SDG711	LOC_Os06g16390	LD repressor	SET domain	RNAi/Overexpression	
SDG718	LOC_Os03g19480	SD activator	SET domain	RNAi	

Table 1: Flowering regulators in rice.

peaks at 35 DAS under SD conditions, transcript levels are higher for the former [23]. Because RFT1 transcripts are more abundant at later developmental stages in Hd3a RNAi plants under SD [23], RFT1 appears to function as a florigen when Hd3a is suppressed. In temperate regions, rice is grown under LD conditions, where RFT1 is preferentially functional as a florigen. An indica cultivar, 'Nona Bokra', flowers extremely late (about 200 DAS) in its natural habitat in Tsukuba, Japan (36°N) but bolts at about 60 DAS under SD conditions

[26] (Uga et al.). Overexpression of the RFT1 allele from 'Nipponbare' or 'Kasalath' induces extremely early flowering while overexpression of that allele from 'Nona Bokra' does not [27] possibly because an amino acid substitution of E to K at the 150th position of RFT1 causes a functional defect in that cultivar. In addition, polymorphisms in the RFT1 promoter region reduce this expression in 'Nona Bokra' [27].

Florigens are produced in leaves and transported to the SAM where they induce reproductive developments [14,19]. After their arrival, they bind to the 14-3-3 proteins GF14b and GF14c [28,29], and the complex interacts with OsFD1 [2]. Crystallographic analyses have also shown that these three proteins form a florigen activation complex [2] that is composed of two Hd3a molecules, the GF14c (14-3-3 protein) dimer, and two OsFD1s. That complex binds to the promoter of OsMADS15 to induce expression of the target gene [2]. Expression of OsMADS14 and OsMADS15 is decreased in the SAM of Hd3a RNAi or RFT1 RNAi plants [19,23,25]. Whereas overexpression of OsMADS14 causes extremely early flowering at the callus induction stage [30], flowering by GF14c overexpression (OX) plants is delayed approximately 20 d. The gf14c/GF14c heterozygous mutants bolt early, by about 14 d under SD [28]. These findings suggest that, although GF14c alone functions as an inhibitor, it becomes an activator when it forms that complex with Hd3a and OsFD1.

Hd1 and Ehd1

Hd3a is controlled by Hd1, which binds to the Hd3a promoter [31]. Hd1 is a rice ortholog of Arabidopsis Constans (CO) [32]. First identified by QTL mapping of a cross between 'Nipponbare' and 'Kasalath' [15,32]. Hd1 is allelic to SE1 (Photoperiodic Sensitivity 1), a major QTL controlling photoperiodic sensitivity [33]. The se1 mutant has been induced by γ-ray irradiation of seeds from the japonica cultivar "Gimbozu". It flowers early under LD conditions [33]. Nonfunctional Hd1 causes late flowering under SD but early flowering under LD [32], indicating that Hd1 can be either a flowering activator or repressor, depending upon light conditions. This behavior is contradictory to that of Arabidopsis CO, which is a constitutive flowering activator under both SD and LD [34]. The hd1-1 mutants generated by Tos17 insertion also show flowering phenotypes similar to se1 [35]. Transcript levels of Hd3a and RFT1 are decreased in hd1 mutants under SD conditions, but increased under LD [28,35]. How Hd1 functions differently between SD and LD is unknown. Overexpression of Hd1 causes late-flowering, but the phenotype is not observed in the osphyB mutant background [36], thereby demonstrating that OsPhyB is necessary for the suppressive functioning of Hd1. As a member of the family of CO-like proteins, Hd1 contains conserved B-box zinc fingers and CCT domains [37,38].

Expression of Hd3a is also controlled by Ehd1, as first revealed by QTL mapping of a cross between 'Taichung 65' and Oryza glaberrima [21]. Introgression of the Ehd1 allele from O. glaberrima causes early flowering in 'Taichung 65' under both SD and LD [39]. Ehd1 protein is a member of the B-type response regulators, which have a receiver domain and a GARP domain [39]. The latter functions to bind DNA; an alteration of Gly in the domain to Arg greatly reduces this binding capacity in 'Taichung 65' [39]. In that cultivar, a single amino acid substitution in the GARP domain disrupts binding to the target DNA [39]. Ehd1 induces transcript levels for Hd3a, RFT1, OsMADS14, and OsMADS15 [39]. Ehd1 RNAi plants flower later, by 7 d under SD [40] and by 10 d under LD [41]. All of these observations indicate that Ehd1 is a flowering activator.

LD-preferential regulatory elements

In temperate zones, rice is grown under LD conditions. When plants are young, flowering is inhibited to allow for sufficient vegetative growth and tillering. After a certain period, a flowering signal is induced to initiate the formation of reproductive organs.

Grain yield and heading date 7 (Ghd7)/Late Heading Date 4 (Lhd4) is an LD-specific repressor identified through QTL mapping of plant height and grain yield traits [42,43]. Introgressing a functional allele of Ghd7 from 'Minghui 63' to 'Zhenshan 97' can cause pleiotropic phenotypes such as late flowering and increases in height, stem diameter, and yield [42,43]. Ghd7 encodes a CO-like protein containing the B-box and CCT domains [38,43]. This gene suppresses flowering by inhibiting the expression of Ehd1 [43]. Ghd7 is strongly expressed in the vascular bundle where Ehd1 is expressed, providing evidence that Ghd7 functions upstream of Ehd1 [43]. Ghd7 is phosphorylated by HEADING DATE 16 (Hd16), a casein kinase-1 protein [24]. It expression is very low under SD but significantly higher under LD [43,44]. Natural variations include Ghd7-1, Ghd7-2, Ghd7-3, Ghd7-0, and Ghd7-0a. Among them, Ghd7-0 and Ghd7-0a are nonfunctional alleles mainly observed in high-latitude cultivars that are grown during shorter seasons [43]. The rest of the alleles are functional and found in low-latitude cultivars [43] (Xue et al.). Therefore, these reports indicate that Ghd7 alleles have played an important role during the progression of rice domestication.

Heading date 5 (Hd5)/QTL for days to heading on chromosome 8 (DTH8)/Grain yield and heading date 8 (Ghd8)/Late Heading Date 1 (LHD1) encodes a putative HAP3 subunit of the CCAAT-box-binding TF [45]. The HAP proteins have three subunits – HAP2, HAP3, and HAP5 – that form trimeric complexes [46,47]. Introgression of the functional Hd5 allele causes late flowering under LD by suppressing expression of Ehd1, Hd3a, and RFT1 [45,48-50]. However, under SD conditions, expression by those genes is not affected by Hd5. Flowering inhibition is independent of activity by Ghd7 and Hd1 [45]. By contrast, the ds9 mutant, another allelic mutant for Hd5, shows an early-flowering phenotype under LD but a late-flowering phenotype under SD [51]. Those flowering patterns are similar to the phenotype of the hd1 mutant [52]. Because the HAP complex forms a trimetric complex with CO protein in Arabidopsis, Hd5 cooperates with Hd1 through protein–protein interactions. The former also controls plant architecture, e.g., height, yield, grain dry weight, and the number of tillers.

Heading date 6 (Hd6) encodes a CK2 α-subunit and was identified by crossing 'Nipponbare' and 'Kasalath' [15,53]. The Hd6 allele from 'Nipponbare' is nonfunctional due to an early stop in the coding region. Introgression of the functional Hd6 allele from 'Kasalath' to 'Nipponbare' results in late flowering [53]. Hd6 functions as a flowering repressor of Hd3a and RFT1 under LD [10]. Although Hd6 has a late-flowering phenotype when functional Hd1 allele is present, it does not phosphorylate Hd1 [10].

Heading date 16 (Hd16)/Early flowering 1 (EL1) is another flowering repressor [50,54]. This gene was identified from a cross between the japonica rice 'Nipponbare' and 'Koshihikari' [55] (Matsubara et al.). 'Koshihikari' flowers later than 'Nipponbare' under SD but earlier under LD [55]. Introgression of a deficient allele of Hd16 from 'Koshihikari' into 'Nipponbare' weakens photoperiodic sensitivity and increases expression of floral activators Ehd1, Hd3a, and RFT1 under LD [24]. Hd16 encodes a casein kinase I that phosphorylates the rice DELLA protein SLR1 [50] and Ghd7 protein [24,50,56]. Therefore, Hd16 might possibly act as a mediator between floral transition and other developmental processes such as gibberellin signaling and tillering.

OsMADS50 is an LD-specific promoter of flowering [57]. Mutation induced by T-DNA causes a flowering delay of more than one month under LD, but has no such effect under SD [57]. OsMADS50 is an ortholog of Suppressor of Overexpression of CO 1 (SOC1), a flowering activator in Arabidopsis [58]. OsMADS50 suppresses expression of OsLFL1 [41] which inhibits Ehd1 by binding to the promoter region [59,60]. Transcript levels of OsMADS50 increase continuously until five weeks after germination [41]. This expression pattern is opposite to that of Ghd7, supporting the idea that OsMADS50 is a floral inducer. OsMADS56, the most homologous to OsMADS50, functions antagonistically when it binds to OsMADS50 [41]. Plants that over-express OsMADS56 flower later, by more than a month, under LD because expression of Ehd1, Hd3a, and RFT1 is repressed [41]. Both OsMADS50 and OsMADS56 are MADS-box TFs that contain a highly conserved MADS-box domain and K domain. The MADS-box domains function in DNA binding and dimerization while the K domains are involved in dimerization. The C domain is the least-conserved in both length and sequence. In some cases, it either possesses transactivation capability or contributes to the formation of multimeric complexes among MADS proteins [61,62]. The C region is shorter in OsMADS56 than in OsMADS50 and their sequences are diverse, suggesting that the former inhibits the latter.

Overexpression of OsDof12 causes early flowering under LD but not under SD [63] (Li et al.). In OsDof12 OX plants, transcript levels of Hd3a and OsMADS14 are increased under LD while those of Hd1, OsMADS51, Ehd1, and OsGI are not changed [63]. These results suggest that OsDof12 induces flowering by a pathway that does not go through Ehd1.

SD-preferential regulatory elements

In addition to the inducer Hd1, several regulatory elements control flowering time under SD. They include OsMADS51, a Type I MADS-box protein [40]. The osmads51 mutant flowers approximately two weeks later under SD, but its timing is unaltered under LD. The OsMADS51 OX plants bolt early, by about 10 d, under SD because expression by Ehd1, Hd3a, and OsMADS14 is enhanced [40]. However, flowering time is not changed under LD. This implies that OsMADS51 induces flowering only under SD, probably because it generates a product that functions preferentially under such conditions. Alternatively, the protein either inhibits an SD-specific repressor or enhances an SD-specific inducer.

Another SD-specific suppressor, OsCO3, is a member of the CO-like proteins and is similar to HvCO3 in barley [38,64]. Transcript levels are high during the daytime and low at night [64]. OsCO3 OX plants show SD-preferential flowering that is delayed by approximately 40 d because Hd3a, FTL, and OsMADS14 are suppressed [65].

Constitutive regulators

OsIndeterminate 1 (OsId1)/Early heading date 2 (Ehd2)/Rice Indeterminate 1 (RID1) is a constitutive activator. RNAi plants of OsId1 have significantly delayed flowering under both SD and LD [65] and knock-out mutants do not flower for more than 365 d [66,67]. This indicates that OsID1 is required for promoting a flowering signal regardless of light conditions. However, no research has shown that overexpression of OsId1 causes early flowering. Therefore, it is possible that OsId1 must form an activation complex with another protein. OsID1 regulates the expression of Ehd1 and its downstream genes Hd3a and RFT1. Transcription of other regulators is not affected in osid1 mutants, suggesting that OsID1 may directly induce Ehd1 expression. OsID1 is the rice ortholog of maize Indeterminate 1 (Id1),

which encodes a TF with an indeterminate domain [65]. Because id1 mutants show phenotypes of prolonged vegetative growth [68], it has been suggested that Id1 is involved in the synthesis of a florigen [69,70]. The presence of Ehd1 homologs in maize implies that the Id1–Ehd1 pathway is conserved among grass species.

Early heading date 4 (Ehd4) is another constitutive activator that encodes a CCCH-type zinc finger protein. The ehd4 mutant was identified as a somaclonal variant in Kitaake [71]. The ehd4 mutant flowered late under both SD and LD [71]. In the mutant, transcript levels of Ehd1, Hd3a and RFT1 are decreased, suggesting that Ehd4 prompts flowering time via Ehd1.

OsCOL4 functions as a constitutive flowering repressor [72]. The oscol4 mutants flower earlier, by approximately two weeks under SD and approximately three weeks under LD because Ehd1 expression is enhanced. Overexpression of OsCOL4 delays flowering by approximately three weeks under SD and approximately six weeks under LD conditions when transcript levels of Ehd1 are suppressed [72]. Expression of OsCOL4 is decreased in osphyB mutants and flowering time is similar between osphyB oscol4 double mutants and osphyB single mutants, indicating that OsCOL4 is positively controlled by OsPhyB [72]. Whereas OsPhyB is involved in the night break (NB) effect [35], OsCOL4 is NB-insensitive, demonstrating that the former controls flowering time through an alternative pathway.

OsLF encodes an atypical bHLH TF that is a constitutive flowering repressor. Overexpression of OsLF causes late flowering regardless of photoperiod length [73] Under SD, expression by OsGI and Hd1 is partially decreased in OX lines [73].

As a flowering activator, miR172 suppresses AP2-like genes in Arabidopsis [74]. Similarly, miR172a and miR172d induce flowering in rice [75]. Transcripts of the AP2-like genes Supernumerary Bract (SNB) and Oryza sativa Indeterminate Spikelet1 (OsIDS1) are targeted by miR172s. Their overexpression delays flowering by suppressing the expression of Ehd1 [75]. Phytochromes inhibit miR172, thereby increasing the expression of those AP2-like genes (Lee et al.). This indicates that the phytochrome–miR172–AP2–Ehd1 pathway also plays a role in controlling flowering time.

Circadian rhythm

GIGANTEA (GI) controls CO, a positive regulator of FT in Arabidopsis [76]. The GI–CO–FT pathway is conserved in rice [31]. Knockout mutants and RNAi-suppressed plants of OsGI flower late under SD but early under LD, exhibiting reduced photoperiodic sensitivity [31,77] OsGI positively controls Hd1 expression, with the latter being reduced in OsGI RNAi plants but increased in OX plants [31]. Transcriptome analyses between WT and osgi mutants have revealed that numerous rhythmic genes are affected [77]. In particular, expression by Ehd1, Hd3a, and Ghd7 is diminished in those mutants [44]. That of OsMADS51 is also decreased in OsGI antisense plants [40], indicating that OsGI controls flowering time via several independent pathways. An osgi knockout mutant generated by T-DNA insertion flowers late by 36 d under SD and by 9 d under LD conditions [78]. This SD-preferential phenotype is similar to that of the mutant by gamma irradiation [79]. Under SD, transcript levels of floral activators such as Hd3a, RFT1, Ehd1, Hd1, OsMADS51 and OsId1 are decreased in the osgi, demonstrating that the delayed flowering is due to decreased expression by the flowering activators [78]. However under LD, transcript level of OsGI is reduced to a low level at later stage of development when the flowering signals are generated. Consequently, effect of OsGI on controlling flowering time under LD conditions is minimal.

Heading date 2 (Hd2) was identified by QTL mapping using crossed progenies between 'Nipponbare'" and 'Kasalath' [15]. Introgressing the Hd2 allele of 'Kasalath' to 'Nipponbare' causes late flowering under SD but early flowering under LD [32]. This is evidence that Hd2 is involved in photoperiodic sensitivity. Hd2 is closely linked to *Oryza sativa* Pseudo Response Regulator 37 (OsPRR37), which encodes a pseudo-response regulator that functions as a major component of the circadian clock [80-83]. In two early-flowering Hokkaido cultivars, 'Kitaake' and 'H143', OsPRR37 is defective [84,85].

QTL mapping has shown that Heading date 3b (Hd3b)/Heading date 17 (Hd17)/Early flowering 7 (Ef7)/*Oryza sativa* Early flowering 3 (OsELF3) is a flowering inducer [16]. Hd3a encodes a homolog of Arabidopsis, EARLY FLOWERING 3 (ELF3), which has crucial roles in maintaining circadian rhythms [16,86-90]. In Arabidopsis, ELF3 modulates re-setting of the circadian clock [91,92]. A mutation in that gene leads to an early and photoperiod-insensitive flowering phenotype in Arabidopsis [93]. Hd17, an allele of Hd3b, has also been identified by QTL mapping of a cross between 'Nipponbare' and 'Koshihikari' [94]. Introgression of the Hd17 allele from 'Koshihikari' to 'Nipponbare' causes late flowering because of enhanced expression by Ghd7. The ef7 mutant, another allelic mutant for Hd3b, was developed via γ-irradiation of seeds from the japonica cultivar 'Gimbozu' [88]. Double-mutant analyses suggest that Hd3b functions as a floral activator by suppressing Ghd7 and Hd1. Hd3b is preferentially expressed in the mesophyll cells of the leaf blade, and also weakly in floral organs such as the lemma, palea, stamen, and pistil [88]. Because Ghd7 is preferentially expressed in the vascular tissue [43] (Xue et al.), further study is needed to determine how the mesophyll preferentially-expressed Hd3b controls Ghd7. Tos17 and T-DNA insertions cause late flowering by enhancing Ghd7 [89,90]. Various circadian clock components, e.g., OsLHY, OsPRR1, -37, -73, -95, and OsGI are changed in the ef7 mutants [89,90]. Therefore, one might conclude that, similar to Arabidopsis ELF3, Hd3b is involved in circadian clock control. In the rice genome, OsEF3 is also highly homologous to ELF3. Mutations in the rice gene cause pleiotropic effects, including late flowering [95].

Photoreception

Photoreception is one of the most important processes by which plants determine flowering time. In rice, day length is recognized by three phytochromes, OsPhyA, OsPhyB, and OsPhyC. Mutations in OsPhyB or OsPhyC result in early flowering, indicating that they function as flowering repressors [96]. OsPhyB suppresses flowering by activating OsCOL4 [35,72] and is also involved in the NB effect, as observed in SD plants [97,98]. Flowering in rice is inhibited by night breaks because of decreased Hd3a expression [35]. However, the NB effect is not seen in the osphyB mutant, and Hd3a transcript levels are not altered there [35].

The flowering time of osphyB osphyC double mutants is similar to that of osphyB single mutants, probably due to the post-translational process of OsPhyC in osphyB mutants, where the amount of protein is significantly lower [79]. In Arabidopsis, PhyC protein levels are also reduced in phyB mutants [99,100]. In addition, phyB and phyC mutants show early flowering phenotypes under SD, and flowering time is similar between phyB and the phyB phyC double mutants [100]. This phenomenon is also found in rice [96,100]. It has been reported that PhyB interacts with PhyC in Arabidopsis [101], suggesting that

OsPhyB and OsPhyC function as flowering repressors by forming a heterodimer, whereas OsPhyB or OsPhyC alone is unstable.

Although mutation of OsPhyA by itself does not affect flowering time, the double mutants osphyA osphyB and osphyA osphyC flower earlier under SD and later under LD when compared with osphyB or osphyC alone [96]. This demonstrates that OsPhyA cooperates with other phytochromes to modulate the expression of flowering regulators [3]. Ghd7 expression is increased in osphyB and osphyB osphyC mutants but decreased in osphyA osphyC and osphyA osphyB, indicating that OsPhyA positively controls Ghd7 expression while OsPhyB and OsPhyC have the opposite effect [79]. Interestingly, Ehd1 expression is not changed despite Ghd7 expression being elevated in the osphyB osphyC mutants. This suggests that OsPhyA controls flowering time through an alternative pathway [79]. In Arabidopsis, PhyA prompts floral induction by stabilizing CO protein whereas PhyB suppresses flowering via CO degradation [102]. In rice, Hd1 protein levels are similar between light and dark conditions in 35S::Hd1:myc plants [36]. This implies that, unlike CO in Arabidopsis, Hd1 protein is not degraded by light in rice.

SE5 encodes a heme oxygenase that is highly homologous to Arabidopsis HY1, which is involved in the biosynthesis of phytochrome chromophore [77,103] (Izawa et al.; Andrés et al.). The se5 mutants exhibit an extremely early flowering phenotype under both SD and LD [103,104] flowering at 44 d after germination, regardless of day length, because of elevated transcript levels of Ehd1 and Hd1 [104]. This phenotype is similar to that of osphyA osphyB mutants [3,77,35,103]

Flowering time is also controlled by a blue light receptor cryptochrome (CRY) in Arabidopsis [102,105-107]. The rice genome has three CRY genes, OsCRY1a, OsCRY1b, and OsCRY2 [108,109]. While OsCRY1a and OsCRY1b are responsible for regulating blue light-mediated de-etiolation [108,109] OsCRY2 functions in controlling flowering time [108,109] OsCRY2 antisense plants flower late under both SD and LD [108] indicating that OsCRY2 is a flowering activator. In Arabidopsis, CRY2 activates flowering by stabilizing CO protein [102]. However, the molecular mechanism for OsCRY2 is not yet known in rice.

In Arabidopsis, GI interacts with several blue light receptors, e.g., Zeitlupe (ZTL) and Flavin-Binding Kelch Repeat F-Box 1 (FKF1) [110,111]. The protein complexes are more stable under blue light conditions [110,111]. In rice, Ehd1 expression is prompted by blue light treatment, but this activation is diminished in osgi mutants, suggesting that OsGI is involved in blue light signaling [3].

Epigenetic regulation

Expression by some flowering time genes is developmentally regulated. For example, transcripts of OsLFL1 are maintained at a high level for the first four weeks but gradually decrease as plants mature. By comparison, transcripts of OsVIL2 (*Oryza sativa* VIN3-like 2)/LC2 (Leaf inclination 2)/OsVIL3 (*Oryza sativa* VIN3-like 3) remain at a low level but increase rapidly as OsLFL1 expression declines. Whereas overexpression of OsLFL1 causes late flowering [59,60] knockout mutations of OsVIL2 delay flowering under both SD and LD [112]. However, a knockout mutation of LC2, an allelic mutation of OsVIL2, causes late flowering only under SD [54,113]. The difference in results reported by those two research groups may be related to the day length that was tested. Whereas experiments by the former group entailed a 14.5-h photoperiod, plants used by the latter group were exposed to 14 h of light.

OsVIL2 protein binds to histone H3 and OsLFL1 chromatin (Yang et al.). Histone H3 lysine 27 trimethyl (H3K27me3) levels of OsLFL1 chromatin are reduced in osvil2 mutants (Yang et al.). This indicates that OsVIL2 suppresses expression of OsLFL1 by mediating the methylation of lysine 27 residue of histone H3. OsVIL2 binds to O. sativa Embryonic Flower 2b (OsEMF2b), a core component of the polycomb repressive complex 2 (PRC2) that represses the expression of target genes by condensing their chromatins [112]. The osemf2b mutants also show late flowering phenotypes [112,114]. Emf2b promotes flowering by directly suppressing the expression of OsLFL1 [115]. OsVIL1–RNAi plants flower late under SD conditions [113]. OsVIL1 engages with OsVIL2, suggesting that the two cooperate as flowering activators through protein–protein interactions.

Transcription of Ghd7 is epigenetically regulated by OsTrx1 (Oryza sativa Trithorax 1), which encodes a histone methyltransferase [75]. OsTrx1 belongs to a Trithorax group (TrxG) of proteins that enhance the expression of target genes by loosening their chromatins. The ostrx1 mutants show significantly delayed flowering that is preferential under LD [75]. In those mutants, Ghd7 transcripts are more abundant under LD [75].

OsTrx1 protein interacts with Early heading date 3 (Ehd3) protein in Plant Homeodomain (PHD) finger regions [75]. The ehd3 mutant was identified as a late-flowering variant of M2 plants from a γ-ray-mutagenized line of O. sativa ssp. japonica 'Tohoku IL9' [116]. Those mutants flower 20 d later than the WT under SD and do not flower for more than 365 d under LD [116]. As with ostrx1 mutants, Ghd7 transcripts are increased in ehd3 mutants, all evidence that OsTrx1 and Ehd3 cooperatively suppress Ghd7.

The rice genome has at least 37 SET domain group (SDG) proteins that may be involved in chromatin remodeling via histone demethylation [117]. For example, SDG724/Long Vegetative Phase 1 (LVP1), which has histone methyltransferase activity, preferentially functions as a flowering activator under LD [118]. In the lvp1 mutant, expression by activators OsMADS50, Hd3a, RFT1, and Ehd1 is significantly decreased under LD [118]. These findings confirm that SDG724 serves as an LD-preferential activator by inducing OsMADS50. Both SDG711 and SDG718 encode the rice enhancer of zeste [E(z)], a subunit of PRC2 [119]. SDG711 is a flowering repressor under LD, as demonstrated by the late flowering phenotype of OX plants, whereas SDG711 RNAi plants display early flowering phenotypes under LD [119]. In the SDG711 OX plants, expression is reduced for Ehd1, Hd3a, RFT1, OsMADS14, OsMADS15, and OsLF while transcript levels of Hd1 are increased under LD [119]. By contrast, in the SDG711 RNAi plants, expression is preferentially increased for OsMADS14 and OsLF but decreased for Hd1 under LD [119]. Those results demonstrate that SDG711 functions as an LD-preferential flowering repressor by inhibiting OsLF. Furthermore, SDG718 RNAi plants flower late preferentially under SD, with expression being decreased for OsMADS14, RFT1, Hd3a, and Hd1 but increased for OsLF [119]. Therefore, SDG718 functions as an SD-preferential flowering activator by suppressing OsLF.

Conclusions

In addition to the conserved GI–Hd1–Hd3a pathway present in all plant species, rice has several other pathways that control flowering time. Most of them are merged to Ehd1 or, occasionally, directly to florigens. The Ghd7 pathway is the major route that functions under LD. Several upstream elements, e.g., phytochromes, GI, and chromatin remodeling factors, modulate Ghd7 expression. The OsMADS50 pathway induces flowering preferentially under LD. By contrast, the OsMADS51 pathway enhances flowering specifically under SD while OsCO3 inhibits flowering only under SD. OsID1 is a constitutive activator that is essential for flowering. The constitutive repressor OsCOL4 is controlled by OsPhyB. Flowering is induced by the miR172 pathway through the degradation of mRNA from AP2 family genes. Although several regulatory genes that control flowering time have been identified, their relationships and the way in which their proteins function are not well understood and must still be addressed. Moreover, the regulatory mechanisms for plant responses to environmental stimuli such as temperature extremes, nutrient deficiencies, and various stresses require further exploration.

Acknowledgements

We thank the members of the Crop Biotech Center at Kyung Hee University for their participation in discussions. We also thank Priscilla Licht for critical proofreading of the manuscript. This work was supported in part by grants from the Next-Generation BioGreen 21 Program (Plant Molecular Breeding Center, No. PJ01108001), Rural Development Administration, Republic of Korea; the Basic Research Promotion Fund, Republic of Korea (NRF-2007-0093862); and Kyung Hee University (20130214).

References

1. Izawa T (2007) Adaptation of flowering-time by natural and artificial selection in Arabidopsis and rice. J Exp Bot 58: 3091-3097.

2. Tsuji H, Taoka K, Shimamoto K (2011) Regulation of flowering in rice: Two florigen genes, a complex gene network, and natural variation. Curr Opin Plant Biol 14: 45-52.

3. Itoh H, Izawa T (2013) The coincidence of critical day length recognition for florigen gene expression and floral transition under long-day conditions in rice. Mol Plant 6: 635-649.

4. Yeang HY (2013) Solar rhythm in the regulation of photoperiodic flowering of long-day and short-day plants. J Exp Bot 64: 2643-2652.

5. Hayama R, Coupland G (2004) The molecular basis of diversity in the photoperiodic flowering responses of Arabidopsis and rice. Plant Physiol 135: 677-684.

6. Greenup A, Peacock WJ, Dennis ES, Trevaskis B (2009) The molecular biology of seasonal flowering-responses in Arabidopsis and the cereals. Ann Bot 103: 1165-1172.

7. Song YH, Ito S, Imaizumi T (2010) Similarities in the circadian clock and photoperiodism in plants. Curr Opin Plant Biol 13: 594-603.

8. Sun C, Chen D, Fang J, Wang P, Deng X, Chu C (2014) Understanding the genetic and epigenetic architecture in complex network of rice flowering pathways. Protein Cell 5: 889-898.

9. Song YL, Luan WJ (2012) Molecular regulatory network of flowering by photoperiod and temperature in rice. Rice Sci 19: 169-176.

10. Ogiso E, Takahashi Y, Sasaki T, Yano M, Izawa T (2010) The role of casein kinase II in flowering time regulation has diversified during evolution. Plant Physiol 152: 808-820.

11. Jeon JS, Lee S, Jung KH, Jun SH, Jeong DH, et al. (2000a) T-DNA insertional mutagenesis for functional genomics in rice. Plant J 22: 561-570.

12. An S, Park S, Jeong DH, Lee DY, Kang HG, et al. (2003) Generation and analysis of end sequence database for T-DNA tagging lines in rice. Plant Physiol 133: 2040-2047.

13. Hirochika H, Guiderdoni E, An G, Hsing YI, Eun MY, Han CD, et al. (2004) Rice mutant resources for gene discovery. Plant Mol Biol 54: 325-334.

14. Tamaki S, Matsuo S, Wong HL, Yokoi S, Shimamoto K (2007) Hd3a protein is a mobile flowering signal in rice. Science 316: 1033-1036.

15. Yano M, Sasaki T (1997) Genetic and molecular dissection of quantitative traits in rice. Plant Mol Biol 35: 145-153.

16. Kojima S, Takahashi Y, Kobayashi Y, Monna L, Sasaki T, et al. (2002) Hd3a, a rice ortholog of the Arabidopsis FT gene, promotes transition to flowering downstream of Hd1 under short-day conditions. Plant Cell Physiol 43: 1096-1105.

17. Monna L, Lin X, Kojima S, Sasaki T, Yano M (2002) Genetic dissection of a genomic region for a quantitative trait locus, *Hd3*, into two loci, *Hd3a* and *Hd3b*, controlling heading date in rice. Theor Appl Genet 104: 772-778.

18. Chardon F, Damerval C (2005) Phylogenomic analysis of the PEBP gene family in cereals. J Mol Evol 61: 579-590.

19. Komiya R, Yokoi S, Shimamoto K (2009) A gene network for long-day flowering activates RFT1 encoding a mobile flowering signal in rice. Development 136: 3443-3450.

20. Hagiwara WE, Uwatoko A, Sasaki A, Matsubara K, Nagano H, et al. (2009) Diversification in flowering time due to tandem FT-like gene duplication, generating novel Mendelian factors in wild and cultivated rice. Mol Ecol 18: 1537-1549.

21. Doi K, Izawa T, Fuse T, Yamanouchi U, Kubo T, et al. (2004) Ehd1, a B-type response regulator in rice, confers short-day promotion of flowering and controls FT-like gene expression independently of Hd1. Genes Dev 18: 926-936.

22. Izawa T, Oikawa T, Sugiyama N, Tanisaka T, Yano M, Shimamoto K (2002) Phytochrome mediates the external light signal to repress FT orthologs in photoperiodic flowering of rice. Genes Dev 16: 2006-2020.

23. Komiya R, Ikegami A, Tamaki S, Yokoi S, Shimamoto K (2008) Hd3a and RFT1 are essential for flowering in rice. Development 135: 767-774.

24. Hori K, Ogiso-Tanaka E, Matsubara K, Yamanouchi U, Ebana K, et al. (2013) Hd16, a gene for casein kinase I, is involved in the control of rice flowering time by modulating the day-length response. Plant J 76: 36-46.

25. Tsuji H, Tamaki S, Komiya R, Shimamoto K (2008) Florigen and the photoperiodic control of flowering in rice. Rice 1: 25-35.

26. Uga Y, Nonoue Y, Liang ZW, Lin HX, Yamamoto S, et al. (2007) Accumulation of additive effects generates a strong photoperiod sensitivity in the extremely late-heading rice cultivar 'Nona Bokra'. Theor Appl Genet 114: 1457-1466.

27. Ogiso-Tanaka E, Matsubara K, Yamamoto S-i, Nonoue Y, Wu J, et al. (2013) Natural variation of the *RICE FLOWERING LOCUS T 1* contributes to flowering time divergence in rice. PLoS One 8: e75959.

28. Purwestri YA, Ogaki Y, Tamaki S, Tsuji H, Shimamoto K (2009) The 14-3-3 protein GF14c acts as a negative regulator of flowering in rice by interacting with the florigen Hd3a. Plant Cell Physiol 50: 429-438.

29. Taoka K-i, Ohki I, Tsuji H, Furuita K, Hayashi K, et al. (2011) 14-3-3 proteins act as intracellular receptors for rice Hd3a florigen. Nature 476: 332-335.

30. Jeon J-S, Lee S, Jung K-H, Yang W-S, Yi G-H, et al. (2000b) Production of transgenic plants showing reduced heading date and plant height by ectopic expression of rice MADS-box genes. Mol Breed 6: 581-592.

31. Hayama R, Yokoi S, Tamaki S, Yano M, Shimamoto K (2003) Adaptation of photoperiodic control pathways produces short-day flowering in rice. Nature 422: 719-722.

32. Lin HX, Yamamoto T, Sasaki T, Yano M (2000) Characterization and detection of epistatic interactions of three QTLs, *Hd1*, *Hd2* and *Hd3*, controlling heading date in rice using nearly isogenic lines. Theor Appl Genet 101: 1021-1028.

33. Inoue H, Tanisaka T, Okumoto Y, Yamagata H (1992) An early-heading mutant gene of a mutant line HS66 of rice (in Japanese with English summary). Rep Soc Crop Sci Breed Kinki 37: 47-52.

34. Putterill J, Robson F, Lee K, Simon R, Coupland G (1995) The *CONSTANS* gene of *Arabidopsis* promotes flowering and encodes a protein showing similarities to zinc finger transcription factors. Cell 80: 847-857.

35. Ishikawa R, Tamaki S, Yokoi S, Inagaki N, Shinomura T, et al. (2005) Suppression of the floral activator gene Hd3a is the principal cause of the night break effect in rice. Plant Cell 17: 3326-3336.

36. Ishikawa R, Aoki M, Kurotani K, Yokoi S, Shinomura T, et al. (2011) Phytochrome B regulates Heading date 1 (Hd1)-mediated expression of rice florigen Hd3a and critical day length in rice. Mol Genet Genom 285: 461-470.

37. Robson F, Costa MMR, Hepworth SR, Vizir I, Pineiro M, Reeves PH, Putterill J, Coupland G (2001) Functional importance of conserved domains in the flowering-time gene CONSTANS demonstrated by analysis of mutant alleles and transgenic plants. Plant J 28: 619-631.

38. Griffiths S, Dunford RP, Coupland G, Laurie DA (2003) The evolution of CONSTANS-like gene families in barley, rice, and Arabidopsis. Plant Physiol 131: 1855-1867.

39. Doi K, Yoshimura A (1998) RFLP mapping of a gene for heading date in an African rice. Rice Genet Newslett 15: 148-149.

40. Kim SL, Lee S, Kim HJ, Nam HG, An G (2007a) OsMADS51 is a short-day flowering promoter that functions upstream of Ehd1, OsMADS14, and Hd3a. Plant Physiol 145: 1484-1494.

41. Ryu CH, Lee S, Kim SL, Lee YS, Choi SC, g HJ, Yi J, Park SJ, Han CD, An G (2009) OsMADS50 and OsMADS56 function antagonistically in regulating long day (LD)-dependent flowering in rice. Plant Cell Environ 32: 1412-1427.

42. Yu SB, Li JX, Xu CG, Tan YF, Li XH, et al. (2002) Identification of quantitative trait loci and epistatic interactions for plant height and heading date in rice. Theor Appl Genet 104: 619-625.

43. Xue W, Xing Y, Weng X, Zhao Y, Tang W, et al. (2008) Natural variation in *Ghd7* is an important regulator of heading date and yield potential in rice. Nat Genet 40: 761-767.

44. Itoh H, Nonouge Y, Yano M, Izawa T (2010) A pair of floral regulators sets critical day length for Hd3a florigen expression in rice. Nat Genet 42: 635-639.

45. Wei X, Xu J, Guo H, Jiang L, Chen S, Yu C, et al. (2010) *DTH8* suppresses flowering in rice, influencing plant height and yield potential simultaneously. Plant Physiol 153: 1747-1758.

46. Thirumurugan T, Ito Y, Kubo T, Serizawa A, Kurata N (2008) Identification, characterization and interaction of HAP family gene in rice. Mol Genet Genom 279: 279-289.

47. Laloum T, De Mita S, Gamas P, Baudin M, Niebel A (2012) CCAAT-box binding transcription factors in plants: Y so many? Trends Plant Sci 18: 157-166.

48. Lin H, Liang Z-W, Sasaki T, Yano M (2003) Fine mapping and characterization of quatitative trait loci *Hd4* and *Hd5* controlling heding data in rice. Breed Sci 53: 51-59.

49. Yan WH, Wang P, Chen HX, Zhou HJ, Li QP, et al. (2011) A major QTL, *Ghd8*, plays pleiotropic roles in regulating grain productivity, plant height, and heading date in rice. Mol Plant 4: 319-330.

50. Dai C, Xue HW (2010) Rice early flowering1, a CKI, phosphorylates DELLA protein SLR1 to negatively regulate gibberellin signalling. EMBO 29: 1916-1927.

51. Xing Q, Zheng Z, Zhou X, Chen X, Guo Z (2015) *Ds9* was isolated encoding as OsHAP3H and its C-terminus was required for interaction with HAP2 and HAP5. J Plant Biol 58: 26-37.

52. Yano M, Katayose Y, Ashikari M, Yamanouchi U, Monna L, et al. (2000) Hd1, a major photoperiod sensitivity quantitative trait locus in rice, is closely related to the arabidopsis flowering time gene CONSTANS. Plant Cell 12: 2473-2483.

53. Takahashi Y, Shomura A, Sasaki T, Yano M (2001) Hd6, a rice quantitative trait locus involved in photoperiod sensitivity, encodes the alpha subunit of protein kinase CK2. Proc Natl Acad Sci USA 98: 7922-7927.

54. Zhao SQ, Hu J, Guo LB, Qian Q, Xue HW (2010) Rice leaf inclination 2, a VIN3-LIKE protein, regulates leaf angle through modulating cell division of the collar. Cell Res 20: 935-947.

55. Matsubara K, Kono I, Hori K, Nonoue Y, Ono N, et al. (2008a) Novel QTLs for photoperiodic flowering revealed by using reciprocal backcross inbred lines from crosses between japonica rice cultivars. Theor Appl Genet 117: 935-945.

56. Kwon CT, Yoo SC, Koo BH, Cho SH, Park JW, et al. (2014) Natural variation in Early flowering 1 contributes to early flowering in japonica rice under long days. Plant Cell Environ 37: 101-112.

57. Lee S, Kim J, Han JJ, Han MJ, An G (2004) Functional analyses of the flowering time gene OsMADS50, the putative SUPPRESSOR OF OVEREXPRESSION OF CO 1/AGAMOUS-LIKE 20 (SOC1/AGL20) ortholog in rice. Plant J 38: 754-764.

58. Lee H, Suh SS, Park E, Cho E, Ahn JH, et al. (2000) The AGAMOUS-LIKE 20 MADS domain protein integrates floral inductive pathways in Arabidopsis. Genes Dev 14: 2366-2376.

59. Peng LT, Shi ZY, Li L, Shen GZ, Zhang JL (2007) Ectopic expression of OsLFL1 in rice represses Ehd1 by binding on its promoter. Biochem Biophys Res Commun 360: 251-256.

60. Peng LT, Shi ZY, Li L, Shen GZ, Zhang JL (2008) Overexpression of transcription factor OsLFL1 delays flowering time in *Oryza sativa*. J Plant Physiol 165: 876-885.

61. Cho S, Jang S, Chae S, Chung KM, Moon YH, et al. (1999) Analysis of the C-terminal region of Arabidopsis thaliana APETALA1 as a transcription activation domain. Plant Mol Biol 40: 419-429.

62. Lim J, Moon YH, An G, Jang SK (2000) Two rice MADS domain proteins interact with OsMADS1. Plant Mol Biol 44: 513-527.

63. Li D, Yang C, Li X, Gan Q, Zhao X, Zhu L (2009) Functional characterization of rice OsDof12. Planta 229: 1159-1169.

64. Kim S-K, Yun C-H, Lee JH, Jang YH, Park H-Y, et al. (2008) OsCO3, a CONSTANS-LIKE gene, controls flowering by negatively regulating the expression of FT-like genes under SD conditions in rice. Planta 228: 355-365.

65. Park SJ, Kim SL, Lee S, Je BI, Piao HL, et al. (2008) Rice Indeterminate 1 (OsId1) is necessary for the expression of Ehd1 (Early heading date 1) regardless of photoperiod. Plant J 56: 1018-1029.

66. Matsubara K, Yamanouchi U, Wang ZX, Minobe Y, Izawa T, et al. (2008b) Ehd2, a rice ortholog of the maize INDETERMINATE1 gene, promotes flowering by up-regulating Ehd1. Plant Physiol 148: 1425-1435.

67. Wu CY, You CJ, Li CS, Long T, Chen GX, et al. (2008) RID1, encoding a Cys2/His2-type zinc finger transcription factor, acts as a master switch from vegetative to floral development in rice. Proc Natl Acad Sci USA 105: 12915-12920.

68. Galinat WC, Naylor AW (1951) Relation of photoperiod to inflorescence proliferation in Zea mays L. Am J Bot 38: 38-47.

69. Colasanti J, Yuan Z, Sundaresan V (1998) The indeterminate gene encodes a zinc finger protein and regulates a leaf-generated signal required for the transition to flowering in maize. Cell 93: 593-603.

70. Colasanti J, Tremblay R, Wong AYM, Coneva V, Kozaki A, et al. (2006) The maize INDETERMINATE1 flowering time regulator defines a highly conserved zinc finger protein family in higher plants. BMC Genomics 7: 158.

71. Gao H, Zheng XM, Fei G, Chen J, Jin M, et al. (2013) Ehd4 encodes a novel and Oryza-genus-specific regulator of photoperiodic flowering in rice. PLoS Genet 9: e10003281.

72. Lee YS, Jeong DH, Lee DY, Yi J, Ryu CH, et al. (2010) OsCOL4 is a constitutive flowering repressor upstream of Ehd1 and downstream of OsphyB. Plant J 63: 18-30.

73. Zhao XL, Shi ZY, Peng LT, Shen GZ, Zhang JL (2011) An atypical HLH protein OsLF in rice regulates flowering time and interacts with OsPIL13 and OsPIL15. Nat Biotechnol 28: 788-797.

74. Aukerman MJ, Sakai H (2003) Regulation of flowering time and floral organ identity by a MicroRNA and its APETALA2-like target genes. Plant Cell 15: 2730-2741.

75. Choi SC, Lee S, Kim SR, Lee YS, Liu C, et al. (2014) Trithorax group protein OsTrx1 controls flowering time in rice via interaction with Ehd3. Plant Physiol 164: 1326-1337.

76. Fowler S, Lee K, Onouchi H, Samach A, Richardson K, et al. (1999) Gigantea: A circadian clock-controlled gene that regulates photoperiodic flowering in Arabidopsis and encodes a protein with several possible membrane spanning domains. EMBO J 18: 4679-4688.

77. Izawa T, Mihara M, Suzuki Y, Gupta M, Itoh H, et al. (2011) Os-Gigantea confers robust diurnal rhythms on the global transcriptome of rice in the field. Plant Cell 23: 1741-1755.

78. Lee YS, An G (2015) OsGI Controls Flowering Time by Modulating Rhythmic Flowering Time Regulators Preferentially under Short Day in Rice. J Plant Biol (in press).

79. Osugi A, Itoh H, Ikeda-Kawakatsu K, Takano M, Izawa T (2011) Molecular dissection of the roles of phytochrome in photoperiodic flowering in rice. Plant Physiol 157: 1128-1137.

80. Kaczorowski KA, Quail PH (2003) Arabidopsis PSEUDORESPONSE REGULATOR7 is a signaling intermediate in phytochrome-regulated seedling deetiolation and phasing of the circadian clock. Plant Cell 15: 2654-2665.

81. Yamamoto Y, Sato E, Shimizu T, Nakamich N, Sato S, et al. (2003) Comparative genetic studies on the APRR5 and APRR7 genes belonging to the APRR1/TOC1 quintet implicated in circadian rhythm, control of flowering time, and early photomorphogenesis. Plant Cell Physiol 44: 1119-1130.

82. Murakami M, Matsushiak A, Ashikari M, Yamashino T, Mizuno T (2005) Circadian-associated rice pseudo response regulators (OsPRRs): Insight into the control of flowering time. Biosci Biotech Biochem 69: 410-414.

83. Farre EM, Kay SA (2007) PRR7 protein levels are regulated by light and the circadian clock in Arabidopsis. Plant J 52: 548-560.

84. Kim SL, Choi M, Jung KH, An G (2013) Analysis of the early-flowering mechanisms and generation of T-DNA tagging lines in Kitaake, a model rice cultivar. J Exp Bot 64: 4169-4182.

85. Koo BH, Yoo SC, Park JW, Kwon CT, Lee BD, et al. (2013) Natural variation in OsPRR37 regulates heading date and contributes to rice cultivation at a wide range of latitudes. Mol Plant 6: 1877-1888.

86. Dixon LE, Knox K, Kozma-Bognar L, Southern MM, Pokhiko A, Millar AJ (2011) Temporal expression of core circadian genes is mediated through early flowering 3 in Arabidopsis. Curr Biol 21: 120-125.

87. Nefissi R, Natsui Y, Miyata K, Oda A, Hase Y, et al. (2011) Double loss-of-function mutation in EARLY FLOWERING 3 and CRYPTOCHROME 2 genes delays flowering under continuous light but accelerates it under long days and short days: An important role for Arabidopsis CRY2 to accelerate flowering time in continuous light. J Exp Bot 62: 2731-2744.

88. Saito H, Ogiso-Tanaka E, Okumoto Y, Yoshitake Y, Izumi H, et al. (2012) Ef7 encodes an ELF3-like protein and promotes rice flowering by negatively regulating the floral repressor gene Ghd7 under both short- and long-day conditions. Plant Cell Physiol 53: 717-728.

89. Yang Y, Peng Q, Chen GX, Li XH, Wu CY (2013b) OsELF3 is involved in circadian clock regulation for promoting flowering under long-day conditions in rice. Mol Plant 6: 202-215.

90. Zhao J, Huang X, Ouyang X, Chen X, Chen W, et al. (2012) OsELF3-1, an ortholog of Arabidopsis EARLY FLOWERING 3, regulates rice circadian rhythm and photoperiodic flowering. PLoS One 7: e43705.

91. McWatters HG, Bastow RM, Hall A, Millar AJ (2000) The ELF3 zeitnehmer regulates light signalling to the circadian clock. Nature 408: 716-720.

92. Covington MF, Panda S, Liu XL, Strayer CA, Wagner DR, et al. (2001) ELF3 modulates resetting of the circadian clock in Arabidopsis. Plant Cell 13: 1305-1315.

93. Zagotta MT, Hicks KA, Jacobs CI, Young JC, Hangarter RP, et al. (1996) The Arabidopsis ELF3 gene regulates vegetative photomorphogenesis and the photoperiodic induction of flowering. Plant J 10: 691-702.

94. Matsubara K, Ogiso-Tanaka E, Hori K, Ebana K, Ando T, Yano M (2012) Natural variation in Hd17, a homolog of Arabidopsis ELF3 that is involved in rice photoperiodic flowering. Plant Cell Physiol 53: 709-716.

95. Fu C, Yang XO, Chen X, Chen W, Ma Y, et al. (2009) OsEF3, a homologous gene of Arabidopsis ELF3, has pleiotropic effects in rice. Plant Biol 11: 751-757.

96. Takano M, Inagaki N, Xie X, Yuzurihara N, Hihara F, et al. (2005) Distinct and cooperative functions of phytochromes A, B, and C in the control of deetiolation and flowering in rice. Plant Cell 17: 3311-3325.

97. Hamner KC, Bonner J (1938) Photoperiodism in relation to hormones as factors in floral initiation and development. Bot Gaz 100: 388-431.

98. Thomas B, Vince-Prue D (1997) Photoperiodism in Plants. (London: Academic Press).

99. Hirschfeld M, Tepperman JM, Clack T, Quail PH, Sharrock RA (1998) Coordination of phytochrome levels in phyB mutants of Arabidopsis as revealed by apoprotein-specific monoclonal antibodies. Genetics 149: 523-535.

100. Monte E, Alonso JM, Ecker JR, Zhang Y, Li X, et al. (2003) Isolation and characterization of phyC mutants in Arabidopsis reveals complex crosstalk between phytochrome signaling pathways. Plant Cell 15: 1962-1980.

101. Sharrock RA, Clack T (2004) Heterodimerization of type II phytochromes in Arabidopsis. Proc Natl Acad Sci USA 101: 11500-11505.

102. Valverde F, Mouradov A, Soppe W, Ravenscroft D, Samach A, et al. (2004) Photoreceptor regulation of Constans protein in photoperiodic flowering. Science 303: 1003-1006.

103. Andrés F, Galbraith DW, Talon M, Domingo M (2009) Analysis of photoperiod sensitivity5 sheds light on the role of phytochromes in photoperiodic flowering in rice. Plant Physiol 151: 681-690.

104. Izawa T, Oikawa T, Tokutomi S, Okuno K, Shimamoto K (2000) Phytochromes

confer the photoperiodic control of flowering in rice (a short-day plant). Plant J 22: 391-399.

105. Guo HW, Yang WY, Mockler TC, Lin CT (1998) Regulation of flowering time by Arabidopsis photoreceptors. Science 279: 1360-1363.

106. Cashmore AR, Jarillo JA, Wu YJ, Liu D (1999) Cryptochromes: Blue light receptors for plants and animals. Science 284: 760-765.

107. Lin C (2002) Blue light receptors and signal transduction. Plant Cell 14: S207-S225.

108. Hirose F, Shinomura T, Tanabata T, Shimada H, Takano M (2006) Involvement of rice cryptochromes in de-etiolation responses and flowering. Plant Cell Physiol 47: 915-925.

109. Zhang YC, Gong SF, Li QH, Sang Y, Yang HQ (2006) Functional and signaling mechanism analysis of rice cryptochrome. Plant J 46: 971-983.

110. Kim WY, Fujiwara S, Suh SS, Kim J, Kim Y, et al. (2007b) ZEITLUPE is a circadian photoreceptor stabilized by GIGANTEA in blue light. Nature 449: 356-360.

111. Sawa M, Nusinow DA, Kay SA, Imaizumi T (2007) FKF1 and GIGANTEA complex formation is required for day-length measurement in Arabidopsis. Science 318: 261-265.

112. Yang J, Lee S, Hang R, Kim SR, Lee YS, et al. (2013a) OsVIL2 functions with PRC2 to induce flowering by repressing OsLFL1 in rice. Plant J 73: 566-578.

113. Wang J, Hu J, Qian Q, Xue HW (2013) LC2 and OsVIL2 promote rice flowering by photoperiod-induced epigenetic silencing of OsLF. Mol Plant 6: 514-527.

114. Conrad L, Khanday I, Johnson C, Guiderdoni E, An G, et al. (2014) The polycomb group gene EMF2B is essential for maintenance of floral meristem determinacy in rice. Plant J 80: 883-894.

115. Xie S, Chen M, Pei R, Quyang Y, Yao J (2015) OsEMF2b acts as a regulator of flowering transition and floral organ identity by mediating H3K27me3 deposition at OsLFL1 and OsMADS4 in rice. Plant Mol Biol Rep 33: 121-132.

116. Matsubara K, Yamanouchi U, Nonoue Y, Sugimoto K, Wang ZX, et al. (2011) Ehd3, encoding a plant homeodomain finger-containing protein, is a critical promoter of rice flowering. Plant J 66: 603-612.

117. Ng DW, Wang T, Chandrasekharan MB, Aramayo R, Kerbundit S, et al. (2007) Plant SET domain-containing proteins: Structure, function and regulation. Biochim Biophys Acta 1769: 316-329.

118. Sun C, Fang J, Zhao T, Xu B, Zhang F, et al. (2012) The histone methyltransferase SDG724 mediates H3K36me2/3 deposition at MADS50 and RFT1 and promotes flowering in rice. Plant Cell 24: 3235-3247.

119. Liu X, Zhou C, Zhao Y, Zhou S, Wang W, et al. (2014) The rice enhancer of zeste [E(z)] genes SDG711 and SDG718 are respectively involved in long day and short day signaling to mediate the accurate photoperiod control of flowering time. Front Plant Sci 5: 591.

14

Genetic Transformation of Rice: Problems, Progress and Prospects

Saroj Kumar Sah*, Amandeep Kaur, Gurwinder Kaur and Gurvinder Singh Cheema

School of Agricultural Biotechnology, Punjab Agricultural University, Ludhiana 141 004, Punjab, India

Abstract

Plant genetic engineering has become one of the most important molecular tools in the modern molecular breeding of crops. Now a days, production of transgenic plants is a routine process in some crop species. Transgenes are delivered into plants to confer novel traits such as improving nutritional qualities, resistance to pests. It is possible to insert genes from plants at evolutionary distant from the host plant, as well as from fungi, viruses, bacteria and even animals. Genetic transformation requires penetration of the transgene through the plant cell wall, facilitated by biological or physical methods. Over the last few decades, a significant progress has been made in the development of new and efficient transformation methods. Despite a variety of available DNA delivery methods, Agrobacterium and Biolistic mediated transformation remain two predominantly applied approaches. The objective of this article is to review the currently used methods for genetic plant transformation, their biological requirements and critical parameters.

Keywords: Genetic transformation rice; *Agrobacterim*; Biolistic transformation.

Introduction

Cereals are the most important source of calories to humans since rice, wheat and maize provide 23%, 17% and 10% calories globally [1]. Rice (*Oryza sativa* L.) is a well-known economical cereal, because it is a staple food included in the diet and most important source of food for more than half of the population. It has been reported that more than 90 per cent of the world's rice is grown and consumed in Asia, where it is a major source of income to rural people [2,3]. Adoption of green revolution varieties led to a drastic change in rice production. Between 1996 and 2011, the population of the low-income countries grew by 110 per cent, but rice production increased by 180 percent only from 257 million tons in 1996 to 718 million tons in 2011. Despite these advances in rice production, still 800 million people is not getting food every day. It is estimated to increase up to 25 per cent of rice production by the year 2030 [1]. This additional rice must be produced from good lands without opening up more fragile lands for rice cultivation. Agriculture production is decreasing due to biotic and abiotic stresses. The major abiotic stresses worldwide which cause threat to food security are high salinity, drought, submergence and cold [4-8]. Among these abiotic stresses, salinity is one of the major factors restricting productivity of crop plant worldwide [9]. Area under salt stress is increasing due to certain factors like climate change, rise in sea levels, excess irrigation without proper drainage in islands and underlying rocks rich in harmful salts,. If current scenario will continue, 50 per cent of present cultivated land will be lost for agriculture by 2050 [10]. Therefore, abiotic stresses are one of the main concerns to fulfill the required food demand [11]. Rendering the problem will entail development of rice varieties, which have higher yields, excellent grain quality, and resistance to biotic and abiotic stresses.

Many breeding programs have been conducted to increase the environmental stress tolerance, pest and disease resistance to the crop. Although conventional breeding programs such as Conventional hybridization, hybrid breeding, wide hybridization and ideo type breeding have developed some rice varieties in addition to several lines have been released in the Philippines, Bangladesh and India [12] but the success rate of conventional breeding is not sufficient [12,13] to meet the requirements. So there is a need to develop rice varieties that can withstand high levels of salt, drought and water stress with optimal yield levels.

Therefore, it is of utmost important to augment the productivity of rice to cope with the increased threat of population boom which is expected to reach 9.1 billion by 2050. To combat these issues, various methods of conventional breeding programs have been deployed, albeit with little progress. In this aspect, genetic transformation can be integral tool in breeding strategies. It permits access to an unlimited gene pool through the transfer of desirable genes [14]. The development of plant transformation techniques during the past decades has made it possible to improve crop plants by introduction of cloned genes. The two most critical steps to be the master for transformation of plants are to transfer of foreign DNA into the plant cell and regeneration of plants from transformed cells [15]. The callus induction and regeneration in tissue culture of rice depend upon different factors like genotype, type of explants and media supplement like basal salts, organic component and growth regulator [16-19]. Development of genetically engineered plants with enhanced tolerance to abiotic and biotic stresses is the important challenges in plant biotechnology research. Rice transformation is a major goal in cereal biotechnology, because rice is world's most important food crop and it is also known as model of cereal genomics [20].

In early 1980s, genetic transformation of crop plants i.e. based on recombinant DNA technology was started and this offered an advantage of transferring novel genes across taxonomic boundaries unlike conventional breeding [3]. Genetic engineering has been used as a prominent tool for rice improvement. Although gene transformation in *japonica* rice is performed routinely in several laboratories, but the system in *indica* rice is more complicated [15]. Until now, the number of copies of a gene(s) inserted and chromosomal locations of the integrated genes are not controllable, the expression of the introduced

*Corresponding author: Saroj Kumar Sah, School of Agricultural Biotechnology, Punjab Agricultural University, Ludhiana 141 004, Punjab, India
E-mail: saroj-biotec@pau.edu

genes varies among individual transformants. Therefore, a relatively large number of transgenic plants must be developed in order to select desirable transformants as well as to study the expression of introduced genes [21].

The most commonly used method for transformation are Biolistic approach and Agrobacterium mediated transformation. This review will summarise various gene delivery methods applied to improve rice traits. Subsequent molecular analysis of the transgenic rice will also be discussed. Additionally, it will consider the future prospects of transgenic researches on the crop.

Genetic Transformation

Purpose of genetic transformation

Main purpose of genetic transformation is to generate plants with useful phenotypes i.e. unachievable by conventional plant breeding, to correct faults in cultivars more efficiently than conventional breeding and to allow the commercial value of improved plant lines to be captured by those investing in the research more fully than is possible under intellectual property laws governing conventionally bred plants. Some reasons for genetic modifications are yield improvement, more resistant to disease and pest resistance, herbicides tolerance, better nutritional value, increased shelf life, better climatic survival by increasing tolerance to drought, flood or frosty conditions to allow the use of previously inhospitable land, higher crop yields, reduced farm costs, increased farm profit and improvement in health and environment..

Biological requirements for transformation

The essential requirements in a gene transfer system for production of transgenic plants are availability of a target tissue including cells competent for plant regeneration, a method to introduce DNA into those re-generable cells and a procedure to select and to re-generate transformed plants at a satisfactory frequency.

Methods of Genetic Transformation

Agrobacterium mediated genetic transformation

The soil pathogen Agrobacterim tumefaciens has been extensively studied since 1907, when it was identified as the causative agent of crown gall disease [22-24] Braun initially proposed the Agrobacterium as a source of a 'tumor inducing principle', possibly DNA, that permanently transformed plant cells from a state of quiescence to active cell division.

A. tumefaciens is a soil dwelling bacteria that naturally infect dicots and causes tumorous growth resulting in crown gall disease. Tumor formation results from incorporation of T-DNA (transfer DNA), a part of small independent DNA molecule outside the bacterial genome called Ti (tumor inducing) plasmid. Phenolic compounds exuded from plant wounds that stimulate the expression of vir genes, located on Ti plasmid and responsible for its excision, transfer and integration into plant genome. The natural capability of Agrobacterium was manipulated in plant transformation by replacing the genes causing tumorous growth by genes of interest [25]. However, Agrobacterium has a natural tendency to infect dicot plants, whereas, monocots including wheat were considered recalcitrant to Agrobacterium transformation. Therefore, most of the Agrobacterium-mediated transformation procedures were established for dicots, whereas, monocots including important cereals such as wheat, rice and maize lagged behind for a considerable time

[26-28]. Low frequency of T-DNA transfer into the target genome was the major limitation. Nonetheless improvements of co-cultivation conditions, use of acetosyringone, selection and regeneration methods for transformed tissues and incorporation of super binary vectors have helped in extending the host range of Agrobacterium to several monocots including important cereals.

Agrobacterium approach is the most popular method to deliver genes to plant cells because of its clean insertion, low-copy number of the inserted genes, easy to handle, higher efficiency, more predictable pattern of foreign DNA integration. Studies on A. tumefaciens provided the basis that has made this soil bacterium as dominant customer for plant transformation.

Critical parameters to increase the efficiency of Agrobacterium-mediated genetic transformation are: Concentration of Agrobacterium broth (in terms of OD), time of inoculation, time of co-cultivation, concentration of acetosyringone.

Optical density of Agrobacterium culture must be in the range of 0.4 to 1.00. During the Agrobacterium infection, OD_{600} of the bacterial suspension must be adjusted within 1.0 because high concentrations of bacteria caused serious injury to callus thus lowering the transformation efficiency.

Co-cultivation duration (5-25min), exposure the explants to 0.4 OD Agrobacterium for 10 min was found to be optimum.

Co-cultivation period should be between 2 to 5 days. Co-cultivation for 2 days showed better infection without browning or Agrobacterium overgrowth.

Co-cultivation with filter paper increased transformation frequency because the filter paper prevented overgrowth of bacteria. In co-cultivation medium, use of cysteine reduces browning of the calli, and Acetosyringone as well as glucose to induce the vir gene activity. Concentration of acetosyringone should be between 100-150 µM. But higher concentration of acetosyringone (200 µM) in co-cultivation media resulted in lower frequency of regeneration likely due to necrotic effect of acetosyringone.

Biolistic Process (Particle Bombardment) mediated genetic transformation

Biolistic transformation was initially invented in 1987 by American Geneticist, Associate Professor John C, Sanford from Cornell University where he and his colleagues showed that small micro projectiles could be delivered into a cell without killing it. In this approach (Also called gene gun or particle bombardment), microscopic particles coated with the DNA fragment representing the desired gene are shot into the plant cells using a special device called gene gun. The key principle behind the gun is to accelerate particles by explosion. A small portion of the DNA which enters the cells becomes incorporated into the chromosomes of the plant cell. The gene gun technique helps to overcome some of the deficiencies of Agrobacterium method such as bacterial contamination, low-efficiency of transfer to cereal crops, and inconsistency of results. The tungsten was first used or DNA coating later on it was replaced by gold particles because tungsten could be oxidized easily and it was potentially harmful to the cell.

It has many advantages than other techniques like rapid gene transfer, efficient, non-specific to tissue, complex cloning strategies with no biological constraints or host limitations and simultaneous multiple gene transfer. There are no intrinsic vector requirements so transgenes of any size and arrangement can be introduced, and multiple gene co-transformation is straightforward. It has bigger advantage that the delivered DNA can be manipulated to influence the quality and structure of the resultant transgene loci. This approach can be used for transfer of more than one gene simultaneously in a host plant. As many as 14 genes have been co-introduced in rice by this approach [29]. Nowadays, particle bombardment is the most efficient way to achieve plastid transformation in plants and is the only method

so far used to achieve mitochondrial transformation [30]. A genotype independent method for rice transformation was originally reported by Christou et al. [31] and it has been widely used throughout the world. Researchers at the International Rice Research Institute, Philippines, have used particle bombardment successfully to transform over 20 different cultivars adapted to different eco-geographic conditions. These cultivars have been transformed with a range of agronomically important genes like *psy, crt1, cry, ferritin, FRO2, Xa21, Bt, Chitinase*,

Genotype	Explants	Promoters	Strain	Plasmid	Transgene	Marker gene	Transformation Efficiency	Transgenic Analysis	References
Ratna (IET411)	Seeds	35S	LBA4404	pCAMBIA1301	hpt	Kanamycin GUS	47%	PCR Analysis	Basu et al. [33]
IRGA424	Seeds	Mpi ubi	LBA4404	pX2.H.C1mpi Cry1Bgene pC1300.ubi Cry1B.nos	Cry1Aa Cry1B	GFP hph	86.4%	PCR Analysis	Pinto et al. [34]
IR36	Mature seeds	35S	EHA105AL	pCAMBL 1301	hpt	Uid A	98% callus induction	PCR Analysis	Krishnan et al. [35]
Indica rice	Mature seeds	35S	LBA4404	pKhg4	Cry1 Ac	Hygromycin resistance gene	-	-	Guruprasad et al. [36]
IR64 Swarna CSR10 PB1	Mature dry seeds	35S	LBA4404	pCAMBL A1304	Glyoxylase 1 (Bigly1)	nptII hptII uidA GFP	Approximately 45%	PCR Analysis	Sahoo et al. [37]
Bg 250	Mature seeds	35S	GV3101	pCAMBL 1303	hptIV	Hygromycin GUS	20%	GUS Analysis	Ratnayaka et al. [38]
Pusa Basmati 1	Mature seeds	35S	LBA4404	pCAMBIA 1301	Am-SOD	Kanamycin resistance genes Hygromycin resistance genes	Very good	PCR Analysis Southern blotting	Sarangi et al. [39]
Kalizira Radhunipagol Tulsimala Pusa basmati-1	Mature embryo with endosperm	35S	EHA105	pIG121-Hm	Uid hpt	nptII	40-75%	GUS assay	Hossain et al. [40]
Heugnam—byeo Daesanbyeo	Scutellum of mature rice seeds	OsCc1 35S	LBA4404	pMJC-GB pMJC-GH	Isoflavone synthase 2 Chalcone reductase	bar hpt	12.8% (Hygromycin At Callus proliferation stage And Phosphinothricin At Shoot regeneration stage) 100% (for vice versa)	RT-PCR	Sohn et al. [41]
HKR-46 HKR-126	Seeds	EHA105		pCAMBIA 1301	Uid Hygromycin resistance gene Carbenicillin resistance gene	Hygromycin resistance gene Carbenicillin resistance gene	28.9%- 44.4%	GUS Assay	Saharan et al. [42]
Taipei 309	Mature zygotic embroys	Glutelin 35S	LBA4404	pAGt1Fe pAGt1Me	Ferritin (pte) Metallathionein-like (rgMT) Phytase (phyA)	hptII hptIV		Western blot Northern blot	Lucca et al. [43]
Senia Tebre Bahia	Mature seeds		LBA4404	pTOK233	GUS gene	hpt npt	5% (Tebra) 3% (Bahia, Senia)	Southern blot	Pons et al. [44]
Taipei 309 Pusa basmati 1 Tinawen	EC		LBA4404	pTOK233		hpt	16-31%(Taipei309) 12-21%(Pusa basmati1) 10-19%(Tinawen)		Azhakanandam et al. [45]

Basmati 122 Tulsi Vaidehi	EC		LBA4404 A281	pNO1		hpt		4.3-5.5%		Datta et al. [46]
Jarah Amoroo	EC		EHA101 AGLO AGL1	p1G121-Hm pTO134 pWBVec10a pBS360 Pbs366		Hpt bar		>10%(Jarah) 2.8%(Amaroo)		Upadhaya et al. [47]
DS20, OMCS96, OMCS97, IR72, IR64	Mature seeds	35S Ubiquitin	LBA4404	PSBbarB-Ubicore And Psb35L-Hyg-L-Gus	barB Cre Hyg gusA	Nos		1.78-13.33%	Southern blot	Hoa et al. [48]
R321	SM		A2260	pGTGUSBAR		bar		15-30% for PPT resistant calluses		Enriquez-Obregon et al. [49]
Pusa basmati	EC		LBA4404	pTOK233		hpt		25%		Mohanty et al. [50]
Seven elite lines including koshihikari	EC		EHA101	P1G12Hm pGFPKH		hpt		25-34%		Hashizume et al. [51]
Tapei 309	EC SDS		LBA4404	pTOK233		hpt		2.3-3.6%		Khanna and Raina [52]
IR64 Karnal local	EC SDS		AGL1	pCAMIAI301		hpt		2.3-3.6%		Khanna and Raina [52]
Nipponbare Zhong8215 ZAU16 91RM T8340 Pin92-528 T90502 Kaybonnet	Mature Or Immature embroys	LBA4404 EHA105	35S Bp10 Ubioquitin	PkUB pKUC pKSB pKBB	Cri 1Ab Cry1Ac	GUS		52-100%	DNA and RNA blot Analysis	Cheng et al. [53]
E-yi105 E-Wan5ZSWG Pusa basmati Tapei 309	EC			pTOK233		Hpt		13.5% (Pusa basmati) 9.1-13% (E-yi105 and E-Wan5ZSWG)		Zhang et al. [54]
	PCIE		LBA4404	pTOK233 pBin93SΩGusint p1G12Hm		hpt		A few events		Uze et al. [55]
Nipponbare Kitaake	Mature seeds	EHA101	Nopaline synthetase Pubi-1	EHA101	PIG121Hm pSMABuba		Hygromycin resistance gene bialaphos resistance gene			Toki et al. [56]
Basmati 370 Basmati 385	Calli derived from scutella	35S	EHA101		hpt gus		Hygromycin			Rashid et al. [57]
Tapei 309	Calli derived from scutella	35S	LBA4404 EHA101		Hpt gus		Hygromycin			Deng et al. [58]
Gulfmont Jefferson Radon	Immature embroys	35S	LBA4404		Hpt gus		Hygromycin			Aldemita and Hodges [59]
TCs10 IR72 Maybelle	Isolated shoot apices	Actin 1	EHA101		bar		PPT			Park et al. [60]
Radon TCS10 IR72	FIIE		LBA4404 At656	pTOK233 pCNL56		hpt		27% (Radon) 1-5% (TCS10 and IR72)		Aldemita and Hodges [59]
Taipei 309	EC		LBA4404	pTOK233		hpt				
Gulfmont Jefferson	EC		EHA101	PIG121Hm		hpt		8.3-13.3%		Dong et al. [58]
Maybelle	ISA		EHA101	pGSFR781 pBARNPT		bar		low		Park et al. [60]
Tsukinohikari Asanohikari Koshihikari	Calli derived from scutella, immature embroys, suspension cells	35S	LBA4404		Hpt gus		Hygromycin		Southern Analysis	Hiei et al. [61]

Tsukinahikari Asanohikari Kashihikari	EC FIIE SC		LBA4404 EHA101	pTOK233 pG121Hm		hpt	12.8-28.6%		Hiei et al. [61]
Shoot apices, root segments from young seedlings,scutella, immatureembroys, call, cells from suspension cultures	Tuskinikari Asanohikari Koshihikari	EHA101 LBA4404	p1G121Hm pTOK233	hpt	GUS Hygromycin	Gus assay	18.2-28.6% (Tuskinikari) 12.8-19.5% (Asanohikari) 17.6-18.8% (Koshihikari)		Hiei et al. [61]
Tannung 62	Precultured immature embroys		A281	pAG8			nptII	A few events	Chan et al. [62]
Taichung native 1	Roots from germinate seeds	Nos 35S	C58C1 LBA4404		nptII gus	G418		Southern Analysis	Chan et al. [62]
Nipponbare Fujisaka 5	Mature embryo from germinate seeds	Nos 35S	LBA4404 A281 A856		NptII Gus onco genes	Kanamycin		Southern Analysis	Raineri et al. [62]
Tainung 62	Cultured immature embroys	nos alpha-amy8	A281		nptII gus	G418			Chan et al. [62]

Table 1: Trends in rice (*Oryza sativa* L.) transformation through Agrobacterium-mediated genetic transformation.

enod12, PEPC, glgC, rolC, sd1 confirming the genotype independence of the transformation method (For details see Table 1) [32]. The main disadvantages of particle bombardment in comparison with *Agrobacterium* i.e. the tendency to generate large transgene arrays containing rearranged and broken transgene copies, are not borne out by the recent detailed structural analysis of transgene loci produced by each of the methods.

Genetic transformation occurs in the two stages: DNA transfer into the cell followed by DNA integration into the genome. The integration stage is much less efficient than the DNA transfer stage, with result of that only a small proportion of the cells that initially receive DNA actually become stably transformed. In many cells, DNA enters but expressed for a short time (transient expression), but it is never integrated and it is eventually degraded by nucleases. Transient expression occurs almost immediately after gene transfer, it does not require the regeneration of whole plants, and it occurs at a much higher frequency than stable integration. Therefore, transient expression can be used as a rapid assay to evaluate the efficiency of direct DNA transfer and to verify the function of expression constructs. In addition, transient expression following particle bombardment with a reporter gene such as gusA or gfp is used routinely to compare different expression constructs and to identify those with most appropriate activities. When the aim is to extract recombinant proteins from transgenic plants, transient expression following particle bombardment may also be used to produce small amounts of protein rapidly for testing [63]. Hoffman et al. [64] used particle bombardment for the mechanical transmission of poleroviruses and particle bombardment is routinely employed for the inoculation of whole plants and leaf tissues with viruses that are difficult to introduce via conventional mechanical infection. It also has an important role in extending virus-induced gene silencing (VIGS) into economically important crop plants [65].

Critical parameters to increase transformation efficiency

Type/physiological and developmental stage of the target tissue: Embryogenic calli of rice have been extensively used in particle bombardment experiments. The calli should be young not more than 30 days old. If we use old calli, the transformation efficiency will be low as calli loses its regeneration capacity ultimately affecting transformation efficiency.

Particle type/size, helium pressure and target distance on bombardment: Efficiency of the DNA delivery to target cells depends largely upon the size of micro-particles and the force in which these DNA coated particles are pushed towards the target cells. The depth of penetration into the target tissues can be controlled by varying particle size, helium pressure and the target distance. The target distance has direct relation with target area i.e. lesser the target distance narrow is the target area and greater the target distance, wider is the target area. So these three parameters are important in determining optimal conditions for efficient DNA delivery to target cells/tissues.

Effect of particle and DNA loads: For obtaining high efficiency of DNA delivery to target tissues, different DNA to particle ratios has been used to determine a suitable combination. A particle load of 3 mg (0.5 mg/shot) is sufficient to produce high level of transient GUS expression during conjunction by 9 μg of DNA (1:3 particle-DNA ratio).

In-Planta transformation

A number of advances in reducing the dependence to tissue culture have been made in rice transformation including, *in planta* transformation method. Cereal transformation via the tissue culture phase has been successful, but involves several limitations. The use of tissue culture allows selection of single transformed cells which are regenerated in a whole plant. However, the tissue culture approach causes somaclonal variation due to both epigenetic effects and chromosomal rearrangements [66,67]. The *in planta* transformation method overcomes the disadvantages of the conventional in vitro *Agrobacterium*-mediated transformation method. The latter requires sterile condition, that is time consuming and causes somatic mutation or somaclonal variation in plant cells during *in vitro* culture, and some plants are recalcitrant to regeneration. In contrast, *in planta* transformation involves no *in vitro* culture of plants cells or tissue, which is its greatest advantage.

Floral dip transformation

Clough and Bent [68] modified the *Agrobacterium* vacuum infiltration method to transform *Arabidopsis thaliana*. This process was eliminated in favor of simple dipping of developing floral tissues into a solution containing *Agrobacterium tumefaciens*, 5% sucrose and 500 μlL^{-1} of surfactant Silwet L-77. Sucrose and surfactant were critical to the success of the floral dip method. Plants inoculated when numerous immature floral buds and few siliques were produced transformed progeny at the highest rate. Plant tissue culture media, the hormone BAP and pH adjustment were unnecessary, and *Agrobacterium* could be applied to plants at a range of cell densities. Repeated application of *Agrobacterium* improved transformation rates and overall yield of transformants approximately two fold. Covering plants for 1 day to retain humidity after inoculation also raised transformation rates two fold.

Desfeux et al. [69] also investigated the mechanisms that underlie the floral-dip method for *Agrobacterium*-mediated transformation in *Arabidopsis* to facilitate its usage in other plant species.In manual outcrossing experiments, application of *Agrobacterium tumefaciens* to pollen donor plants did not produce any transformed progeny, whereas application of *Agrobacterium* to pollen recipient plants yielded transformants at a rate of 0.48%. Their results suggested that ovules were the site of productive transformation in the floral-dip method, and further suggested that *Agrobacterium* must be delivered to the interior of the developing gynoecium prior to locule closure.

Rod-in et al. [70] reported *Agrobacterium-mediated* transformation method for rice carrying the *gusA* gene to infect rice spikelets via the floral-dip method. The tip-cut spikelets of the rice inflorescence stage 51 (beginning of panicle emergence: tip of inflorescence emerged from sheath) were dipped in the Agrobacterium and co-cultivated at 25 °C for 3 d. Floral dip approach transform primarily anthers and to a small extent the ovary. The highest transformation efficiency in case of anther was 89.16% whereas in case of ovary 7.23% Their results suggest that floral-dip transformation is a simple potential tool for production of transgenic rice with no requirement of tissue culture.

Naseri et al. [71] demonstrated the inoculation of *A. tumefaciens* into embryonic apical meristem of the soaked seeds, a region on the seed surface where a shoot would later emerge and was pierced twice up to depth of about 1 to 1.5 mm with a needle (0.70 mm) dipped in the *A. tumefaciens* inoculums. The inoculated seeds were then placed on filter papers on wet per liter in flasks covered with aluminum foil and incubated at 23°C in dark for nine days that 70 to 75% of inoculated seeds are germinated to seedlings. PCR results showed that 24% plants inoculated with *Agrobacterium* integrated transgene.

Critical parameters: During the *agrobacterium* infection, OD$_{600}$ of the bacterial suspension must be 1.0 for higher transformation.

Source of Explants

Tissue culture which involves manipulating the totipotent nature of plant cells is one of the necessary technology for production of transgenic plants. For genetic transformation in different experiments, many types of explants has been used by researchers likewise mature seeds, mature zygotic embryos, immature embryos, mature embryos, calli, inflorescence, embryogenic apical meristem, spikelet, roots, isolated shoot apices (Tables 1-3).

Vectors and Markers in Gene Transfer

In plant research, binary vectors are standard tools for delivery of a wide range of genes into the cells of higher plants [72] This vector has certain important characters like multiple cloning sites, bacterial origin of replication, unique restriction sites and selectable markers gene cassettes enabling the recognition of untransformed cells. A selectable marker gene encodes a product that allows the transformed cell to survive and grow under conditions that kill or restrict the growth of non-transformed cells. The most used genes in rice are dominant selectable markers that confer resistant to antibiotics or herbicides i.e. *ble* (glycopeptides binding protein), *dhfr* (dihydrofolate reductase), *hpt* (hygromycin phosphotransferase), *nptII* (neomycin phosphotransferase), bar and pat (phosphinothricin acetyltransferase), *csr1-1* (acetolactate synthase), *tms2* (indolaacetic acid hydrolase) (For details see Table 1 [20]. In addition to selectable marker genes, plant researchers also used reporter genes known as screenable marker, a visual marker or a scorable marker,such as *gus A* (ß-glucuronidase, GUS), *luc* (firefly luciferase), *gfp* (green fluorescent protein) (For details see Table 1) [20]. It generates a product that can be detected using a simple and often quantitative assay. It is mainly used for confirming transformation, determining transformation efficiency and monitoring gene or protein activity. It has ability to form fusion genes at the transcriptional level and it can be used to assay the activity of regulatory elements. It also forms translational fusion products, which allows them to be used, to monitor protein localization in the cell or at a whole-plant level. ß-glucuronidase activity can be easily monitored within 1-2 days post co-cultivation in infected plant tissues by enzymatic conversion of colourless X-Gluc substrate to blue precipitates [73] so, it is a powerful assay for detecting transformation events.

Next major aspects of binary vector is the usage of constitutive promoters providing transgene expression in majority of the plant tissues. Studies have been done on rice transformation by using CaMV35s constitutive promoters. By using this promoter Brietler et al. [74] reported 70-80% genetic transformation for *yfp* and *hph* genes. Besides this promoter, there are many others has been used for genetic transformation in rice such as Actin 1, ubi-1, ubiquitin, OSCc1 (Tables 1-3).

Transgenic Analysis

To confirm the real transgenic, efforts should be undertaken with a long term goal rather than just the analysis of the initial transformants. Development of transgenic plants is not only limited to the confinement of laboratory, it should have wider scope of application. So, to test the real transgenic we have to go for different complementary tests like GUS Assay, PCR, Semi-quantitative PCR, Quantitative Real Time PCR, Southern blotting, Western blotting, ELISA, Northern blotting. Most of the studies conducted molecular analysis of the primary transformants, rather than proceeding further. GUS analysis revealed transformation efficiency in wide range (7.23% to 95%) that has been shown by different studies by using different explants in various cultivars of rice [21,75-

Genotype	Explant	Promoter	plasmid	Transgene	Marker gene	Efficiency	Analysis	References
MR219	Mature seeds	35S	pCAMBIA1304	hptII	gusA mgfp5	Regenerability 100%	Southern Analysis GUS staining RT-PCR Flouroscence microscopy Stastical Analysis	Htwe et al. [78]
Swarna Mahsuri	Mature seeds	35S	pCAMBIA1301	Cry1Ac hpt	GUS hygromycin	Regeneration efficiency Swarna (79.23%) Mahsuri(86.07%)	GUS assay Southern blot and PCR Analysis	Pravin et al. [79]
PAU201	Mature seeds	35S	pWRG1515	GUS	GUS	32-39%	GUS Assay	Wani et al. [77]
Rice cv. MR81 Taipei 309	Mature seeds	35S	pRQ6	hph	GUS	>95%	GUS Assay	Rahman et al. [76]
BMS (black mexican sweet) Maize	Mature embryo	35S	pRESQ70	HPH	GUS		PCR and southern analysis	Sivamani et al. [81]
L1345 L3	Immature embryos	35S	Pcambia3301	PAT	GUS	L1345 (20.23-49.38%) L3 (25.19-69.02%)	GUS Analysis	Petrillo et al. [82]
Xiushui04 Jia59	Immature embryo	35S	pCB1	bar			Southern Analysis	Zhao et al. [83]
Rice cv. Chainat 1	Mature seeds	35S	PCAMBIA 1305.1	Chitinase	GUS Hygromycin resistance	100% at 9cm distance from stopping screen to callus and 35.5%at 12 cm distance	PCR Analysis	Maneewan et al. [84]
Rasi Taipei309	Mature seeds	35S	PUCGUS pHX4	hph	GUS	Rasi (15.5-24.2%) Taipei309 (33.3-37.55)	GUS assay	Ramesh et al. [86]
Taipei309 Nippanbare	Mature seeds embryos	35S	pYEP p1LTAB227	Yfp hph		70-80%	Southern blot analysis	Breitler et al. [74]
Eyi105 Ewan 5	Mature seeds	35S	pJIMB15 pRSSGNA1 pAHXA21	hpt gusA Xa21 gna	Hygromycin	Eyi105 (16%) Ewan 5 (18%)	PCR analysis Dot and southern blot analysis RT-PCR Analysis Genetic Analysis	Tang et al. [88]
IR64 IR72 Minghui 63 BG90-2	seeds	35S	pC822 pHX4	Xa21 hph	Hygromycin	IR64(91%) IR72(100%) Minghui 63(54%) BG90-2 (100%)	Southern blot	Zhang et al. [54]
Jingyin 119 Zhongbai 4	Immature embroys	35S Act1	pCB1	Cecropin Bgene	bar		PCR analysis Dot and southern blot analysis Northern blot analysis	Danian et al. [89]
Taipei 309 77125 Tetep TN1 8706	Immature emryo Callus	Actin1 35S	pAct1D pNG3 PMON410	hph	GUS	Taipei 309 (25.6%) 8706 (31.0-42.3%)	GUS assay	Li et al. [21]

Table 2: Trends in rice (*Oryza sativa* L.) transformation through particle bombardment.

77]. PCR analysis showed various range of transformation efficiency i.e. 16%-100% [48-53]. Molecular analysis were carried out by most of the workers to check the integration pattern and copy number by southern blot analysis, northern and real time pcr [78,79,81-84].

Future Prospects

As we look to the future of rice biotechnology, we have many transformation technique but till now no transgenic line of rice has been released successfully. Near about 100 genotypes have been used by different groups for transformation (Tables 1-3). Among them, Taipei 309 (*Japonica* rice) and Basmati (*Indica* rice) has been widely used in *Agrobacterium*-mediated genetic transformation with an efficiency of 31% and 40-75%,respectively.Though some reports has been shown that some of the transgenic line has been released. Gupta et al. [85] reported that Huaghi-1 Rice for insect resistance has been released by Huazhong Agricultural University, China. SO, we have to focus on particular assay of that transgene which has been transferred along with series of confirmatory test. Ultimately we need high yield so, our main objective should be to check our transgene is working or not, which may not affect the yield. As for example, if we transfer salinity tolerance gene, we must go for salinity screening technique, in which transgenic line must show higher tolerance level in respect to wild. In most of the research this assay are lacking. So, to check or confirm real transgenic we must go for particular assay.

Conclusion

In the present scenario, population growth is increasing day by day along with demand for food supply. Genetic transformation must be implemented to bridge the gap between production and human need. Whatever achievement has been done so far in transgenic, is not sufficient to fill the demand. The full realization of the plant biotechnology revolution depends on successful and innovative

Genotype	Explants	Promoters	Strain	Plasmid	Transgene	Marker gene	Efficiency	Analysis	References
RD41	Rice spikelets	35S	AGL1	pCAMBIA1304	hptII	gusA gfp	Anther-89.16% Ovary- 7.23%	Histochemical Gus assay	Rodin et al. [47]
Zhong Zuo321 Nippanbare IR64	Antepenultimate emerged leaf of plantlets at tillering stage	Ubi-1 CsVMV	EHA101 AGL1 LBA4404 GV2200 GV3101	pC5300 PB10S738	UidA Sc4A	GUS	-	Real time PCR Analysis	Andrieu et al. [87]
Oryza sativa	Seeds	35S	EHA101	pAJ21	TLPD34		24%	PCR Analysis	Naseri et al. [71]
Koshihikari	Embryonic apical meristem	35S	M-21 LBA4404	P1G121Hm	Hpr nptII	GUS	40-43%	PCR	Supartana et al. [90]
Taipei 309	Infloroscence	35S	LBA4404	PJD4	Bialaphos resistance gene	GUS	-	GenomeDNA blot analysis PCR analysis	Dong et al. [91]

Table 3: Trends in rice (*Oryza sativa* L.) transformation through in-planta transformation.

research, as well as on favorable regulatory guidelines and public acceptance. Thus, all the strategies discussed in the present review will definitely contribute to biotechnological breeding programs of rice for its improvement.

References

1. Khush GS (2003) Productivity improvement in rice. Nutritional Review 61: 114-116.

2. Datta SK (2004) Rice biotechnology: A need for developing countries. Ag Bio Forum 7: 31-35.

3. Bakshi S, Dewan D (2013) Status of Transgenic Cereal Crops: A Review. Clon Transgen 3: 1.

4. Epstein E, Norlyn JD, Rush DW, Kingsbury RW, Kelley DB, et al. (1980) Saline culture of crops: a genetic approach. Science 210: 399-404.

5. Boyer JS (1982) Plant productivity and environment. Science 218: 443.

6. Grover A, Minhas D (2000) Towards production of abiotic stress tolerant transgenic rice plants: Issues, progress and future research needs. Proeedings of Indian National Science Academy (PINSA) B66: 13-32.

7. Thakur P, Kumar S, Malik JA, Berger JD, Nayyar H (2010) Cold stress effects on reproductive development in grain crops: An overview. Environmental and Experimental Botany 67: 429-443.

8. Mantri N, Patade V, Penna S, Ford R, Pang E (2012) Abiotic stress responses in plants: present and future. In: Ahmad P, Prasad MNV (eds.) Abiotic stress responses in plants: metabolism, productivity and sustainability. Springer, New York, pp. 1-19.

9. Munns R, Tester M (2008) Mechanisms of salinity tolerance. Annual Review of Plant Biology 59: 651- 681.

10. Wang W, Vincur B, Altman (2003) Plant responses to drought, salinity and extreme temperature. es: Towards genetic engineering for stress tolerance. Planta 218: 1-14.

11. Shanker A, Venkateswarlu B (2011) Abiotic stress Response in Plants-Physiological, Biochemical and Genetic perspectives. Intech Publishers.

12. Ismail AM, Heuer S, Thomson MJ, Wissuwa M (2007) Genetic and Genomic approaches to develop rice germplasm for problem soils. Plant Molecular Biology 65: 547-570.

13. Singh AK, Ansari MW, Preek A, Singla-pareek SL (2008) Raising salinity tolerant rice: Recent progress and future prospectives. Physiology and Molecular Biology of Plant 14: 23-32.

14. Hiei Y, Ohta S, Komari T, Kubo T (1997) Transformation of rice mediated by Agrobacterium tumefaciens. Plant Molecular Biology 35: 205-218.

15. Yookongkaew N, Srivatanakul M, Narangajavana J (2007) Development of Genotype independent regeneration system for transformation of rice (Oryza sativa ssp. Indica). Journal of Plant Research 120: 237-245.

16. Abe T, Futsuhara Y (1984) Varietal differences in plant regeneration from root callus tissues of rice.

17. Abe T, Futsuhara Y (1986) Genotypic variability for callus formation and plant regeneration in rice. Theoretical Applied Genetics 72: 3-10.

18. Kavi kishor PB, Reddy GM (1986) Retention and revival of regenerating ability by osmotic adjustment in long-term cultures of four varieties of rice. Journal of Plant Physiology 126: 49-54.

19. Mikami T, Kinoshita T (1988) Genotypic effects on the callus formation from different explants of rice, Oryza sativa L. Plant Cell and Tissue Organ Culture 12: 311-314.

20. Twyman RM, Stoger E, Kohli A, Capell T, Christou P (2002) Selectable and Screenable Markers for Rice Transformation. Molecular Methods ofPlant Analysis, Vol. 22 Testing for Genetic Manipulation in Plants Edited by J.F. Jackson, H.F. Linskens, R.B. Inman © Springer-Verlag Berlin Heidelberg 2002.

21. Li L, Qu R, Kochko AD, Fauquet C, Beachy RN (1993) An improved rice transformation system using the biolistic method. Plant Cell Reports 12: 250-255.

22. Smith EF, Townsend CO (1907) A plant tumor of bacterial origin. Science 25: 671-673.

23. Braun, A C (1947) Thermal studies on the factors responsible for tumor initiation in crown gall. American Journal of Botany 34: 234-240.

24. Braun AC, Mandle RJ (1948) Studies on the inactivation of the tumor-inducing principle in crown gall. Growth 12: 255-269.

25. Horsch RB, Fraley RT, Rogers SG, Sanders PR, Lloyd A, Hoffman N (1984) Inheritance of functional genes in plants. Science 223: 496-498.

26. Barro F, Rooke L, Bekes F, Gras P, Tatham AS, et al. (1997) Transformation of wheat with high molecular weight subunit genes resulted in improved functional properties. National Biotechnology 15: 1295-1299.

27. Dobrzańska M, Krysiak C, Kraszewska E (1997) Transient and stable transformation of wheat with DNA preparations delivered by a biolistic method. Acta Physiologiae Plantarum 19: 277-284.

28. Takumi S, Shimada T (1997) Genetic transformation of durum wheat (Triticum durum Desf.) through particle bombardment of scuteller tissues. Plant Biotechnology 14: 151-156.

29. Cheng X, Sardana R, Kaplan H, Altosaar I (1998) Agrobacterium-transformed rice plants expressing synthetic cryIA(b) and cryIA(c) genes are highly toxic to striped stem borer and yellow stem borer. Proceedings of National Academy and Sciences USA 95: 2767-2772.

30. Johnston SA, Anziano PQ, Shark K, Sanford JC, Butow RA (1988) Mitochondrial transformation in yeast by bombardment with microprojectiles. Science 240: 1538-1541.

31. Christou P, Ford TL, Kofron M (1991) Genotype independent stable transformation of rice (Oryza sativa) plants. Bio/Technology 9: 957-962.

32. Alpeter F, Baisakh N, Beachy R, Bock R (2005) Particle bombardment and the genetic enhancement of crops: myths and realities. Molecular Breeding 15: 305-327.

33. Basu A, Ray S, Sarkar S, chadhuri RT, Kundu S (2014) Agrobacterium mediated genetic tansformation of popular indica rice Ratna (IET1411). African journal of biotechnology 13: 4187-4197.

34. Pinto LMN, Fiuza LM, Ziegler D, Oliveira JVD, Menezes VG, et al. (2013) Indica Rice Cultivar IRGA 424, Transformed with cry Genes of B. thuringiensis,

Provided High Resistance Against Spodoptera frugiperda (Lepidoptera: Noctuidae). Journal of Economic Entomology 106: 2585-2594.

35. Radhesh Krishnan SR, Priya AM, Ramesh M (2013) Rapid regeneration and ploidy stability of 'cvIR36' indica rice (Oryza Sativa. L) confers efficient protocol for in vitro callus organogenesis and Agrobacterium tumefaciens mediated transformation. Botanical Studies 54: 47.

36. Guruprasad M, Raja DS, Jaffar SK, Shanthisri KV, Naik MS (2012) An Efficient And Regeneration Protocol for Agrobacterium mediated transformation of indica rice. International journal of Pharmacy And Technology 4: 4280-4286.

37. Sahoo KK, Tripathi AK, Pareek A, Sopory SK, Singla-Pareek SL(2011) An improved protocol for efficient transformation and regeneration of diverse indica rice cultivars. Plant Methods 7: 49.

38. Ratnayake RMKL, Hettiarachchi GHCM (2010) Development of an Efficient Agrobacterium Mediated Transformation Protocol for Sri Lankan Rice Variety (Bg 250). Tropical Agricultural Research 22: 45-53.

39. Sarangi S, Ghosh J, Bora A, Das S, Mandal AB (2009) Agrobacterium mediated genetic transformation of Indica rice varieties involving Am-SOD gene. Indian Journal Of Biotechnology 10: 9-18.

40. Hossain MR, Hassan L, Patwary AK, Ferdous MJ (2009) Optimization of Agrobacterium tumefaciens mediated genetic transformation protocol for aromatic rice. Journal of Bangladesh Agricultural. University 7: 235-240.

41. Sohn SI, Kim YH, Cho JH, Kim JK, Lee JY (2006) An Efficient Selection Scheme for Agrobacterium-mediated Co-transformation of Rice Using Two Selectable Marker Genes hpt and bar. Korean Journal of Breeding 38: 173-179.

42. Saharan V, Yadav RC, Yadav NR, Ram K (2004) Studies on improved Agrobacterium-mediated transformation in two indica rice (Oryza sativa L.). African Journal of Biotechnology 3: 572-575.

43. Lucca P, Hurrell R, Potrykus I (2001) Genetic engineering approaches to improve the bioavailability and the level of iron in rice grains. Theoretical Applied Genetics 102: 392-397.

44. Pons MJ, MarfaV, Mele E, Messeguer J (2000) Regeneration and genetic transformation of Spanish rice cultivars using mature embryos. Euphytica 114: 117-122.

45. Azhakanandam K, McCabe MS, Power B, Lowe KC, Cocking EC and Davey MR (2000) T-DNA transfer, integration, expression and inheritance in rice: effects of plant genotype and Agrobacterium super-virulence. Journal of Plant Physiology 157: 429-439.

46. Datta K, Koukolıkova-Nicola Z, Baisakh N, Oliva N, Datta SK (2000) Agrobacterium-mediated engineering for sheath blight resistance of indica rice cultivars from different ecosystems. Theoretical Applied Genetics 100: 832-839.

47. Upadhyaya NM, Surin B, Ramm K, Gaudron J, Schnmann PHD, Taylor W Waterhouse PM and Wang MB (2000) Agrobacterium mediated transformation of Australian rice cultivars Jarrah and Amaroo using modified promoters and selectable markers. Austratlian Journal of Plant Physiology 27: 201-210.

48. Hoa THC, Bahagiawati A, Hodges TK (1999) Agrobacterium-Mediated Transformation of Indica Rice Cultivars Grown in Vietnam. Omonrice 7.

49. Enrıquez-Obregon GA, Prieto-Samsonov DL, de la Riva GA, Perez M, Selman-Housein G, et al. (1999) Agrobacterium mediated Japonica rice transformation: a procedure assisted by an antinecrotic treatment. Plant Cell Tissue and Organ Culture 59: 159-168.

50. Mohanty A, Sarma NP, Tyagi AK (1999) Agrobacterium-meditated high frequency transformation of an elite indica rice variety Pusa Basmati 1 and transmission of the transgenes to R2 progeny. Plant Science 147: 127-137.

51. Hashizume F, Tsuchiya T, Ugaki M, Niwa Y, Tachibana N, et al. (1999) Efficient Agrobacterium-mediated transformation and the usefulness of a sythetic GFP reporter gene in leading varieties of japonica rice. Plant Biotechnol. 16: 397-401.

52. Khanna HK, Raina SK (1999) Agrobacterium-mediated transformation of indica rice cultivars using binary and superbinary vectors. Australian Journal of Plant Physiology 26: 311-324.

53. Cheng X, Sardana R, Kaplan H, Altosaar I (1998) Agrobacterium-transformed rice plants expressing synthetic cryIA(b) and cryIA(c) genes are highly toxic to striped stem borer and yellow stem borer. Proceedings of National Academy and Sciences USA 95: 2767-2772.

54. Zhang J, Xu RJ, Elliott MC, Chen DF (1997)Agrobacterium-mediated transformation of elite indica and japonica rice cultivars. Molecular Biotechnology 8: 223-231.

55. Uze M, Wunn J, Puonti-Kaerlas J, Potrykus I, Sautter C (1997) Plasmolysis of precultured immature embryos improves Agrobacterium mediated gene transfer to rice (Oryza sativa L.). Plant Science 130: 87-95.

56. Toki S (1997) Rapid and Efficient Agrobacterium-Mediated Transformation in Rice. Plant Molecular Biology Reporter 15.

57. Rashid H, Yokoi S, Toriyama K, Hinata K (1996) Transgenic plant production mediated by Agrobacterium in indica rice. Plant Cell Reporter 15: 727-730.

58. Dong J, Teng W, Buchholz WG, Hall TC (1996) Agrobacterium-mediated transformation of javanica rice. Molecular Breeding 2: 267-276.

59. Aldemita RR, Hodges TK (1996) Agrobacterium tumefaciens-mediated transformation of japonica

60. Park SH, Pinson SR, Smith RR (1996) T-DNA integration into genomic DNA of rice following Agrobacterium inoculation of isolated shoot apices. Plant Molecular Biology 32: 1135-1148.

61. Hiei Y, Ohta S, Komari T, Kumashiro T (1994) Efficient transformation of rice (Oryza sativa L.) mediated by Agrobacterium and sequence analysis of the boundaries of the T-DNA. The Plant Journal 6: 271-282.

62. Chan MT, Chang HH, Ho SL, Tong WF, Yu SM (1993) Agrobacterium-mediated production of transgenic rice plants expressing a chimeric a-amylase promoter/b-glucuronidase gene. Plant Molecular Biology 22: 491-506.

63. Twyman RM, Stoger E, Schillberg S, Christou P, Fischer R (2003) Molecular farming in plants: host systems and expression technology. Trends in Biotechnology 21: 570-8.

64. Hoffmann K, Verbeek M, Romano A, Dullemans A M, vanden Heuvel JF, et al. (2001) Mechanical transmission of polerovirus. J Virol Methods 91: 197-201.

65. Fofana IBF, Sangare A, Collier R, Taylor C, Fauquet CM (2005) A geminivirus-induced gene silencing system for gene function validation in Cassava. Plant Molecular Biology 56: 613-624.

66. Kaeppler SM, Kaeppler HF, Rhee Y (2000) Epigenetic aspects of somaclonal variation in plants. Plant Molecular Biology 43: 179-188.

67. Mohan Jain S (2001) Tissue culture-derived variation in crop improvement. Euphytica 118: 153-166.

68. Clough SJ, Bent AF(1998) Floral dip: a simplified method for Agrobacterium mediated transformation of Arabidopsis thaliana. Plant Journal 16: 735-743.

69. Desfeux C, Clough SJ, Bent AF (2000) Female reproductive tissues are the primary target of Agrobacterium mediated transformation by the Arabidopsis floral dip method. Plant Physiology 123: 895-904.

70. Rod-in W Sujipuli K, Ratanasut K (2014) The floral-dip method for rice (Oryza sativa) transformation. Journal of Agricultural Technology 10(2): 467-474.

71. Naseri G, Sohani MM, Pourmassalehgou A, Allahi S (2012) In planta transformation of rice (Oryza sativa) using thaumatin-like protein gene for enhancing resistance to sheath blight. African Journal of Biotechnology 11(31): 7885-7893.

72. Dafny-Yelin M, Tzfira T (2007) Delivery of multiple transgenes to plant cells. Plant Physiology145: 1118-1128.

73. Jefferson RA (1987) Assaying chimeric genes in plants : The GUS gene fusion system. Plant Mol Biol Rep 5: 387-405.

74. Breitler JC, Labeyrie A, Meynard D, Legavre T, Guiderdoni E (2002) Efficient microprojectile bombardment-mediated transformation of rice using gene cassettes. Theoretical Applied Genetics 104: 709-719.

75. Petrillo CP, Carneiro NP, Antanio Álvaro, Carvalho CHS, Alves JD, et al. (2008) Otimizaco dos parametros de bombardeamento de partículas para a transformação genética de linhagens brasileiras de milho. Pesq. agropec. bras., Brasília 43: 371-378.

76. Rahman ZA, Seman ZA, Roowi S, Basirun N, Subramaniam S (2010) Production of transgenic Indica rice (Oryza sativa L.) Cv. MR 81 via particle bombardment system. Emir. J. Food Agric. 22 (5): 353-366.

77. Wani SH, Sanghera GS, Haribhushan A, Singh NB, Gosal SS (2011) Bio-physical parameters affecting transient GUS expression in Indica rice variety PAU 201 via particle bombardment. Elixir Bio. Tech. 34: 2496-2501.

78. Htwe NN, Ling HC, Zaman FQ, Mazizh M (2014) Plant genetic Transformation efficiency of selected Malaysian Rice Based on Selectable Marker Gene (hptII). Pakistan Journal of Biological Sciences 17: 472-481.

79. Pravin J, Mahendra D, Saluja T, Sarawgi AK, Ravi S, et al. (2011) Assessment of critical factors influencing callus induction, in vitro regeneration and selection of bombarded indica rice genotypes. Cell & Plant Science 2: 24-42.

80. Rahman ZA, Seman ZA, Roowi S, Basirun N, Subramaniam S (2010) Production of transgenic Indica rice (Oryza sativa L.) Cv. MR 81 via particle bombardment system. Emir. J. Food Agric. 22: 353-366.

81. Sivamani E, DeLong RK, Qu R (2009) Protamine-mediated DNA coating remarkably improves bombardment transformation efficiency in plant cells. Plant Cell Reporter 28: 213-221.

82. Petrillo CP, Carneiro NP, Antanio Álvaro Corsetti Purcino AAC, Carvalho CHS, Alves JD, et al. (2008) Otimizaco dos parametros de bombardeamento de partículas para a transformação genética de linhagens brasileiras de milho. Pesq. agropec. bras., Brasília 43: 371-378.

83. Zhao Y, Qian Q, Wang H, Huang D (2007) Hereditary Behavior of bar Gene Cassette is Complex in Rice Mediated by Particle Bombardment. Journal of Genetics and Genomics 34: 824-835.

84. Maneewan K, Bunnag S, Theerakulpisut P, Kosittrakun M, Suwanagul A (2005) Transformation of rice (Oryza sativa L.) cv. Chainat 1 using chitinase gene. Songklanakarin Journal of Science and Technology 27: 1151-1162.

85. Gupta SM, Grover A, Nasim M (2013) Transgenic Technologies In Agriculture: From Lab To field to market, CIBTech Journal of Biotechnology 3: 20-47.

86. Ramesh M, Gupta AK (2005) Transient expression of β-glucuronidase gene in indica and japonica rice (Oryza sativa L.) callus cultures after different stages of co-bombardment. African Journal of Biotechnology 4: 596-600.

87. Andrieu A, Breitler JC, Sire C, Meynard D, Gantet, et al. (2012) An in planta, Agrobacterium-mediated transient gene expression method for inducing gene silencing in rice (Oryza sativa L.) leaves. Rice 5: 23.

88. Tang K, Tinjuangjun P, Xu Y, Sun X, Gatehouse JA, et al. (1999) Particle-bombardment-mediated co-transformation of elite Chinese rice cultivars with genes conferring resistance to bacterial blight and sap-sucking insect pests. Planta 208: 552-563.

89. Danian H, Bing Z, Wei Y, Rui X, Han X, et al. (1996) Introduction of cecropin B gene into rice (Oryza sativa L.) by particle bombardment and analysis of transgenic plants. Science in China 39: 652-661.

90. Supartana P, Shimizu T, Shioiri H, Nogawa M, Nozue M, et al. (2005) Development of Simple and Efficient in Planta Transformation Method for Rice (Oryza sativa L.) Using Agrobacterium tumefaciens. Journal of bioscience and bioengineering 100: 391-397.

91. Dong J, Kharb P, Teng W, Hall TC (2001) Characterization of rice transformed via an Agrobacterium-mediated inflorescence approach. Molecular Breeding 7: 187-194.

15

Genetic Analysis and Traits Association in F2 Intervarietal Populations in Rice Under Aerobic Condition

Farhad Kahani and Shailaja Hittalmani*

Marker Assisted Selection Laboratory, Genetics and Plant Breeding, University of Agricultural Sciences, GKVK, Bangalore 560065, India
Professor and University Head, Genetics and Plant Breeding, University of Agricultural Sciences, GKVK, Bangalore-560065, India

Abstract

Genetic variability studies provide basic information concerning the genetic properties of the population based on which breeding methods could be formulated for further improvement of the crop. The estimates of heritability, coefficients of variability and genetic advance was computed in F_2 segregating populations of 15 crosses for 14 characters including drought and yield contributing traits under aerobic condition during dry season 2012. The estimates of phenotypic coefficients of variation (PCV) were high and moderate for days between flowering and maturity (10.71%), number of tillers plant (19.70%), number of panicles plant (21.44%), 100 grain weight (g) (27.01%), panicle exertion (14.75%), panicle length (cm) (12.90%), leaf width (cm) (20.96%), straw weight (g) (30.99%) and yield (g) (34.04%)in different crosses. High heritability coupled with high and moderate genetic advance was observed for all the plant traits observed. Correlation studies revealed that Grain yield plant[-1] was positively and significantly correlated with number of tillers, number of panicles, grain length and straw weight. However Grain yield plant[-1] was negatively significantly correlated with Days to flowering, Days to maturity, Plant Height, 100-grain weight, Grain width and Leaf width [1,2].

Keywords: Rice; Aerobic; Genetic variability; Heritability; Genetic advance

Introduction

Rice is the life and the prince among cereals as this unique grain helps to sustain two thirds of the world's population. Asia is the biggest rice producer and consumer, accounting for 90 per cent of the world's production and consumption of rice. Except Antarctica, it is grown in all the continents, occupying 159 million hectare area and producing 683 million tones (equivalent to 456 million tons of milled rice); In Asia, more than 50% of all water used for irrigation is for rice. About 55% of the rice area is irrigated and accounts for 75% of the total rice production in the world. Being an extravagant consumer of water, rice uses around 5000 liters of fresh water to produce 1 kg of rice. The increase in depletion of fresh water resources is a major threat to the traditional way of rice cultivation. Efforts are therefore underway to develop water saving technologies such as alternate wetting and dry continuous soil saturation, irrigation at fixed soil moisture tensions varying from 0 to 40 kPa, or irrigation at an interval of 1-5 days after disappearance of standing water. To combat this problem water-saving rice production system like, aerobic rice cultivation is being popularized to obtain optimum yield with less water consumption. True aerobic rice system (ARS) is a new production system in which rice is grown under non-puddled, non-flooded, and non-saturated soil conditions. Thus in ARS, soils are kept in aerobic situation almost throughout the rice growing season. As already pointed out, ARS aims at growing rice without puddling and flooding under non-saturated soil conditions as other upland crops. The driving force behind ARS is water economy and reported a saving of 73% in land preparation and 56% during crop growth. The work on developing varieties suitable for ARS started only recently and is generally restricted to screening available varieties [3-15].

Moreover before launching any breeding programme, a breeder should have a thorough knowledge on nature and magnitude of genetic variability, heritability, genetic advance, genetic divergence and character association in a crop species. Since yield is inherited in a complex way and is influenced by the environment path coefficient analysis will be an added advantage to the breeder in crop improvement programme. The study on above aspects is essential to identify superior genotypes. The crosses between the genotypes with maximum genetic divergence would be responsible for improvement as they are likely to yield desirable recombinants in the progeny. Keeping in view the genetic studies in aerobic rice were undertaken to estimate the genetic component of variance for yield and yield components involving aerobic cultivars and to compute the heritability, coefficients of variability and genetic advance in F_2 segregating populations of the 15 crosses for 14 characters.

Material and Methods

Crossing, development of f_1 and f_2 populations

Fifteen F2 populations were generated using di allel mating design of rice by crossing 6parental genotypes viz., Moroberekean, Gandhasala, IR-50, IM-90, IM-114 and OYC-145. Staggered sowing of the selected parental genotypes was done to achieve synchronization in the flowering for effective crossing programme to generate F1. The seedlings were raised during dry season 2011 following all the recommended agronomic practices. At panicle emergence and flowering stage, the florets of female parents were hand emasculated early in the morning, before 7 a.m. and later the pollen was collected from male parent and dusted on to the stigma around 11 a.m which is the ideal time for effective pollination. To avoid contaminations from

***Corresponding author:** Shailaja Hittalmani, Professor and University Head, Genetics and Plant Breeding, University of Agricultural Sciences, GKVK, Bangalore 560065, India, E-mail: shailajah_maslab@rediffmail.com

foreign pollens, emasculated panicle were covered with butter paper packet. The seeds set on female plants were harvested, around 25-27 days after crossing event. Fifteen crosses were effected in a pair wise combination during dry season 2011, at K block, UAS, Bangalore.

Evaluation of F_2 generation population

The F_2 seeds of all the 15 crosses of rice were directly sown in the field under moisture scarce aerobic situation during dry season-2012 with single seeds per hill at spacing of 25 × 25 cm. On an average of 300 population size for each cross was maintained along with two rows parental lines. Mean values were utilized for statistical analysis and the characters observed for eliciting the information are: days to flowering, days between flowering and maturity, days to maturity, plant height (cm), number of tillers plant, number of Panicles plant[-1], productive tillers plant[-1], mother panicle weight (g), 100 grain weight (g), panicle length (cm), panicle exertion (± cm), leaf length (cm), leaf width (cm), grain length (mm), grain breadth (mm), straw weight (g) and grain yield plant-1 (g).

Statistical analysis

The observations recorded in respect of all the above quantitative traits were subjected to following standard statistical analysis: Descriptive statistics Sunderaraj et al. (1972), Mean, Range, Standard error, Variance, Skewness and Kurtosis. The mean values of quantitative traits were used to estimate coefficients of skewness and kurtosis using 'SPSS' software program, Genetic variability parameters, Coefficients of Variation Burton and, Phenotypic coefficient of variation (PCV), Heritability (h^2)Johnson et al., (1955), Genetic advance (GA) [16-18].

$$PCV\ (\%) = \frac{\sigma_p^2}{\overline{X}} \times 100$$

$$h_{(bs)}^2 = \frac{\sigma_g^2}{\sigma_p^2} \times 100$$

$$GA = K\ h2\ \sigma p$$

$$GAM = \frac{GA}{\overline{X}} \times 100$$

Results and Discussion

Genotypic coefficient of variation, phenotypic coefficient of variation, heritability in narrow sense, genetic advance and genetic advance as percentage of means were estimated for yield and quality in F_2 generation as presented in Table1. Expectedly phenotypic coefficient of variation (PCV) was higher than genotypic coefficient of variation (GCV) in all the characters studied. The difference between PCV and GCV is probably due to environmental effects. High heritability estimates for all the characters except panicle excretion and straw weight suggesting that the environmental factors did not affect greatly the phenotypic performance of these traits. Highest PCV (34.04%) and GCV (33.58%) were observed for grain yield plant[-1]. PCV generally ranged between 2.67% for grain width to 34.04 % for grain yield plant[-1]. Similarly, GCV ranged between 2.12% for grain width to 33.58% for grain yield plant[-1]. A similar finding of higher PCV than GCV was also reported by [2,10,15]. Generally, heritability in broad sense estimate varied from 19.67% for panicle exertion and 99.95% for leaf width, respectively. Similarly, genetic advance was ranged between 0.16% for panicle excretion and 21.36% for plant height. A joint consideration of GCV, heritability broad sense and genetic advance revealed that 100-grain weight (27.01, 99.20 and 16.18%), leaf width (20.96, 99.95 and 11.46%) and grain yield plant[-1] (33.58, 96.12 and 16.04%) combined high GCV, high heritability broad sense and moderate genetic advance. Thus, high estimates of GCV and heritability could be good predictors of seed yield in rice. Hence selection based on the phenotypic performance will be reliable and effective. Furthermore, moderate to high heritability, GCV and GA% in a mean could be explained by additive gene action and their improvement could be achieved through mass selection [19-21] (Table 1).

Correlation studies revealed that Grain yield plant[-1] was positively significantly correlated with number of tillers, number of panicles, grain length and straw weight. These results suggest that selection to improve rice yield directed by phenotype of these traits may be effective [20]. However Grain yield plant[-1] was negatively significantly correlated with days to flowering, days to flowering, days to maturity, plant height, 100-grain weight, grain width and leaf width. Similar findings were also recorded by [22] for days to flowering (Table 2).

Characters	PCV	GCV	h^2	GA	GA% MEAN
DF	4.42	4.38	97.43	9.06	8.92
DFM	10.71	10.30	88.72	8.47	20.49
DM	6.17	6.11	97.69	17.78	12.48
PHT	14.49	14.33	97.45	21.36	29.22
NT	19.70	18.88	90.55	6.40	37.40
NP	21.44	20.73	93.16	6.13	41.33
100-GW	27.01	24.71	99.20	16.18	50.75
GL	15.31	11.03	72.57	0.24	19.08
GW	2.67	2.12	69.36	0.26	3.07
LW	20.96	20.96	99.95	11.46	43.16
PL	12.90	12.46	92.88	4.99	24.86
PE	14.75	16.88	19.67	0.16	5.48
SW	30.99	16.71	39.57	1.80	13.49
GYLD	34.04	33.58	96.12	16.04	68.30

Table 1: Estimates of PCV, GCV, heritability (broad sense) GA and GA%Mean in F_2 population.

Characters	DF	DFM	DM	PHT	NT	NP	100 GW	GL	GW	LW	PL	PE	SW	GYLD
DF	1													
DFM	0.17**	1												
DM	0.87**	0.63**	1											
PHT	0.21**	0.43**	0.38**	1										
NT	0.14**	-0.60**	-0.19**	-0.70**	1									
NP	-0.05	-0.67**	-0.37**	-0.72**	0.98**	1								
100 GW	0.25**	0.17**	0.28**	0.38**	-0.34**	-0.41	1							
GL	-0.14**	-0.58**	-0.40**	-0.33**	0.62**	0.66**	-0.22**	1						
GW	0.42**	0.00	0.33**	0.55**	-0.14**	-0.20	0.02	-0.07	1					
LW	-0.13**	0.45**	0.12	0.15**	-0.21**	-0.23**	0.05	-0.22**	0.22**	1				
PL	-0.21**	0.27**	-0.03	-0.11	-0.04	-0.03	0.38**	0.00	-0.16**	0.23**	1			
PE	0.28**	-0.05	0.19**	-0.05	0.07	0.02	-0.19**	0.36**	0.16**	-0.03	0.05	1		
SW	-0.50**	-0.21**	-0.50**	-0.44**	0.35**	0.41**	-0.37**	0.20**	-0.22**	0.45**	-0.11	-0.26**	1	
GYLD	-0.40**	-0.87**	-0.75**	-0.66**	0.67**	0.75**	-0.19**	0.58**	-0.22**	-0.17**	-0.02	-0.03	0.58**	1

*significant @ p=0.01, **significant @ p=0.05.

Table 2: Phenotypic correlations among traits in F_2 generations.

DF=Days to flowering	PHT=Plant Height	GW=Grain width	LW=Leaf width	SW=Straw weight
DFM=Days to flowering and maturity	NT=Number of tillers Plant[-1]	GL=Grain length	PL=Panicle length	GYLD=Grain yield plant[-1]
DM=Days to maturity	NP=Number of panicles	100-GW=hundred grain weight	PE=Panicle excretion	

Characters	DF	DFM	DM	PHT	NT	NP	100-GW	GL	GW	LW	PL	PE	SW	'r' values
DF	0.06	-0.10	-0.24	-0.03	0.04	0.01	0.04	0.01	0.02	0.01	-0.03	0.03	-0.21	-0.40**
DFM	0.01	-0.56	-0.17	-0.07	-0.18	0.15	0.03	0.02	0.01	-0.04	0.04	0.01	-0.09	-0.86**
DM	0.05	-0.35	-0.28	-0.06	-0.06	0.08	0.04	0.01	0.01	-0.01	0.01	0.02	-0.21	-0.74**
PHT	0.01	-0.24	-0.10	-0.15	-0.21	0.16	0.06	0.01	0.02	-0.01	-0.01	0.01	-0.19	-0.65**
NT	0.01	0.34	0.05	0.11	0.30	-0.22	-0.05	-0.02	-0.01	0.02	-0.01	0.01	0.15	0.67**
NP	0.01	0.38	0.10	0.11	0.29	-0.22	-0.06	-0.02	-0.01	0.02	0.01	0.01	0.17	0.75**
100-GW	0.01	-0.10	-0.08	-0.06	-0.10	0.09	0.16	0.01	0.01	0.01	0.05	-0.02	-0.16	-0.19**
GL	-0.01	0.33	0.11	0.05	0.18	-0.15	-0.03	-0.03	0.01	0.02	0.01	0.03	0.08	0.57**
GW	0.02	0.01	-0.09	-0.09	-0.04	0.04	0.01	0.01	0.04	-0.02	-0.02	0.01	-0.09	-0.19**
LW	-0.01	-0.25	-0.03	-0.02	-0.06	0.05	0.01	0.01	0.01	-0.08	0.03	0.01	0.19	-0.17**
PL	-0.01	-0.15	0.01	0.02	-0.01	0.01	0.06	0.01	-0.01	-0.02	0.13	0.01	-0.05	-0.01
PE	0.02	0.03	-0.05	0.01	0.02	0.01	-0.03	-0.01	0.01	0.01	0.01	0.09	-0.11	-0.03
SW	-0.03	0.12	0.14	0.07	0.10	-0.09	-0.06	-0.01	-0.01	-0.04	-0.01	-0.02	0.42	0.57**

Residual Effect=0.0356.

Table 3: Genotypic direct (diagonal) and indirect effects of different quantitative traits in F_2 generation.

DF=Days to flowering	PHT=Plant Height	GW=Grain width	LW=Leaf width	SW=Straw weight
DFM=Days to flowering and maturity	NT=Number of tillers Plant[-1]	GL=Grain length	PL=Panicle length	GYLD=Grain yield plant[-1]
DM=Days to maturity	NP=Number of panicles	100-GW=hundred grain weight	PE=Panicle excretion	

The path analysis is a useful parameter to understand more clearly the association among different variables as recorded by simple correlation values. It helps to partition the overall association of particular variables with dependent variable into direct and indirect effects. While dealing with a more complex character like grain yield, it enables the breeder to specifically identify the important component trait of such a nature and differential emphasis can be laid on those characters for selection. Table 3 shows the results of the path analysis for the examined traits. Path coefficient analysis of yield and its component traits revealed that Straw weight (0.42) had the highest positive direct effect on grain yield. The Number of tillers Plant[-1](0.30) was the second most important character

it showed highest positive direct effect on grain yield [23] reported that indirect selection based on number of productive tillers per plant, harvest index and biomass per plant can be adopted for enhancement of grain yield as it showed positive direct effect on grain yield per plant. The number of tillers Plant[-1] had negative indirect effect on grain yield via number of panicles (-0.22), similar results were also reported by [24,25]. The traits that have direct affect on grain help to select them directly to enhance the yield. The lower residual effect indicated that different characters other than the characters considered in this study influence the grain yield considerably (Table 3 and Figure 1).

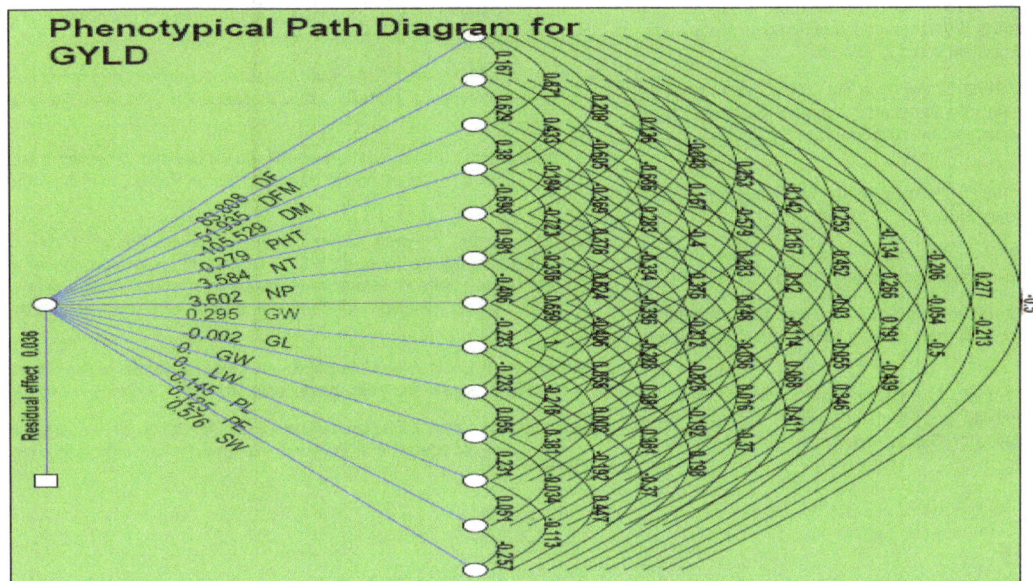

DF=Days to flowering	PHT=Plant Height	GW=Grain width	LW=Leaf width	SW=Straw weight
DFM=Days to flowering and maturity	NT=Number of tillers Plant⁻¹	GL=Grain length	PL=Panicle length	
DM=Days to maturity	NP=Number of panicles	100-GW=hundred grain weight	PE=Panicle excretion	

Figure 1: Path diagram for F$_2$ population.

Conclusion

An overall consideration of the results revealed that grain yield plant⁻¹ could be improved through selection for traits number of tillers, number of panicles, grain length and straw weight which were positively correlated with grain yield plant⁻¹. High heritability estimates for all the characters except for panicle excretion and straw weight revealed that selection for these characters will have to be carried in advance segregating generations.

Acknowledgements

We express the opportunity to thank various funding agencies including DBT, New Delhi, NICRA, Hyderabad, India for carrying this work at University of Agricultural Sciences, GKVK, Bangalore, India.

References

1. Sala M, Ananda Kumar CR, Geetha S (2015) Variability Studies for Quality Traits in Rice with High Iron and Zinc Content in Segregating Population. Rice Genomics and Genetics 6: 1-5.

2. Ratnakar MS, Rajanna MP, Ramesh S, Sheshshayee MS, Mahadevu P (2012) Genetic variability, correlation and path coefficient studies in F2 generation of aerobic rice (Orzyasativa L). Electronic J of Plant Breed 3: 925-931.

3. Barker R, Dave D, Tuong TP, Bhuuiyan SI, Guerra LC (1999) The outlook for resources in the year 2020: Challenges for research on water management in rice production. In "Assessment and Orientation towards 21st Century. Proceedings of 19th Session, International Rice Commission, Cairo, Egypt, 7-9 September 1998, FAO, Rome," 96-109.

4. Bouman BAM (2001) Water efficient management strategies in rice production.

Int. Rice Res.Notes 16: 17-22.

5. Bouman BAM, Tuong TP (2001) Field water management to save water and increase its productivity in irrigate rice. Agric. Water Manage 49: 11-30.

6. Vanitha RJ, UshaKumari KA, Robin S (2015) Genetic evaluation of rice genotypes for zinc deficiency tolerance and yield traits under aerobic condition in rice. Electronic J of Plant Breed 6: 191-195.

7. Borell A, Garside A, Fulai S (1997) Improving efficiency of water for irrigated rice in a semi-arid environment. Field Crops Res. 52, 231-248.

8. Sharma PK, Bhushan L, Ladha JK, Naresh PK, Gupta RK (2002) Crop water relations in rice–wheat cropping under different tillage systems and water management practices in a marginally sodic, medium textured soil. In "Water-Wise Rice Production. International Rice Research Institute, Los Banos, Philippines 223-235.

9. Singh AK, Choudhary BU, Bouman BAM (2002) Effect of rice establishment methods on crop performance, water and mineral nitrogen. In "Water-Wise Rice Production". International Rice Research Institute, Los Banos, Philippines 237-246.

10. Venkataravana P, S Hittalmani (1999) Genetic variability on some important traits in two F2 segregants of rice (Oryza sativa L.) under non-submergence conditions. Crop Res 18: 53-56.

11. Hittalmani S (1986) Genetics of Quantitative traits in Rice (Oryza sativa L) in rice under unfavorable conditions Ph. D thesis submitted to the University of Agricultural sciences, GKVK, Bangalore, India.

12. Chaudhary TN (1997) Water Requirement of Rice for Efficient Production. Directorate of Water Management Research, Patna, India.

13. Tuong TP, Bouman BAM (2003) Rice production in water-scarce environments. In "Water Productivity in Agriculture: Limits and Opportunities for Improvements"

CABI Publishing, UK 53-67.

14. Murthy KBC, Kumar A, Hittalmani S (2011) Response of rice genotypes under aerobic conditions, Electronic J of Plant breed 2: 194-96.

15. Castaneda AR, Bouman BAM, Peng S, Visperas RM(2003) The potential of aerobic rice to reduce water use in water-scarce irrigated low-lands in the tropics. In "Water-Wise Rice Production" International Rice Research Institute, Los Banos, Philippines.

16. Snedecor GW, Cochran WG (1994) Statistical Methods. Fifth Edition. Iowa State University Press. Ames. Iowa, USA.

17. Burton GW, Devane EM(1953) Estimating heritability in tall Fescue (*Festucaarundinaceae*) from replicated Clonal-material Agron J 51: 515-518.

18. Johnson DA, Rumbaugh MD (1995) Genetic variation and inheritance characteristics for Carbon isotope discrimination in alfalfa J Range Manage 48: 126-131.

19. Tuhina-Khatun, Mohamed M, Hanafi, Mohd Rafii Yusop, Wong MY, et al. (2015) Genetic Variation, Heritability, and Diversity Analysis of Upland Rice (*Oryza sativa* L.) Genotypes Based on Quantitative Traits. BioMed Research International P 1-7.

20. Ogunbayo SA, Sie M, Ojo DK, Sanni KA, Akinwale MG, et al. (2014) Genetic variation and heritability of yield and related traits in promising rice genotypes (*Oryza sativa* L) J Plant breed Crop Sci 6: 153-159.

21. Ibharim MM, Hussein RM (2006) Variability, heritability and genetic advance in some genotypes of Roselle (*Hibiscus sabdariffa* L) World J Agric Sci 2: 340-345.

22. Chakraborty Subrata, Chaturvedi HP (2014) Genetic Variability in Upland Rice (*Oryzasativa* L) Genotypes of Nagaland Indian Res J Genet & Biotech 6: 374-378.

23. Rakhi Soman, Naveen kumar Gande, Rajeswari Ambati, Pavan J Kundur, Ashwathanarayana R, et al. (2014) Studies for grain iron concentration and yield related traits in recombinant inbred lines of rice (*Oryza sativa* L) grown under aerobic condition. International Journal of Current Research 6: 5869-5874.

24. Shantala J, Latha J, Shailaja Hittalmani, (2006) Heterosis of rice (*Oryza sativa* L) hybrids for growth and yield components. Research on Crops 7: 143-146.

25. Shantala J, Latha J, Shailaja Hittalmani (2004) Path co-efficient analysis for grain yield with yield components in hybrid rice. Environment and Ecology (Spl 2): 734-736.

16

Optimization of Processing Conditions for Aqueous Pigmented Rice Extracts as Bases for Antioxidant Drinks

Adyati Putriekasari Handayani, Roselina Karim and Kharidah Muhammad*

Universiti Putra Malaysia, Serdang 43400, Selangor Darul Ehsan, Malaysia

Abstract

Pigmented rice can be categorized as a functional food due to its various health benefits, mainly from its polar antioxidant content which consists of anthocyanins in black rice and proanthocyanidins in red rice. This rice is usually cooked in excess water and removal of the water will be a waste as it can be further utilized as a base for antioxidant drink. Therefore, the objective of this study was to determine the optimum processing conditions (extraction temperature, time, and water/rice (W/R) ratio) for minimum 1,1-diphenyl-2-picrylhydrazyl (DPPH) radical scavenging activity IC_{50}, maximum total flavonoid (TFC), and Maximum Total Phenolic Content (TPC) in the pigmented rice extracts using response surface methodology (RSM). The optimum hot water extraction conditions for black rice were W/R ratio of 20 ml/g at 95.6°C for 40 minutes, while that for red rice are W/R ratio of 20 ml/g at 97°C for 30 minutes. It can be concluded that RSM is a useful method in optimizing the processing conditions for production of antioxidant drink from pigmented rice and hot water extraction showed great potential in extracting antioxidants from pigmented rice.

Keywords: Red rice; Black rice; Antioxidant drink; Response surface methodology; Optimization

Introduction

The color pigments on the bran layer of black rice are anthocyanins [1] while that of red rice are proanthocyanidins [2]. Previous investigations have shown that anthocyanins and proanthocyanidins contained possessed anti-oxidative and anti-inflammatory activities [3], anti-cancer [4], and various other healthful advantages. These compounds, however, are water soluble and can leach into cooking water. Finocchiaro et al. [5] found that the cooking water of red rice contained 57.66% of the total antioxidant content of red rice, showing that the majority of the antioxidants in red rice were lost into the cooking water. Compared to fruits and vegetables which were commonly used as raw materials for antioxidant beverages, pigmented rice is less perishable, can be stored for longer time due to its low moisture [6], and is cheaper in price especially compared to anthocyanin-rich berries or blackcurrants [7]. Therefore, it can be seen that the cooking water of pigmented rice, which is in the form of an aqueous extract, has the potential to be the base for an antioxidant drink.

Hot water can be used in the extraction of antioxidant compounds from pigmented rice to be further processed into antioxidant drinks. Hot Water Extraction (HWE) has better extraction efficiency compared to cold water extraction due to the change in diffusivity characteristics, viscosity, permittivity, and surface tension [8]. Furthermore, usage of solvents such as ethanol to extract antioxidants from pigmented rice for application in beverages is not recommended as it may affect the Halal status of the beverages.

Response Surface Methodology (RSM) is an effective statistical technique useful for developing, improving, and optimizing processes [9]. The principle of the approach has been reviewed previously by Henika [10,11] and Giovanni [12]. It is a designed regression analysis used to predict the value of a dependent variable based on the controlled values of the independent variables [13,14]. RSM is preferred in this study for the process of optimization because it is able to lead to the need for an experimental design which can generate multiple samples for consumer evaluation in a short period of time, and thus laboratory level tests are more efficient [15]. From the parameter estimates, it can

be determined which variable contributes the most to the prediction model, which enables the user to focus on the variables which are most important to the product [16].

The objective of this study was to determine the optimum processing conditions (extraction temperature, time, and water/rice ratio) for minimum DPPH radical scavenging activity IC_{50}, maximum Total Flavonoid Content (TFC), and Maximum Total Phenolic Content (TPC) in pigmented rice extracted by HWE for further application in antioxidant drinks.

Materials and Methods

Source of materials

Two types of Malaysian commercial pigmented rice were purchased from local supermarkets. Both rice samples were from indica subspecies, non-glutinous type, and contained intermediate level of amylose (20-25%).

Reagents and chemicals

Gallic acid, sodium acetate (CH_3COONa), sodium carbonate (Na_2CO_3), sodium nitrate ($NaNO_2$), (+)-catechin, and 1,1-diphenyl-2-picrylhydrazyl (DPPH) were from Sigma-Aldrich (St. Louis, MO, USA). Aluminium chloride ($AlCl_3$), Folin-Ciocalteu phenol reagent, iron (III) chloride hexahydrate ($FeCl_3.6H_2O$), and hydrochloric acid (HCl) were from Merck (Germany). NaOH was from R&M Chemicals

---

*****Corresponding author:** Kharidah Muhammad, Faculty of Food Science and Technology, Universiti Putra Malaysia, Serdang 43400, Selangor Darul Ehsan, Malaysia, E-mail: kharidah@upm.edu.my

(Essex, UK). Acetic acid glacial ($C_2H_4O_2$) was from JT Baker (New Jersey, USA).

Preparation of aqueous pigmented rice extracts

Optimization of processing conditions for production of aqueous pigmented rice extracts was conducted by mixing 5 g of pigmented rice with different volumes of distilled water once the temperature had reached the specified temperature based on the experimental design in a Julabo SW23 shaking waterbath (Julabo Labortechnik GmbH, Seelbach, Germany). After the required time, the aqueous pigmented rice extracts were filtered and centrifuged at $3000 \times g$ for 10 minutes. The supernatant was used for analyses.

Determination of DPPH radical scavenging activity IC_{50}

The free radical scavenging activity of samples on DPPH radical was carried out according to the procedure described by Brand-Williams et al. [17]. Concentration of the sample which is required to scavenge 50% of the DPPH free radicals (IC_{50}) was estimated using a nonlinear regression algorithm.

Determination of total phenolic content (TPC)

Total phenolic content of samples was measured by Folin Ciocalteu reagent assay conveyed by Slinkard and Singleton [18] and Singleton et al. [19]. Results were expressed as milligrams of Gallic Acid Equivalent (GAE) per ml of extract.

Determination of total flavonoid content (TFC)

Determination of total flavonoid content was conducted by using aluminium chloride colorimetric assay as described by Shams Ardekani et al. [20]. Total flavonoid content was expressed as mg catechin equivalents (CE) per one ml of extract.

Experimental design

Central Composite Design (CCD) with three factors (X_1, X_2, X_3) and three levels (-1, 0, +1) was used to design the experiments (Table 1). The factors selected based on the preliminary study were water/rice ratio (X_1, ml/g), extraction time (X_2, min), and extraction temperature (X_3, °C), while the response variables were TPC (mg GAE/ml of aqueous pigmented rice extracts), TFC (µM TE), and DPPH radical scavenging activity IC_{50}.

Each experiment was performed in triplicates and the average values were taken as the response, Y. Experimental data were fitted to second order polynomial model and regression coefficients were obtained. The generalized second-order polynomial model proposed for the response surface analysis was as follows:

$$Y = \beta_0 + \sum_{i=1}^{3} \beta_i x_i + \sum_{i=1}^{3} \beta_{ii} x^2_i + \sum_{i=1}^{2} \sum_{j=i+1}^{3} \beta_{ij} X_i X_j + \varepsilon \qquad (1)$$

where y is the response variable and β_0, β_i, β_{ii}, and β_{ij} are the regression coefficients for intercept, linear, quadratic, and interaction terms, respectively. X_i and X_j are coded values of the independent variables while k equals to the number of the tested factors (k=3).

All analyses were carried out in triplicates and the experimental values were reported as means ± standard deviations. Statistical analysis was performed using the Minitab 14.0 (Minitab Inc., State College, PA, USA) software and fitted to a second-order polynomial regression model containing the coefficient of linear, quadratic, and interaction terms. An Analysis Of Variance (ANOVA) with 95%

confidence level was then carried out for each response variable in order to test the significance and suitability. The predicted values were obtained according to the recommended optimum conditions and compared with experimental values in order to determine the validity of the model.

Results and Discussion

Optimization of extraction conditions using RSM

The results for DPPH radical scavenging activity, TPC, and TFC of each run of the experimental design are presented in Table 2. The TPC values for aqueous black rice extract ranged from 0.169 to 0.220 mg GAE/ml, while that for aqueous red rice extract ranged from 0.098 to 0.185 mg GAE/ml. The results for DPPH radical scavenging activity IC_{50} of aqueous black rice extract ranged from 0.216 to 0.404, and 0.369 to 0.549 for aqueous red rice extract. Additionally, TFC values for aqueous black rice and red rice extracts varied from 0.056 to 0.077 mg CE/ml and 0.028 to 0.059 mg CE/ml, respectively.

Regression coefficients of the second-order polynomial models for DPPH radical scavenging activity, TFC and TPC are summarized in Table 3. It can be seen that the regression parameters of the surface response analysis which had significant effects on the response variables of aqueous black rice extract were those of full quadratic model for DPPH radical scavenging activity, linear+interaction model for TPC, and linear+square model for TFC of aqueous black rice extract. The ones which had significant effects on the response variables of aqueous red rice extract were those of full quadratic model for TPC and DPPH radical scavenging activity, while linear+square model gave significant effects to the TFC of aqueous red rice extract. The models were used to construct three dimensional response surface plots to predict the relationship between process and response variables.

Effects of processing variables on DPPH radical scavenging activity

The analytical results for DPPH radical scavenging activity IC_{50} of

Run order	Factor 1 (X_1) Water/rice ratio (ml/g)	Factor 2 (X_2) Temperature (°C)	Factor 3 (X_3) Time (min)
1	30 (+1)[a]	87 (-1)	50 (+1)
2	20 (-1)	87 (-1)	30 (-1)
3	20 (-1)	97 (+1)	50 (+1)
4	30 (+1)	97 (+1)	30 (-1)
5[b]	25 (0)	92 (0)	40 (0)
6[b]	25 (0)	92 (0)	40 (0)
7	25 (0)	87 (-1)	40 (0)
8	30 (+1)	92 (0)	40 (0)
9[b]	25 (0)	92 (0)	40 (0)
10[b]	25 (0)	92 (0)	40 (0)
11	20 (-1)	92 (0)	40 (0)
12	25 (0)	92 (0)	30 (-1)
13	25 (0)	92 (0)	50 (+1)
14	25 (0)	97 (+1)	40 (0)
15	30 (+1)	97 (+1)	50 (+1)
16	20 (-1)	87 (-1)	50 (+1)
17[b]	25 (0)	92 (0)	40 (0)
18	20 (-1)	97 (+1)	30 (-1)
19[b]	25 (0)	92 (0)	40 (0)
20	30 (+1)	87 (-1)	30 (-1)

[a] Values in parentheses are the coded levels [b]Center points.

Table 1: Combinations of three-variable, three-level central composite design (CCD) used in RSM.

Run order	Aqueous black rice extract			Aqueous red rice extract		
	DPPH IC_{50}	TPC (mg GAE/ml)	TFC (mg CE/ml)	DPPH IC_{50}	TPC (mg GAE/ml)	TFC (mg CE/ml)
1	0.363 ± 0.002	0.177 ± 0.001	0.056 ± 0.000	0.549 ± 0.000	0.098 ± 0.000	0.028 ± 0.000
2	0.228 ± 0.001	0.205 ± 0.002	0.076 ± 0.000	0.422 ± 0.000	0.156 ± 0.003	0.053 ± 0.002
3	0.242 ± 0.001	0.220 ± 0.002	0.075 ± 0.000	0.432 ± 0.000	0.142 ± 0.003	0.045 ± 0.001
4	0.362 ± 0.003	0.215 ± 0.001	0.057 ± 0.000	0.458 ± 0.001	0.112 ± 0.002	0.036 ± 0.000
5	0.277 ± 0.002	0.195 ± 0.001	0.067 ± 0.000	0.469 ± 0.001	0.119 ± 0.002	0.037 ± 0.001
6	0.276 ± 0.001	0.196 ± 0.001	0.067 ± 0.000	0.469 ± 0.001	0.119 ± 0.001	0.037 ± 0.000
7	0.295 ± 0.001	0.191 ± 0.000	0.061 ± 0.001	0.505 ± 0.001	0.117 ± 0.001	0.045 ± 0.000
8	0.343 ± 0.002	0.178 ± 0.000	0.058 ± 0.000	0.496 ± 0.000	0.104 ± 0.001	0.033 ± 0.000
9	0.276 ± 0.001	0.199 ± 0.001	0.066 ± 0.000	0.470 ± 0.000	0.118 ± 0.001	0.038 ± 0.000
10	0.277 ± 0.001	0.194 ± 0.001	0.066 ± 0.001	0.426 ± 0.000	0.147 ± 0.002	0.045 ± 0.001
11	0.216 ± 0.001	0.207 ± 0.002	0.077 ± 0.001	0.444 ± 0.000	0.130 ± 0.001	0.042 ± 0.000
12	0.288 ± 0.002	0.195 ± 0.001	0.061 ± 0.000	0.473 ± 0.001	0.119 ± 0.001	0.034 ± 0.000
13	0.296 ± 0.002	0.192 ± 0.000	0.066 ± 0.000	0.454 ± 0.001	0.121 ± 0.001	0.036 ± 0.001
14	0.294 ± 0.002	0.197 ± 0.001	0.061 ± 0.000	0.474 ± 0.002	0.103 ± 0.002	0.029 ± 0.000
15	0.356 ± 0.003	0.176 ± 0.000	0.058 ± 0.000	0.463 ± 0.000	0.142 ± 0.002	0.049 ± 0.001
16	0.264 ± 0.001	0.205 ± 0.001	0.069 ± 0.001	0.470 ± 0.001	0.119 ± 0.001	0.037 ± 0.000
17	0.278 ± 0.001	0.196 ± 0.001	0.066 ± 0.000	0.369 ± 0.000	0.185 ± 0.002	0.059 ± 0.002
18	0.232 ± 0.001	0.216 ± 0.002	0.077 ± 0.000	0.471 ± 0.001	0.118 ± 0.001	0.038 ± 0.000
19	0.276 ± 0.001	0.195 ± 0.000	0.066 ± 0.000	0.504 ± 0.002	0.114 ± 0.000	0.036 ± 0.000
20	0.404 ± 0.003	0.169 ± 0.000	0.057 ± 0.000			

Table 2: Experimental data for the antioxidant activity of aqueous pigmented rice extracts based on the design showed in Table 1.

both aqueous black and red rice extracts are displayed in Table 2. As described by the three dimensional surface plots in Figures 1(a) and 1(d), DPPH radical scavenging activity IC_{50} increased as water/rice ratio increased, which was due to increasing volume of water as solvent for the antioxidants. Cooking time also showed the same pattern, but the increase in IC_{50} was more noticeable after around 40 minutes. Less time and lower water/rice ratio were favored since water exhibits thermodynamic properties favorable to heat transfer, resulting in better heat transfer with increasing water quantity. Hence, anthocyanins and proanthocyanidins might be chemically degraded when in longer contact with heat [21].

Response surface plot of aqueous black rice extract for interaction between cooking temperature and time (Figure 1(b)) displayed a negative quadratic function at fixed water/rice ratio. At water/rice ratio of 25 ml/g, minimum DPPH scavenging activity IC_{50} could be obtained by applying the middle values of optimized cooking time and cooking temperature, which were approximately at 92°C in 40 minutes. The plot showed that at cooking temperature and time before the middle values, antioxidants were probably not fully extracted yet, while the increase in IC_{50} at cooking temperature and time after the middle values indicated interference of antioxidant stability due to, as previously explained, chemical degradation from longer exposure to heat and application of increasing temperature. In contrast with black rice, at fixed water/rice ratio (25 ml/g), DPPH scavenging activity IC_{50} of aqueous red rice extract decreased as cooking time decreased and cooking temperature increased (Figure 1(e)). This might be an indication that higher temperature was required to break the matrix of red rice bran in order to extract the antioxidants in red rice.

At a fixed time of 40 minutes, the results indicated that a decrease in water/rice ratio decreased the DPPH radical scavenging activity IC_{50} for aqueous black rice extract (Figure 1(c)). However, cooking temperature had no significant effect on the IC_{50} at the range of 87-97°C, with only a slight decrease in IC_{50} as the temperature reached approximately 92°C. This illustrated that cooking time had more significant effect on the IC_{50} than cooking temperature in terms of the

interaction with heat and the breaking of bran matrix to release the antioxidants. This might relate to the interaction of antioxidants in the rice with other food constituents which led to different time required to reach equilibrium between the solution in the bran matrix and in the water [22]. For aqueous red rice extract, minimum DPPH radical scavenging activity IC_{50} could be obtained by applying lower water/rice ratio and higher cooking temperature (Figure 1(f)). Contrary with aqueous black rice extract, cooking temperature influenced the release of antioxidants in the aqueous red rice extract.

Effects of processing variables on total phenolic content (TPC)

The analytical results for TPC of both aqueous black and red rice extracts are displayed in Table 2. For the interaction between water/rice ratio and cooking time, both surface plots (Figure 2(a) and 2(d)) showed increasing TPC as water/rice ratio decreased and cooking time decreased. The pattern of effects from interaction between water/rice ratio and cooking time for TPC of both aqueous black and red rice extracts was similar to that of DPPH radical scavenging activity of both extracts as discussed previously.

At fixed water/rice ratio (25 ml/g), increased TPC for both aqueous pigmented rice extracts was obtained by increasing cooking temperature at less cooking time (Figure 2(b) and 2(e)). This pattern was also seen in the interaction between cooking temperature and cooking time for DPPH radical scavenging activity of aqueous red rice extract. However, TPC of aqueous black rice extract showed a contradictory pattern to the DPPH radical scavenging activity since increasing cooking temperature and less cooking time was required to obtain maximum TPC. This might be because the Folin-Ciocalteu reagent used in this assay was also reactive towards other non-antioxidant compounds such as proteins, carbohydrates, or inorganic ions which might be released from the rice bran matrix as temperature increased [23].

At fixed cooking time (40 minutes), it was found that for both aqueous black and red rice extracts, a combination of lower water/rice ratio and higher temperature was required to achieve maximum

Model parameter	Term	Aqueous black rice extract	Aqueous red rice extract
Total Phenolic Content (TPC)			
Intercept	β_0	-0.13^{ns}	0.14^{ns}
Water/rice ratio, x_1	β_1	-0.78×10^{-2ns}	-0.96×10^{-2ns}
Time, x_2	β_2	$0.12 \times 10^{-1*}$	-0.25×10^{-2ns}
Temperature, x_3	β_3	0.36×10^{-2ns}	0.42×10^{-2ns}
$x_1 x_2$	β_{12}	$-0.86 \times 10^{-4*}$	$0.80 \times 10^{-4*}$
$x_1 x_3$	β_{13}	0.92×10^{-4ns}	$-0.13 \times 10^{-3*}$
$x_2 x_3$	β_{23}	$-0.11 \times 10^{-3*}$	-0.55×10^{-4ns}
x_1^2	β_{11}	-	$0.27 \times 10^{-3*}$
x_2^2	β_{22}	-	$0.58 \times 10^{-4*}$
x_3^2	β_{33}	-	0.11×10^{-4ns}
R^2		0.94	0.99
DPPH radical scavenging activity			
Intercept	β_0	5.06^*	2.91^*
Water/rice Ratio, x_1	β_1	$0.35 \times 10^{-1*}$	$0.47 \times 10^{-1*}$
Time, x_2	β_2	-0.69×10^{-2ns}	$0.15 \times 10^{-1*}$
Temperature, x_3	β_3	-0.12^*	$-0.70 \times 10^{-1*}$
$x_1 x_2$	β_{12}	$-0.23 \times 10^{-3*}$	$-0.11 \times 10^{-3*}$
$x_1 x_3$	β_{13}	-0.15×10^{-3ns}	$-0.18 \times 10^{-3*}$
$x_2 x_3$	β_{23}	0.20×10^{-4ns}	-0.20×10^{-4ns}
x_1^2	β_{11}	0.30×10^{-4ns}	$-0.36 \times 10^{-3*}$
x_2^2	β_{22}	$0.13 \times 10^{-3*}$	$-0.11 \times 10^{-3*}$
x_3^2	β_{33}	$0.63 \times 10^{-3*}$	$0.38 \times 10^{-3*}$
R^2		0.99	0.99
Total Flavonoid Content (TFC)			
Intercept	β_0	-0.86^*	1.10^{ns}
Water/rice Ratio, x_1	β_1	$-0.87 \times 10^{-2*}$	$-0.44 \times 10^{-2*}$
Time, x_2	β_2	0.39×10^{-3ns}	$-0.66 \times 10^{-3*}$
Temperature, x_3	β_3	$0.23 \times 10^{-1*}$	0.21×10^{-1ns}
$x_1 x_2$	β_{12}	-	-
$x_1 x_3$	β_{13}	-	-
$x_2 x_3$	β_{23}	-	-
x_1^2	β_{11}	$0.14 \times 10^{-3*}$	$0.5 \times 10^{-4*}$
x_2^2	β_{22}	-0.11×10^{-4ns}	0.1×10^{-5ns}
x_3^2	β_{33}	$-0.5 \times 10^{-5*}$	0.11×10^{-3ns}
R^2		0.95	0.97

ns - not significant. *Significant at 5%. R^2: Coefficient of multiple determinations.

Table 3: Estimated regression coefficients of the fitted second-order polynomial model of the antioxidant activity of aqueous pigmented rice extracts

TPC (Figure 2(c) and 2(f)). Both aqueous pigmented rice extracts also showed the same pattern for their DPPH radical scavenging activity.

Effects of processing variables on total flavonoid content (TFC)

The analytical results for TFC of both aqueous black and red rice extracts are displayed in Table 2. It was found that at fixed temperature (92 °C), a combination of minimum cooking time and minimum water/rice ratio was desired to obtain maximum TFC of both extracts (Figure 3(a) and 3(d)). This pattern was also observed in the effect of interaction between cooking time and water/rice ratio at fixed temperature on DPPH radical scavenging activity and TPC of both aqueous pigmented rice extracts.

The parabolic shape of the surface plot for the effect of cooking temperature and time on TFC of aqueous black rice extract showed that the middle values for cooking temperature and time at fixed water/

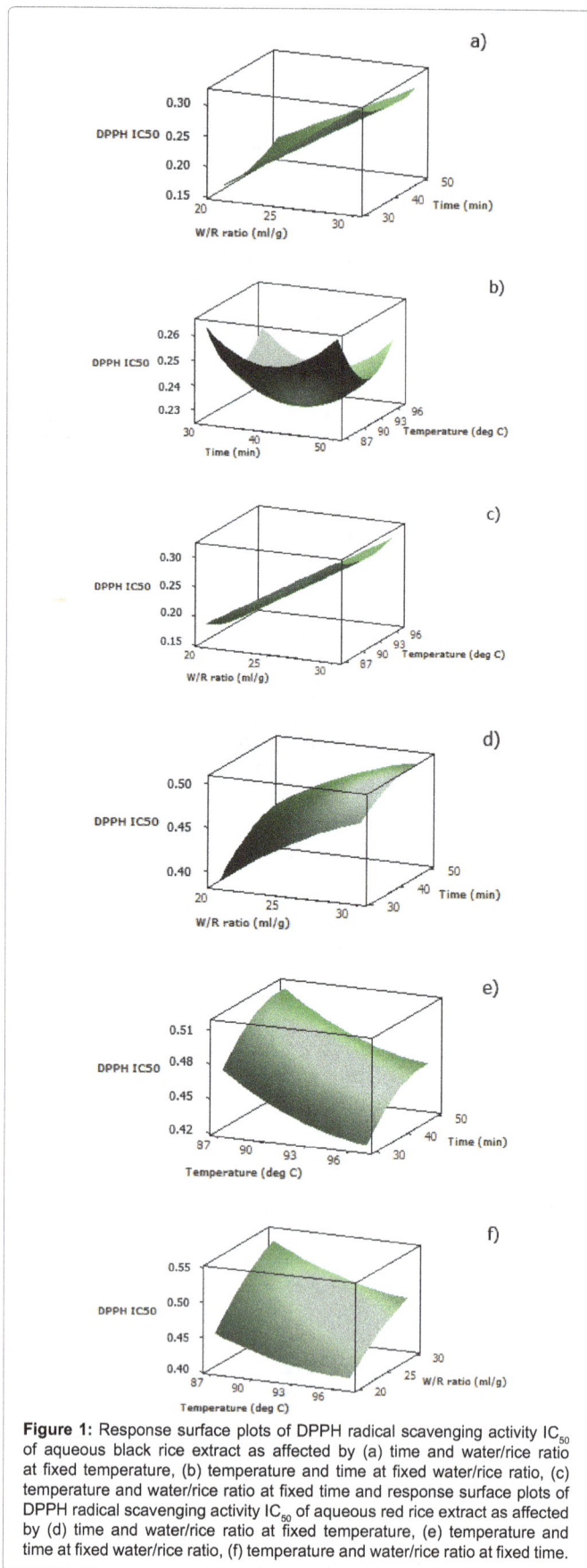

Figure 1: Response surface plots of DPPH radical scavenging activity IC_{50} of aqueous black rice extract as affected by (a) time and water/rice ratio at fixed temperature, (b) temperature and time at fixed water/rice ratio, (c) temperature and water/rice ratio at fixed time and response surface plots of DPPH radical scavenging activity IC_{50} of aqueous red rice extract as affected by (d) time and water/rice ratio at fixed temperature, (e) temperature and time at fixed water/rice ratio, (f) temperature and water/rice ratio at fixed time.

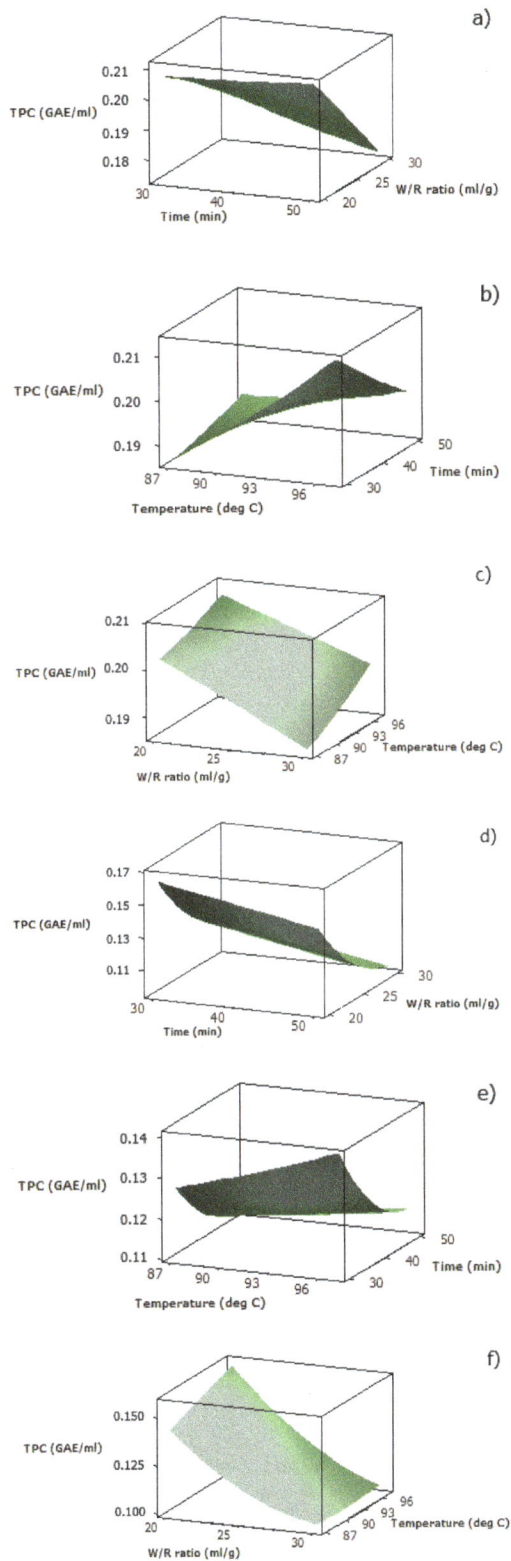

Figure 2: Response surface plots of TPC of aqueous black rice extract as affected by (a) time and water/rice ratio at fixed temperature, (b) temperature and time at fixed water/rice ratio, (c) temperature and water/rice ratio at fixed time and response surface plots of TPC of aqueous red rice extract as affected by (d) time and water/rice ratio at fixed temperature, (e) temperature and time at fixed water/rice ratio, (f) temperature and water/rice ratio at fixed time.

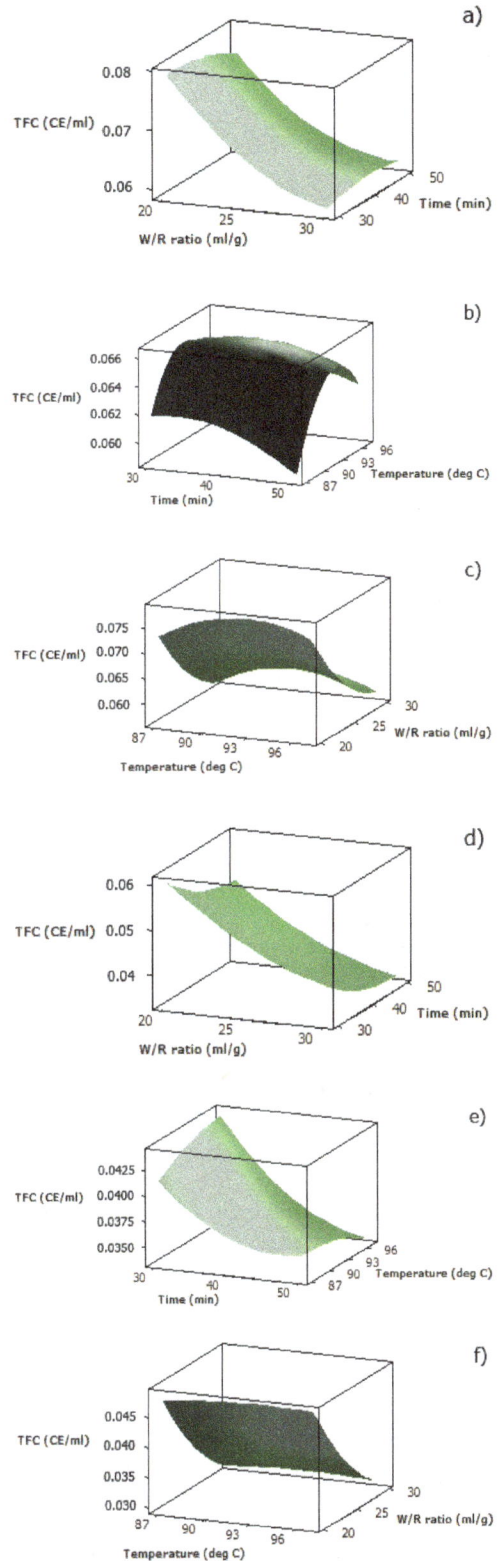

Figure 3: Response surface plots of TFC of aqueous black rice extract as affected by (a) time and water/rice ratio at fixed temperature, (b) temperature and time at fixed water/rice ratio, (c) temperature and water/rice ratio at fixed time and response surface plots of TFC of aqueous red rice extract as affected by (d) time and water/rice ratio at fixed temperature, (e) temperature and time at fixed water/rice ratio, (f) temperature and water/rice ratio at fixed time.

Response variables	Optimum processing conditions			Optimum values	
	Water/rice ratio (ml/g)	Time (min)	Temperature (°C)	Predicted	Actual
TPC (mg GAE/ml)	20	30	97	0.220[a]	0.223 ± 0.004[a]
DPPH IC$_{50}$	20	39.64	92.35	0.213[b]	0.234 ± 0.010[b]
TFC (mg CE/ml)	20	39.88	92.11	0.079[c]	0.081 ± 0.001[c]

Means followed by the same letter(s) within a column are not significantly different at 5% level of probability.

Table 4: Estimated optimum processing conditions, predicted, and experimental values of responses for aqueous black rice extract.

Response variables	Optimum processing conditions			Optimum values	
	Water/rice ratio (ml/g)	Time (min)	Temperature (°C)	Predicted	Actual
TPC (mg GAE/ml)	20	30	97	0.178[a]	0.187 ± 0.004[a]
DPPH IC$_{50}$	20	30	97	0.376[b]	0.381 ± 0.003[b]
TFC (mg CE/ml)	20	30	97	0.056[c]	0.060 ± 0.001[c]

Means followed by the same letter(s) within a column are not significantly different at 5% level of probability.

Table 5: Estimated optimum processing conditions, predicted, and experimental values of responses for aqueous red rice extract.

rice ratio (25 ml/g) gave maximum TFC (Figure 3(b)), similar to DPPH radical scavenging activity when subjected under the same variables. This further proved that the major contributors for radical scavenging activity of aqueous black rice extract were flavonoid compounds, specifically anthocyanins, which are the major flavonoid compounds in black rice, together with other non-anthocyanin flavonoid compounds such as quercetin, isorhamnetin, apigenin, and kaempferol [24,25]. The surface plot for effect of cooking temperature and time on TFC of aqueous red rice extract (Figure 3(e)) also showed the same pattern as the one for DPPH radical scavenging activity and TPC, in which maximum TFC was obtained by applying higher temperature at less cooking time.

At fixed cooking time (40 minutes), TFC of aqueous black rice extract increased as water/rice ratio decreased (Figure 3(c)). However, comparable to the trend showed by DPPH radical scavenging activity, a slight increase in TFC was obtained as cooking temperature reached around 92°C. The surface plot of aqueous red rice extract also showed an increase in TFC as the water/rice ratio decreased (Figure 3(f)), with insignificant effect from cooking temperature towards TFC at fixed cooking time, which differed from the one of its DPPH radical scavenging activity. This explains the nature of proanthocyanidins which are heat-stable antioxidant compounds [26].

Optimum processing conditions of aqueous pigmented rice extracts

Tables 4 and 5 show the optimum conditions for the production of aqueous pigmented rice extracts with regard to the antioxidant content based on response surface methodology. For aqueous black rice extract, maximum TPC (0.215 mg GAE/ml) (d=0.91553), maximum TFC (0.076 mg CE/ml) (d=0.95599), and minimum DPPH scavenging activity IC$_{50}$ (0.216) (d=1.0000) was achieved by applying the optimum conditions (d=0.95655) which consist of cooking temperature of 92 °C, cooking time of 40 minutes, and water/rice ratio of 20 ml/g. On the other hand, combination of cooking temperature of 97 °C, cooking time of 30 minutes, and water/rice ratio of 20 ml/g were the optimum conditions (d=0.93555) required to obtain maximum TPC (0.178 mg GAE/ml) (d=0.92480), maximum TFC (0.056 mg CE/ml) (d=0.92281) and minimum DPPH scavenging activity IC$_{50}$ (0.376) (d=0.95948) for aqueous red rice extract. The fitted results had no significant difference (p>0.05) with the experimental results, showing that the RSM model was theoretically validated. The practical validation of the RSM model also showed that the predicted results had no significant difference (p > 0.05) from the experimental results obtained using optimum extraction conditions.

Conclusions

The water attained from cooking pigmented rice in excess water was found to have high potential to be developed into antioxidant drinks because antioxidant compounds in the rice bran tend to seep out into the cooking water. The optimum hot water extraction conditions for black rice were water/rice (W/R) ratio of 20 ml/g at 92 °C for 40 minutes, while that for red rice were W/R ratio of 20 ml/g at 97 °C for 30 minutes. Hence, it can be concluded that RSM is a useful method in optimizing the processing conditions for production of antioxidant drink from pigmented rice and Hot Water Extraction (HWE) showed great potential in extracting antioxidants from pigmented rice, making it suitable for the production of antioxidant drinks.

Acknowledgments

Authors thank Padiberas Nasional Berhad (BERNAS) and Faculty of Food Science and Technology UPM for the financial support of this research.

References

1. Escribano-Bailon MT, Santos-Buelga C, Rivas-Gonzalo JC (2004) Anthocyanins in cereals. Journal of Chromatography A 1054: 129-141.

2. Oki T, Masuda M, Kobayashi M, Nishiba Y, Furuta S, et al. (2002) Polymeric procyanidins as radical-scavenging components in red-hulled rice. Journal of Agricultural and Food Chemistry 50: 7524-7529.

3. Hu C, Zawistowski J, Ling WH, Kitts DD (2003) Black rice (Oryza sativa L. indica) pigmented fraction suppresses both reactive oxygen species and nitric oxide in chemical and biological model systems. Journal of Agricultural and Food Chemistry 51: 5271-5277.

4. Kamei H, Hashimoto Y, Koide T, Kojima T, Hasegawa M (1998) Anti-tumor effect of methanol extracts from red and white wines. Cancer Biotherapy and Radiopharmacology 13: 447-452.

5. Finocchiaro F, Ferrari B, Gianinetti A, Dall'Asta C, Galaverna G, et al. (2007) Characterization of antioxidant compounds of red and white rice and changes in total antioxidant capacity during processing. Molecular Nutrition & Food Research 51: 1006-1019.

6. Handayani AP, Ramakrishnan Y, Karim R, Muhammad K (2014) Antioxidant properties, degradation kinetics and storage stability of drinks prepared from the cooking water of pigmented rice. Advance Journal of Food Science and Technology 6: 668-679.

7. Jang SJ, Xu ZM (2009) Lipophilic and hydrophilic antioxidants and their antioxidant activities in purple rice bran. Journal of Agricultural and Food Chemistry 57: 858-862.

8. Teo CC, Tan SN, Yong JHW, Hew CS, Ong ES (2010) Pressurized hot water extraction (PHWE). Journal of Chromatography A 1217: 2484-2494.

9. Myers R, Montgomery DC (2002) Response Surface Methodology. John Wiley, New York, USA.

10. Henika RG (1972) Simple and effective system for use with response surface methodology. Cereal Science Today 17: 309-314, 334.

11. Henika RG (1982) Use of response surface methodology in sensory evaluation. Food Technology 36: 96-101.

12. Giovanni M (1983) Response surface methodology and product optimization. Food Technology 37: 41-45.

13. Meilgaard M, Civille GV, Carr BT (1991) Sensory evaluation techniques, (2ndedn) CRC Press, Boca Raton, Florida.

14. Resurreccion AVA (1998) Quantitative of quality attributes as perceived by the consumer. In Consumer Sensory Testing For Product Development. Aspen Publishers, Gaithersburg, MarylandL: 203-228.

15. Moskowaitz HR (1994) Product optimization approaches and applications. In Measurement of Food Preferences; Macfie, HJH and Thomson DMH, Eds. Blackie Academie & Professional, Glasgow, UK. Pp. 97-136.

16. Schutz HG (1983) Multiple regression approach to optimization. Food Technology 37: 46-48, 62.

17. Brand-Williams W, Cuvelier ME, Berset C (1995) Use of a free radical method to evaluate antioxidant activity. Lebensmittel-Wissenschaft und Technologie 28: 25-30.

18. Slinkard K, Singleton VL (1977) Total phenols analysis: Automation and comparison with manual methods. American Journal of Enology and Viticulture 28: 49-55.

19. Singleton V, Orthofer R, Lamuela-Raventos RM (1999) Analysis of total phenols and other oxidation subtrates and antioxidants by means of Folin-Ciocalteu reagent. Methods in Enzymology 299: 152-175.

20. Shams Ardekani MR, Hajimahmoodi M, Oveisi MR, Sadeghi N, Jannat B, et al. (2011) Comparative antioxidant activity and total flavonoid content of Persian pomegranate (Punica granatum L.) cultivars. Iranian Journal of Pharmaceutical Research 10: 519-524.

21. Goto-Yamamoto N, Mori K, Numata M, Koyama K, Kitayama M (2010) Effects of temperature and water regimes on flavonoid contents and composition in the skin of red-wine grapes. Journal International des Sciences de la Vigne et du Vin 43: 75-80.

22. Thoo YY, Ho SK, Liang JY, Ho CW, Tan CP (2010) Effects of binary solvent extraction system, extraction time, and extraction temperature on phenolic antioxidants and antioxidant capacity from mengkudu (Morinda citrifolia). Food Chemistry 120: 290-295.

23. Everette JD, Bryant QM, Green AM, Abbey YA, Wangila GW, et al. (2010) Thorough study of reactivity of various compound classes toward the Folin-Ciocalteu reagent. Journal of Agricultural and Food Chemistry 58: 8139-8144.

24. Sriseadka T, Wongpornchai S, Rayanakorn M (2012) Quantification of flavonoids in black rice by liquid chromatography-negative electrospray ionization tandem mass spectrometry. Journal of Agricultural and Food Chemistry 60: 11723-11732.

25. Kim JK, Lee SY, Chu SM, Lim SH, Suh SC, et al. (2010) Variation and correlation analysis of flavonoids and carotenoids in Korean pigmented rice (Oryza sativa L.) Cultivars. Journal of Agricultural and Food Chemistry 58: 12804-12809.

26. Satoshi K, Madoka T, Teruyoshi M (2001) Effect of heat and pH on the radical-scavenging activity of proanthocyanidin-rich extract from grape seeds and production of konjac enriched with proanthocyanidin. Journal of the Japanese Society for Food Science and Technology 48: 591-597.

17

Sustainability of Rice Cultivation: A Study of Manipur

Komol Singha* and Sneha Mishra

Department of Economics, Sikkim University, Gangtok, India

Abstract

When many states in India are gradually moving away from their traditional agriculture-based to industry or service-oriented economy, Manipur continues to depend on agricultural sector, especially the rice crop. The crop's yield level is found to be much higher than that of the national level, but its demand in the state is much higher that the supply having known this importance, the present paper attempts to assess sustainability of rice crop in Manipur With the help of primary data collected from 152 farmers and using PCA, the study found that 20 percent of farmer as sustainable, 64 percent as moderate and 16 as vulnerable. This implies that the rice cultivation in Manipur has not been very impressive, despite its favorable agro-climatic condition of the five dimensions included in measuring sustainability, social factor was igured out at the top, followed by the economic factor.

Keywords: Manipur; PCA; Rice cultivation; Sustainability

Introduction

Rice or paddy (Oryzasativa L) has not only been the staple food for more than half of the humanity [1] but also shaped the culture, diet and economy of the majority of the world's population, especially the east and south-east Asian continents. Its production primarily depends on good agronomic practices, and the most consistent and the highest yields of the crop can be harvested in irrigated systems [2]. Good agronomic practices include the effective fertilization, water and weed management, lower plant densities and sustainability of the farmers [3,4]. In the countries where rice is the dominant crop and staple food, the livelihood of the people depends on the crop's availability, quality and sustainability [1]. Consequently, sustainability of this crop has become one of the important issues in the world now. Sustainable rice farming is very important in any developing countries/region, reason being not only the staple food of the majority of the people, but also the country's food security, poverty alleviation and rural employment depend largely on rice production [4,5]. While many states in India are gradually moving away from their traditional agriculture-based to industry or service-oriented economy, Manipur (one of the north eastern states of India covers an area of 22,327 km2. It is bounded by Nagaland to the north, Mizoram to the south, and Assam to the west; Burma lies to its east) continues to be heavily dependent on the agricultural sector. More than half (52.19 percent) of its population depend upon the agriculture sector, especially the rice cultivation. Gross Cropped Area (GCA) is 3.3 lakh hectares; it is around 12.98 per cent. Around 90 percent of GCA of the state is covered by rice and average cropping intensity is accounted at 143.26 per cent in 2011-12 [6]. This implies that the average ratio of rice area sown is found to be 1.5 times annually. In term of rice yield level, as per the Directorate of Rice Development, Patna, Government of India, , Manipur ranked 8th in the country with 2369 kg/ha in 2006-07.

Though rice area had slightly declined from 1.6 lakh hectares in 1995-96 to 1.4 lakh hectares in 2002-03 [7] after 2002-03, it increased and recorded 2.2 lakh hectares in 2011-12. Despite this, demand for rice crop has increased significantly in the recent past, much higher than its supply capacity [2]. For instances, the shortage of rice in 2009-10 was around 116 thousand tonnes in the state. Growth of technology or improvement in yield performance did not show any respite to the ever increasing demand for it.

Conceptual Framework

Though the idea of sustainable development was conceived in

the United Nation's (UN) first conference on Human Development, held in 1972 at Stockholm, it came to be prominence through the Brundtland Commission's report Our Common Future 1987. Thereafter, the UN has made a consciousness in the world on ecology, environment and poverty. These themes have been brought to the centre stage in the new development policies. In a broader sense, [8] defined sustainable development as a pattern of social and structural economic transformations which optimise the economic and other societal benefits available in the present without jeopardizing the likely potential for similar benefits in the future. Nevertheless, till today, concrete progress towards the goal of sustainable development has not been satisfactory. It is not so easy to specify and quantify true indicator of sustainable level and no specific statistical system or tool has been designed to quantify and assign weights to often conflicting indicators for operationalising the measurement and overall monitoring of the sustainability [5,9].

While conceptualising agricultural sustainability, the Department for International Development, University of Essex, UK viewed it as the resilience (capacity of the systems to adjust in buffer shocks and stresses) and persistence, means the capacity of systems to carry on [10]. It implies that the capacity to adjust as external and internal conditions gets changed. As of the rice farming sustainability, it may be understood as the process by which farmers manage soil, water and other basic inputs to enhance productivity and maintain it to meet farm and family needs, without adversely affecting the production environment and future resources [11]. Before them, [12] had reflected on two basic characteristics of sustainability - one, it is a people-centred concept that aims to improve the quality of human life and conservation-based concept that is conditioned by the need to respect nature's ability to provide resources and life-support services. In this perspective, sustainable development means improving the quality

*Corresponding author: Komol Singha, Associate Professor, Department of Economics, Sikkim University, Gangtok, India, E-mail: ksingha@cus.ac.in

of human life while living within the carrying capacity of supporting ecosystems. Secondly, sustainable development is also a normative concept that embodies standard of judgement and behaviour to be respected as the human community and the society seeks to satisfy its needs of survival and well-being. It can also be interpreted in many different ways, but at its core, an approach to development that looks to balance different, and often competing, needs against an awareness of the environmental, social and economic limitations we face as a society.

However, contemporary theories of agricultural development [4,5,13,14] broadened the basic parameters of sustainable development that had focussed more on environmental aspects, incorporated economic, social and political dimensions. In agriculture, it can also be defined as the common face of agronomic, ecologic, economic and social factors. Production is an important component for sustainability of agro-ecosystems, and improving agricultural production would result in increasing system of sustainability [15]. Moreover, a good and sustainable rice farming model requires a clear structural framework that includes essential indicators, strategy and goal definition of dimension, and should be supported by the social, political, economic (including technical) and environmental institutions [5,16]. Nevertheless, the central question raised by [1] on rice research is– how to balance the need for ever-greater food production at prices that poor consumers can afford and get reasonable profit without jeopardising natural resources and the environment for generations to come. Similarly, [11] also opined that it links to a range of problems that farmers face and should be addressed without disturbing or disrupting the usually healthy rhythms of rural life. Therefore, a devised strategy for sustainable rice farming should be a regional/country specific and that should address each region/country's economic, political, environmental and social conditions and objectives.

Objectives and Methodology

This study tries to analyse overall sustainability of rice farming in Manipur. While measuring sustainability, parameters include the resources available with the farmers, not necessarily the ecological or environmental component alone. How have the institutions–social, economic, political, technology and environment in a synergetic manner helped in making rice farming sustainable in Manipur is discussed. The specific objectives of the study are given as:

To measure overall sustainability of rice farmers in Manipur.

To identify institutional dimensions that impacted the most in rice farming sustainability.

Methodologically, the present study is based on primary data. Based on agro-climatic conditions and growth pattern of rice crop, two districts (Senapati and Thoubal) of Manipur have been selected and altogether 152 rice farmers were interviewed through a well-structured questionnaire. The primary field survey was conducted in the month of September to November 2013. To satisfy the two objectives mentioned above, the Principal Component Analysis (PCA) method was used. In the process, a combination of single indicators into a meaningful composite indicator was made under a fitness-for-purpose principle. Twenty-six indicators have been selected on the basis of their analytical soundness, measurability, relevance to the phenomenon and relationship to each other. Some proxy variables have also been used to make up data limitation. The normalized values

$$(\text{Normalisation} = 1 - \frac{ei - Min}{Max - Min} \; ; \; \text{Where, } ei = \text{Actual Value}) \text{ of the}$$

variables were multiplied with their respective PCA scores and the

summation of the entire individual variables yielded an index value. The Composite Index of an individual dimension (institution) was arrived at by adding all individual index values social, economic, political, technology and environment. While normalising index value, all the indicators have been converted in a range of 0 to 1. The larger value influences more on sustainability of rice farming. Finally, using the criterion of "Mean ± Standard Deviation" of the index value [17-20] farmers have been categorized into three groups– sustainable, moderately sustainable and vulnerable (Table 1).

Result and Discussion

As of the sustainability, according to [21], it needs to define with respect to systems rather than doing singular analyses of inputs and outputs, because crop varieties and inputs produce nothing in isolation. The most relevant issue today is to design suitable technologies, as well as compatible strategies from the social, economic and ecological viewpoints that will bring about the necessary behavioural changes to achieve the objectives of sustainable agriculture. As [15] mentioned, rice sustainability is a composite effort of ecologic, economic and social factors, consequently, it can primarily be explained more by a composite index. As given in the methodology, the study categorised the paddy farmers of Manipur into three as:

Sustainable=House-holds with an index value ≥ Mean + SD;

Moderately Sustainable=House-holds with index value between Mean ± SD;

Vulnerable=House-holds with index value ≤ Mean – SD

Table 1 shows overall sustainability of rice farmers in Manipur. Using the formula of Mean± SD of the index value, of the total 152 sample, 30 rice farming households have been categorised as sustainable, 97 as moderate and 25 as vulnerable. This implies that the rice cultivation has been at the cross-road and majority of them are mainly cultivating at the subsistence level. As of the institutional impact, [11] endorsed that agricultural sustainability; particularly the rice farming requires technological support to social, economic, political and ecological realities. As an evidence of technological supports and modernisation, fertilizer consumption, irrigation and HYV seeds were found to have positively correlated with rice farming development [2,22,23]. Apart from the physical or economic variables, [23] identified the role of social institution and attitudinal environment of the society for agricultural development in Manipur. Another study in Assam by [2] found that institutional credit was found to be an important factor for enhancing paddy cultivation. According to [13] despite rice being a major staple crop, commonly grown in the NER, rice-based agriculture system has failed to provide required household income-security primarily due to weak institutional systems. They are– adherence to traditional agricultural practices, low adoption of modern rice varieties (HYVs), poorly defined of property right, small size of operational holdings, weak institutional credit facility, high vulnerability to natural calamities and over-dependence on monsoon rain with poor irrigation infrastructure (Table 2).

Table 2 explains the impact of various institutions and their indicators on Sustainability Index construction. Of the total, the share of social institution has been very great, composed of 30.80 per cent. This was followed by the economic institution with the share of 28.75 per cent. The institutions of technology and environmental system have also impacted more on the rice farming sustainability in Manipur, accounted for 13.11 per cent and 18.31 per cent respectively. As expected, the role of political institution has been very negligible

Dimension	Indicators	Sign	Definition and Measurement
Social– to develop the quality of life of society at large	Education	+	Education of the farmers is one of the key indicators of farm sustainability. Level of education is coded as: 0=Illiterate, 1=Primary(1-5), 2=Secondary/Higher Secondary (6-12), 3= Graduate and above
	Family size	+	Number of family member or size of the family is also one of the important indicators of agricultural development, as labour supply, social security. It is taken as absolute number of the family.
	Stability of the Farmer in the village	+	Recent settlers have some disadvantages. Stability of the farmers in a place or village has got certain advantages, like approach to agriculture department, local fellow farmers and leaders for agricultural assistance. If the farmer is a recent settler, it is coded as: 1= Yes; 2=No
	No of family (female) labour engaged	+	Engage family female labour force in the rice cultivation indicates that the importance and need of the crop in the family.
	Importance of the crop	+	How many times the farmer visited the paddy field or farm after transplantation for better harvest of rice/paddy
	Use of traditional implements	+	Traditional implements like wooden plough, wooden thresher, hoe, etc. are widely used by the farmers and it is a part of rural social culture: 1=Yes, 2=No.
Economic– to achieve economic feasibility in rice cultivation	Land productivity	+	Measurement of the productivity of rice per unit of land in physical units (collected during the survey) can reflect economic condition of the farmer. It is measured as kilogram of paddy per hectare of land. Kg/ha
	Net Sown Area	+	Net Sown Area under the crop in physical units, as measured in hectares during the field survey. It also determines the size of the farmer's economy.
	Availed/want to avail loan	+	To avail loan from the bank or any other financial institution by the farmers for cultivation is a sign of agricultural development: 1=Yes, 2=No.
	Insurance policy	+	Any of the insurance policy of the farmer indicates the safety. It is coded as 1=Yes; 2=No
	Irrigation	+	Irrigation or the assured water supply is one of the most important inputs in rice cultivation. Percentage of area irrigated of the farmer implies improvement of cultivation.
	Land brought	+	Land bought in the last five years for rice cultivation is considered as one of the indicators of development or sustainability of the sector.
	Usage of seeds	+	Good seed means good output. Seed usage is coded as: 1=Farm saved/own seed saved last year, 2=Local seeds, 3=HYV seeds/new varieties.
Technology– to identify the role of technology in agriculture/rice cultivation	Modern implements/power operated machines	+	Adoption modern techniques in cultivation imply development of agriculture. Modern implements like threshers, tractors, sprayers, etc. used by the farmers is coded as: 1=Used, 2=No.
	Availing facility of community warehouse	+	Getting community warehouse is one of the benefits of the farmers. As warehouse is a kind of public good and its investment is practically not possible by individual farmers, it is coded as: 1=Yes; 2=No
	Harvesting method	+	Using modern technology in harvesting is one of the indicators of advancement of the sector. It is coded as: 1= modern; 2= traditional method.
	Mode of transportation and harvesting	+	Application of modern techniques of harvesting and transportation in paddy cultivation imply development of the sector and is coded as: 1=Yes, 2=No
	Experience & knowledge of rice cultivation	+	Knowledge of the sector or occupation also depends on the year of experience. Number of years engage in rice cultivation–
Environment– to assess the awareness of environmental degradation	Flood/draught resistant variety	+	Whether the farmer can protect their crop from the flood or draught situation. Do they have flood resistant crop or different seed for different season, so as to survive from the calamities?
	Knowledge of application of fertilisers	+	Fertilizer is one of the most important inputs to developing agricultural sector. Knowledge about the application of fertiliser by the farmer indicates the usage of it. Whether farmer knows the appropriate dose of fertilizer: 1=Yes, 2=No.
	Amount of chemical fertilizer used	–	Larger the amount of chemical fertilizer used in the field more is degraded the environment, coded as: Kilogram per hectare of land.
	Diversification of crop	+	Whether the land is used only for one crop or other crops too along with the main crop (rice in this study)? It reflects the importance of the crop grown by the farmer. For measurement, crop diversification (from rice to other): 1=Yes, 2=No
Political– to build equitable access to the agricultural resources	Membership	+	If the farmer is a member of any organisation related to agriculture, definitely the farmer is considered as well informed or agriculturally alert farmer.
	Joint effort with fellow farmers or in group	+	Making an effort to work together with the fellow farmers is one of the political or group forces that makes in achieving unachievable individual effort made earlier: 1= Yes; 2= No
	Benefits received from the agriculture Department.	+	Ministry of agriculture has initiated a lot of provision and benefits for the rural farmers. But, often than not, many of the farmers do not aware of the benefits provided by the department. If benefited in the last one year (in cash or kind): 1= yes; 2= No
	Attended agriculture related meetings	+	Many of the rural farmers do not aware of the vital information and basic agricultural provisions provided by the department. If farmer attended agriculture related meetings in the last one year.

Table 1: Indicators of sustainable rice cultivation in Manipur.

impact on rice farming sustainability, estimated at 9.03 percent of the total score. Within the social institution, caste category and continuous cultivation of the crop played a major role in the farmer's sustainability. In the case of economic institution, net sown area and irrigation have influenced the most. Under the environment institution, amount of chemical fertiliser used has been very significant. The application of chemical fertiliser has been very influential in rice cultivation, though it affects environmentally. Within the political institution, joint effort with fellow farmers has impacted more in farmers' sustainability in Manipur (Table 3).

Conclusion

To conclude, of the total 152 sample, 30 rice farmers have been categorised as sustainable, 97 as moderate and 25 as vulnerable.

Categorisation	Status	Index Value	Households
Index Value ≥ Mean + SD	Sustainable	2.10 and above	30
Index value between Mean ± SD	Moderate	1.60 to 2.10	97
Index Value ≤ Mean - SD	Vulnerable	Below 1.60	25
Total Household (N)	-	-	**152**

Source: Authors' estimation

Table 2: Sustainability Categorisation of the Farmers.

Indicators	Effect (%) of Indicators on Index
1. Social	**30.80**
a) Education	3.50
b) Family Size	2.51
c) Caste	11.48
d) Credibility of the Farmer (social respect)	2.66
e) Importance of Crop (continuity of rice cultivation)	10.24
f) Usage of Traditional Implements	0.41
2. Economic	**28.75**
a) Land Productivity (fertility)	5.47
b) Net Sown Area	8.74
c) Availed/Want to avail Loan	0.07
d) Insurance Policy	2.50
e) Irrigation	6.09
f) Usage of Seeds	5.89
3. Technology	**13.11**
a) Usage of Modern Implements	0.57
b) Harvesting Method/post-harvest care	6.55
c) Mode of Transportation & Harvesting	0.05
d) Experience & Knowledge of Rice Cultivation	5.93
4. Environment	**18.31**
a) Flood/Draught resistant variety	0.28
b) Knowledge of Application of Fertilizers	5.70
c) Amount of Chemical Fertilizer used	11.08
d) Diversification of Crop	1.24
5. Political	**9.03**
a) Membership in social organisation	4.02
b) Joint effort with Fellow Farmers	4.22
c) Attended agricultural related meetings	0.79
Total	**100.00**

Source: Authors' estimation

Table 3: Influence of the Indicators/Dimensions on Sustainability.

In percentage term, it is 20 per cent, 64 per cent and 16 per cent respectively. This implies that the rice cultivation in Manipur has not been very impressive, despite its favourable agro-climatic condition. Of the different dimensions/institutions that influenced rice farming sustainability the most has been the social institution, followed by the economic institution. The institution of technology has been very negligible role on rice farming sustainability in Manipur. Nevertheless, the role of environment institution on rice farming economy still occupies some significant impacts, while the role political institution has been very negligible.

Acknowledgement

This paper is an outcome of the major research project titled "Institutional Structure and Performance of Agriculture in Northeast India" funded by Indian Council of Social Science Research, New Delhi.

References

1. Fischer KS (1998) Challenges for Rice Research in Asia. In Dowling NG; SM Greenfield and Fischer KS (eds), Sustainability of rice in the Global Food System. Manila (Philippines): International Rice Research Institute 95-98.

2. Singha K (2013) Growth of Paddy Production in India's North Eastern Region: A Case of Assam. Anvesak 42: 193-206.

3. FAO (2006) Report of the International Rice Commission-Twenty-First Session Rome: Food and Agriculture Organization.

4. Hossain M (1998) Sustaining Food Security in Asia Economic, Social and Political Aspects. In Dowling NG; SM Greenfield and KS Fischer (eds), Sustainability of rice in the Global Food System. Manila (Philippines): International Rice Research Institute 19-43.

5. Roy R, Chan NW, Rainis R (2014) Rice Farming Sustainability Assessment in Bangladesh, Sustainability Science 9: 31-44.

6. GoM (2012) Economic survey of Manipur 2010-11. Imphal: Directorate of economics and statistics, Government of Manipur.

7. GoI (2013) State of Indian agriculture 2012-13. Department of Agriculture and Cooperation, Ministry of Agriculture, New Delhi: Government of India.

8. Goodland R, Ledec G (1987) Neo-classical Economics and Principles of Sustainable Development, Ecological Economics 38: 19-46.

9. Acharya SS (2006) Sustainable Agriculture and Rural Livelihoods. Agricultural Economics Research Review 19: 205-217.

10. DFID (2002) Agricultural Sustainability, Department for International Development, Department of Biological Sciences, University of Essex, UK.

11. Najim MMM, Lee TS, Haque MA, Esham M (2007) Sustainability of Rice Production: A Malaysian Perspective. The Journal of Agricultural Sciences 3: 1-12.

12. Abrahamsson KV (1997) Paradigms of Sustainability. In Sörlin S (ed) The Road Towards Sustainability-A Historical Perspective, A Sustainable Baltic Region Programme, Sweden: Uppsalla University 30-35.

13. Barah BC (2006) Agricultural Development in North East India–Challenges and Opportunities, Policy Brief No 25, New Delhi: National Centre for Agricultural Economics and Policy Research.

14. (2005) Dynamics of rice Economy in India– Emerging Scenario and Policy Options. Occasional Paper No 47, Mumbai: National Bank for Agriculture and rural Development, Department of Economic Analysis and Research.

15. Damghani AM, Khoshbakht K, Veisi H (2009) Sustainability of Rice-Based Agro-ecosystems in Mazandaran, Iran: Agro-technical Characteristics. Environmental Sciences, 6: 135-144.

16. (2013) Development of an Empirical Model of Sustainable Rice Farming– A Case Study from Three Rice-Growing Ecosystems in Bangladesh, American-Eurasian Journal of Agricultural & Environmental Science 13: 449-460.

17. Tesso G, Emana B, Ketema M (2012) Analysis of Vulnerability and Resilience to Climate Change Induced Shocks in North Shewa, Ethiopia. Agricultural Sciences 3: 871-888.

18. Sherbinin AD (2003) The Role of Sustainability Indicators as a Tool for Assessing Territorial Environmental Competitiveness. Presented at the International Forum for Rural Development on 4–6 November 2003 at Brasilia, Brazil.

19. Bigazzi AY, Bertini RL (2009) Adding Green Performance Metrics to a Transportation Data Archive, Transportation Research Record: Journal of the Transportation Research Board No 2121, 30-40.

20. (2012) Regional Disparity of Rice Cultivation: A Case of Assam. Economic Affairs 57: 29-36.

21. Herdt RW, Steiner RA (1995) Agricultural Sustainability: Concepts and Conundrums. In Barnett V, Payne R, Steiner R (eds): Agricultural Sustainability - Economic, Environmental, and Statistical Considerations. London: John Wiley and Sons Ltd.

22. IHD (2010) Manipur State Development Report, New Delhi: Institute for Human Development.

23. Pou K (2010) Ethnic Impediments to Agricultural Development in the North-Eastern Region of India-An Institutional Perspective with Reference to PoumaiNaga, Quarterly Bulletin 1: 2-16.

18

Influence of water depth and seedling rate on the performance of late season lowland rice (*Oryza sativa* L) in a Southern Guinea Savanna ecology of Nigeria

U Ismaila[1]*, MGM Kolo[2], JA Odofin[2] and AS Gana[2]

[1]*National Cereals Research Institute, Badeggi, P.M.B. 08, Bida, Nigeria*
[2]*Federal University of Technology P.M.B. 65, Minna, Nigeria*

Abstract

A three year field experiment was conducted between 2010 and 2012 at the irrigated lowland experimental field of National Cereals Research Institute in Edozhigi (9°04N, 6°7E) of the Southern Guinea savannah ecological zone of Nigeria, to determine the effect of different water depths and seedling rates on yield and yield components of lowland rice. The trial was laid out in a split plot and arranged in a randomized complete block design by six regimes of water depths (5 cm, 10 cm, 15 cm, 20 cm, saturated soil and continuous flow of water at 2 cm depth) was accommodated in the main plot while the seedling rate of 2, 4 and 6 per stand constituted the sub-plots. The results indicated that both grain yield and yield components of rice were enhanced while the water level increased to 20 cm, although both tiller and height were negatively affected by higher water level at the early stage of growth but it was later compensated at later stage. Water depth of 15 - 20 cm revealed higher grain yield of 5051.8, 4700.4 and 4066.0 kg ha⁻¹ which was 84.4%, 85.2% and 84.7% higher than yields obtained from saturated plot in 2010, 2011 and 2012 respectively. Rice yield and yield components were significantly affected by different seedling rates and six seedlings per stand gave grain yield that is 13.1%, 27.8% and 14.4% higher than 2 seedling rate in 2010, 2011 and 2012, respectively. It is therefore concluded that maintaining water depth between 15 to 20 cm and seedling rate of 4 to 6 enhanced yield and yield components of lowland rice

Keywords: Water depth; Seedling rate; Performance; Lowland rice

Introduction

Rice is a semi aquatic plant that can be grown in standing water and dry land conditions. It was reported by Fogleman [1] that rice is grown under many different conditions and production systems, but submergence in water is the most common method worldwide. Most varieties of rice grow well and produce better yield when those are grown in flooded soils rather than non-flooded soils [2].

In farming systems of low land rice, water control is the most important practice that determines efficacy of production inputs such as nutrients, herbicides, pesticides, farm machineries, microbial activity and mineralization rate. Water plays a pivotal role in the management of rice systems. Different rice agro-ecosystems are mostly classified based on their hydrology and extent of water availability. Juraimi *et al.* [3] considered water as the most important component for sustainable rice production, especially in the traditional growing areas. Therefore, reduction or large removal of water from field can significantly decrease sustainability of rice production.

Irrigated lowland systems comprise 55% of the total rice area and provide 70% of the global rice production while rain-fed lowland and flood-prone areas constitute 35% of the total rice area, covering 4.7 million hectares in Asia but providing only 25% of global rice production because of various abiotic challenges associated with rain-fed ecosystems [4]. Chopra and Prakash [5] reported that 57% of rice is grown on irrigated lands, 25% on rain-fed lowlands, 10% on the uplands, 6% in deepwater and 2% in tidal wetlands while the report of Alan [2] indicated that irrigated rice ecosystem accounts for 75% of the global rice production.

Although, rice is a water loving crop, but continuous flooding could be stressful to the crop particularly during the seedling establishment stage when increase in water depth could jeopardize the performance of the crop. It is therefore very imperative to determine the safe limit for water depth that will give better rice performance, hence this study.

The study was conducted with the main objectives of determining the response of lowland rice yield and yield components to different depths of ponded water and appropriate seedling rate that gives better rice grain yield.

Materials and Methods

The experiment was conducted at Edozhigi lowland rice research field of National Cereals Research Institute, Badeggi, Bida, (Latitude 09° 45' N and Longitude 6° 07' E at an elevation of 75 meters above sea level). Niger State is in the Southern Guinea Savannah ecological zone of Nigeria between 2010 to 2012 late seasons. The average annual rain fall was 1287.5, 1158.3 and 1158.6 mm in 2010, 2011 and 2012 respectively, while the peak of rain fall was between July to September annually (Figure 1). During the three years of field experimentation, rain fall began in April and ended in October except for 2010 when rain fall extended to November.

The trial was laid out using a split plot design by six levels of water (5 cm, 10 cm, 15 cm, 20 cm, saturated soil and free flow of water) as the main plots while three seedling rates (2, 4, and 6 seedlings per stand) constituted the sub-plots. The main plot size was 10 m x 4 m while the sub – plot was 3 m x 4 m, the net plot was 2.8 m x 4 m by

*Corresponding author: U Ismaila, National Cereals Research Institute, Badeggi, P.M.B. 08, Bida, Nigeria, E-mail: ismailaumar72@yahoo.com

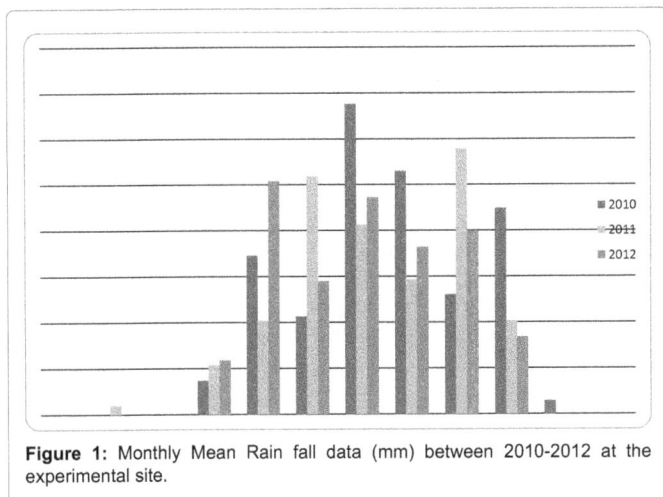

Figure 1: Monthly Mean Rain fall data (mm) between 2010-2012 at the experimental site.

three replicates. The experiment was conducted between September to December which represents the late cropping season in Nigeria.

Water management

Irrigation water was supplied through a channel that had its source from the perennial River Kaduna. The water was let into the field through the alley way and 3-inch PVC pipe was connected from the alley way to each plot serving as water inlet pipe. White plastic indicator was fixed at the middle of each plot to monitor the water level, while 10 cm plastic hose of 3-inches diameter was connected to each plot to drain the excess water when the maximum water level was attained.

Agronomic practices

The land was mechanically ploughed, harrowed and leveled but the bonds round the perimeter of the plots were manually constructed with a hoe. The FARO 57 rice seeds used in the study were obtained from the Seed Unit of National Cereals Research Institute, Badeggi. A nursery was established on August 20th, 30th and 21th of 2010, 2011 and 2012 respectively. Rice seedlings were transplanted 30 days after seeding (DAS) according to expatiate seedling treatments at the spacing of 20 x 20 cm. Each sub-plot received a uniform application of 40 kg/ha N, 40 kg/ha P_2O_5 and 40 kg/ha K_2O one week before transplanting. Additional 40 kg/ha N was applied at panicle initiation stage. The source of N, P_2O_5 and K_2O was urea (46% N), single super phosphate (18% P_2O_5) and muriate of potash (60% K_2O) respectively. The field was flooded to various heights as dictated by the treatments at 15 days after transplanting (DAT).

Fertilizer application was by broadcast after proper drainage of water from the plots. The plots were then flooded immediately after fertilizer application. The plots were finally drained one week before harvest and harvesting was done when the grains were hard and turned yellowish/brown, which occurred at between 30-45 days after flowering or one month after 50% flowering.

Rice parameters

The rice plant height was taken at four intervals (30, 45, 60 and 75 DAT) using metric rule which was measured from the ground level to apex of the tallest leaf. Rice tiller count was taken from 1m2 quadrant at four intervals (30, 45, 60 and 75 DAT). Number of panicles per m2 was also taken from 1m2 quadrant (averaged over 3 - quadrants thrown at random at 100 DAS). Days to 50 % flowering were determined when 50

% of the rice in each plot had flowered.

Rice grain yield was obtained from the net plot; the chaff was separated from the grain by soaking in water for two minutes. After proper steering, the floating chaff and the grain were collected and dried separately and weighed using weighing balance. Obtained grain was converted into tonnes per hectare. Percentage chaff was determined using the following formula.

$$\% \text{ Chaff} = \frac{\text{Chaff weight}}{\text{Total harvest}} \times 100$$

The weight of 1000 grains was determined by taking the measurement of 100 grains using an electrical digital weighing balance and the results were multiplied by 10 to give 1000 grains weight.

Data Analysis

Experimental data were subjected to Analysis of Variance (ANOVA) using the M-Stat-C version 1.3 statistic package and significant means were compared using Duncan Multiple Range Test at 5% probability.

Physical and chemical properties of soil in the experimental site

Composition of soil samples were collected from 0 to 30 cm of each block before transplanting. The samples were air-dried, ground to be passed through a 2 mm sieve. The samples were analyzed for selected physicochemical properties, namely, organic carbon, total nitrogen (N) soil pH, available phosphorus (P), available K, cation exchange capacity (CEC) and texture.

Organic carbon content of the soil was determined using the volumetric method Walkley and Black [6] as described by guide to laboratory establishment for plant nutrient analysis in Food and Agriculture Organization of the United Nations [7]. Total soil nitogen was analyzed by Micro-Kjeldahl digestion method with sulphuric acid. The pH of the soil was determined using 1:2.5 (weight/volume) soil samples to water ratio using a glass electrode attached to a digital pH meter.

Cation exchange capacity was measured after saturating the soil with 1M ammonium acetate (NH4OAc) and displacing it with 1M NaOAc [8]. Available phosphorous was extracted by the Olsen method, and phosphorous analyzed using spectrophotometer.

Particle size distribution was done by the hydrometer method according to FAO [7]. To determine exchangeable potassium in the soil, the soil samples were extracted with 0.5 N ammonium-acetate at pH 7.0 and the exchangeable potassium was determined with a flame photometer according to Hesse [9] (Table 1).

Soil characteristics measurement	
Texture	
Sand %	94.2
Silt %	1.2
Clay %	4.6
Textural class	Sand loam
pH (H_2O)	4.6
Organic matter (g kg^{-1})	1.2
Available P (ppm)	6.3
Exchangeable bases (cmo1kg^{-1})	
ECEC (mg kg^{-1})	13.1

Table 1: Physcio - Chemical properties of soil of the experimental site, Edozhigi.

	2010				2011				2012			
Days after transplanting												
Treatment	30	45	60	75	30	45	60	75	30	45	60	75
Water depth cm												
5	48.9e	67.7c	91.5b	114.8c	49.0f	67.5d	92.8b	114.4b	48.4d	65.3e	86.2c	113.4d
10	51.2d	70.2c	91.8b	111.4d	50.7e	70.2c	92.4b	114.6b	50.5c	68.9c	91.1b	115.6c
15	53.7c	71.9b	93.7b	117.3b	53.7c	72.7b	104.2a	119.6a	53.8b	74.3b	105.2a	120.5b
20	55.4b	90.6a	110.6a	119.2a	54.7b	73.5a	103.9a	120.0a	54.6b	76.7a	104.9a	125.8a
Cont. flow	60.3a	65.8d	92.9b	113.0c	61.7a	66.9d	80.4c	108.2c	60.9a	67.5d	76.3d	99.8e
Saturated	48.0f	72.2b	78.8c	112.6d	52.0d	60.9e	72.2d	108.3c	49.3d	56.5f	67.0e	91.1f
CV	1.9	13.7	14.4	3.8	1.7	1.6	1.4	1.0	1.5	3.2	2.7	1.7
SE	0.3	3.3	4.5	1.5	0.6	0.4	0.4	0.4	0.4	0.5	1.8	0.7
Seedling rate (S)												
2	43.8c	55.5c	60.6b	111.6	45.1c	54.9c	70.1c	108.8a	45.1c	58.3c	73.3c	104.4c
4	55.7b	70.8b	108.7a	114.8	56.7b	69.9b	100.4b	116.9b	55.6a	67.4b	90.3b	111.8b
6	59.2a	92.8a	110.3a	117.8	59.1a	81.1a	102.5a	116.8b	58.1b	78.9a	101.8a	116.9a
CV	1.8	13.7	14.4	3.8	1.7	1.6	1.4	1.0	1.5	3.2	2.7	1.69
SE	0.2	2.4	3.2	1.0	0.2	0.3	0.3	0.3	0.2	0.5	0.6	0.4
W X S	S	S	NS	NS	S	S	S	S	S	S	S	S

Means followed by the same letter (s) within a column are not significantly different at 5% level of probability. (DMRT). NS –not significant, * - significant, at P ≤ 0.05

Table 2: Effects of water depth and seedling rate on plant height (cm) in late season sown rice in 2010- 2012.

	2010				2011				2012			
Days after transplanting												
Treatment	30	45	60	75	30	45	60	75	30	45	60	75
Water depth cm												
5	104c	242d	348d	370d	109.9c	239d	338d	381c	106c	248d	305d	322d
10	93c	339c	476c	426c	93.2d	332c	480c	412b	92d	315c	486c	526c
15	89d	431a	553b	497b	89.0e	441a	563a	574a	85e	423b	553b	573b
20	87e	419b	561a	549a	88.8e	419b	555b	578a	86e	454a	558a	587a
Cont. flow	112b	237e	324e	347e	117.1b	221e	314e	269d	129a	192e	152e	137e
Saturated	126a	216f	285f	332f	121.7a	194f	220f	186e	123b	140f	84f	76f
CV	3.8	3.6	4.1	3.4	7.0	2.5	1.8	2.4	3.5	9.5	2.5	4.4
SE	1.3	3.7	5.8	4.8	2.4	2.3	3.6	3.9	2.1	7.9	3.3	6.5
Seedling rate (S)												
2	130a	402a	472a	532a	394a	394a	454a	472a	122a	343a	388a	402a
4	98b	345b	418b	390c	322b	322b	408b	387b	100b	290b	351b	364b
6	78c	196c	383c	339d	207c	207c	373c	342c	89c	252c	331c	345c
CV	3.9	3.6	4.1	3.4	2.6	2.6	1.8	2.4	3.5	9.5	2.5	4.2
SE	0.9	2.6	4.1	3.4	1.9	1.8	1.7	3.9	0.8	6.6	2.1	3.7
W X S	S	S	S	S	S	S	S	S	S	NS	S	S

Means followed by the same letter (s) within a column are not significantly different at 5% level of probability. (DMRT). NS –not significant, S - significant, at P ≤ 0.05

Table 3: Effect of water depth and seedling rate on tiller m-2 in late season sown rice in 2010 - 2012.

Results

The rice plant height was significantly affected by both water depth and seedling rate at all sampling periods in the three years of study (Table 2). At the early stage of growth (30 DAT), the highest plant height was obtained when the field was under continuous flow of water, but as plant growth progressed toward maturity, the water depths of 15–20 cm produced significantly taller plants than the other treatments. This result was consistent in the three years of study.

For seedling rates, higher seedling rate of six gave significant higher plant height at all the sampling periods (Table 2).

Water depth and seedling rate had significant effect on rice tillering at all the sampling periods (Table 3). Saturated plots gave better rice tillers at the early stage of growth (30 DAT) than the other treatments (Table 3). However, as growth progressed, significant tiller number was attained along with increased water depth.

The water depth and seedling rate have decisive bearing on rice yield components considered in the current study (Table 4). Days to 50 % of flowering were attained earlier in the plots not flooded and the days increased by elevation of water depth while the higher seedling rate of six showed significant lower days to 50 % flowering than the rest. The number of panicles per m² was significantly higher in 20 cm water depth while seedling rate of six revealed higher number of panicles per m² (Table 4).

The percentage chaff was lower in flooded plots than that of unflooded (Table 4). The 20 cm water depth showed significant lower % chaff which was consistent throughout the three years of study

Treatment	Days 50% flowering			Panicle m^{-2}			% chaff		
	2010	2011	2012	2010	2011	2012	2010	2011	2012
Water depth cm (W)									
5	69.9^c	70.8^c	71.1^d	191.0^d	187.8^d	185.3^d	7.9^c	8.0^c	7.8^c
10	74.6^b	74.4^b	76.0^c	290.3^c	285.7^c	269.0^c	8.0^c	7.7^c	7.9^c
15	78.9^a	80.9^a	80.7^b	398.1^b	384.8^b	368.7^b	6.8^d	7.0^d	6.5^d
20	79.6^a	81.6^a	83.4^a	464.7^a	441.7^a	406.6^a	6.9^d	6.7^d	6.2^d
Cont. flow	67.0^d	66.7^d	67.2^e	121.8^e	110.7^e	90.6^e	12.7^b	13.4^b	12.1^b
Saturated	65.4^e	64.4^e	66.2^f	101.8^f	53.1^f	34.3^f	14.6^a	15.6^a	14.6^a
CV	2.1	0.7	1.1	5.7	1.9	3.0	7.9	4.5	4.6
SE	0.5	1.1	0.2	4.9	3.0	1.7	0.5	0.4	0.4
Seedling rate (S)									
2	79.5^a	79.6^a	79.6^a	200.3^c	193.6^c	184.8^c	8.4	7.7	7.9
4	72.1^b	72.1^b	72.1^b	267.7^b	255.3^b	237.7^b	7.9	8.3	7.5
6	66.1c	67.7^c	70.7^c	315.9^a	282.9^a	254.7^a	8.0	7.9	8.0
CV	2.1	0.8	1.1	5.6	1.9	3.0	7.9	4.5	4.6
SE	0.4	0.1	0.2	3.4	1.1	1.6	0.8	0.6	0.4
W X S	S	S	S	S	S	S	S	S	S

Means followed by the same letter (s) within a column are not significantly different at 5% level of probability. (DMRT). NS –not significant, S - Significant

Table 4: Effect of water depth and seedling rate on yield components in late season sown rice in 2010 – 2012.

	2010				2011				2012			
Days after transplanting												
Treatment	30	45	60	75	30	45	60	75	30	45	60	75
Water level cm (W)												
5 cm	85.5^c	78.4^c	72.8^c	68.1^c	52.9^c	94.1^c	83.9^c	73.1^c	62.1^c	95.8^c	85.8^c	83.7^c
10 cm	63.5^d	58.4^d	50.2^d	64.5^d	50.7^c	79.4^d	62.1^d	51.2^d	49.5^d	79.9^d	76.6^d	72.1^d
15 cm	38.7^e	34.7^e	30.9^e	38.6^e	34.7^d	42.0^e	34.7^e	28.4^e	30.1^e	52.1^e	42.4^e	39.7^e
20 cm	26.9^f	20.2^f	17.7^f	15.4^f	32.1^e	32.1^f	30.1^f	25.4^f	24.9^f	43.9^f	37.4^f	38.5^f
Con't flow	121.4^b	110.1^b	130.8^a	145.2^b	90.6^b	126.1^b	113.5^b	114.6^b	103.1^b	224.3^b	3204.1^b	4283.8^b
Saturated	152.9^a	122.0^a	110.6^b	220.1^a	127.7^a	152.7^a	159.5^a	148.1^a	171.91^a	277.4^a	4574.6^a	4919.2^a
CV	2.6	7.1	54.8	11.2	11.5	4.09	14.7	5.9	3.43	5.9	5.9	4.4
SE	0.7	1.7	12.6	3.4	2.7	0.72	3.9	1.0	1.60	4.8	36.1	47.2
Seedling rate (S)												
2	143.7^a	124.6^a	118.0^a	164.7^a	88.7^a	140.1^a	127.0^a	111.1^a	97.6^a	169.5^a	1474.0^a	1691.6^a
4	67.6^b	60.7^b	49.5^b	70.4^b	58.0^b	71.7^b	67.7^b	59.7^b	66.3^b	125.8^b	1338.9^a	1562.8^b
6	33.1^c	26.6^c	39.0^c	40.7^c	47.7^c	51.4^c	47.2^c	49.6^c	56.9^c	91.5^c	1197.6^c	1464.0^c
CV	2.6	7.1	54.8	11.2	11.5	4.09	14.7	5.9	3.43	5.9	5.9	4.4
SE	0.5	1.2	8.9	2.4	1.7	0.85	2.8	1.0	0.59	1.8	18.5	16.3
W X S	S	S	S	S	S	S		S	S		S	S

Means followed by the same letter (s) within a column are not significantly different at 5% level of probability. (DMRT). NS –not significant, *

Table 5: Effect of water level and seedling rate on weed dry matter (g m-1) in late season sown rice in 2010 - 2012.

(Table 4). The result of this study indicated that seedling rate have no significant effect on % chaff. Weight of Rice grain was significantly higher at 20 cm water depth and decreased with decrease in water depth. Fewer seedling rate of two gave significantly higher grain weight than either four or six (Table 5). The plots that were left unflooded as in free flow and saturated plots, weed competition tended to be higher which could be attributed to low grain weight in those plot (Table 6).

Production of rice grains was significantly affected by both water depth and seedling rate so that the highest grain yield was recorded when the field was under continuous 20 cm of water depth from 15 days after transplanting till maturity (Table 5). The saturated plot had the lowest grain yield across the three years of study.

Weed dry weight was significantly affected by different water depth and seedling rate in the three studies. Lowest weed dry weight was recorded in 20 cm water depth and increased as the water depth decreased. The highest weed dry weight accumulation was recorded in saturated plot, where no ponding was done.

Discussions

Ponding of rice plant at the early stage of development had negative impact on plant height which might be due to the fact that the rice seedlings are too young to stand against stress of flood at this stage. Williams *et al.* [10] observed that growth is always stressful during seedling establishment in standing water so that deep water can jeopardize the rice crop. It is important then, to determine a safe limit for water depths in order avoid unacceptable risks. The result is in agreement with the finding of Tadesse *et al.* [11], who observed shorter rice plant in continuous flooding condition.

The taller rice plants associated with the higher numbers of

Treatment	1000 grain weight (g)			Grain yield (kg ha-1)		
	2010	2011	2012	2010	2011	2012
Water depth cm (W)						
5	29.1[d]	29.1[d]	28.6[b]	3289.1[d]	3245.2[d]	3141.0[d]
10	30.0[c]	30.0[c]	29.7[a]	3551.7[c]	3550.0[c]	3311.3[c]
15	31.3[b]	32.0[b]	32.3[a]	4702.3[b]	4493.1[b]	4066.0[a]
20	31.8[a]	32.5[a]	32.7[a]	5051.8[a]	4700.4[a]	4033.1[b]
Cont. flow	28.3[e]	28.2[e]	28.0[b]	2812.3[e]	2534.3[e]	2103.7[e]
Saturated	27.7[f]	27.6[f]	27.8[b]	990.0[f]	696.2[f]	607.2[f]
CV	2.5	1.8	25.9	3.0	1.3	2.5
SE	0.2	0.3	2.7	34.1	12.5	19.3
Seedling rate (S)						
2	31.3[a]	31.1[a]	31.2[a]	3128.3[c]	3245.2[c]	2616.9[c]
4	29.0[b]	29.3[b]	29.5[b]	3468.9[b]	3550.0[b]	2956.2[b]
6	28.8[c]	29.3[c]	28.6[b]	3601.4[a]	4493.1[a]	3058.1[a]
CV	2.5	1.8	25.9	3.0	1.3	2.5
SE	0.2	0.1	1.9	24.1	9.9	16.60
W X S	S	NS	NS	S	S	S

Means followed by the same letter (s) within a column are not significantly different at 5% level of probability. (DMRT). NS –not significant, S - significant, at P ≤ 0.05

Table 6: Effect of water depth and seedling rate on yield in late season sown rice in 2010 - 2012.

seedling/stand might be due to intra-plant competition for sunlight energy. Similar results were recorded by Gupta [12] who observed that higher seedling/hill gave significant higher plant height. The experiment conducted by Prabha et al. [13] to compared one and three seedlings/hill; it was also observed that the higher seedlings/hill gave significant taller plant.

This study indicated that the fewer seedling rates of two lead to significantly better tillers than those of four and six. Williams et al. [10] also observed that rice developed more tillers in shallow water depth than deeper water at the early stages of growth. However, Balasubramanian and Palaniappan [14] reported that drainage at maximum tillering stage stimulates vigorous root growth and reduced the development of unproductive tillers. Shrirame et al. [15] also found that planting of two seedlings/hill lead to significantly higher number of total tillers over single seedling/hill. Results of the current study are also in line the finding of Prabha et al. [13] and Sen et al. [16]. The greater number of tillers produced in the 15 – 20 cm water depth and two seedlings/hill might be due to less competition among rice and its weeds in deeper water situation (Table 6) [17].

Delay in 50% flowering in the water depth of 15 – 20 cm could be due to luxury availability of water that tended to enhance vegetative growth rather than reproductive stage of rice plant. This is not in line with the work of William et al. [10] who reported that despite the slow start, rice in deep water attained maturity about 4 days before the rice in shallow water. They observed that different temperatures had no effects since the deep water was generally cooler than the shallow water. The earlier heading in the saturated soil condition might be due to a a stress reaction.

The higher panicle, grain weight and low percentage chaff observed in higher water depth plots might be due to suppression of weed growth. Since weed is a known enemy of these yield attributes, less competition from weed will definitely enhance their productivity. Several workers Juraimi et al. [3], Sariam [18] indicated that production of rice panicles was significantly influenced by flooding. According to Sariam [18] and Siti Mardina [19], production of panicles was significantly higher when the field was under continuous flooding than saturated one. This study is in line with the report of the above authors. The higher number of

panicles m-2 in all flooded regimes is also observed to be due to the high number of tillers in the same flooding treatments which could also be attributed to low weed infestation. The report by Mohd Razi Ismail et al, [20] on growth and yield response of rice variety on different water regimes indicated that rice panicle and 1000 grain weight was higher under flooding condition than saturated condition.

The high percentage chaff in saturated plots might be due to reduced water availability at the flowering stage or the effect of weed infestation in the treatment. In fact, the finding could be caused by shortage of water at the anthesis (flowering) stage, which restricted rice pollination process leading to produce infertile and empty rice grain.

In the current study, flood depth of 20 cm effectively suppressed weed growth (Table 6), which resulted in lower competition between the rice and weed. Kakade and Soner [21] observed that continuous flood of rice field up to flowering significantly increased grain yield over alternation between floods and drying. Yakan and Surek [22] found no significant differences in grain yield among the relevant treatments. On the other hand Borrell [23] compared different irrigation regimes in dry seeded rice production of Australia. The authors found that flooded irrigation from sowing to maturity resulted in the highest grain yield.

Beser [24] conducted an experiment on different irrigation methods and observed that highest rice grain yield in continuous flooding irrigation. Also, the highest values of total biological yield and harvest index were achieved in continuous flooding irrigation. Although Tabbal et al. [25] observed no significant difference in the yield between rice grown in flooded condition, alternate flooded conditions and saturated condition.

Generally, grain yield was better with increased water depth which resulted to decreased weed infestation as a result of impact of ponding water on the weed seed and the killing of weed seedlings. This result is in line with the work of Sen [16], who observed that increasing the yield of lowland rice required reducing crop losses from weeds, which in turn require good water management to minimize weed infestation. Bhagat et al [26], Pane [27] and Zainal et al. [28] indicated that germination of weed seeds was significantly reduced by flooding. Zainal et al [28] concluded that flooding is the most effective cultural practice to weed control in lowland rice ecology, particularly when fields are continuous flooded to water depth of 7.5 – 15 cm.

The higher grain yield recorded in higher seedling rate of 6 could be due to the fact this rate suppressed weed growth much more than it counterpart.

Conclusion

Results of the current study indicated that performance of lowland rice is enhanced by continuous flood water depth. At the early stage of rice growth, both the height and tillers were negatively affected by flood. It is therefore, advisable to delay introduction of continuous flooding in the rice field beyond 15 DAT when the rice would be strong enough to stand the stress of flooded water. At this stage, either post emergent herbicide application or manual weeding could be carried out before introduction of flooded water to control weeds. The study also showed that transplanting four to six seedling/hill enhanced fast canopy cover hence weed suppression which might likely result to better grain yield of rice.

References

1. Fogleman M (2010) Introduction to Rice.

2. Alan KW (2002) Flooding (weed control in rice) IRRI Makati City, Philippines: In encyclopedia of pest management CRC press.

3. Juraimi AS, Muhammad MY, Begum M, Anuar AR, Azmi M, et al. (2009) Influence of flood intensity and duration on rice growth and yield . Pertanika journal of Tropical Agricultural science 32: 195- 208.

4. Ismail AM, Johnson DE, Ella ES, Vergara GV, Baltazar AM (2012) Adaptation to flooding during emergence and seedling growth in rice and weeds, and implications for crop establishment.

5. Chopra VL, Prakash S (2002) Evolution and Adaptation of Cereal Crops. Science Publishers Inc, NH, USA.

6. Walkley AJ, Black IA (1984) Estimation of soil organic carbon by the chromic acid titration method. Soil Sci, 37: 29-38.

7. FAO (2008) FAO fertilizer and plant nutrition bulletin: Guide to laboratory establishment for plant nutrient analysis. Rome, Italy. Pp. 203.

8. Chapman HD (1965) Cation-exchange capacity. InMethods of SoilAnalysis; Black CA: American Society of Agronomy: Madison, WI, Pp. 891-901

9. Hesse PR (1971) A Text Book of Soil Chemical Analysis. John Murray, London Pp 520

10. Williams JF, Roberts SR, Hill JE, Scardaci SC, Tibbits G (1990) Managing water for weed control in rice. California Agric. 441: 7-10.

11. Tadesse TF, Decchassa NR, Bayu W, Gebeyehu S (2013) Impact of rainwater management on growth and yield of rainfed lowland rice. Woodpecker journal of Agricultural research 2: 108-114.

12. Gupta SK (1996) Effect of planting, seedlings and nitrogen on yield of rice (Oryza sativa). Indian J. Agron, 41: 581-583.

13. Prabha ACS, Thiyagarajan TM, Senthivrlu (2011) System of Rice Intensification Principles on Growth Parameters, Yield Attributes and Yields of Rice (Oryza sativa L.)

14. Balasubramanian P, Palaniappan SP (2007) Principles and Practices of Agronomy, 2nd ed. Agrobis, India. Pp. 291-292.

15. Shrirame MD, Rajgire HJ, Rajgire AH (2000) Effect of spacing and seedling number per hill on growth attributes and yield of rice hybrids under lowland condition. J. Soils Crops. 10: 109-113.

16. Sen LTH, Ranamukhaarachchi SL, Zoebisch MA, Hasan MM, Meskuntavon W (2002) Effects of early-inundation and water depth on weed competition and grain yield of rice in the Central Plains of Thailand. Conference on International Agricultural Research for Development.

17. Kolo MGM, Umaru I (2012) Weed competitiveness and yield of inter-and intraspecific upland rice (Oryza sativa L.) under different weed control practices at Badeggi, Niger State, Nigeria. African Journal of Agricultural Research 7: 1687-1693.

18. Sariam O (2004) Growth of non-flooded rice and its response to nitrogen fertilization. Ph.D Thesis, Universiti Putra Malaysia. Pp. 260.

19. Siti Mardina I (2005) Kajian paras air berbeza ke atas populasi rumpai dan hasil padi. Final Year Project Paper, Universiti Putra Malaysia. Pp. 74.

20. Mohd Razi I, Kamal Uddin MD, Wan Ahmad Z, Maziah M, Ismail CH (2013) Growth and yield reposnse of rice variety MR220 to different water regimes under direct seeded conditions. WFL Publisher volume 11: 367-371.

21. Kakade BV, Soner KR (1983) Nutrient uptake and yield of rice as influenced by moisture regimes. In : International Rice Commission Newsletter, 32: 38-40

22. Yakan H, Sürek H (1990) Edirne yöresinde çeltik sulaması. Köyhizmetleri Genel Müdürlüsü, Kırklareli Atatürk Aras. Ens. Müd. Yayınları.

23. Borrell AK (1991) Irrigation methods for rice in tropical Australia. Inter. In: Rice Research Notes. 16: 28.

24. Beser N (1997) Trakya Bölgesi'nde degisik ekim ve sulama yöntemlerinin Çeltikte (Oryza sativa L.) verim ve verim unsurları ile kalite karakterlerine etkisi. Ph D Thesis, Trakya University, Edirne, Turkey. Pp. 160.

25. Tabbal DF, Lampayan RM, Bhuiyan SI (1992) Water-efficient irrigation technique for rice. In Proceedings of International Workshop on Soil and Water Engineering for Paddy Field Management, Asian Institute of Technology, Pp. 146-159.

26. Bhagat RM, Bhuiyan SI, Moody K (1996) Water, tillage and weed interactions in lowland tropical rice: A review. Agric. Water Manage. 31: 165-184.

27. Pane H (1997) Studies on ecology and biology of red sprangletop and its management in direct seeded rice. Ph.D. thesis, Universiti Sains Malaysia, Malaysia.

28. Zainal AAH, Mashhor M, Azmi M (2007) The effect of flooding on seed germination of weedy rice (oryza sativa complex, locally called padi angin) under plant house condition.

19

Characterization and QTL Analysis of *Oryza longistaminata* Introgression Line, pLIA-1, derived from a Cross between *Oryza longistaminata* and *Oryza sativa* (Taichung 65) under Non-fertilized Conditions

Emily Gichuhi[1], Eiko Himi[2], Hidekazu Takahashi[3] and Masahiko Maekawa[2]*

[1]*Graduate School of Environmental and Life Sciences, Okayama University, Japan*
[2]*Institute of Plant Science and Resources, Okayama University, Japan*
[3]*Bioresource Sciences, Akita Prefectural University, Japan*

Abstract

To meet and sustain the food demands of an ever-increasing world population, improving the yield of major cereal crops such as rice is necessary with sustainable cultivation harmonized with the environment. It is useful to utilize wild rice species as reservoirs of novel traits for breeding low-input adaptable (LIA) crops. *Oryza longistaminata*, a wild species of rice native only to Africa, possesses the vigorous biomass needed under low-input conditions. Thus, a potential LIA (pLIA) candidate, pLIA-1, showing large biomass, tall culm, large panicle with many primary and secondary branches and thick culms was selected from a selfed progeny of the cross between *O. longistaminata* and Taichung 65 (T-65), a japonica variety, under non-fertilized conditions. The pLIA-1 performance was superior to that of Koshihikari, Norin 18, T-65 and Nipponbare under fertilized and non-fertilized conditions suggesting that pLIA-1's characteristics might be useful for breeding low-input adaptable varieties. QTL analysis in F2 of the cross between pLIA-1 and Norin 18 detected 31 QTLs for yield-elated traits under non-fertilized conditions. The pLIA-1 allele had a positive contribution in 20 of the QTLs detected. Importantly, many of the QTLs were identified around regions where *O. longistaminata* chromosome segments were introgressed into pLIA-1. These results suggests that the QTLs detected in the F2 are important to improve modern varieties for adaptability to low-input conditions.

Keywords: Low-input adaptable; *Oryza longistaminata*; Wild rice; *Oryza sativa*

Introduction

In the near future, food insecurity will be a serious global problem, due to an explosively growing population that is predicted to reach 9 billion by 2050 [1]. Therefore, increasing crop productivity of major cereal crops such as rice; to meet the rising demand of an increasing population is mandatory. Rice is a staple food and it accounts for more than 21% of the calorific needs of the world's population [2].

A major milestone during the green revolution was the development and extensive adoption of semi-dwarf rice cultivars that almost tripled worldwide rice production and greatly enhanced food security [3]. Their high yields were primarily due to their improved harvest index and responsiveness to high inputs, especially nitrogen and water [4]. The extensive usage of semi-dwarf cultivars was, therefore, accompanied by increased inputs of chemical fertilizers and pesticides. This tripled the global grain yield, yet at a high environmental cost. The negative impacts of the current high input agriculture include soil acidification; pollution of rivers, lakes and ground water and the emission of greenhouse gases that may influence global warming [5].

To secure enough food without environmental degradation and excess dependency on inputs, it is necessary to breed new varieties with comparatively high yields and adaptability to low-input conditions. This can be achieved by maximizing crops' ability to produce biomass and tolerance of various abiotic and biotic stresses through the utilization of genetic resources and distantly related wild relatives. Wild species are reservoirs of latent useful traits for domesticated rice improvement [6]. Although many desirable alleles in wild relatives have, however, not yet been fully exploited, several QTLs for yield-related traits have been identified from *Oryza rufipogon* [6,7] and *Oryza glumaepatula* [8]. Of the wild species carrying the AA genome in rice, *Oryza longistaminata*, native only in Africa, is most distantly related to *Oryza sativa* [9]. It is a perennial species with resistance to bacterial leaf blight [10-13]

and is characterized by strong rhizomes, long anthers and allogamy which are important traits for hybrid seed production [14]. Further, it shows a vigorous biomass under low-input conditions [15,16]. The large biomass of *O. longistaminata* is considered to be an important trait for breeding low-input adaptable rice. However, its utilization in breeding programs for rice improvement has been very limited, due to developed crossing barriers and hybrid sterility [15,17-19].

We successfully crossed *O. longistaminata*, with Taichung 65 (T-65), an *O. sativa*, and successive self-fertilized plants of the progeny were selected based on large biomass production at a non-fertilized paddy field and called potential low-input adaptable (pLIA) rice line. In this study, we report the characterization of the introgression line, pLIA-1, carrying chromosome segments of *O. longistaminata* and QTLs for some important yield-related traits detected in F2 of the cross between pLIA-1 and Norin 18, a japonica variety, under non-fertilized conditions.

Materials and Methods

Breeding of the introgression line

Oryza longistaminata, locally known as Mpunga wa Majani (MwM),

---

*Corresponding author: Masahiko Maekawa, Institute of Plant Science and Resources, Okayama University, Japan, E-mail: mmaekawa@rib.okayama-u.ac.jp

collected in a valley 50 km away from Mombasa, Kenya, was crossed with T-65, a japonica variety, as a female parent. A single unmatured crossed seed was then subjected to half-strength MS medium [20] culture. F2 population of 169 plants was grown in a greenhouse at the Institute of Plant Science and Resources (IPSR), Okayama University, Kurashiki, Japan. The progeny from the F3 generation was then grown at a non-fertilized paddy field that had been kept without any fertilizer application for more than 20 years at IPSR. Plants showing a large biomass were selected at the F3 generation at the non-fertilized paddy field. Twelve vigorous plants selected at F5 generation were continuously self-fertilized every year. At the F11 generation, a line showing a large biomass under non-fertilized conditions was bred and called potential low-input adaptable (pLIA) rice line, pLIA-1.

To evaluate the contribution of introgressed segment of O. longistaminata (MwM) into pLIA-1 to yield-related traits, F2 plants and F3 plants derived from the cross between pLIA-1 and Norin 18 were grown under non-fertilized conditions and used for QTL mapping.

Characterization of pLIA-1 and its progeny

The agronomic performance of pLIA-1, under both non-fertilized and fertilized conditions, was evaluated with two replicates and compared to T-65, Norin 18, Nipponbare and Koshihikari in 2012 and T-65 and Norin 18 in 2013. In both experiments, 3-week-old seedlings were transplanted with a spacing of 40 cm between rows and 15 cm between plants. Fifty kg/ha of N, P and K were supplied in fertilized field as basal dressing. The following agronomic traits were measured: the number of panicles per plant, total weight of panicles, total weight of the biomass of the plant shoot, 100-grain weight and grain yield. Culm length, flag leaf length, panicle length, fresh panicle-base diameter, fresh culm-base diameter (culm diameter at 5 cm above ground), number of primary branches per panicle, number of secondary branches per panicle, number of spikelets per panicle and spikelet fertility were measured on the tallest culm of a given plant. Grain length and grain width were measured in 30 grains with 3 replicates. The rough harvest index was calculated as the panicle weight/total plant weight in 2012. Days to heading were from sowing date to the emergence of the first panicle.

F2 and F3 plants derived from the cross between pLIA-1 and Norin 18 were only grown under non-fertilized conditions in 2011 and 2012, respectively. A total of 14 yield-related traits were measured and evaluated.

Analysis of variance (ANOVA) was performed using R software. The Tukey's test ($P=0.05$) was used to test the differences between means, where significant differences were detected by ANOVA.

Graphical genotyping of pLIA-1

Leaf samples from seedlings of MwM, Norin 18, T-65 and pLIA-1 were collected and dried at 50°C overnight. Genomic DNA was extracted from the leaf tissues using a modified procedure as described by Kawasaki [21]. PCR reaction was prepared by mixing 3.5 μl of distilled water, 20 pmol of 0.5 μl forward primer, 20 pmol of 0.5 μl reverse primer, 5 μl of Quick Taq (TOYOBO, Japan) and 5 ng of the extracted DNA. Amplification was performed as follows: the initial denaturing step at 95°C for 7 min, 30 cycles for 45 sec at 95°C, 30 sec at 55°C and 30 sec at 72°C. Electrophoresis was done in a 3% agarose gel for 90 minutes. The band pattern of the samples was observed in UV-lighting after staining with Ethidium bromide.

QTL analysis

The F2 plants derived from the cross between pLIA-1 and Norin

18 were genotyped using 35 SSR markers found to be polymorphic between pLIA-1 and Norin 18, out of 111 SSR markers genome-widely distributed. To validate the QTLs identified in the F2 population, 230 F3 plants were genotyped using additional 8 SSR markers in the vicinity of the QTL identified on the distal end of the long arm of chromosome 8. The genetic linkage map was constructed using Mapmaker/Exp version 3.0. Composite interval mapping was performed for QTL analysis using the software Window QTL Cartographer 2.5 [22]. For each of the traits, the LOD threshold was determined at significant probability level of 5% by computing 1000 permutations, in both populations.

Results and Discussion

The breeding process

Crossing barriers and hybrid sterility highly developed between O. longistaminata and O. sativa [17] hindered the direct analysis of some important traits in the F2 population. Therefore, many researchers bred backcrossed populations for genetic analyses. Maekawa et al. [23] reported that O. longistaminata from Ethiopia has comparatively good crossing ability with O. sativa and an F2 population could be produced. Iwamoto et al. [24] demonstrated that O. longistaminata from Kenya is closely related to that from Ethiopia, based on the Catalase gene structure. In this study, O. longistaminata from Kenya was successfully crossed with T-65 as a pollen-donor parent and an F2 population was obtained. A single F1 plant of the cross between MwM and T-65 showed low spikelet fertility of 27%. Although 107 of the 169 plants showed extremely high sterility in the F2, a few highly fertile plants segregated, as shown in Figure 1A. In order to obtain large biomass plants derived from the cross, descendants from fertile F2 plants were grown in the non-fertilized field. Thus, one promising plant was selected based on its large biomass production and large panicle. The relationship between the culm length and the number of spikelets per panicle of F4 plants selected from the F3 plant demonstrated that plants carrying large numbers of spikelets per panicle had relatively long culm lengths (Figure 1B). Furthermore, the segregation of F5 plants was examined and the segregation pattern was found to be similar to that of F4 plants (Figure 1C). Twelve plants (#1 to #12) showing large numbers of spikelets per panicle were, therefore, selected. These selected plants were characterized by large panicles, large numbers of spikelets per panicle, long culms, many elongated internodes, thick culm base diameters and few tillers, compared with those of Norin 18 (Figures 1D and 1E). Although plants carrying large panicles and long culms tend to lodge easily, these selected plants were highly tolerant to lodging due to the very thick culm-base diameter. Of the 12 selected plants, #1 plant with more spikelets per panicle was selectively grown for further self-fertilization at the non-fertilized field. After 11 generations of self-fertilization, this line was named potential low-input adaptable (pLIA) line, pLIA-1.

Characterization of the pLIA-1

Recently, it has been reported that most of the semi-dwarf rice varieties have high numbers of unproductive tillers, small panicles and are susceptible to lodging in directly seeded conditions [25]. These traits are identified as the major constraints on improving yields in these varieties; whose yield potential has stagnated due to their plant type. Therefore, the concept of ideal plant architecture (IPA) has been proposed and demonstrated using plants with low tiller numbers with few unproductive tillers, more grains per panicle than in the currently cultivated varieties and thick and sturdy stems [26,27]. As shown in Figure 2, pLIA-1 had a very large number of spikelets per panicle with a large number of primary and secondary branches, compared with

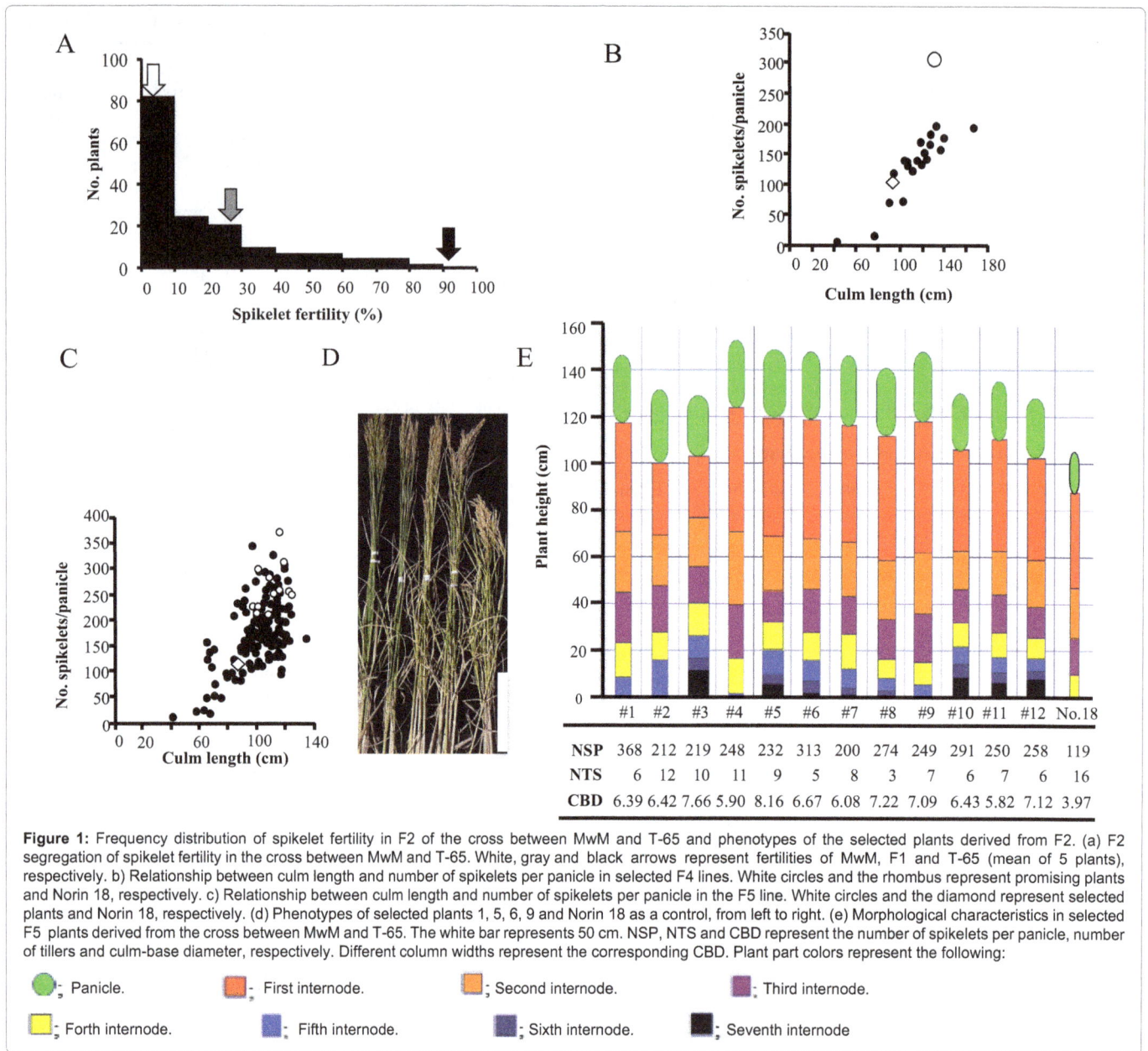

Figure 1: Frequency distribution of spikelet fertility in F2 of the cross between MwM and T-65 and phenotypes of the selected plants derived from F2. (a) F2 segregation of spikelet fertility in the cross between MwM and T-65. White, gray and black arrows represent fertilities of MwM, F1 and T-65 (mean of 5 plants), respectively. b) Relationship between culm length and number of spikelets per panicle in selected F4 lines. White circles and the rhombus represent promising plants and Norin 18, respectively. c) Relationship between culm length and number of spikelets per panicle in the F5 line. White circles and the diamond represent selected plants and Norin 18, respectively. (d) Phenotypes of selected plants 1, 5, 6, 9 and Norin 18 as a control, from left to right. (e) Morphological characteristics in selected F5 plants derived from the cross between MwM and T-65. The white bar represents 50 cm. NSP, NTS and CBD represent the number of spikelets per panicle, number of tillers and culm-base diameter, respectively. Different column widths represent the corresponding CBD. Plant part colors represent the following:

●; Panicle.   ■ - First internode.   ■; Second internode.   ■; Third internode.

■; Forth internode.   ■; Fifth internode.   ■; Sixth internode.   ■; Seventh internode

T-65 and Norin 18 (Figures 2A and 2B). Significantly larger numbers of primary and secondary branches per panicle resulted in a larger number of spikelets per panicle (Figures 2C-2E). Therefore, the agronomic traits of pLIA-1 were compared to those of T-65, Nipponbare and Koshihikari together with Norin 18 in 2012 and Norin 18 and T-65 in 2013, under both fertilized and non-fertilized conditions (Figures 3 and 4). Significant differences in the culm length, panicle length, culm-base diameter, panicle-base diameter, and number of secondary branches, number of spikelets per panicle and grain length between pLIA-1 and other varieties were observed under both conditions, in both years (Figures 3A-3D, 3H, 3I, 3R, 4A-4D, 4H, 4I and 4P). T-65 showed a significant decrease in panicle weight, total grain yield, flag leaf length and biomass in 2012 (Figures 3F, 3K, 3N and 3O) and number of panicles in 2013 (Figure 4E) under non-fertilized conditions. On the other hand, Norin 18 showed significantly decreased panicle weight, number of primary branches and number of grains/plant in 2012 (Figures 3F, 3G and 3K) and number of panicles, days to heading and

flag leaf length under non-fertilized conditions in 2013 (Figures 4E, 4M and 4N). Significant increases in the culm-base diameter, number of primary branches and days to heading were observed under non-fertilized conditions for pLIA-1 in 2012 (Figures 3C, 3G and 3M). Increases in the panicle-base diameter, number of secondary branches and number of spikelets per panicle were also observed under non-fertilized conditions in both years; however, the increase was not significant (Figures 3D, 3H, 3I, 3L, 4D, 4H, 4I and 4L). Although pLIA-1 showed significantly shorter grain length under non-fertilized conditions in 2013 (Figure 4P), it did not show a significant reduction in any other trait under non-fertilized conditions. These results suggest that pLIA-1 performance is superior under non-fertilized conditions, compared to other varieties. The characteristics of the pLIA-1 selected under non-fertilized conditions are, hence, comparable to IPA. Overall, in comparison to other varieties, pLIA-1 was significantly superior in most of the traits measured. However, the spikelet fertility of pLIA-1 was found to be significantly lower than in other varieties. The spikelet

Figure 2: Morphological characteristics of pLIA-1, Norin 18 and T-65. (a) pLIA-1, Norin 18 and T-65 phenotypes at maturity. (b) Panicles of Norin 18 and pLIA-1. (c) Number of primary branches. (d) Number of secondary branches. (e) Number of spikelets/panicle. Data represent the mean values of 16 plants, with the standard error. Different letters show significant differences at the 5% level by Tukey's test.

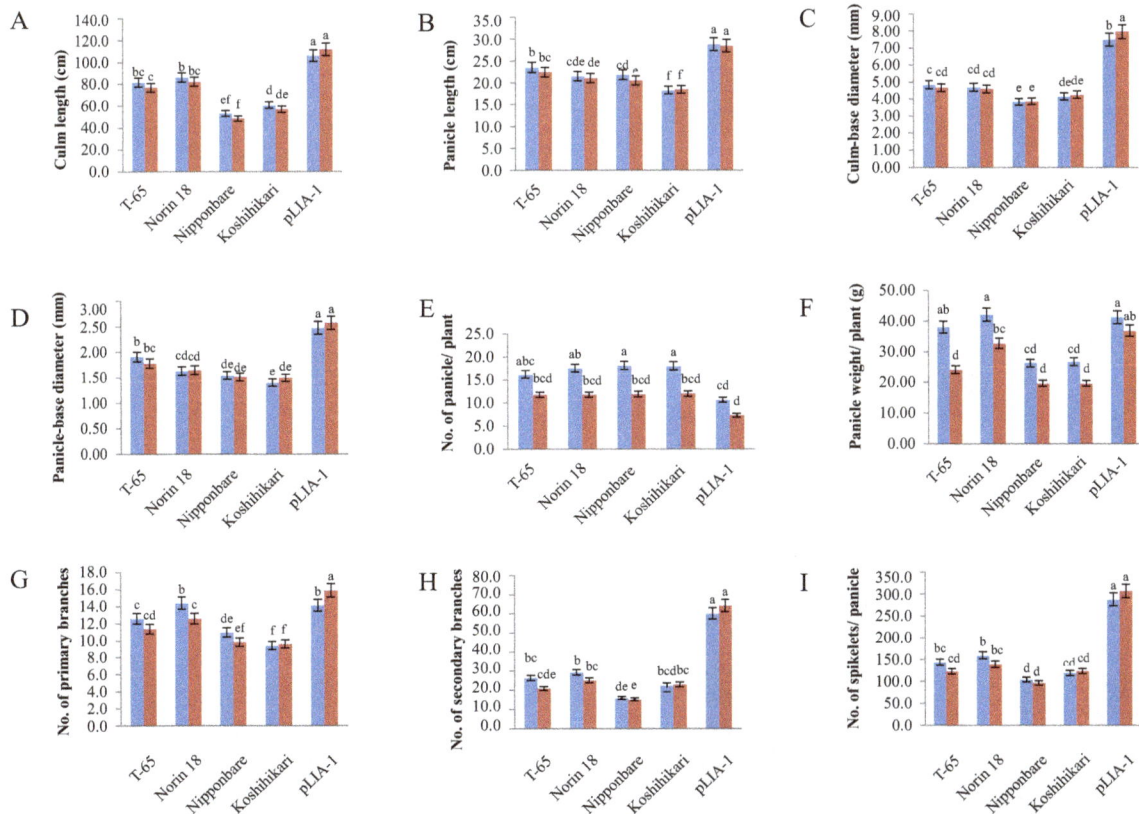

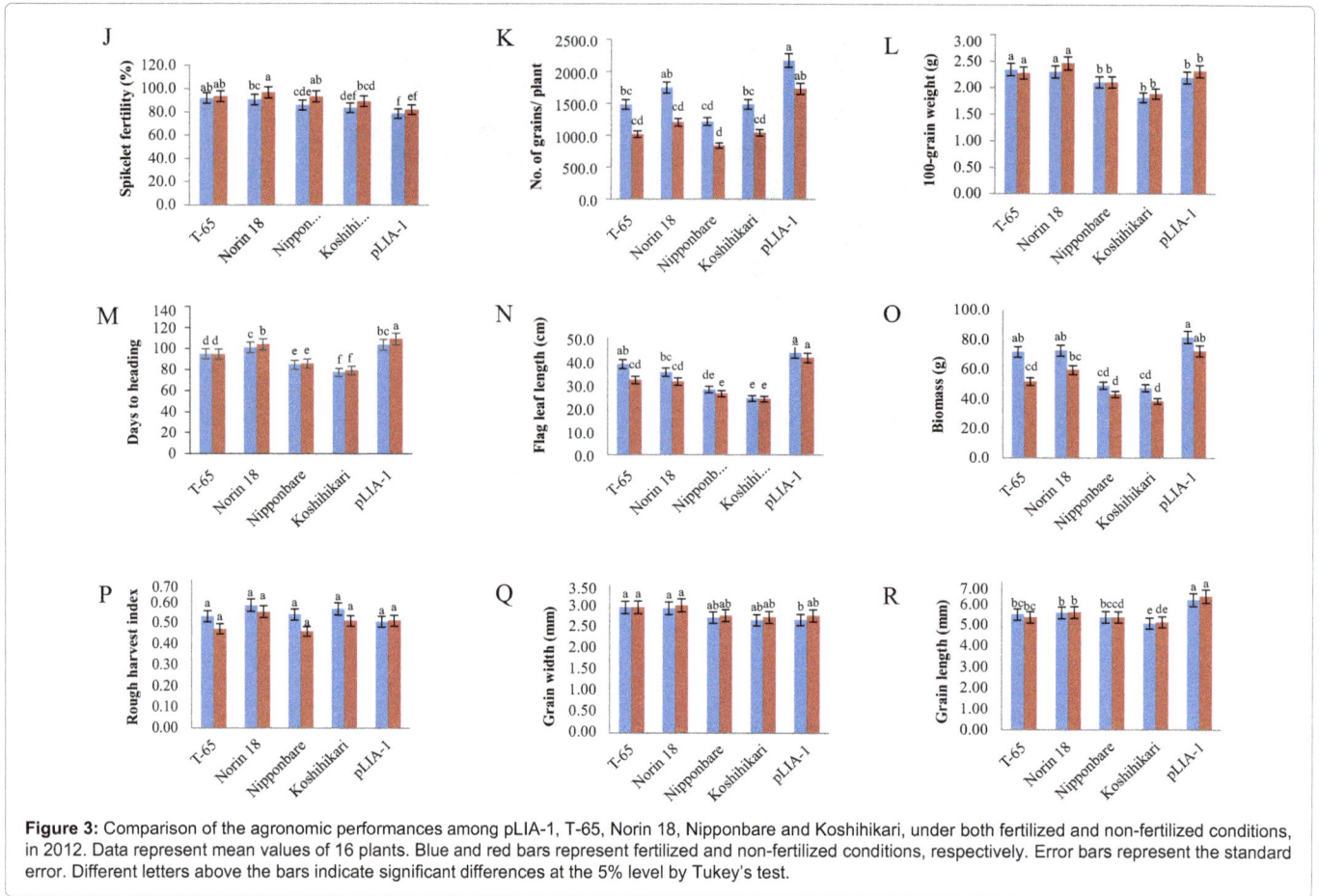

Figure 3: Comparison of the agronomic performances among pLIA-1, T-65, Norin 18, Nipponbare and Koshihikari, under both fertilized and non-fertilized conditions, in 2012. Data represent mean values of 16 plants. Blue and red bars represent fertilized and non-fertilized conditions, respectively. Error bars represent the standard error. Different letters above the bars indicate significant differences at the 5% level by Tukey's test.

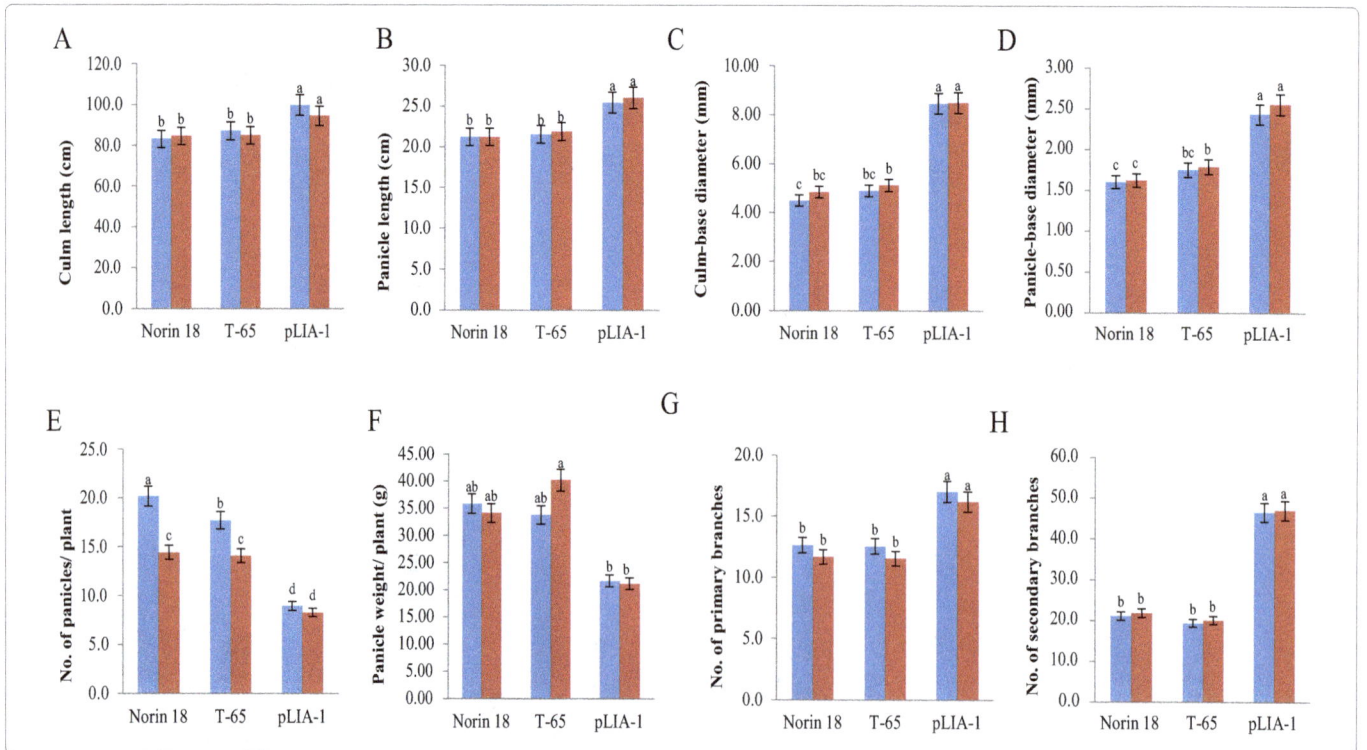

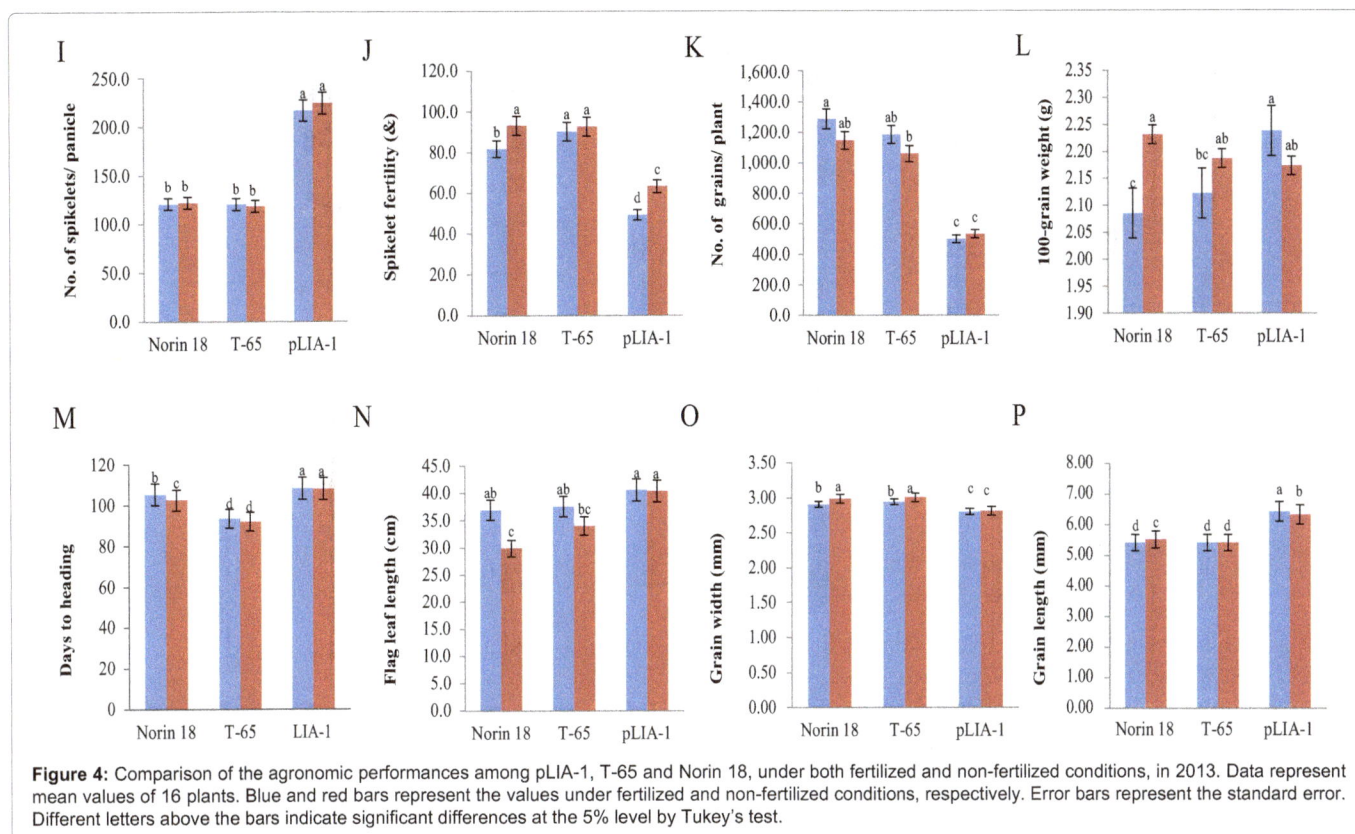

Figure 4: Comparison of the agronomic performances among pLIA-1, T-65 and Norin 18, under both fertilized and non-fertilized conditions, in 2013. Data represent mean values of 16 plants. Blue and red bars represent the values under fertilized and non-fertilized conditions, respectively. Error bars represent the standard error. Different letters above the bars indicate significant differences at the 5% level by Tukey's test.

fertility recorded in 2013 was markedly low and may have had a direct effect on the total grain yield. The inconsistence observed in spikelet fertility and a few other traits between 2012 and 2013 may be due to environmental effect.

Graphical genotyping of pLIA-1

In order to utilize pLIA-1's characteristics in breeding programs, introgressed segments in pLIA-1 were needed to be revealed. Tian et al. [28] found that QTLs derived from *O. rufipogon* introgressed segments in a set of 159 introgressed lines were usually associated with improvement of the target trait (panicles per plant, grains per panicle and filled grains per panicle). In this study, pLIA-1 was found to possess introgressed chromosome segments of *O. longistaminata* and T-65 by using genome-wide SSR markers. As shown in Figure 5, MwM showed specific polymorphisms in 85 (76.6% frequency) of 111 markers used. Thirteen markers were not amplified in MwM. These non-amplified markers were distributed in 10 chromosomes, except chromosomes 5 and 12 in MwM. Since *O. longistaminata* is predicted to carry a highly rearranged DNA sequence (Maekawa, unpublished), it is likely that 13 of the markers were not amplified. Additionally, other 13 markers were not polymorphic among the markers used. However, it was observed that pLIA-1 showed MwM-specific band patterns in 18 of 85 markers in 7 chromosomes, except chromosomes 4, 5, 7, 9 and 12 (Figure 5). Ten markers showed T-65-specific band patterns on chromosomes 1, 4, 5, 7, 8 and 10. Consequently, pLIA-1 was found to carry 20 MwM-specific markers including 2 non-amplified markers on chromosomes 1, 2, 3, 6, 8, 10 and 11 with a frequency of 18.0%. In particular, the short arm of chromosome 6 of *O. longistaminata* is presumed to be introduced into pLIA-1. Relatively few segments of *O. longistaminata* were introgressed into pLIA-1 chromosomes. Of 111 genome-1 wide SSR markers used, only 18 *O. longistaminata*

specific markers were observed with a frequency of 16.2%. However, pLIA-1 was shown to exhibit the large biomass characteristic under non-fertilized conditions. This suggests that the large biomass and low-input tolerance characteristics may be controlled by small segments of the *O. longistaminata* chromosome segments. On the other hand, novel introgressed chromosome segments observed on chromosomes 1, 5, 8, 9, 10 and 11 may have been caused by changed short sequence repeat numbers through successive self-fertilization or more likely outcrossing during the early stages of the breeding process of pLIA-1.

Correlations between yield-related traits

To reveal the important QTLs for yield-related traits, segregation patterns of the traits and correlations between them were examined in F2 of the cross between pLIA-1 and Norin 18. In most of the traits measured, segregation patterns of normal distribution were observed (Supplementary Figure 1) and transgressive segregations were found in all the traits. In order to understand QTL cluster, correlations among agronomic traits were examined in the F2. It was found that culm-base diameter was significantly positively correlated with panicle traits (panicle length, panicle weight, number of primary branches, number of secondary branches and number of spikelets per panicle) and flag leaf length (Table 1). The panicle traits were further observed to be positively correlated to each other. The number of spikelets per panicle was strongly correlated to the number of secondary branches (Table 1). Further, significant positive correlations between flag leaf length and number of primary branches, number of secondary branches and number of spikelets per panicle were observed (Table 1).

QTL analysis

To explore the genetic resources from wild rice, several populations

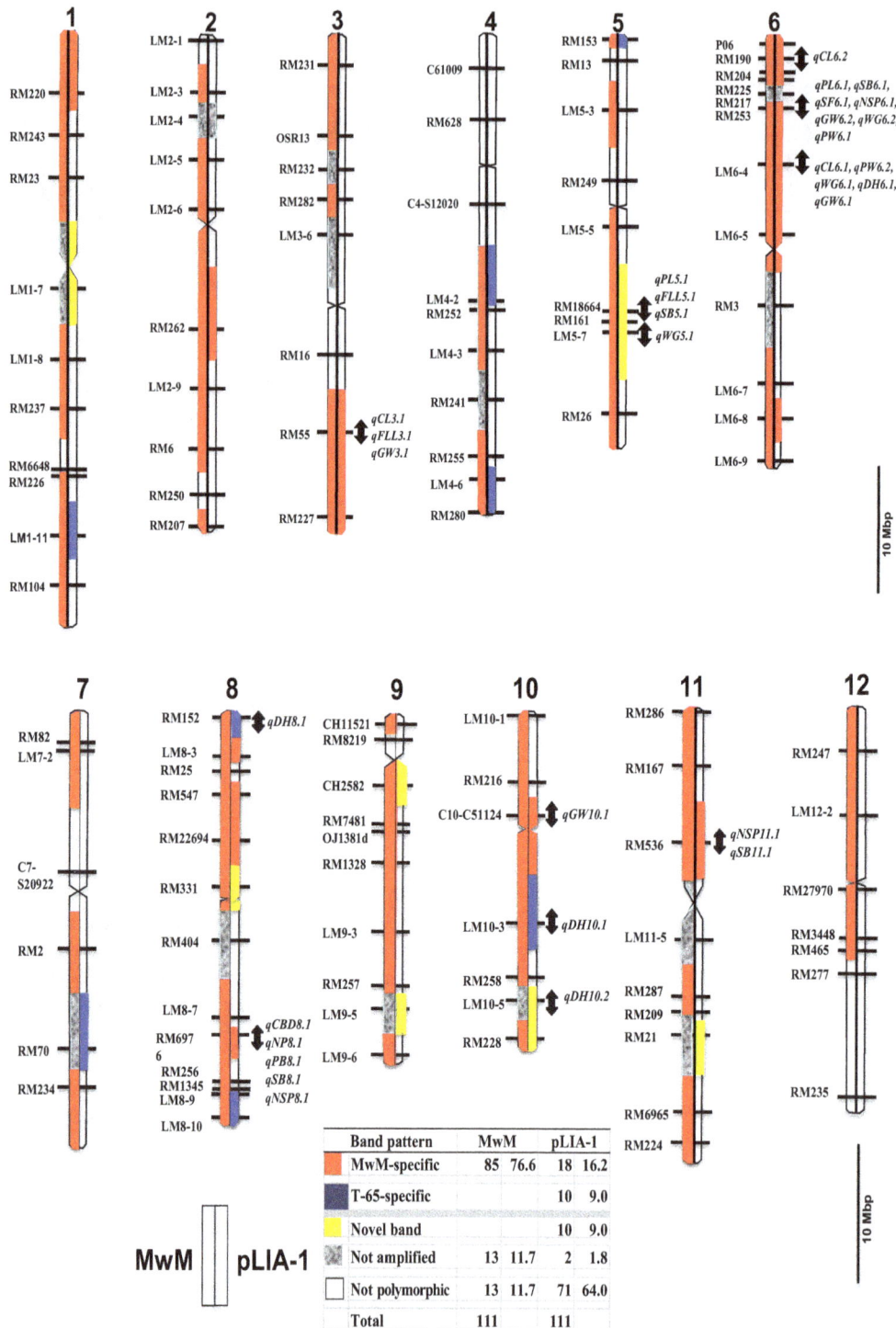

Figure 5: Graphical genotype of pLIA-1 and identified QTLs for yield-related traits. Chromosomes showing polymorphisms observed between MwM, pLIA-1, Norin 18 and T-65. The left side of each chromosome represents the MwM chromosome and the right side represents the pLIA-1 chromosome. On the left, red sections indicate that MwM-specific allele was observed and white sections indicate that the allele was similar to that of Norin 18. On the right chromosome, red, blue and white chromosome sections show the MwM, T-65 and Norin 18 alleles, respectively. Yellow chromosome sections indicate novel alleles not observed in MwM, T-65 or Norin 18. Shaded chromosome sections represents non-amplified allele. Two headed black arrows indicate tentative position of detected QTLs for the yield-related traits in the F2 of the cross between pLIA-1 and Norin 18.

derived from crosses between various cultivars and wild rice have been used for identification of important QTLs for agronomic traits [7,8,25]. In particular, Xiao et al. [6] detected a total of 68 QTLs for 12 traits

using a backcross population derived from a wild rice (O. rufipogon) and cultivated rice. In this study, 31 QTLs for 13 yield-related traits were detected on chromosomes 3, 5 6, 8, 10 and 11 in the F2 of the cross

Trait	Culm length (cm)	Panicle length (cm)	No. of panicles	Culm-base diamter (mm)	Flag leaf length (cm)	One panicle weight (g)	No. of primary branches	No. of secondary branches	No. of secondary spikelets/ panicle	Spikelet fertility (%)	100-grain weight (g)	Grain length (mm)	Grain width (mm)
Panicle length (cm)	-0.118												
No. of panicles	0.231**	0.137											
Culm-base diamter (mm)	0.153	0.237**	0.058										
Flag leaf length (cm)	-0.307**	0.407**	-0.003	0.225**									
One panicle weight (g)	0.095	0.19*	0.021	0.34**	0.154								
No. of primary branches	0.043	0.124	0.007	0.477**	0.248**	0.265**							
No of secondary branches	0.081	0.376**	0.167*	0.508**	0.368**	0.499**	0.379**						
No. of spikelets/ panicle	0.072	0.319**	0.173*	0.521**	0.349**	0.494**	0.534**	0.847**					
Spikelet fertility (%)	0.113	0.068	0.064	-0.02	-0.105	0.546**	0.002	0.079	0.041				
100-grain weight (g)	-0.218**	0.006	-0.019	-0.032	0.052	0.004	-0.027	-0.163*	-0.097	-0.275**			
Grain length (mm)	-0.155*	0.125	-0.051	0.077	0.104	0.114	0.021	0.039	-0.025	-0.108	0.583**		
Grain width (mm)	-0.117	-0.106	-0.06	0.046	-0.057	-0.098	0.056	-0.123	0.04	-0.317**	0.47**	-0.165*	
Days to heading	0.063	-0.171*	-0.151	-0.148	-0.121	-0.25**	-0.184*	-0.169*	-0.220**	-0.267**	-0.024	0.088	-0.066

*,**, Significant at 5% and 1% levels, respectively.

Table 1: Correlation coefficient of yield-related traits in F2 of the cross between pLIA-1 and Norin 18 under non-fertilized conditions.

between pLIA-1 and Norin 18 (Table 2 and Figure 5). These included 3 QTLs for culm length, 2 QTLs for panicle length, 1 QTL for number of panicles per plant, 1 QTL for culm-base diameter, 2 QTLs for flag leaf length, 2 QTLs for panicle weight, 1 QTL for number of primary branches per panicle, 4 QTLs for number of secondary branches per panicle, 3 QTLs for number of spikelets per panicle, 1 QTL for spikelet fertility, 3 QTLs for 100 grain weight, 4 QTLs for days to heading and 4 QTLs for grain width (Table 2). The QTLs were distributed on chromosomes 3, 5, 6, 8, 10 and 11(Figure 5). In 20 of the QTLs identified, pLIA-1 had a positive contribution to the trait. QTLs for strongly correlated traits were observed to be localized near the same region on the chromosome. These clusters of QTLs were observed on chromosomes 3, 5, 6 and 8 (Figure 5). Further analysis using an F3 population revealed that the region around RM6976 on chromosome 8 carried a crucial QTL cluster for culm-base diameter, panicle-base diameter and number of primary branches (Figure 5 and Table 3). This is considered to be caused by a strong positive correlation observed between the number of primary branches, number of secondary branches and number of spikelets per panicle. Hence, it is plausible that QTLs for number of primary branches and number of spikelets per panicle were identified in the same QTL cluster in the F2 population. These results strongly suggest that O. longistaminata has great potential for utilization in yield improvement and especially under low-input conditions despite the low spikelet fertility. Among the traits transferred in interspecific crosses using wild rice relatives, spikelet sterility is especially a serious constraint.

A QTL for spikelet fertility was identified on chromosome 6 with a very high LOD score of 30.9 and 60% contribution to the total phenotypic variation (Table 2), suggesting that this QTL might be the major cause of the low spikelet fertility of pLIA-1. In fact, this QTL was located near the same chromosome region where a QTL for pollen and spikelet fertility was previously identified in a cross using O. longistaminata [29].

QTL clusters of functionally related genes are of great interests in crop improvement. A total of 4 QTL clusters on chromosome 3, 5, 6 and 8 were observed (Figure 5). Highly significant correlations were also observed between the yield-related traits whose QTLs' were observed to cluster in the same chromosome locations (Table 1). In previous QTL analysis it has been observed that QTL for significantly correlated traits usually had same chromosome location [8,28]. The QTL cluster on chromosome 8 was identified near the same chromosome region where the WFP (Wealthy Farmer's Panicle) was found to be located [30]. This result suggest that these traits are either controlled by strongly linked genes or are as a result of pleiotropism of a single gene locus located at the regions where QTL clusters were observed. Previously, Ookawa et al. [31] reported that the APO1 gene of Habataki on chromosome 6 increased spikelet number together with thicker culm through increased size of inflorescence meristem, hence, the higher spikelet number induced culm thickness pleiotropically. The WFP gene of ST-12 which encodes OsSPL14 (Squamosa Promoter Binding Protein-Like14) drastically increases primary branch number, resulting in increased number of spikelets per panicle [30]. Since in the report, inflorescence meristem of ST-12 was found to be larger than that of Nipponbare, ST-12 was presumed to have thick culms. Thus, it is plausible that the QTL of culm-base diameter was located near the QTL for primary branch number on chromosome 8 based on significantly positive correlation between culm-base diameter and panicle traits observed in this study. Further, there was no correlation between culm-base diameter and culm length in this study suggesting that different culm length plants could be bred with thick culm and larger panicles. Taken together, O. longistaminata is therefore suggested to carry several useful traits under low-input conditions.

The pLIA-1 line reported here is considered to have high potential for low-input adaptability. Identification of QTLs for yield-related traits under non-fertilized conditions further proves its potential for utilization in yield improvement of rice. This line could therefore be

Trait	QTL	Chr.	Marker	LOD	Additive effect	r2
Culm length	qCL3.1	3	RM55	3.4	-3.69	0.07
	qCL6.2	6	LM6_4	6.4	6.45	0.16
	qCL6.1	6	RM190	6.3	5.75	0.12
Panicle length	qPL5.1	5	RM18664	5.9	1.24	0.25
	qPL6.1	6	RM253	6.7	-1.29	0.18
No. of panicles/ plant	qNP8.1	8	RM6976	4.6	-1.35	0.11
Culm-base diameter	qCBD8.1	8	RM6976	7.1	0.44	0.15
Flag leaf length	qFLL3.1	3	RM55	3.2	2.83	0.07
	qFLL5.1	5	RM18664	3.0	2.42	0.14
Panicle weight	qPW6.1	6	RM253	9.3	-0.23	0.29
	qPW6.2	6	LM6_4	3.7	-0.24	0.08
No. of primary branches	qPB8.1	8	RM6976	15.3	1.74	0.29
No. of secondary branches	qSB5.1	5	RM18664	5.1	2.67	0.18
	qSB6.1	6	RM253	4.9	-4.48	0.15
	qSB8.1	8	RM6976	3.2	2.63	0.06
	qSB11.1	11	RM536	3.5	3.77	0.07
No. of spikelets/ panicle	qNSP6.1	6	RM253	3.2	-12.77	0.10
	qNSP8.1	8	RM6976	6.0	17.57	0.12
	qNSP11.1	11	RM536	3.0	13.55	0.06
Spikelet fertility	qSF6.1	6	RM253	30.9	-1.57	0.60
Weight of 100-grains	qWG5.1	5	LM5_7	3.7	-0.16	0.39
	qWG6.2	6	LM6_4	9.1	-0.11	0.23
	qWG6.1	6	RM253	7.9	0.00	0.16
Days to heading	qDH6.1	6	LM6_4	24.2	2.95	0.33
	qDH8.1	8	RM152	8.7	-1.92	0.14
	qDH10.1	10	LM10_3	10.6	2.17	0.22
	qDH10.2	10	LM10_5	3.5	0.94	0.04
Grain width	qGW3.1	3	RM55	13.7	-0.06	0.23
	qGW6.2	6	LM6_4	7.5	-0.05	0.12
	qGW6.1	6	RM253	5.5	0.00	0.07
	qGW10.1	10	C51124	3.9	0.03	0.06

Table 2: Marker position, LOD score, additive effect and contribution rate of QTL for yield-related traits identified in the F2 population.

Trait	QTL	Marker	LOD	Additive effect	r2
Culm-base diameter	qCBD8.1	RM6976	8.8	0.39	0.16
Panicle-base diameter	qPBD8.1	RM210	2.2	0.06	0.04
No. of primary branches	qPB8.1	RM6976	18.2	1.16	0.30
		RM210	17.1	1.12	0.29

Table 3: QTL for yield-related traits validated in the F3 population on chromosome 8 QTL cluster region.

utilized to introgress high productivity under low-input conditions to elite rice varieties. Improvement of nitrogen use efficiency has been proposed as a target for the Second Green Revolution [32]. Hence, "low-input and high output" agriculture is required for sustainability. Further analysis of this line and identification of genes governing the QTLs of agronomic importance is necessary for better understanding of its tolerance to low-input conditions.

Conclusion

In order to utilize *O. longistaminata* as gene resources, some selected plants which showed vigorous biomass under non-fertilized conditions were developed from the F2 of the cross between MwM, *O. longistaminata* collected in Kenya and T-65, *O. sativa*. Out of the selected plants, pLIA-1 showed tolerance to non-fertilized conditions compared to the other varieties, hence was named potential Low-input Adaptable-1. The pLIA-1 was subjected to polymorphic analysis against Norin 18 using SSR markers. Although pLIA-1 carried 20 MwM-specific segments, very important QTLs for panicle-related traits were especially found to be intensively located on the distal region of long arm of chromosome 8 in F2 of the cross between pLIA-1 and Norin 18.

Acknowledgements

This research was funded by the Japan Science and Technology Agency (JST)/Japan International Cooperation Agency (JICA), the Science and Technology Research Partnership for Sustainable Development (SATREPS) and the Ministry of Education, Culture, Sports, Science and Technology (MEXT), Japan.

References

1. Alexandratos N, Bruinsma J (2012) World agriculture towards 2030/2050: The 2012 revision. ESA Working Paper No. 12-03, Rome.

2. Fitzgerald MA, McCouch SR, Hall RD (2009) Not just a grain of rice: the quest for quality. Trends Plant Sci 14: 133-139.

3. Tilman D (1998) The greening of the green revolution. Science 396: 211-212.

4. Matson PA, Parton WJ, Power AG, Swift MJ (1997) Agricultural intensification and ecosystem properties. Science 277: 504-509.

5. Ravishankara AR, Daniel JS, Portmann RW (2009) Nitrous Oxide (N2O): The dominant ozone-depleting substance emitted in the 21st Century. Science 326: 123-125.

6. Xiao JH, Li JM, Grandillo S, Ahn SN, Yuan LP, et al. (1998) Identification of trait-improving quantitative trait loci alleles from a wild rice relative, *Oryza rufipogon*. Genetics 150: 899-909.

7. Reddy CS, Babu AP, Swamy BPM, Sarla N (2007) Insight into genes underlying yield enhancing QTLs from *O. rufipogon*. Rice Genetics Newsletter 23: 53-55.

8. Brondani C, Rangel PHN, Brondani RPV, Ferreira ME (2002) QTL mapping and introgression of yield-related traits from Oryza glumaepatula to cultivated rice (*Oryza sativa*) using microsatellite markers. Theor Appl Genet 104: 1192-1203.

9. Causse MA, Fulton TM, Cho YG, Ahn SN, Chunwongse J, et al. (1994) Saturated molecular map of the rice genome based on an interspecific backcross population. Genetics 138: 1251-1274.

10. Khush GS, Bacalango E, Ogawa T (1990) A New Gene for Resistance to Bacterial Blight from *O. longistaminata*. Rice Genetics Newsletter 7: 121-122.

11. Khush GS, Mackill DJ, Sidhu GS (1989) Breeding rice for resistance to bacterial blight. In: Bacterial blight of rice. Proceeding of international workshop on bacterial blight of rice, IRRI, Manila, Philippines. pp: 207-217.

12. Ronald PC, Albano B, Tabien R, Abenes L, Wu KS, et al. (1992) Genetic and physical analysis of the rice bacterial-blight disease resistance locus, Xa21. Molecular & General Genetics 236: 113-120.

13. Song WY, Wang GL, Chen LL, Kim HS, Pi LY, et al. (1995) A receptor kinase-like protein encoded by the rice disease resistance gene, Xa21. Science 270: 1804-1806.

14. Virmani SS, Aquino RC, Khush GS (1982) Heterosis breeding in rice (*Oryza sativa* L). Theoretical and Applied Genetics 63: 373-380.

15. Sacks EJ, Roxas JP, Cruz MTS (2003) Developing perennial upland rice II: Field performance of S-1 families from an intermated Oryza sativa/*O. longistaminata* population. Crop Sci 43: 129-134.

16. Yang H, Hu L, Hurek T, Reinhold-Hurek B (2010) Global characterization of the root transcriptome of a wild species of rice, *Oryza longistaminata*, by deep sequencing. BMC Genomics 11: 705.

17. Chu YE, Oka HI (1970) The genetic basis of crossing barriers between Oryza Perennis Subsp. barthii and its related taxa. Evolution 24: 135-144.

18. Hu F, Wang D, Zhao X, Zhang T, Sun H, et al. (2011) Identification of rhizome-specific genes by genome-wide differential expression analysis in *Oryza longistaminata*. BMC Plant Biol 11: 18-31.

19. Zong Y, Huang L, Zhang T, Qin Q, Wang W, et al. (2014) Differential microRNA expression between shoots and rhizomes in *Oryza longistaminata* using high-throughput RNA sequencing. The Crop J 2: 102-109.

20. Murashige T, Skoog FK (1962) A revised medium for rapid growth and bio-assays with tobacco tissue cultures. Physiologia Plantarum 15: 473-497.

21. Kawasaki T (1997) In: Shimamoto K, Sasaki T (ed.) Simplified extraction method of rice genomic DNA for PCR analysis. PCR-based Experimental Protocol in Plants, Shujyunsya, Tokyo pp: 67-68.

22. Wang S, Basten CJ, Zeng ZB (2007) Windows QTL Cartographer 2.5. Department of Statistics, North Carolina State University, Raleigh, USA.

23. Maekawa M, Rikiishi K, Matsuura K, Noda K (1996) Genetic analysis of rhizomatous trait of wild species (*Oryza longistaminata*) in rice (in Japanese). Breed Sci 46: 323

24. Iwamoto M, Maekawa M, Saito A, Higo H, Higo K (1998) Evolutionary relationship of plant catalase genes inferred from exon-intron structures: Isozyme divergence after the separation of monocots and dicots. Appl Genet 97: 9-19.

25. Peng S, Khush GS, Virk P, Tang Q, Zou Y (2008) Progress in ideotype breeding to increase rice yield potential. Field Crops Res 108: 32-38.

26. Khush GS (1995) Modern varieties: Their real contribution to food supply and equity. Geo J 35: 275-284.

27. Jiao Y, Wang Y, Xue D, Wang J, Yan M, et al. (2010) Regulation of OsSPL14 by OsmiR156 defines ideal plant architecture in rice. Nat Genet 42: 541-544.

28. Tian F, Li DJ, Fu Q, Zhu ZF, Fu YC, et al. (2006) Construction of introgression lines carrying wild rice (*Oryza rufipogon* Griff.) segments in cultivated rice (*Oryza sativa* L.) background and characterization of introgressed segments associated with yield-related traits. Theor Appl Genet 112: 570-580.

29. Chen Z, Hu F, Xu P, Li J, Deng X, et al. (2009) QTL analysis for hybrid sterility and plant height in interspecific populations derived from a wild rice relative, *Oryza longistaminata*. Breed Sci 59: 441-445.

30. Miura K, Ikeda M, Matsubara A, Song X, Ito M, et al. (2010) OsSPL14 promotes panicle branching and higher grain productivity in rice. Nat Genet 42: 545-549.

31. Ookawa T, Hobo T, Yano M, Murata K, Ando T, et al. (2010) New approach for rice improvement using a pleiotropic QTL gene for lodging resistance and yield. Nat Commun 1: 132.

32. de Ribou SDB, Douam F, Hamant O, Frohlich MW, Negrutiu J (2013) Plant science and agricultural productivity: Why are we hitting the yield ceiling? Plant Sci 210: 159-176.

20

Effects of Citric Acid and Butylated Hydroxytoluene Alone or In Combination on Prevention of Rancidity of Rice Bran during Storage

NSBM Atapattu[1]*, KP Wickramasinghe[1], Thaksala serasinghe[1], SP Gunarathne[2]

[1]Faculty of Agriculture, University of Ruhuna, Sri Lanka
[2]Faculty of Veterinary Science, University of Peradeniya, Sri Lanka

Abstract

Rice bran (RB), a byproduct of rice milling process is a valuable feed resource for the livestock industry. Both lipolytic and oxidative rancidity starts right after bran is removed from the grain during the milling process thereby making long term storage of RB difficult. Anti oxidants such as butylated hydroxytoluene (BHT) control the oxidative rancidity only for a short period of time. Therefore, effective rancidity control measures need to be adopted to maintain the nutritive and feeding value of RB. The objective of this study was to evaluate whether rancidity of RB could be curtailed by storing RB with citric acid (CA) and BHT alone or in combination.

RB was collected immediately after milling and stored with three levels of CA (0, 2, and 4%) and three levels of BHT (0, 200, and 400 ppm) in a 3×3 factorial arrangement. Each treatment combination had three replicates. Treated RB was stored in plastic containers with a closed lid for 82 days and analyzed for peroxide value (PV) and pH on 7, 25, 55, 82 days. The mean PV of un-treated RB on day 7 (2.85 meq kg^{-1}) increased to 14.42meq kg^{-1} during 82 days of storage. When BHT or CA was not used, a significant ($p<0.05$) negative correlation was found between PV and pH [PV=(7.3291–pH x 0.112; R^2=0.901]. Peroxide values were not significantly ($p>0.05$) affected by BHT or CA on day 7. The use of 0, 200 or 400 ppm BHT significantly reduced the PV during 25 days of storage time, but not thereafter. The use of 2% or 4% CA significantly ($p<0.05$) reduced the PV up to day 82. It can be concluded that BHT was effective in reducing rancidity of RB only for about month whereas CA controls both lipolytic and oxidative rancidity for about three months.

Keywords: Rice bran; Rancidity; Citric acid; BHT; Storage

Introduction

Rice bran (RB) is a major by-product in the milling industry of raw paddy (rough rice) into rice. A total of 40-45 million metric tons of RB are produced annually mainly in East and South Asia [1]. RB is a valuable feed resource for all the classes of livestock. The lipid content of RB is high and varies from 15% to 23% and three major fatty acids; palmitic (12%-18%), oleic (40%-50%) and linoleic (30%-35%) account for up 90% of total fatty acids [2]. One of the major restriction for the use of RB as an animal feed ingredient is its high susceptibility to rancidity during storage. It has been reported [3] that upto 50% of the fat in the bran was converted into free fatty acids within 6 weeks after milling due to rancidity. Nutritional quality of rice bran deteriorates rapidly due to rancidity [2].

Oil in un-milled rice is stable because lipolytic enzymes of intact kernel are located in the cross cells of the seed coat while most of the oil is stored in the aleurone layer and the germ [4]. After the bran layer is removed from the endosperm, the individual cells are disrupted and the bran lipids come into contact with lipases. Oxidative rancidity involves a reaction between free oxygen and lipids. The reaction attacks the double-bonds of the fatty acids and is accelerated by oxygen, free radicals, metal ions (iron, cobolt and copper), light, radiation and enzyme lipoxygenase [5]. The reaction of oxygen with unsaturated lipids involves free radical initiation, propagation, and termination processes [6].

Stabilization of RB just after milling is of utmost importance to mitigate the adverse effects of rancidity. Many experimental procedures such as lowering pH [7], use of ethanol vapour [8] and heating [4] have been used to inactivate lipases and thereby to stabilize RB and prolong its shelf-life. Microwave heating, extrusion cooking, roasting, pelleting are the other methods that can be used to stabilize RB [2]. Though physical treatments such as heating right after milling can substantially reduce the rancidity particularly the lipolytic rancidity, these methods are costly, demands high amount of energy and cumbersome [9].

Due to the inherent instability of natural antioxidants, several synthetic antioxidants have been used to stabilize fats and oils [10,11]. Butylated hydroxyanisole (BHA) and BHT compounds have been used as antioxidants in human foods since 1954 [12]. However, potential health hazards of synthetic antioxidants in foods, including possible carcinogenicity, have been reported times [11]. Prabhaker and Venkatash [7] concluded that chemical methods are not very effective in RB stabilization. Therefore, socio-economically viable rancidity prevention methods to stabilize RB are of importance.

Citric acid is a relatively cheap, readily available organic acid. CA is used in poultry and swine diets, mainly to increase the availability of phytate phosphorus [13]. The food and pharmaceutical industries utilize CA extensively because of its general recognition of safety, pleasant acid taste and chelating and buffering properties [14]. CA or citrates have been used to control oxidative deterioration of flavor or colour of a wide range of foodstuffs [15]. CA has been shown to play a synergetic role with primary antioxidants and oxygen scavengers during vegetable oil storage [16]. The objective of this study was to

*Corresponding author: NSBM Atapattu, Faculty of Agriculture, University of Ruhuna, Sri Lanka, E-mail: mahindaatapattu@gmail.com

evaluate whether rancidity of RB could be curtailed by storing RB with CA and BHT alone or in combination.

Materials and Methods

Citric acid (2-hydroxypropane-1, 2, 3-tricarboxylic acid) and butylated hydroxytoluene were purchased from SIGMA chemicals private limited, Sri Lanka. RB was collected (un-known variety) from a local rice mill and immediately transported to the research laboratory. The experiment was a completely randomized design in 3×3 factorial arrangement. Experimental factors were three BHT (0 ppm, 200 ppm and 400 ppm) and three CA (0%, 2%, and 4%) levels. Each treatment combination had three replicates. Treated RB samples were stored in the 2 dm^3 volume plastic containers with closed lids, under room temperature (27-31°C) and RH conditions (mean 85%).

Treated RB in each container was mixed well before collecting the samples for analysis. Three random samples from each treatment combination were taken on day 7, 25, 55 and 82 (Approximately one week, one month, two month and three months after storage) and were analyzed for peroxide value [17] and pH value. Data were analyzed using the GLM procedure of SAS [18]. Effects were considered significant when $p < 0.05$. When interactions were significant LS mean comparison procedure was used to compare the means.

Determination of pH value

The pH values were obtained by using a pH meter of a glass electrode (Corning pH meter 430). Sample preparations for pH determination were performed according to method described by Dev et al. [19]. 10g of samples were mixed with 10 ml of distilled water to keep fluidity. Mixture was kept still for about 30 minutes after shaking well. Three replicates of each solution were analyzed for pH and mean value was taken as the pH value of the samples.

Determination of the Peroxide Value

10 g of RB sample was taken from each replicated. Oil was extracted by soxhlet extraction method before being subjected to chemical analyses. The peroxide value was determined using the method of AOAC [17]. The peroxide value of RB oil was determined by dissolving the oil in a solvent mixture of acetic acid and carbon tetrachloride, warmed with potassium iodide and, then titrated with sodium thiosulphate solution using starch as the indicator.

Results and Discussion

The effects of three levels of CA and BHT on the PV of RB on storage are shown in Table 1. The interaction between different levels of CA and BHT was not significant ($p > 0.05$). The main effects; BHT and CA levels had significant ($p < 0.05$) effect on PV of RB during storage. Both BHT and CA levels had no significant ($p > 0.05$) effect on PV after a week of storage. The mean PV of un-treated RB on day 7 (2.85 meq kg^{-1}) increased to 14.42 meq kg^{-1} during 82 days of storage. The PVs observed in the present study were slightly higher than those values reported by Mujahid et al. [20]. Njobeh et al. [21] reported that the tropical climatic conditions (temperature above 25°C and RH greater than 70%) are ideal for feed spoilage. Therefore differences in experimental conditions may be the reason for higher PVs observed in the present study.

BHT reduced the PV of RB until day 25 but not thereafter. On day 25, both 200 ppm (3.88 meqkg^{-1}) and 400ppm (3.54 meqkg^{-1}) of BHT significantly ($p < 0.05$) lowered the PV compared to 0% BHT (4.6 meqkg^{-1}). Several authors [4,12,22,23] have also reported that oxidative rancidity can be controlled by adding antioxidants such as Pyrocatechol, Homocatechol, Propylgallate, BHA and BHT. Interestingly, on day 25, 4% CA also reduced PV (3.13 meqkg^{-1}) significantly ($p < 0.05$), compared to 2% (4.19 meqkg^{-1}) and 0% (4.69 meqkg^{-1}) CA. Similar effects of CA have been reported by others as well [15,16].

Results of the present study suggest that BHT is effective in controlling oxidative rancidity of RB only for about 25 days. Mujahid et al. [20] have also reported that anti oxidants did not prevent the free fatty acid formation and increase of PV during long storage of RB and

BHT (ppm)	CA (%)	Peroxide value (mean ± SE)			
		on day 07	on day 25	on day 55	on day 82
0	0	2.85 ± 0.01	5.55 ± 0.02	6.48 ± 1.21	14.42 ± 0.35
0	2	2.16 ± 0.23	4.80 ± 0.17	5.97 ± 2.19	7.10 ± 0.69
0	4	2.46 ± 0.12	3.46 ± 0.01	4.53 ± 0.40	5.15 ± 0.29
200	0	2.20 ± 0.98	4.15 ± 0.01	7.48 ±0.69	12.33 ± 0.87
200	2	2.09 ± 0.23	4.26 ± 0.06	4.22 ± 0.52	5.82 ± 0.12
200	4	2.53 ± 0.29	3.22 ± 0.35	2.99 ± 0.17	7.37 ± 0.01
400	0	2.23 ± 0.23	4.38 ± 0.29	6.26 ± 0.06	10.72 ± 1.56
400	2	2.39 ± 0.01	3.53 ± 0.40	4.37 ± 0.06	5.47 ± 0.98
400	4	2.29 ± 0.35	2.72 ± 0.12	5.33 ± 0.52	6.72 ± 0.06
Main effects means ± SE					
0		2.49 ± 0.1	4.60^a ± 0.3	5.66 ± 0.8	8.89 ± 1.8
200		2.27 ± 0.3	3.88^b ± 0.2	4.89 ± 0.8	8.51 ± 1.2
400		2.31 ± 0.1	3.54^b ± 0.3	5.32 ± 0.3	7.64 ± 1.1
	0	2.43 ± 0.3	4.69^a ± 0.2	6.74^a ± 0.5	12.49^a ± 0.8
	2	2.22 ± 0.1	4.19^a ± 0.2	4.86^{ab} ± 0.8	6.13^b ± 0.5
	4	2.43 ± 0.1	3.13^b ± 0.1	4.28^b ± 0.4	6.41^b ± 0.4
Level of significance					
BHT		NS	0.0038	NS	NS
CA		NS	0.0003	0.05	0.0001
BHT×CA		NS	NS	NS	NS

NS=Non significant ($p > 0.05$)
$^{a-c}$Means within a column with no common superscript differ significantly ($P < 0.05$)

Table 1: Peroxide value of rice bran during storage as affected by three levels of CA and BHT.
SE = standard error.

thus addition of antioxidant to rice bran was not an effective mean to stabilize rice bran. Several other studies [21,22] have also shown that anti oxidants are effective only for about a month. Rapid auto oxidation of antioxidants and/or leaching with time and low penetration of antioxidants in fibrous RB are suggested as the possible reasons for shorter period of effectiveness of BHT in RB.

In contrast to BHT, CA significantly ($p<0.05$) reduced the oxidative rancidity as determined by PV of RB over a longer period of storage. When measured after 25, 55 and 82 days of storage, 4% CA significantly ($p<0.05$) reduced the PV compared to the 0% CA. 2% CA did not significantly ($p>0.05$) reduce the PV compared to 0% CA on 25 days. But, but later on day 82, the PV of the RB treated with 2% was significantly lower than that of RB with 0% CA.

The changes of PV of RB stored with BHT or CA are shown in Figures 1 and 2, respectively. The PVs of the RB treated both with BHT and CA increased over the storage period. The rate of increase in PV of the RB stored with 4% CA was lower than those of the BHT treated RB. Furthermore, throughout the storage period, the PVs of the RB stored with CA were lower than those of the RB stored without CA (Figure 2). Above findings suggest that CA controls the rancidity of RB and anti-oxidative properties of CA in controlling oxidative rancidity were more powerful than BHT. This hypothesis is further supported by the findings of Tao [24] who reported that 0.1 to 2% acetic acid having anti-oxidative properties, maintained the stability of the parboiled RB for at least 6 months.

The effect of different BHT and CA levels on pH value of RB is shown in Table 2. The interaction between different levels of CA and BHT was not consistent. Significant ($p<0.05$) interactions were found on day 25 and day 82 but not on day 55. A clear explanation could not be found for this inconsistency. When BHT or CA was not used (Control samples), a significant ($p<0.05$) negative correlation (Figure 3) was found between PV and pH [PV=(7.3291-pH×0.112; R^2=0.901]. This suggests that when facilities are not available for PV determination,

PV can be predicted fairly accurately by measuring the pH value of the RB, stored for 82 days.

During 82 days of storage period, RB mixed with CA had significantly ($p<0.05$) lower pH values than the control (RB without CA) whereas

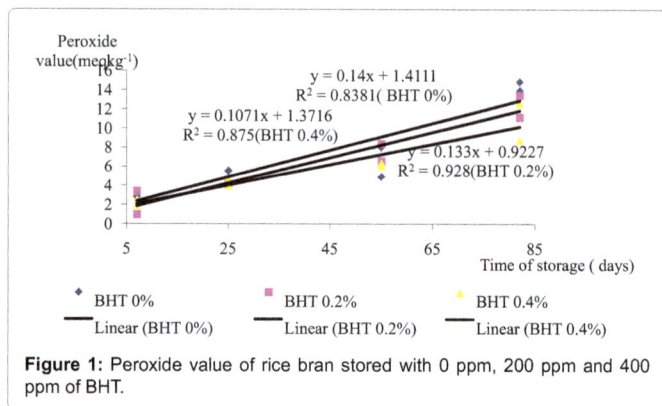

Figure 1: Peroxide value of rice bran stored with 0 ppm, 200 ppm and 400 ppm of BHT.

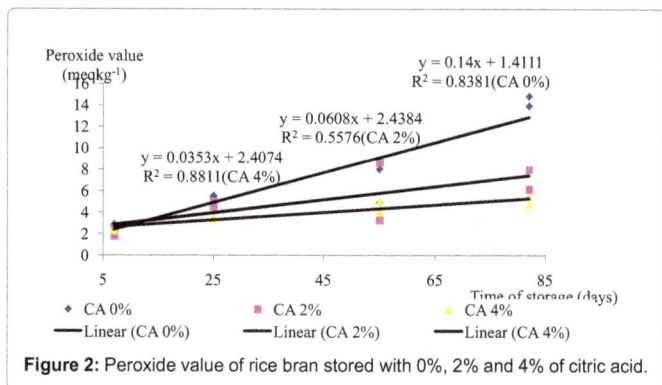

Figure 2: Peroxide value of rice bran stored with 0%, 2% and 4% of citric acid.

BHT (ppm)	CA (%)	Mean pH ± SE			
		on day 07	on day 25	on day 55	on day 82
0	0	7.02 ± 0.001	6.86 ± 0.006	6.57 ± 0.023	5.68 ± 0.052
200	2	5.93 ± 0.006	5.82 ± 0.012	5.88 ± 0.012	5.34 ± 0.023
400	4	5.17 ± 0.046	5.34 ± 0.029	5.26 ± 0.040	4.81 ± 0.012
0	0	7.06 ± 0.046	6.85 ± 0.023	6.63 ± 0.006	5.86 ± 0.012
200	2	5.98 ± 0.040	6.14 ± 0.092	5.90 ± 0.069	5.30 ± 0.040
400	4	5.38 ± 0.040	5.26 ± 0.012	5.18 ± 0.012	4.82 ± 0.017
0	0	7.04 ± 0.006	6.89 ± 0.023	6.66 ± 0.023	5.98 ± 0.001
200	2	5.74 ± 0.069	5.74 ± 0.069	5.78 ± 0.001	5.16 ± 0.001
400	4	5.36 ± 0.069	5.28 ± 0.001	5.10 ± 0.001	4.84 ± 0.001
Main effects means ± SE					
0		6.04 ± 0.34	6.01[ab] ± 0.28	5.90 ± 0.24	5.28 ± 0.16
200		6.14 ± 0.31	6.08[a] ± 0.29	5.90 ± 0.26	5.32 ± 0.19
400		6.04 ± 0.32	5.97[b] ± 0.30	5.84 ± 0.28	5.32 ± 0.21
	0	7.04[a] ± 0.01	6.87[a] ± 0.01	6.62[a] ± 0.02	5.84[a] ± 0.05
	2	5.88[b] ± 0.05	5.90[b] ± 0.08	5.85[b] ± 0.03	5.26[b] ± 0.03
	4	5.31[c] ± 0.05	5.29[c] ± 0.01	5.18[c] ± 0.03	4.82[c] ± 0.01
level of significance					
BHT		NS	NS	NS	NS
CA		0.0001	0.0001	0.0001	0.0001
BHT×CA		NS	0.0080	NS	0.0005

SE = standard error
NS= non significant ($p>0.05$)
[a-c]Means within a column with no common superscript differ significantly ($P < 0.05$)

Table 2: The pH values of rice bran during storage as affected by three levels of CA and BHT.

BHT had no effect on pH (Table 2). Suggesting a formation of free fatty acids due to lipolytic rancidity, the pH of the RB stored with BHT and those stored without BHT or CA decreased drastically over the storage period (Figures 4 and 5). This suggests that BHT does not control the lipolytic rancidity. Meanwhile, throughout the storage period, the pH values of the RB stored with 2 and 4% of CA were significantly ($p<0.05$) lower than those of control samples. Apparently this may primarily be due to the acidification due to the presence of CA (Figure 5).

The changes of pH and PV suggest that BHT, being an anti-oxidant, has controlled the rancidity, but without inhibiting the formation of free fatty acids. Coppen [25] has also found that anti-oxidants could reduce the formation of peroxides but not the free fatty acids. Even though the pH value of CA treated RB also decreased over the storage period, the rate of pH reduction was lower than that of the RB stored without CA and also the RB stored with BHT. This observation indicates that apart from controlling the oxidative rancidity, CA has controlled the lipolytic rancidity that results in free fatty acid formation.

The rancidity controlling capacity of may be due to a number of reasons. Hydrolytic rancidity of RB is initiated when lipases and lipids in bran come into contact upon milling. RB lipases have pH optima of 7.5-8.0 and either increase or decrease of pH may reduce the rancidity [2]. It has been shown that complete cessation of lipase activity cannot be achieved even at pH 4 [7]. RB stored with CA always had lower pH values than the control. Therefore, it may be a possibility that reduced pH due to CA might have slowed the activity of lipase and thus the lipolytic rancidity.

Oxidative rancidity involves a reaction between free fatty acids and oxygen. The strong anti-oxidant properties of the CA may be the reason for the oxidative-rancidity controlling capacity of CA. The oxidative rancidity is accelerated by oxygen, free radicals and metallic ions such as iron, cobalt, copper as well as light and enzyme lipoxigenase [26]. CA is a strong chelating agent and thus can bind heavy metals [27]. Therefore, it can be suggested that CA binds with heavy meals in the RB, thereby preventing them acting as a catalyst in oxidative rancidity reactions. It is concluded that BHT was effective in reducing rancidity of RB only for about month whereas CA controls both lipolytic and oxidative rancidity for about three months.

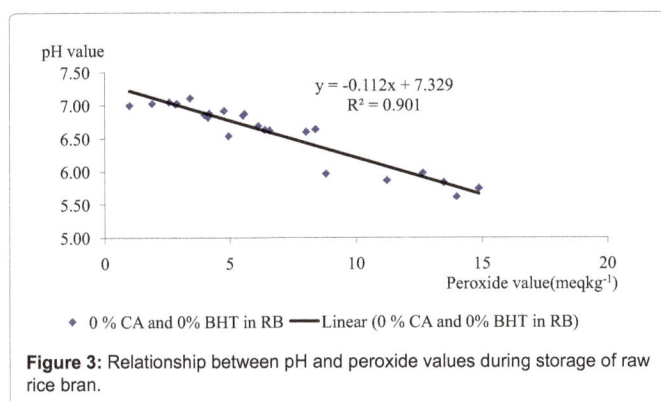

Figure 3: Relationship between pH and peroxide values during storage of raw rice bran.

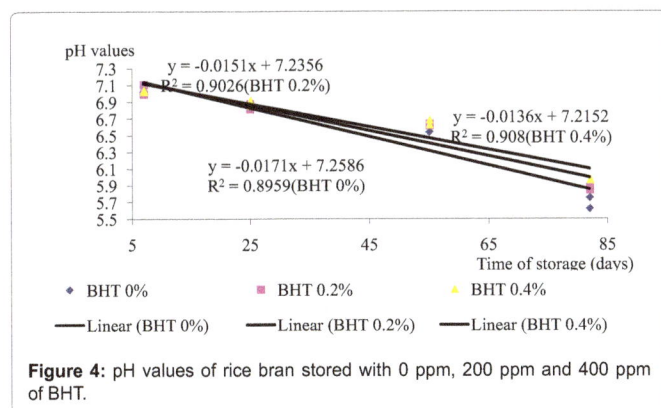

Figure 4: pH values of rice bran stored with 0 ppm, 200 ppm and 400 ppm of BHT.

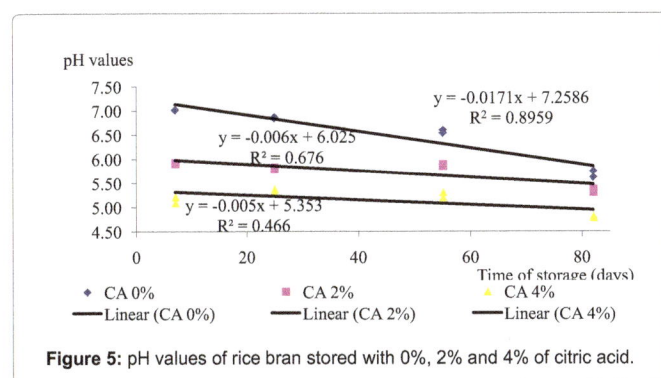

Figure 5: pH values of rice bran stored with 0%, 2% and 4% of citric acid.

References

1. Farrell DJ (1994) Utilization of rice bran in diets for domestic fowl and ducklings. World's Poultry Science Journal 50: 115-131.

2. Malekian F, Ramu M Rao, Prinyawiwatkul et al. (2000) Lipase and lipoxigenase activity, functionality, and nutrient losses in rice bran during storage. Lousiana State University Agriculture Centre, USA.

3. Warren BE, Ferrell DJ (1990) The nutritive value of full-fat and defatted Australian rice bran. I. Chemical composition. Animal Feed Sci Technol 27: 219-228.

4. Sounders RM (1985) Rice bran: Composition and potential food sources. Food Rev Int 1: 465-495.

5. Takahama U (1985) Inhibition of lipoxygenase-dependent lipid peroxidation by quercetin: Mechanism of antioxidative functions. Phytochem 24: 1443-1446.

6. Frankel EN (1984) Lipid oxidation: Mechanisms, products and biological significance. J Am Oil Chemists Soc 61: 1908-1916.

7. Prabhakar JV,Venkatesh KVL (1986) A simple chemical method for stabilization of rice bran. J Am Oil Chemists Soc 63: 644-646.

8. Champagne ET, Robert J. Hron Sr, Abraham G (1992) Utilizing ethanol to produce stabilized brown rice products. J Am Oil Chemists Soc 69: 205-208.

9. Atapattu NSBM (2005) Rice bran: potential and constraints as a poultry feed ingredient. Proceedings of the 1st international symposium of Sabaragamuwa University.

10. Hilton JW (1989) Antioxidants: function, types and necessity of inclusion in pet foods. Can Vet J 30: 682-684.

11. Sharafi SM, Rasooli I, Owlia P, Taghizadeh M, Astaneh SD (2010) Protective effects of bioactive phytochemicals from Mentha piperita with multiple health potentials. Pharmacogn Mag 6: 147-153.

12. Hilton JW (1989) Antioxidants: function, types and necessity of inclusion in pet foods. Can Vet J 30: 682-684.

13. Kopecky J, Hrncar C, Weis J (2012) Effect of organic acids supplement on performance of broiler chickens. Scientific Papers Animal Science and Biotechnologies 45.

14. Soccol CR, Vandenberghe LP, Rodrigues C, Pandey A (2006) New perspectives for citric acid production and application. Food Technol Biotechnol 44: 141-149.

15. Milsom PE (1987) Organic acids by fermentation, especially citric acid. In Food Biotechnology-1. Springer Netherlands 273-307.

16. Santiago PA, Francisco PA, Jose MG (2004) Studies on rancidity inhibition in frozen horse mackerel (Trachurus trachurus) by citric and ascorbic acids. Eur J Lipid Sci Technol 100: 232-240.

17. Association of Official Analytical Chemists (1980) Official Methods of Analysis. 14th ed. Association of Official Analytical Chemists. Washington, DC, USA.

18. SAS Institute (1986) SAS user's guide statistics, SS Institute, USA.

19. Dev CN, Carol C, Donna D, Yvon-louis T (2003) Determination of the pH of foods including foods in hermitically sealed containers. Food section, Ottawa laboratory (Carling) Canadian food inspection Agency.

20. Mujahid A, Ikram ul Haq, Asif M, Gilani AH (2005) Effect of various processing techniques and different levels of antioxidant on stability of rice bran during storage. J Sci Food Agric 85: 847-852.

21. Njobeh PB, Iji PA, Nsahlai IV (2006) Influence of composition and storage conditions on the concentrations of free fatty Acids and peroxides in broiler diets. Int J Poult Sci 5: 279-283.

22. Connor TP, Brien NM (1991) Significance of lipoxygenase in fruits and vegetables. In: Food Enzymology, Fox, P.F. Elsevier Science Publishing, USA.

23. Cabel MC, Waldroup PW (1989) Research note; Ethoxyquin and ethylenediaminetetra acetic acid for the prevention of rancidity in rice bran stored at elevated temperature and humidity for various lengths of time. Poult Sci 68: 438-442.

24. Tao J (2001) Method of stabilization of rice bran by acid treatment and composition of the same. U.S. Patent No. 6,245,377. Washington, DC: U.S. Patent and Trademark Office.

25. Coppen PP (1989) The use of antioxidants. In: Rancidity in Foods. Allen JC, Hamilton J (Eds.), Elsevier Appl Sci, London, UK.

26. Ramezanzadeh FM, Rao RM, Windhauser M, Prinyawiwatkul W, Tulley R, et al. (1999) Prevention of hydrolytic rancidity in rice bran during storage. J Agric Food Chem 47: 3050-3052.

27. Boling-Frankenbach SD, Snow JL, Parsons CM, Baker DH (2001) The effect of citric acid on the calcium and phosphorus requirements of chicks fed corn-soybean meal diets. Poult Sci 80: 783-788.

21

Effect of *Polygonum persicaria* Agglutinin on Digestive α-amylase of Rice Striped Stem Borer

Arash Zibaee*

Department of Plant Protection, Faculty of Agricultural Sciences, University of Guilan, Rasht, Iran

Abstract

Results of the current study revealed that a lectin extracted from *Polygonum persicaria* L. had significant inhibition on digestive α-amylase of *Chilo suppressalis* Walker. Concentrations 0.05 to 2 mg/ml of PPA significantly decreased enzymatic activity from 5 to 72%. The highest inhibition of α-amylase was obtained at pH 9 while the optimal pH for amylolytic activity was found to be 8 and 9. Besides, optimal temperature of α-amylase activity was at 30˚C but the highest inhibition of the enzyme was found at temperatures of 30-40˚C. Kinetic parameters of α-amylase revealed lower and higher amounts of V_{max} and K_m in PPA treated enzyme indicating mixed inhibition. In vitro inhibition of α-amylase clearly showed that PPA could intervene in digestive process of *C. suppressalis* and decrease nutrient income of larvae. Further experiments are required to fully understand negative effects of PPA against rice striped stem borer to reach an efficient pest control via providing resistant varieties.

Keywords: Rice striped stem borer; α-amylase; Polygonum persicaria; Agglutinine

Introduction

Rice striped stem borer, *Chilo suppressalis* Walker (Lepidoptera: Crambidae) is the major pest of rice in north of Iran [1]. Larvae intensively utilize inner parts of stems and cause dead heart and white head of rice leading to severe loss of rice production [1]. *C. suppressalis* has 2-3 generations per year and causes significant damages to almost all rice varieties. Control procedures are basically to use synthetic insecticides like diazinon, padan and fenitrotione [2]. Although, biological control using *Trichogramma* spp. demonstrated positive results but insecticides still remains as the first and main tactic [2].

Digestive carbohydrases contain α-amylases, glycosidases, xylanases, chitinases and pectinases but α-amylases have crucial role in digestion of dietary starch and glycogen as the initial energy sources of insects. α-Amylases (EC 3.2.1.1) are hydrolyzing enzymes that catalyze breaking-down of inner and long α-1,4-glucan chains in starch and glycogen [3]. Several studies revealed that the enzyme has molecular weight of 48–60 kDa, pI values of 3.5–4.0, and K_m values by using soluble starch ~0.1% [4]. Also, it has been reported that α-amylases are calcium-dependent and those are activated by chloride with displacement of pH optimum [4]. Zibaee et al. [5] reported that activities of α-amylase in the midgut and salivary glands of *C. suppressalis* were 0.06 and 0.036 μmol/min/mg protein, respectively. The optimal pH and temperature of the enzyme found to be 9 and 35-40°C which is consistent with reports of other lepidopteran insects. Finally, the authors demonstrated that the α-amylase was inhibited by addition of NaCl, KCl, $MgCl_2$, Urea, EDTA, and SDS while $CaCl_2$ enhanced enzymatic activity [5].

Since *C. suppressalis* utilizes inner parts of rice stems, no chemical could reach the larvae. So, insecticide spraying has been limited to wide spraying against 1st larval instar. Zibaee et al. [2] reported that rice striped stem borer has been resistant to diazinon in four different regions of northern Iran. It means that using diazinon could not be more efficient. There are six classes of α-amylase inhibitors including lectin-like, knottin-like, cereal-type, Kunitz-like, c-purothionin-like, and thaumatin-like [6]. These inhibitors show structural diversity, differ¬ent modes of inhibition and different specificity profiles against diverse range of α-amylases [7]. Lectins are the heterogeneous molecules (glycoprotein) that have been extensively distributed in nature and plays several functions in physiology of living organisms [8]. They have shown entomotoxic and inhibitory properties against insects [8]. Hence, aims of the current study were to purify a lectin from *Polygonum persicaria* and determine its inhibition on digestive α-amylase of *C. suppressalis*.

Materials and Methods

Insect rearing

Egg patches of *C. suppressalis* were collected from rice fields and reared in containers provided by rice seedlings. In¬sects were reared based on a modified method of Zibaee et al. [2] at 28 ± 2°C, 80% relative humidity (RH) under conditions of 16 h light: 8 h dark. Laboratory conditions were checked, containers were cleaned and fresh stems were daily provided for larvae.

Purification of PPA

Stems of *P. persicaria* were incubated in phosphate buffer (0.1 M pH 7.1) for approximately 72 h at 4°C. Then, stems were grounded in the buffer to completely destroy the tissues prior to additional incubation for 24 h. The mixture was filtrated by a layer of cheesecloth, then it was centrifuged at 5000 rpm for 20 min. Remaining debris was removed by passing the supernatant through filter paper (Whatmann No.4) [9,10]. Supernatant was precipitated by 0-60% concentrations of ammonium sulfate and centrifuged at 5000 rpm for 20 min. Debris was eluted in Tris-HCl buffer (0.1 M, pH 7) and dialyzed in the same buffer overnight [10]. Dialyzed samples were loaded into Sepharose 4B-galactose column equilibrated with Tris-HCl buffer (0.1 M, pH 7)

---

*Corresponding author: Arash Zibaee, Department of Plant Protection, Faculty of Agricultural Sciences, University of Guilan, Rasht, Iran
E-mail: arash.zibaee@gmx.com, arash.zibaee@guilan.ac.ir

as affinity chromatography. The affinity column was washed with Tris-HCl buffer and buffer containing 20 mM 1,3-diaminopropane (DAP) [9,10]. Fractions showing the highest protein content were pooled and used for forthcoming step. Fractions obtained after the first affinity chromatography were loaded on DEAE-Cellulose fast flow equilibrated with DAP [9,10]. Finally, the lectin was eluted using Tris–HCl (0.1 M, pH 7.0) containing 0.5 M NaCl after DAP washing. Again, the samples showing the highest protein content were dialyzed against water and purified molecule was analyzed by SDS-PAGE stained with commassie brilliant blue [11].

Sample preparation

Larvae of *C. suppressalis* were randomly selected and their midguts were removed by dissection under a stereomicroscope in ice-cold saline buffer (NaCl, 10 mM). The midguts were rinsed in ice-cold distilled water, placed in a pre-cooled homogenizer and grounded before centrifugation. Equal portions of larval midgut and distilled water were used to have desirable concentration of the enzyme (W/V). Homogenates were separately transferred to 1.5 ml tubes and centrifuged at 13,000 rpm for 20 min at 4°C. The supernatants were pooled and stored at −20°C for subsequent analyses.

α-Amylase assay

The method using dinitrosalycylic acid (DNS) was used to assay α-amylase activity [12]. Ten microlitres of the enzyme was incubated for 30 min at 35°C with 50 μl of phos¬phate buffer (0.02 M, pH 7.1) and 20 μl of soluble starch as substrate. The reaction was stopped by adding 80 μl of DNS and heating in boiling wa¬ter for 10 min prior to read the absorbance at 545 nm. One unit of α-amylase activity was defined as the amount of enzyme required to produce 1 mg maltose in 30 min at 35°C. The negative control contained all reaction mix¬tures with pre-boiled enzyme (for 15 min) to prove the enzyme presence in the samples.

Inhibition of α-amylase by different concentrations of PPA

Inhibition experiment was carried out using 50 μl of PBS (phosphate buffer solution) (0.02 M, pH 7.1) 20 μl of starch 1% and 20 μl of different concentrations of PPA (0, 0.1, 0.5, 1, 1.5, and 2 mg/ml) Then, 10 μl of the enzyme was added and the reaction continued as described earlier. The blank con¬tained PBS, starch 1%, and each concentration of PPA.

Effect of pH on α-amylase inhibition by PPA

pH dependency of the PPA inhibition was deter¬mined at different pH values using Tris-HCl buffer (20 mM) with pH set at 5, 6, 7, 8, 9, 10, 11, and 12. A-Amylase activity was assayed after incubation of the reac¬tion mixture containing Tris-HCl buffer (in the given pH value), starch 1%, PPA (2 mg/ml) and enzyme.

Effect of temperature on α-amylase inhibition by PPA

Effects of temperature regimes on α-amylase inhibi¬tion by PPA were made by the reaction mixture containing Tris-HCl (20 mM pH, 9), Starch 1%, PPA (2 mg/ml) and enzyme at different temperatures set at 15, 20, 25, 30, 35, 40, 45, 50, and 60°C.

Kinetic Studies

Kinetic parameters of inhibition were carried out using different concentrations of starch as substrate (0.5–2.0%) in the presence of PPA (2 mg/ml). Lineweaver-Burk plot analysis was made to find Km and Vmax values by Sigma-Plot software (version 12).

Protein Assay

Protein concentrations were assayed according to the method described by Lowry et al. [13]. The method recruits reaction of Cu^{2+}, produced by the oxidation of peptide bonds with Folin–Ciocalteu reagent. In the assay, 20 μl of sample was added to 100 μl of reagent, and incubation was made for 30 min prior to read the absorbance at 545 nm (Recommended by Ziest Chem. Co., Tehran-Iran).

Statistical Analysis

All data were compared by one-way analysis of vari¬ance (ANOVA) followed by Tukey's studentized test when significant differences were found at $p \leq 0.05$, and marked in figures with letters.

Results and Discussion

Lectins are the heterogeneous proteins that are able to recognize carbohydrates in physiological media [14]. In facts, lectins reversibly bind to carbohydrates without altering their covalent structure [15]. These molecules may have several functions in living organisms but plant lectins have been shown entomotoxic effects and enzymatic inhibition [8]. Certain tissues of plants like seeds, bark and bulbs contains lectins that might indicate their roles as storage proteins [8]. In the current study, different concentrations of PPA significantly affected digestive amylolytic activity of *C. suppressalis* (Figure 1). Although the highest inhibition was found to be 72% but concentrations of 1, 1.5 and 2 mg/ml show somehow same inhibition (Figure 1). Among six types of α-amylase inhibitors, lectin-like one has demonstrated various effects. For example, α-AI1 from *Phaseolus vulgaris* L. inhibited digestive α-amylases of *Callosobruchus maculatus* Fabricius (Coleoptera: Bruchidae) and *C. chinensis* while α-AI2 had no inhibitory effects on these insects [16,17]. In a current study, different concentrations of extracted lectin, *Citrullus colocynthis* agglutinin (CCA), inhibited digestive amylolytic activity in Ectomyelois ceratoniae Zeller (Lepidoptera: Pyralidae) by 22-49% [18]. Direct inhibition of digestive α-amylase by PPA and well-known function of lectins as binding moelcules to epithelial cells of midgut point out entomotoxic and inhibitory effects that definitely disrupt digestive processes of insects lead to mal-nutrition and death of pest without other effects on non-target organisms and environmental pollutions.

Although optimal pHs of the enzyme were found at pHs 8 and 9 but the highest inhibition occurred at pH 9 (Figure 2). In case of temperature, the highest inhibition obtained at temperatures of 15, 35, 45 and 50˚C (Figure 3). Although these temperatures showed the

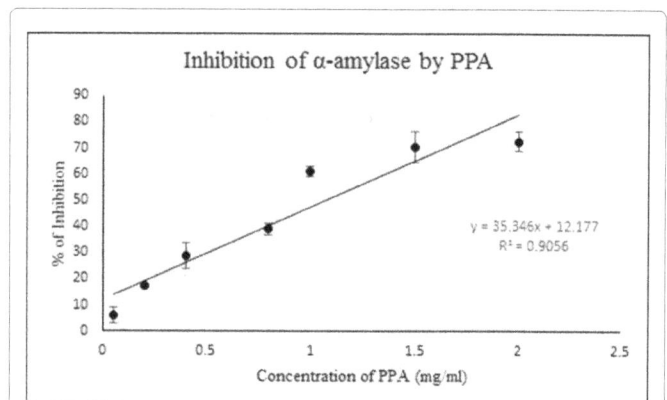

Figure 1: Inhibition of digestive α-amylase from *C. suppressalis* by different concentrations of PPA.

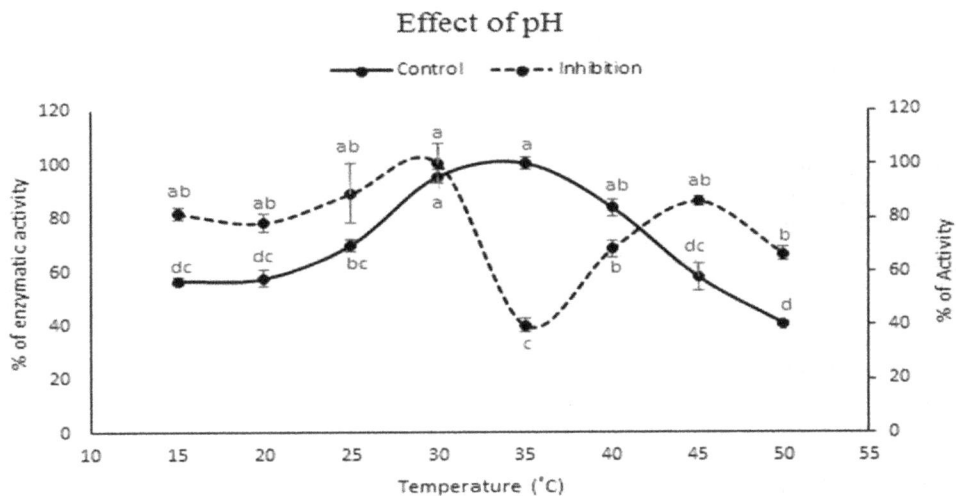

Figure 2: pH dependency of PPA inhibition on α-amylase from *C. suppressalis* by PPA. Statistical differences have been shown by various letters (Tukey tes, $p \leq 0.05$).

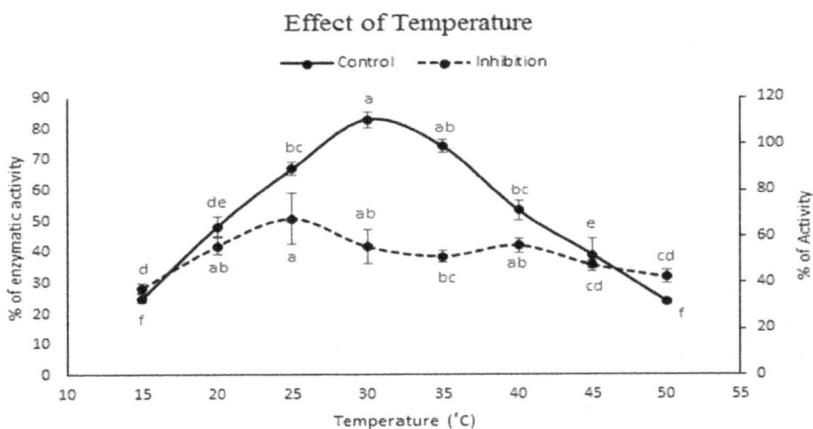

Figure 3: Effect of temperature on PPA inhibition on α-amylase from *C. suppressalis*. Statistical differences have been shown by various letters (Tukey tes, $p \leq 0.05$).

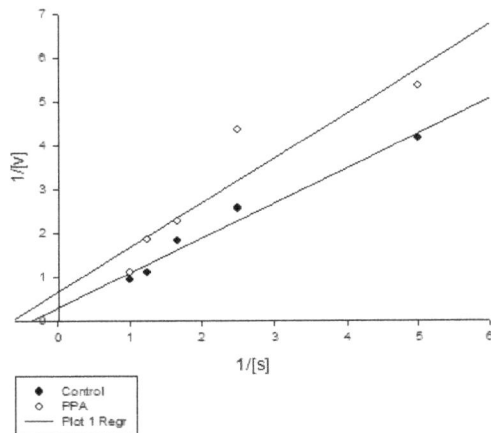

Figure 4: Effect of PPA on α-amylase inhibition from *C. suppressalis* in Lineweaver-Burk plot.

Treatment	V_{max} (U/mg protein)	K_m (%)
Control	3.44 ± 0.87a	2.72 ± 0.61a
PPA	1.53 ± 0.24b	1.55 ± 0.33b

*. Statistical differences have been shown by various letters (Tukey tes, $p \leq 0.05$)

Table 1: Kinetic parameters of α-amylase from *C. suppressalis* in control and PPA treatment.

highest inhibitions but the correct value could be 35°C because other values have been shown lower amylolytic activity in both control and PPA treated enzymes (Figure 3). On the other hand, lower activity in 15, 45 and 50°C might be due to lower enzymatic velocity or its denaturation not PPA inhibition. Other reports imply on correlation between optimal values of pH and temperature for enzymatic activity versus inhibition. For example, Mehrabadi et al. [7] reported pHs of 5 and 6 for inhibition of salivary and midgut α-amylase in *Eurygaster integriceps* Puton (Hemiptera: Scutelleridae). Barbosa et al. [19] showed that αAI extracted from P. vulgaris inhibited porcine pancreatic α-amylase at pH 5.5 depending on the strain of bean used. Finally, Ramzi and Sahragard [18] demonstrated the highest inhibition of *E. ceratoniae* α-amylase by CCA at pH values of 8 and 9. Meanwhile, inhibition of the enzymes was found at 40°C as control optimal value.

Enzyme kinetic parameters revealed lower and higher amounts of V_{max} and K_m in PPA treated enzyme that is a mixed inhibition (Figure 4 and Table 1). This kind of inhibition is considered as a combination of competitive and uncompetitive processes leading to decrease in both K_m and V_{max} values. V_{max} and K_m are the two basic parameters to calculate enzyme behavior in biochemical media. In details, K_m shows affinity of an inhibitor or a substrate to an enzyme. Mixed inhibition has been reported in pancreatic amylase inhibition P. *vulgaris* seeds [20], and α-amylase of *Rhyzopertha dominica* by wheat α-amylase inhibitors [21].

Synthetic chemicals mainly organocholorous and orhanophosphorous have been made several concerns on environmental pollutions, residue in non-target organisms, resistance and etc. Since 1972 that rice stripes stem borer was introduced to Iran, chemical insecticides, mainly diazinon, were the unique control against the pest [2]. In 2009, resistant of the pest has been reported indicating necessity of using an alternative control procedure. Resistant varieties have shown well results against herbivorous insects when biotechnological approaches were adopted to express an entomotoxic protein or inhibitor in host plants. A traditional example could be *Galanthus nivalis* L. agglutinin that has been expressed in several crops showing well results. Since, a low concentration of PPA showed inhibitory effects against α-amylase of rice striped stem borer, it could be a nice candidate to provide resistant varieties of rice expressing the molecule. Although this objective requires comprehensive studies in both biotechnological and pest control aspects.

References

1. Khanjani M (2006) Crop pests of Iran. Boali Sina university press 717.

2. Zibaee A, Sendi JJ, Ghadamyari M, Alinia F, Etebari K (2009) Diazinon resistance in different selected strains of Chilo suppressalis (Lepidoptera: Crambidae) in northern Iran. J Econ Entomol 102: 1189-1196.

3. Nation JL (2008) Digestion. Insect Physiology and Biochemistry CRC Press 27-63.

4. Terra WR, Ferriera C (2012) Biochemistry of digestion. In: Gilbert LI, Editor. Insect molecular biology and biochemistry 365-418.

5. Zibaee A, Bandani AR, Kafil M, Ramzi S (2008) Characterization of a-amylase in midgut and salivary glands of Chilo suppressalis Walker (Lepidoptera: Pyralidae), rice striped stem borer. J Asia-Pacific Entomol 11: 201-205.

6. Franco OL, Rigden DJ, Melo FR, Bloch C Jr, Silva CP, et al. (2000) Activity of wheat alpha-amylase inhibitors towards bruchid alpha-amylases and structural explanation of observed specificities. Eur J Biochem 267: 2166-2173.

7. Mehrabadi M, Bandani AR, Mehrabadi R, Alizadeh H (2012) Inhibitory activity of proteinaceous a-amylase inhibitors from Triticale seeds against Eurygaster integriceps salivary a-amylases: interaction of the inhibitors and the insect digestive enzymes. Pestic Biochem Physiol 102: 220-228.

8. Michiels K, Van Damme EJ, Smagghe G (2010) Plant-insect interactions: what can we learn from plant lectins? Arch Insect Biochem Physiol 73: 193-212.

9. de Oliveira CFT, Luz LA, Paiva PMG, Coelho LCBB, Marangoni S, et al. (2011) Evaluation of seed coagulant Moringa oleifera lectin (cMoL) as a bioinsecticidal tool with potential for the control of insects. Proc Biochem 46: 498-504.

10. Hamshou M, Van Damme EJ, Smagghe G (2010) Entomotoxic effects of fungal lectin from Rhizoctonia solani towards Spodoptera littoralis. Fungal Biol 114: 34-40.

11. Laemmli UK (1970) Cleavage of structural proteins during the assembly of the head of bacteriophage T4. Nature 227: 680-685.

12. Bernfeld P (1955) Amylases, a and ß. Meth Enzymol 1: 149-158.

13. Lowry OH, Rosebrough NJ, Farr AL, Randall RJ (1951) Protein measurement with the Folin phenol reagent. J Biol Chem 193: 265-275.

14. Carlini CR, Grossi-de-Sá MF (2002) Plant toxic proteins with insecticidal properties. A review on their potentialities as bioinsecticides. Toxicon 40: 1515-1539.

15. Pusztai A, Bardocz S (2009) Biological Effects of Plant Lectins on the Gastrointestinal Tract:Metabolic Consequences and Applications. Trends. Glycosci. Glycotechnol 8: 149-165.

16. Da Silva MC, de Sá MF, Chrispeels MJ, Togawa RC, Neshich G (2000) Analysis of structural and physico-chemical parameters involved in the specificity of binding between alpha-amylases and their inhibitors. Protein Eng 13: 167-177.

17. Yamada T, Hattori K, Ishimoto M (2001) Purification and characterization of two alpha-amylase inhibitors from seeds of tepary bean (Phaseolus acutifolius A. Gray). Phytochemistry 58: 59-66.

18. Ramzi S, Sahragard A (2013) A lectin extracted from Citrullus colocynthis L.(Cucurbitaceae) inhibits digestive a-amylase of Ectomyelois ceratoniae Zeller (Lepidoptera: Pyralidae). J Entomol Acarol Res 45: 110-116.

19. Barbosa AE, Albuquerque EV, Silva MC, Souza DS, Oliveira-Neto OB, et al. (2010) Alpha-amylase inhibitor-1 gene from Phaseolus vulgaris expressed in Coffea arabica plants inhibits alpha-amylases from the coffee berry borer pest. BMC Biotechnol 10: 44.

20. Le Berre-Anton V, Bompard-Gilles C, Payan F, Rougé P (1997) Characterization and functional properties of the alpha-amylase inhibitor (alpha-AI) from kidney bean (Phaseolus vulgaris) seeds. Biochim Biophys Acta 1343: 31-40.

21. Priya S, Kaur N, Gupta AK (2010) Purification, characterization and inhibition studies of a-amylase of Rhyzopertha dominica. Pestic Biochem Physiol 98: 231-237.

22

Bacteria-Soil-Plant Interaction: This Relationship to Generate can Inputs and New Products for the Food Industry

Jeremias Pakulski Panizzon[1,2], Harry Luiz Pilz Júnior[1], Neiva Knaak[1*], Denize Righetto Ziegler[2], Renata Cristina Ramos[2] and Lidia Mariana Fiuza[1]

[1]Microbiology and Toxicology Laboratory; PPG Biology / University Valley River Unisinos Bells, São Leopoldo-RS, Brazil
[2]Technological Institute of Food for Health - ITT NUTRIFOR, São Leopoldo-RS, Brazil

Abstract

A thorough study of microbial communities that inhabit aquatic agro-ecosystems is crucial to a better understanding of what happens in the soil, since these microorganisms play important roles for the maintenance of the habitat. Irrigated rice culture is very common in Brazil. However, the incorrect use of these environments can affect the soil and cause damage. The bacteria-soil-plant interaction has been used in order to support biotechnology, as the rhizosphere possesses a different microbial ecology from the rest of the soil. Microorganisms from this region are directly related to plant growth. Bacterial diversity in soil is extremely diverse, with its population changed rapidly as new nutrients are made available or the existing ones are exhausted. The bacteria that live there receive both the plant nutrients and antimicrobial agents, which are selective and inhibit certain undesirable microorganisms. Aiming at this data, this article has an objective to review existing literature on the interaction between microorganisms and their relationship with the plants, which can be transmitted to food through minerals and/or enzymes, thus enabling the generation of inputs or new products.

Keywords: Bacteria; Soil plant; Food microbial comunity

Introduction

Rice is one of the most consumed foods in the world, being one of the most important grains in global economic terms. It is also an important product in the economy of many Latin American countries because it is a staple in the diet of the population [1].

Brazil is among the 10 countries with the highest production of rice, reaching the significant amount of 13 million tons per year, representing about 82% of production in the Mercosul bloc. Rio Grande do Sul is the largest producing state in Brazil, accounting for 61% of total production. Along with the state of Santa Catarina, the southern region accounts for 70% of total production in the country, ensuring the supply of this grain to the entire population of Brazil. The average consumption in the country is 45 kg per person per year [2].

The irrigated rice culture is the largest agricultural user of water worldwide. It is believed that it may cause environmental impacts and this not completely known. Some of the possible impacts of rice culture may be linked to phytosanitary treatments, water, and culture management in a general way [3]. New rice varieties were introduced in the second half of the 20th century that provided increased yields, and these yield gains were almost doubled when synthetic fertilizers were used [4]. Therefore, anthropogenic activities, such as the expansion of cities, pesticides and pollution may directly affect the microbiota of water and soil. However, little is known about how these factors may influence the action of these microorganisms in the ecosystems [5].

The key aspect of this review is to obtain answers to the following questions; (I) how the literature shows the interaction between soil, plant and bacteria; and (II) how the bibliography relates microorganisms with the generation of inputs for the food industry.

Rice

Rice is a plant of the genus *Oryza* belonging to the Gramineae family, who most cultivated species are *O. sativa* (Asian rice) and *O. glaberima* (African rice). The rice domestication happened about 10,000

years ago in Asia and, in Brazil, the plant came through Portuguese colonization [2].

Rice is an annual monocot and can be planted in wetland or highland ecosystems, but has variations in its yield, since the plant needs to adapt to the environment [6]. The plant grows in a period of 100 to 140 days, depending on the chosen crop [7]. The rice development cycle is separated into three phases: the vegetative stage; the reproductive phase; and the grain filling stage [8].

Because it is a culture of easy adaptation, which develops in many soils and climate conditions, the rice is being cultivated more, as the world population is in broad expansion and demand in food production should continually increase to meet the daily nutritional needs of the people, since rice is a mandatory component in the Brazilian meal [9]. The soil microbiota helps in the development of the rice plant. Such as the establishment of plant-microorganism association in the root system that is critical to the chemotactic response of the endophyte to root exudates.

The rice is a cereal that is recommended to people with coeliac disease for its lack of gluten, which is the element of some flour that gives softness to the food. Brown rice is a food that, for each 100 g, consists of 79.3 g of starch, 6.61 g mg of protein and some mg of vitamins such as vitamin B1, B2, B6, niacin, phosphorus, magnesium, iron, potassium and zinc [10,11].

*Corresponding author: Neiva Knaak, Microbiology and Toxicology Laboratory; PPG Biology / Universidade do Vale do Rio dos Sinos, São Leopoldo -RS- Brazil, E-mail: neivaknaak@gmail.com

The rice grain is divided into three parts: Bark, containing the embryo, the endosperm and a fiber complex; - Bran, a thin layer of differentiated tissues such as fibers, proteins, fats, and vitamin B; - Embryo, containing starch, amylose and amylopectin [2].

Wet Lands

Irrigated agriculture shows significant growth in Brazil [12]. The rice fields may be considered wetlands because, according to Maltchik et al. [13], wetlands are environments created by men that help promote the proliferation of microorganism populations on substrates such as sand, gravel or other material in association with plants.

Each cultivation system also presents a different demand of irrigation water. Minimum tillage and no-till systems have similar methods of irrigation [14]. The pre-germinated system presents a different irrigation system of soil, with a flooding prior to their preparation, remaining for 20 days before sowing.

Soil

The soil is a complex mixture of chemicals, including inorganic such as minerals, and the changes taking place in its composition are called biogeochemical transformations. The rate of mineralization grade depends on the availability of oxygen. Compared with the anaerobic metabolism, aerobic is more versatile. Many organic materials are mineralized only if there is oxygen. When the soil is dry and loose, oxygen penetrates more easily, up to about 30 cm depth. Yet, small soil particles are anaerobic, since microorganisms that use oxygen consume it quickly. The mineralization happens slowly in flooded soils because of this oxygen deficit. Soil fertility depends on adequate supply of oxygen, nitrogen, phosphorus and potassium whereas the inorganic forms of these minerals are produced by microorganisms as they mineralize organic material. Natural fertilizers are added to enrich the soil with these elements [15,16]. Necessary the maintain of vegetation, because it avoids the loss of soil nutrients and contamination, and the rational use of pesticides, as it preserves the food chain and also prevents contamination of water resources in the region. It also requires soil analysis before planting to make proper fertilization, since the answer will be different because the cultivation system chosen requires specific conditions [17]. Other important aspects are associated with the use of agrochemicals, once they can quickly reach surface water sources through surface drainage, lateral seepage, surface and subsurface draining, drift, volatilization and also reaching underground water sources by leaching and facilitated flow, reaching non-target organisms and causing environmental contamination [17].

The most widespread cropping systems for irrigated rice are: no-till, conventional, pre-germinated and minimum tillage. Each system is characterized by different types of soil management and of culture itself. In conventional cultivation, soil preparation is done with plowing, in other words, occurs very intense movement of topsoil. For no-till, there is little soil movement, which is held far in advance of planting period, in order to use an herbicide of total action against competing plants that germinate in the period between soil preparation and planting. The pre-germinated planting is characterized by the sowing of pre-germinated seeds in soil flooded beforehand. The soil preparation is done with machinery and implements, by working within the fully flooded plantation frame. Minimum tillage cultivation uses less soil mobilization. When compared to the conventional system, the seeding is performed directly on the vegetation cover previously desiccated with herbicide without soil disturbance [2].

In paddy fields, the first fertilization occurs with nitrogen coverage in a dry ground. After the development of 3 or 4 leaves in the plants, irrigation starts. Another detail that should be noted is that early irrigation, up to two days after herbicide application, results in greater weed control [2]. Knowing how to handle the soil to preserve or even improve their characteristics in sustainable systems is one of the challenges to the current agriculture [18]. Agricultural practices affect physical and chemical characteristics of the soil, influencing diverse populations and bacterial communities [19]. The frequent use of the soil, over time, tends to lead to a reduction of its heterogeneity [20], resulting in the decrease of nutrients available to the microorganisms and consequently the plants. In addition to the common use of soil, the use of chemicals may also have antimicrobial effects [21].

The continuous use of the soil, through the planting of rice, can cause loss of important resources for maintaining this culture, like any other culture that uses it, and the loss of nutrients ends up being inevitable. The soil microbiota promotes important processes for their maintenance and less impacted soil have positive influence on crop development and demand for fewer inputs, which represent costs in the production of grain [22].

Microorganisms

In the soil, it is possible to find various types of microorganisms, such as fungi, protozoa, bacteria and yeasts, which live in symbiosis with others. Some of these relationships are mutualistic, also with plants [23]. The bacteria are differentiated by size and cell structure, and may be spherical, rod-shaped, helical and others. Some obtain energy by processing organic compounds (food), others get nutrients from the environment where they live, using intermediates of glycolysis and other degradation routes [24]. Microorganisms have great importance in biochemical and geochemical cycles and have a great chemical and molecular arsenal. In the context of rice cultivation, bacterial abundance can be considered wide by its variety of microhabitats, caused by constant irrigation, resulting in complex bacterial communities [25].

Because it is a heterogeneous environment, the soil system may harbor the development of many important groups (diazotrophs, nitrifying bacteria, decomposer fungi and antagonists) that can be affected by several factors such as physical disturbances, use of fertilizers and plant species [26]. Most bacteria in the rhizosphere are highly dependent on associations with plants. Therefore, the bacterial diversity plays a key role in the agricultural environment. The result set shows that the irrigated rice culture influences the bacterial density present in water samples used for soil irrigation [27].

From an ecological point of view, decomposer organisms are the most important, which act degrading organic molecules and turning them into inorganic, so that can be used by plants. They are also protagonists in nitrogen and sulfur cycles [28]. Microorganisms are also of great importance in biotechnology, such as the use of bacteria in the bioremediation of pollutants and toxic wastes generated by the industry. Some bacteria are able to use pollutants as energy sources or to convert toxins into less harmful substances. The effect of these bacteria in the environment is highly positive, since the toxins may be removed from different environments, thus being possible their use in oil spills and locations with the presence of toxic waste. Microbial communities are particularly affected by the management and impact in the ground. Agricultural practices such as soil alterations during preparation and irrigation can modify bacterial communities. Water stress is a physiological state of the bacterial community. This stress induces osmotic shock that may result in cell lysis and release of intracellular solutes. Therefore, bacteria can survive under water since they remain in soil particles adhered to the roots, and thus can provide the ideal moisture to the roots to keep them alive [1]. There are groups of microorganisms that produce endospores, which stay dormant and viable in the environment for years, remaining so during adverse situations waiting for a favorable environment for their development, when they return to active form. Some organisms use the energy of chemical interactions, others are even able to photosynthesize and produce their own energy [29]. Most studies of microbial communities in irrigated rice are focused on mass of populations from soil experimentally developed, as many microorganisms are closely linked to the ground by removing nutrients and interfering in its composition [30].

Diazotrophic Bacteria

The endophytic bacteria play a fundamental role in plants and do not cause disease symptoms in which they are associated. These species are able to invade the internal tissues providing a systemic dissemination. The population of viable endophytic diazotrophs in cultivated rice varies with the type of soil, the growth phase of rice culture, and plant tissue. In general, bacterial populations are larger in the roots, compared with stems and leaves [30]. The rice roots harbor endophytes equivalent to 10^8 cultivable nitrogen-fixing bacteria by root gram of dry weight, and an even larger number of non-cultivable bacteria [31]. Nitrogen is one of the most important nutrients for achieving high productivity of annual crops due to high demand of the plant for this nutrient. Therefore, the low availability of this nutrient limits the productivity of the crop. Most of the nitrogen fixation from air takes place through diazotrophs such as *Azospirilium*, *Herbaspirilium* and *Burlkoderia* [32]. The diazotrophs, which are inserted in nitrogen utilization in the soil, are important organisms that can be used as an alternative to nitrogen fertilization [33].

The fact of being able to process organic and inorganic substrates successfully, bacterias become critical for the dynamics of aquatic ecosystems. The interaction between plant and microorganism is little explored in agriculture, despite having global and local importance in the dynamic equilibrium of ecosystems.

The plant and microorganism association in the root system is essential for the chemotactic response of the endophyte to root exudates. In the colonization process, several stages occur, beginning with the displacement of the microorganism into the root system, clinging and distributing through roots [34].

The endophytic bacteria, in order to penetrate the roots, first need the formation of intra and intercellular microcolonies. The different associations of endophytic bacteria can cause changes in plant colonization processes. Accordingly, the microorganisms migrate to the rhizosphere in response to root exudates, which are rich in amino acids, organic acids, sugars, vitamins, purines/pyrimidines, among others. In addition to providing nutritious substances, the plants can also eliminate secretions that facilitate colonization of specific groups of bacteria [35,36]. Microorganisms allow the recycling of nutrients such as lost carbon that can be reintroduced into the food web. Aquatic environments differ in physical and chemical aspects as there are microbial differences. Fungi and bacteria are mainly responsible for the decomposition process in aquatic ecosystems by converting organic matter into inorganic substances. Rainfall, when high, appears to contribute to the high bacterial rates [37].

Heterotrophic Bacteria

Bacteria that inhabit the rhizosphere promote the growth of host plants through the production of phytohormones such as auxins, the phosphate solubilization, the production of iron chelators (siderophores), the release of antimicrobial metabolites and for competition for nutrients [38]. The bacteria found in soil are highly diversified. In hot soils, for example, there is the presence of thermophilic microorganisms and microbial population changes very quickly as the available nutrients are modified [39].

The soil has great spatial variability composed of many micro habitats that may differ in their physicochemical properties [40], (Figure 1). These characteristics provide a diverse composition of microorganisms, which accomplish the primary decomposition, cycling and regulation of nutrients and minerals retention. Bacteria

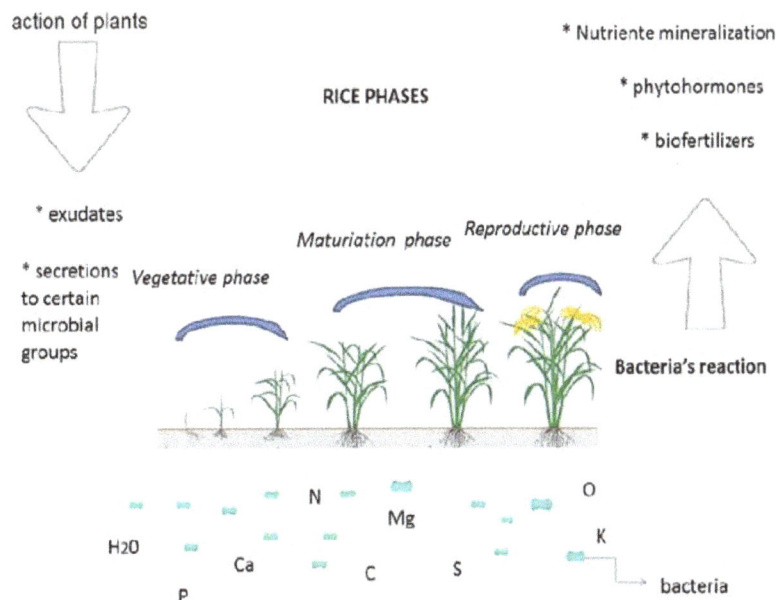

Figure 1: Interactions between bacteria and the rice plants in their different development stages.

also secrete various enzymes such as catalase, urease, cellulose, among others [41], also important for the plants. The cultivars and also the stage of the irrigated rice crop development can influence the microbial populations associated with these agro-ecosystems [17], as well as the different soil management systems, the application of pesticides and the fertilizers used for better development of this culture (Figure 1).

The composition of a community hardly will remain static, so any change in the habitat can cause changes in the composition of populations. Species such as *Bacillus thuringiensis*, *B. subtilis*, *B. cereus* and *Lysinibacillus sphaericus* may have insecticidal activity to different orders of insects [42] and are widely used in biological pest control. Some species are also associated with plant roots, aiding in the absorption and retention of nutrients in the soil.

They also have biotechnological applications and produce antibiotics [43], among other important functions. Therefore, understanding the impact of different rice cultivation systems in the bacterial community in the soil is important for the knowledge of lower environmental impact soil technologies and, eventually, the ones that can contribute positively to the improvement of ecological interactions that take place under sunshine [39]. Cosmopolitan microorganisms isolated from soil, water and vegetables such as bacteria of the Enterobacteriaceae family can also be found in the digestive tract of animals and humans, although it is also possible to find them in transient or normal microbiota. Among them include: *Escherichia* spp., *Klebsiella* spp., *Salmonella* spp., *Enterobacter* spp., *Serratia* spp., *Hafnia* spp., *Citrobacter* spp., *Yersinia* spp., *Proteus* spp., *Rhanella* ssp., *Providencia* spp., *Morganella* spp., *Shigella* spp., *Edwarsiella* spp., *Ewingella* ssp., *Budvicia* ssp., *Tatumella* spp., *Erwinia* spp., *Koserella* ssp., *Kluyvera* ssp., *Hoganella* ssp., *Moellenella* ssp., *Leminorella* ssp., *Buttiauxella* spp. and *Pantoea* spp. They can be pathogenic or opportunistic, occasionally forming components with physicochemical and biological properties beneficial to humans and the environment in which they operate [44]. Pereira et al. [45] points out that there is a wide variety of bacteria in the soil that, when associated with plant hosts, stimulate their growth, such as the rhizobacteria [46]. Stimulation in plants growth is mainly due to increased availability of mineral nutrients [47] and production of growth hormones such as gibberellins and auxin (induced resistance to disease and suppression of harmful microorganisms from the rhizosphere of plants). These direct effects provide a high gain to producers, allowing them to lower their use of inputs in farming or monoculture, reducing costs and also potential environmental problems such as contamination of soil and water sources, caused by the use of chemicals for pest control [48].

Pseudomonas fluorescens, *P. putida*, *Azospirillum brasiliense*, *Serratia marcecens*, *Bacillus subtilis*, *B .megaterium*, *Rizobium* sp., *Bradyrhizobium* sp., *Arthrobacter* sp., *Enterobacter* sp., *Azotobacter* sp. among others are heterotrophic bacteria genres most cited in the literature [49]. Microorganisms that inhabit the soil, along with biological processes, have been investigated as indicators of the sustainability of agriculture and / or soil quality [18]. They also influence the biological quality of products and the productivity achieved [19].

Food

Many diseases that affect plants and impair food production in crops can be caused by bacteria. Some examples well known for rice farmers are the brown spot and the stem rot caused by the plant pathogens *Helminthosporium oryzae* and *Sclerotium oryzae*, respectively [2]. However, besides being studied by the negative aspects, recent scientific research has given great importance to microorganisms for

other functions. As an example in the food industry, the fermentation, this allows the manufacture of cheeses, dairy products, breads, meat products, alcoholic beverages and plants. Microorganisms present in the soil can influence the nutrients the food provides being ingested by humans, thus being able to bring beneficial changes to health. Most food is chemically stable, rotting only when contaminated by microorganisms. Factors such as pH, temperature, water availability and addition of chemical substances help in the preservation of food [24].

In the case of rice, its processing ensures quality control. It undergoes different processes leading to each type of rice: whole, parboiled, polished and white rice. White rice is polished and the starch is what predominantly stays in the grain. Parboiled rice is sterile in which the nutrients of the husk migrate into the grain. Whole rice has a medium and long grain, being darker than the processed, because it retains the film and the germ, which is more nutritious and rich in vitamins [11].

The rice is responsible for 27% of the daily calorie and 20% of the protein needs of the population, containing thiamine, riboflavin and niacin. It is easily digestible, being a good fiber regulator. The rice-based diet selects fermentative bacteria that are resistant to pathogens. By not containing gluten, it does not cause eating disorders. It has low sodium content, 5 mg in 100 g, and high potassium content, 92 mg in 100 g. It is also useful for osteoporosis because it has selenium (an antioxidant that acts against free radicals and unstable molecules of high reactivity) and silicon that helps forming collagen, elastin and protein of the cognitive tissue [50]. Foods considered new are resulted from new techniques of biotechnology and plant breeding. There are few genetically modified plants being commercialized. The genetic traits that are changed confer resistance to insects, herbicides etc. In order to arouse consumer interest in the products, there must be a higher quality of these that improves health or that has a greater shelf life. Biotechnology can improve organoleptic properties such as flavor and pigmentations [51].

Molecular Biology in Food

At present, the microorganisms have been used in the production of enzymes, antibiotics, solvents, amino acids and dietary supplements. Microbial diversity is a source of genetic resources, with each microorganism playing a unique and particular function. Therefore, first the decoded gene from the specific enzyme is identified and then isolated and transferred by recombinant DNA techniques into a known microorganism. The main bacteria used are: *Escherichia coli* and *Bacillus* sp, The main fungi used is: *Aspergillus* sp. and the main yeast is: *Saccharomyces cerevisiae*. Enzymes are biological catalysts present in all beings, therefore are used in various fields, including the bioremediation. The biocatalysts are protein molecules with an associated catalytic power. These biocatalysts degrade the molecules present in the environment such as cellulose, starch, lignin etc. [52,53].

Application of molecular biology techniques has greatly contributed to the exploration of new enzymes and new enzymatic properties and certainly continued to promote and develop production. Enzymes generate added value to the product in line with the demand of technological, market and environmental preservation characters. The use of catalysis is critical in Brazil, as it helps in environmental renewal, with the increase of enzymes used in pharmacy and food [38].

Conclusion

Agriculture is one of the most important activities, but it is necessary to observe some points, as food production is directly linked to environmental quality. Rice is one of the most consumed foods in the world, one of the most important grains in global economic terms. Rice is a culture that besides being simple is a great resource for human consumption. Rice, although not a food rich in vitamins, has a differential, it is easily assimilated by the starch, which provides power, serving as fuel for the operation body. Currently, crops cultivation methods seek to optimize the potential of agricultural production through the application of fertilizers and pesticides and which consequently cause human health problems and an imbalance in agricultural ecosystems, especially in the communities that inhabit the soil. The rice crops suffer and benefit from various microbial actions, including interactions between plants and microorganisms [54,55]. Rice agroecosystems consist of several micro-habitats and provide the suitability of a wide variety of microorganisms. The management of rice promotes changes of physical and chemical characteristics of the water and due to changes in pH, turbidity, temperature, radiation and amount of organic matter that may be related to the dynamics of microbial communities in the soil. Soil is a habitat full of living microorganisms that directly influence the development of the plant. The bacteria that act in it are inserted in the process of chemical transformations that facilitate nutrient cycling and can be added to the food, generating inputs that provide functionality and well-being to the human being. The challenge of scientists in the area is still the search for a better quality of life, environmental protection and sustainable development, because rice is one of the grains produced in the world and extremely important for the world economy.

Acknowledgement

"Give thanks to the Lord, for he is good." Psalm 136:1. The authors thank CNPq, FAPERGS, NUTRIFOR and UNISINOS for the financial support. I thank my friends! Thanks to my wife Ana Paula.

References

1. Arshad R, Farooq S, Farooq-E-Azam (2006) Rhizospheric Bacterial Diversity: Is it Partly Responsible for Water Deficiency Tolerance in Wheat? Pak. J Bot 38: 1751-1758.

2. SOSBAI (2014) Irrigated Rice: Technical Research Recommendations Southern Brazil. Technical Report, XXX Technical Meeting of Rice Culture Irrigated, Bento Goncalves, Brazil pp: 9-192.

3. Reche MHLR, Fiuza LM (2012) Bacterial diversity in rice-field water in Rio Grande doSul. Brazilian J Microbiol 36: 253-257.

4. Pittol M, Durso L, Valiati VH, Fiuza LM (2015) Agronomic and environmental aspects of diazotrophic bacteria in rice fields. Ann Microbiol p1-20.

5. Kirk MD, Little CL, Lem M, Fyfe M, Genobile D, et al. (2004) An outbreak due to peanuts in their shell caused by Salmonella enterica serotypes Stanley and Newport-sharing molecular information to solve international outbreaks. Epidemiol Infect 132: 571-577.

6. Pinto LFS, Laus-Neto JA, Pauletto EA (2004) Solos de várzea do Sul do Brasil cultivados com arroz irrigado. Technical Report, Arroz Irrigado no Sul do Brasil p 75-95.

7. Macedo VRMM, Menezes VG (2004) Influence of production systems and management of water use by the rice plant. Use of water in agriculture pp: 1-20.

8. Streck NA, Bosco LC, Michelon S, Walter LC, Marcolin E (2006) Duration development cycle of rice cultivators as a function of emission leaves on the main stem. Rural Science 36: 1086-1093.

9. Bera S, Ghosh R (2013) Soil Physico-Chemical Properties and Microflora as Influenced by Bispyribac Sodium 10% SC in Transplanted Kharif Rice. Rice Science 20: 298-302.

10. Pamplona J (2008) The healing power of food. Publishing house Brazil 155-158.

11. Franco, J (2003) Table of chemical composition of the feed . Atheneu , they are Paul, pp: 7-64.

12. De Bona FD, Bayer C, Bergamaschi H, Dieckow J (2006) organic carbon in soil irrigated by sprinkling systems under no-tillage and preparation conventional. J of Soil Science 30: 911-920.

13. Maltchik L, Rolon AS, Guadagnin DL, Stenert C (2004) Wetlands of Rio Grande do Sul, Brazil: a classification with emphasis on plant communities. Acta Limnol Bras 16: 137-151.

14. Vernetti-Junior FJ, Gomes AS (2004) conventional system of rice. Technical Report, irrigated rice in Southern Brazil pp: 339-348.

15. Zhao S, Li K, Zhou W, Qiu S, Huang S, et al. (2015) Changes in soil microbial community, enzyme activities and organic matter fractions under long-term straw return in north-central China. Agric Ecosyst Environ 216: 82-88.

16. Faccin GL, Miotto LA, Vieira LN, Barreto PLM, Amante ER (2009) Chemical, Sensorial and Rheological Properties of a New Organic Rice Bran Beverage. Rice Science 16: 226-234.

17. Silva DRO, Avila LA, Agostinetto D, Primel EG, Bundt ADC (2009) Occurrence of pesticides used in irrigated rice crops in water sources underground in southern Brazil. Brazilian Congress of Irrigated Rice 6: 167-170.

18. Reis Junior FB, Mendes IC (2007) Biomassa microbiana do solo Planaltina.

19. Sessitsch A, Hardoim P, Döring J, Weilharter A, Krause A et al. (2012) Functional Characteristics of an Endophyte Community Colonizing Rice Roots as Revealed by Metagenomics Analysis. Mol Plant Microbe In 25: 28-36.

20. Pereira JC, Neves MCP, Gava CAG (2000) soybean cultivation effect on the dynamics the bacterial population in cerrado soils. Agricultural research Brazilian 35: 1182-1190.

21. Cavalcante EGS, Alves MC, Souza ZM, Pereira GT (2011) Variability Spatial physical attributes of soil under different uses and management . Rev Bras AGR Eng 15: 237-243.

22. Griffiths BS, Ritz K, Wheatley R, Kuan HL, Boag B, et al. (2001) An examination of the biodiversity–ecosystem function relationship in arable soil microbial communities. Soil Biol Biochem 33: 1713-1722.

23. Giri R, Dudeja SS (2013) Root Colonization of Root and Nodule Endophytic Bacteria in Legume and Non Legume Plants Grown in Liquid Medium. J of Microbiology Research and Reviews 1: 7582.

24. Tian W, Wang L, Li Y, Zhuang K, Li G, et al. (2015) Responses of microbial activity, abundance, and community in wheat soil after three years of heavy fertilization with manure-based compost and inorganic nitrogen. Agric Ecosyst Environ 213: 219-227.

25. Panizzon JP, Pilz Júnior HL, Knaak N, Ramos RN, Ziegler DR, et al. (2015) Microbial Diversity: Relevance and Relationship Between Environmental Conservation and Human Health. Braz Arch Biol Technol 58: 137-145.

26. Reali C, Fiuza LM (2012) Ecology of Bacillus sp and Lysinibacillus sp in rice field soils from Southern Brazil, in: Microbes in Applied Research: Current Advances and Challenges. Fulsland Offset Printing 89-93.

27. EMBRAPA (2014) Irrigated rice cultivation in Brazil: Economic Importance agricultural and food.

28. Lindow SE, Leveau, JHJ (2002) Phyllosphere microbiology. Curr Opin Biotech 13: 238-243.

29. Neves MCP, Rumjanek NG (1998) Ecology of nitrogen fixing bacteria in tropical soils. Microbial Ecology 1: 15-60.

30. Curtis H, Barnea NS (2004) Medical Biology Panamericana De Bona F, Bayer C, Bergamaschi H, Dieckow J (2006) not only organic carbon in systems Irrigated for spraying and tillage under conventional prepared. Magazine only the Brazilian Science 30: 911-920.

31. Prakamhang J, Minamisawa K, Teamtaisong K, Boonkerd N, Teaumroong N (2009) The communities of endophytic diazotrophic bacteria in cultivated rice (Oryza sativa L). Appl Soil Ecology 42: 141-149.

32. Sessitsch A, Hardoim P, Döring J, Weilharter A, Krause A, et al. (2012) Functional Characteristics of an Endophyte Community Colonizing Rice Roots as Revealed by Metagenomic Analysis. MolPlantMicrobe in 25: 28-36.

33. Guimarães SL, Baldani JI, Baldani VLD (2003) Effect of inoculation of diazotróficasendofíticas bacteria in upland rice. Agronomy 37: 25-30.

34. Moreira APB, Teixeira TFS, Peluzio GM, Alfenas CG (2012) Gut microbiota and the development of obesity. Nutr Hosp 27: 1408-1414.

35. Jha PN, Gupta G, Jha P, Mehrotra R (2013) Association of Rhizospheric/ Endophytic Bacteria with Plants: A Potential Gateway to Sustainable Agriculture. Greener J of Agricultural Sciences 3: 73-84.

36. Bais HP, Weir TL, Perry LG, Gilroy S, Vivanco JM (2006) The role of root exudates in rhizosphere interactions with plants and other organisms. Ann Rev of Plant Bio 57: 233-266.

37. Compant S, Mitter B, Colli-Mull JG, Gangl H, Sessitsch A (2011) Endophytes of Grapevine Flowers, Berries, and Seeds: Identification of Cultivable Bacteria, Comparison with Other Plant Parts, and Visualization of Niches of Colonization. Microbial Ecol 62: 188-197.

38. Araújo MFF, Costa IAC (2004) Comunidades Microbial (Bacterioplancton and protozooplankton) In Reservoirs Do Brazilian semiarid region. Oecologia Brasiliensis 11: 422-432.

39. Berg G (2009) Plant-microbe interactions promoting plant growth and health: Perspectives for controlled use of microorganisms in agriculture. Appl Microbiol Biot 84: 11-18.

40. Godoy SG, Stone LF, Ferreira EPB, Cobucci T, Lacerda MC (2015) Correlation between rice productivity in tillage and soil properties. Rev Bras Eng Agr Amb 19: 119-125.

41. Carson JK, Campbell L, Rooney D, Clipson N, Gleeson DB (2009) Correlation between rice productivity in tillage and soil properties. Rev Bras Eng Agr Amb 19: 119-125. FEMS Microbiol Ecol 67: 381-388.

42. Kunito T, Saeki K, Goto S, Hayashi H, Oyaizu H, et al. (2001) Copper and zinc fractions affecting microorganisms in long-term sludge-amended soils. Bio resource Technol. 79: 135-146.

43. Lindow SE, Leveau, JHJ (2002) Phyllosphere microbiology. Curr Opin Biotech 13: 238-243.

44. Apaydin Ö, Yenidünya F, Harsa Ş, Güneş H (2005) Isolation and characterization of Bacillus thuringiensis strains from different grain habitats in Turkey. World j microb biot 21: 285-292.

45. Garbeva P, Veen JD, Elsas, JD (2003) Predominat Bacillus spp. In agricultural Soil Under Different Management Regimes Detected via PCR-DGGE. Microbial Ecol 45: 302-316.

46. Cortés-sánchez AJ, Diaz-ramirez M, Hernández-álvarez AJ, García-ochoa F, Villanueva-carvajal A, et al. (2015) Bio surfactants produced by enterobacterial-Glo Adv Res. J Microbiol 4: 103-112.

47. Pereira RM, Silveira EL, Carareto-Alves LM, Lemos EGM (2008) Evaluation populations of possible rhizobacteria in soils under forest species. J of Soil Science 32: 1921-1927.

48. Sottero AN, Freitas SS, Melo AMT, Trani PE (2006) Rhizobacteria and lettuce: root colonization, plant growth promotion and biological control. Magazine Soil Science Brazilian 30: 125-134.

49. Peña HB, Reyes I (2007) Isolation and evaluation of bacteria sfijadoras nitrogen and phosphates enlapromocióndelcrecimiento solvents lalechuga (Lactuca sativa L). Interciencia 32: 560-565.

50. Hardoim PR, Hardoim CCP, Overbeek LS, Elsas JD (2012) Dynamics of seed-borne rice endophytes on early plant growth stages. PloSone 7: e30438.

51. Melo IS (1998) Promoting rhizobacteria of planatas growth: Description and potential use in agriculture. Ecology Microbial 3: 87-116.

52. Walter M, Marchezan E, Avila LA (2008) Rice: Composition and Characteristics Nutrition. Rural Science 38: 1184-1192.

53. Itani T, Tamaki M, Arai E, Horino T (2002) Distribution of Amylose, Nitrogen, and Minerals in Rice Kernels with Various Characters. J Agr Food Chem 50: 5326-5332.

54. Walters D (2009) Are plants in the field already induced? Implications for practical disease control. Crop Prot 28: 459-465.

55. Kumar KVK, Yellareddygari SKR, Reddy MS, Kloepper JW, Lawrence KS et al. (2012) Efficacy of Bacillus subtilis MBI 600 against health blight caused by Rhizoctoniasolani and on growth and yield of rice. Rice Science 19: 55-63.

23

Pregerminated Brown Rice Enhanced NMDA Receptor/CaMKIIα Signaling in the Frontal Cortex of Mice

Takayoshi Mamiya[1]*, Keiko Morikawa[2] and Mitsuo Kise[2]

[1]Department of Chemical Pharmacology, Faculty of Pharmacy, Meijo University, Japan
[2]FANCL Institute, FANCL Corporation, Yokohama, Japan

Abstract

We previously reported that the continuous feeding of mice with pellets of pregerminated brown rice (PGBR; *Hatsuga genmai* in Japanese) enhances their spatial learning. Here, we show the possible relationships of the enhancement of learning and memory with the glutamatergic system in the brain of PGBR-pellet-fed mice. The enhancement of learning and memory in the novel object recognition and Y-maze tests after 28-day-feeding of PGBR pellets was inhibited by dizocilpine (10 μg/kg s.c.), an N-methyl-D-aspartate (NMDA) receptor antagonist, whereas the extracellular glutamate level and the glutamate content were not affected in the frontal cortex and hippocampus. In the frontal cortex of mice fed PGBR pellets, the phosphorylation of calcium calmodulin kinase IIα (CaMKIIα), one of the important events after NMDA receptor activation, was facilitated compared with that of mice fed control pellets. This facilitation was inhibited by dizocilpine (10 μg/kg s.c.), whereas the phosphorylation of extracellular signal-regulated protein kinases (ERKs), another index of memory formation was not affected by PGBR pellets. On the other hand, in the hippocampus, there was no significant difference in the phosphorylation of CaMKIIα and ERKs between the control and PGBR pellets-fed mice. Taken together, these results suggest that PGBR enhances the NMDA receptor/CaMKIIα signaling in the frontal cortex, leading to enhanced learning and memory in mice.

Keywords: Learning and memory; Pregerminated brown rice; NMDA receptor/CaMKIIα cascade; Frontal cortex

Introduction

A rice grain consists of the endosperm, bran layer, and germ layer, and is an important energy source for humans. Polished rice (PR), the staple food of Asians, is produced by eliminating the fiber-rich bran layer from unpolished rice, namely, brown rice (BR). Pregerminated brown rice (PGBR) is produced by soaking BR in water to induce partial germination. PGBR contains abundant amino acids, dietary fiber, vitamins, and minerals in the bran layer and embryo, and it tastes better than BR. We have reported the beneficial effects of PGBR on learning and memory in mice and raised a possibility that the glutamatergic system is involved in the enhancement of learning and memory by PGBR [1].

The glutamatergic system in the central nervous system is one of most important neuroregulatory systems in learning and memory [2]. For example, the dysfunction of the N-methyl-D-aspartate (NMDA) receptor induced by pharmacological or genetic techniques impairs learning and memory [3,4]. In contrast, overexpression of NMDA receptor 2B in the forebrain leads to enhanced activation of the NMDA receptor and better learning and memory in mice [5]. These findings suggest that continuous feeding with PGBR pellets may regulate learning and memory via the glutamatergic system in the brain, but the details remain unclear.

In this study, firstly we examined whether the glutamatergic system plays a significant role in the PGBR-induced enhancement of learning and memory using an NMDA receptor antagonist in two behavioral tests. Next, we assessed the extracellular glutamate level by *in vivo* brain microdialysis experiments and glutamate content in the hippocampus and frontal cortex. Additionally, we examined the phosphorylation of calcium calmodulin kinase IIα (CaMKIIαα and extracellular signal-regulated protein kinases (ERKs), because NMDA receptor activation leads to their phosphorylation, which is an important intracellular event related to learning and memory [6-8].

Materials and Methods

Animals and foods

Five-week-old male ICR mice (Nihon SLC Co., Shizuoka, Japan) were purchased. We received the mice from 11h00 and 13h00, then put them into the home cage and gave adequate food pellets (day 0). The animals were housed in a controlled environment (23 ± 1°C, 50 ± 5% humidity) and given access to water ad libitum. Room lights were on between 7:30 and 19:30. We used commercial pellets (AIN-93G; Oriental Yeast, Tokyo, Japan) as the control, the ingredients of which are as follows: 39.7% cornstarch, 20% casein, 0.3% L-cystine, 13.2% α-cornstarch, 10% sucrose, 7% bean oil, 5% cellulose powder, 3.5% minerals, 1% vitamins, 0.25% choline bicitrate, and 0.0014% butylhydroquinone. The rice we used was grown in Hokkaido area in Japan [*Oryza sativa subsp. japonica* (Hoshino-yume)]. PGBR was prepared at 25–30% water content to induce germination and dried to 15% according to a patented procedure (Patent No. 3738025, JP, November 4, 2005). PGBR was manufactured as powdered feed by Oriental Yeast. The nutrients in the PGBR pellets were the same as those in AIN-93G, except that cornstarch was replaced with PGBR, as reported previously [1,9]. All experiments were conducted in accordance with the Guidelines for Animal Experiments of Meijo University (Approval number: Yaku-Jitsu-12) and the Guiding

*Corresponding author: Takayoshi Mamiya, Department of Chemical Pharmacology, Faculty of Pharmacy, Meijo University, 150 Yagotoyama, Tempaku-ku, Nagoya 468-8503, Japan, E-mail: mamiya@meijo-u.ac.jp

Principles for the Care and Use of Laboratory Animals approved by the Japanese Pharmacological Society (2007). Each experiment was carried out independently on the 29th day.

Y-maze test: Short-term memory was examined by monitoring spontaneous alternation behavior in the Y-maze [1,9,10]. We used 13 mice in each group. The maze was made of wood painted black and each arm was 40 cm long, 12 cm high, 3 cm wide at the bottom, and 10 cm wide at the top. The arms converged at an equilateral triangular central area that was 4 cm at its longest axis. The apparatus was placed on the floor of the experimental room and illuminated with a 100 W bulb from 200 cm above. Each mouse was placed at the end of one arm and allowed to move freely through the maze during an 8-min session, and the series of arm entries was recorded visually. The alternation behavior was defined by the successive entry into the three arms, on overlapping triplet sets, and such behavior (%) was expressed as the ratio of actual alternations to possible alternations (defined as the total number of arm entries minus two), multiplied by 100. Dizocilpine (0-30 µg/kg) was dissolved in 0.9% physiological saline and administered subcutaneously to mice 20 min before the test.

Novel object recognition test: For the novel object recognition test, a mouse was habituated to a black plastic cage (30 x 30 x 50 cm) for 15 min for consecutive 3 days. In the training trial, two same objects (white film cases, Ø 3 x 6 cm) were placed in the cage, and the mouse was allowed to explore freely for 5 min. The time spent on approaching each object was recorded manually. In the retention trial immediately after the training trial, the mouse was removed once and then placed back in the same cage, in which one of the familiar objects used in the training trial was replaced by a novel object (black dry battery cell, Ø3 x 6 cm), and allowed to explore for 5 min. Exploratory preference, the ratio of time spent on approaching one of the two same objects in the training trial or the novel one in the retention trial over the total time spent approaching both objects, was used to measure recognition memory [approach duration (%)] [9,11]. We used 10 mice in each group. Dizocilpine (10 µg/kg) was administered subcutaneously to mice 20 min before the retention trial.

***In vivo* brain microdialysis:** One day before the microdialysis, mice were anesthetized with pentobarbital-Na (50 mg/kg, i.p.) and the dialysis probe (D-I-3-01, Eicom Corp., Kyoto, Japan) was implanted into the prefrontal cortex (+1.7 mm anteroposterior, -0.3 mm mediolateral from the bregma, -2.5 mm dorsoventral from the skull) and hippocampus (-1.6 mm anteroposterior, +1 mm mediolateral, -1.5 mm dorsoventral from the skull) according to the brain atlas [12]. The dialysis probe was perfused with artificial cerebrospinal fluid (aCSF; mM: NaCl, 127.6; KCl, 2.5; CaCl2, 1.3, pH 7.4) at a rate of 1 µL/min and connected to a microinfusion pump (Microsyringe Pump, ESP-64, Eicom) via a single-channel liquid swivel. The mice were placed in individual acrylic cages (20 × 30 × 25 cm high) and allowed to adapt for at least 60 min before the experiment was started. About 3 h after the probe was inserted, samples (10 µL) were collected at 10-min intervals. The locations of dialysis probes were confirmed after the experiments.

Glutamate in the dialysate was quantified using an HPLC-ECD system (HTEC-500, Eicom) with an immobilized enzyme column (E-Enzympak, Eicom). The mobile phase consisted of 0.1 M ammonium chloride buffer (pH 7.2) containing 10% ammonia solution, and 685 µM hexadecyltrimethylammonium bromide was delivered at a flow rate of 400 µL/min. To protect the analytical column from impurities in the mobile phase and samples, a precolumn (CH-GEL, Eicom) was placed between the pump and the injector, and the column temperature was maintained at 25°C. Aliquots (10 µL) of the perfusate samples were injected into the HPLC system and separated by a column of Eicompak Glu-Gel (Ø 4.6 × 150 mm, Eicom). The column containing glutamate oxidase catalyzed the formation of hydrogen peroxide from glutamate and oxygen. The resultant hydrogen peroxide was detected by ECD with a platinum working electrode at +450 mV.

Glutamate content: Mice were sacrificed by focused microwave irradiation for 1.4 sec at 5 kW. The brains were quickly removed, and the hippocampus and frontal cortex were dissected out on an ice-cold glass plate as in our previous report [13]. Each frozen brain sample was weighed and homogenized with an ultrasonic processor in 0.1 M ammonium chloride buffer (pH 7.2) containing 10% ammonia solution and 685 µM hexadecyltrimethylammonium bromide. The homogenates were centrifuged at 15,000 x g for 15 min at 4°C. The supernatants were centrifuged by ultrafiltration (Ultrafree-MC, 5000 NMWL, Millipore, USA) at 5,000 x g for 20 min at 4°C, then aliquots (10 µL) were injected into the HPLC-ECD system (HTEC-500, Eicom). The HPLC-ECD system condition was similar to that in the microdialysis analysis.

Western blot analysis of CaMKIIα and ERK phosphorylation: The mice were decapitated, and the hippocampus and frontal cortex were dissected (5 mice in each group). Each brain tissue was homogenized by sonication in ice-cold buffer (50 mM Tris-HCl, 150 mM NaCl, 10 mM NaF, 10 mM EDTA, 1 mM sodium orthovanadate, 2 µg/mL pepstatin, 2 µg/mL leupeptin, 2 µg/mL aprotinin, 1 mM PMSF, and 1 mM DTT, pH 7.4). The samples were boiled in Laemmli sample buffer, separated on 7.5% polyacrylamide gel and subsequently transferred to PVDF membranes (Millipore). The membranes were blocked with an ECL Blocking Agent (ECL Plus Western Blotting Detection Reagents, GE Healthcare, USA) for 2 h at room temperature and probed with a monoclonal antibody to the phospho-CaMKIIα subunit at Thr^{286} [0.2 µg/ml, Affinity BioReagents, Inc. (Thermo Fisher Scientific), USA] or a polyclonal antibody to phospho-p44/42 MAP kinase at Thr^{202}/Tyr^{204} (0.1 µg/mL, Cell Signaling Technology, USA) overnight at 4°C. The membranes were washed with TBST buffer (10 mM Tris-HCl, 150 mM NaCl, and 0.1% Tween 20, pH 7.4) and subsequently incubated with a goat anti-rabbit horseradish peroxidase-conjugated secondary antibody for 2 h at room temperature. The immune complexes were detected by chemiluminescence immunoassay (GE Healthcare) and exposed to an X-ray film. The band intensities of the film were analyzed by densitometry (GelDoc 2000, BioRad, USA). To confirm equal loading of each protein, the membranes were stripped with stripping buffer (62.5 mM Tris-HCl, 100 mM 2-mercaptoethanol, and 2% SDS, pH 6.7) at 50°C for 5 min. They were then incubated with a monoclonal antibody to the CaMKIIα subunit [0.2 µg/mL, Affinity BioReagents, Inc. (Thermo Fisher Scientific)] or an anti-p44/42 MAP kinase polyclonal antibody (0.1 µg/mL, Cell Signaling Technology) and analyzed as described above [7,14].

Data analysis: All results were expressed as means ± SEM for each group. All data were analyzed by one-way ANOVA, followed by Tukey's or Student's t-tests. The level of significant difference was taken at $P < 0.05$.

Results

Y-maze test

As shown in Figure 1A, there was a significant enhancement of alternation behavior in the PGBR- pellet-fed mice compared with the control pellets-fed mice ($F_{5,77} = 8.04$, $P < 0.001$, Tukey's test; $P < 0.01$). In the control-pellet-fed mice, dizocilpine (10 µg/kg) did not affect this

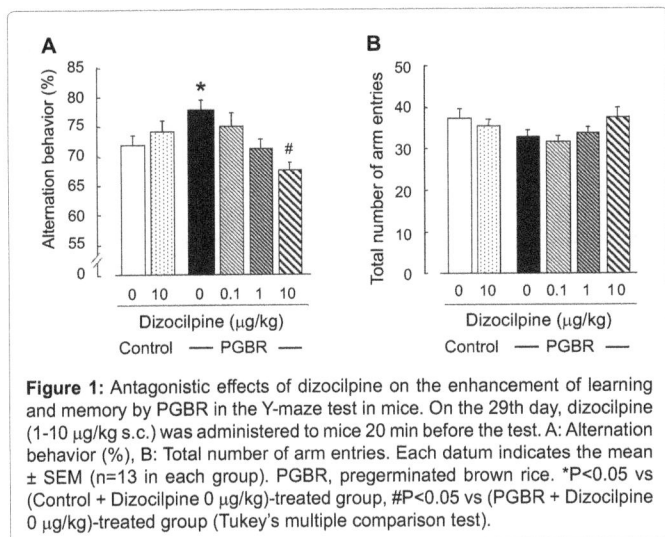

Figure 1: Antagonistic effects of dizocilpine on the enhancement of learning and memory by PGBR in the Y-maze test in mice. On the 29th day, dizocilpine (1-10 µg/kg s.c.) was administered to mice 20 min before the test. A: Alternation behavior (%), B: Total number of arm entries. Each datum indicates the mean ± SEM (n=13 in each group). PGBR, pregerminated brown rice. *P<0.05 vs (Control + Dizocilpine 0 µg/kg)-treated group, #P<0.05 vs (PGBR + Dizocilpine 0 µg/kg)-treated group (Tukey's multiple comparison test).

behavior (Figure 1A), but at 30 µg/kg, it induced remarkable reduction (data not shown), as previously reported [10]. In the PGBR-pellet-fed mice, dizocilpine antagonized the effects of PGBR pellets in a dose-dependent manner (Tukey's test; P<0.05). On the other hand, statistical analysis revealed that the total number of arm entries was not affected by PGBR pellets or dizocilpine ($F_{5,77}$=1.58, P=0.176, Figure 1B).

Novel object recognition test

There was no significant difference in approach duration (exploratory preference) in the training trial among the groups (white columns; $F_{4,49}$=0.265, P=0.89), indicating that these groups essentially had the same levels of curiosity and/or motivation to explore the two same objects. In the retention trial, however, all groups exhibited greater preference toward the novel object than the familiar object (dotted columns). Approaching a novel object for a longer duration was inhibited by dizocilpine (10 µg/kg) in the PGBR-fed mice but not in the control-pellet-fed mice ($F_{2,29}$=3.92, P=0.031, Tukey's test; P<0.05) (Figure 2).

Extracellular glutamate level

The average basal levels of glutamate were 1.45 ± 0.19 and 6.97 ± 0.96 pmol/min in the hippocampus and frontal cortex of the control-pellet-fed mice respectively. There were no significant differences in accumulation for 60 min between the control- and PGBR-pellets-fed mice in both brain regions (Figures 3A and 4A).

Glutamate content

There were no significant differences in glutamate content between the control- and PGBR- pellet-fed mice in both brain regions [hippocampus: Control, 1912.9 ± 77; PGBR, 1803.9 ± 52 nmol/g wet tissue ($F_{1,9}$=1.61, P>0.05); frontal cortex: Control, 1674.3 ± 43; PGBR, 1560.7 ± 22 nmol/g wet tissue ($F_{1,9}$=4.38, P>0.05)].

Western blot analysis of CaMKIIα and ERK phosphorylation

CaMKIIα: There were no significant differences in CaMKIIα phosphorylation in the hippocampus between the groups we examined ($F_{3,19}$=0.12, P=0.95, Figure 3B). On the other hand, in the frontal cortex of the Sal-treated mice, PGBR enhanced significantly CaMKIIα phosphorylation compared with that of the control-pellet-fed mice ($F_{3,19}$=14.3, P<0.01, Tukey's test; P<0.01, Figure 4B). In the dizocilpine-treated mice, the enhanced CaMKIIα phosphorylation was inhibited

to the same level as that in the control treated with saline, whereas dizocilpine did not affect the phosphorylation in the control.

ERK: On the other hand, we observed no marked changes in ERK phosphorylation in both the hippocampus ($F_{3,19}$=1.58, P=0.23) and frontal cortex ($F_{3,19}$=0.68, P=0.58) of the four groups (Figure 3C and 4C).

Discussion

Here, we confirmed the effects of continuous PGBR feeding on learning and memory in mice in two independent behavioral tests. We also investigated the implication of the glutamatergic system in PGBR-induced enhancement of learning and memory in the brain.

As expected from our previous study [1], the PGBR-pellet-fed mice showed a more enhanced learning and memory in the two tests than in the control-pellet-fed mice. Therefore, to investigate whether the NMDA receptor is involved in those enhancements, we injected

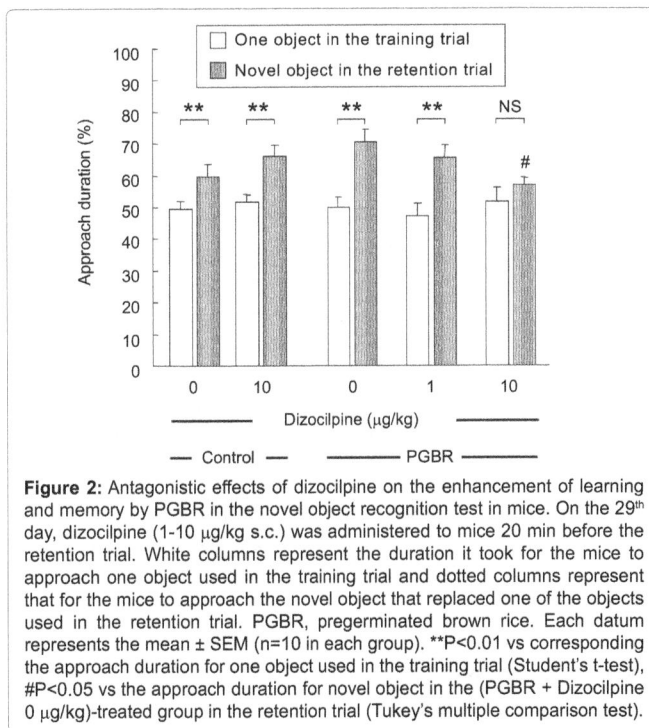

Figure 2: Antagonistic effects of dizocilpine on the enhancement of learning and memory by PGBR in the novel object recognition test in mice. On the 29th day, dizocilpine (1-10 µg/kg s.c.) was administered to mice 20 min before the retention trial. White columns represent the duration it took for the mice to approach one object used in the training trial and dotted columns represent that for the mice to approach the novel object that replaced one of the objects used in the retention trial. PGBR, pregerminated brown rice. Each datum represents the mean ± SEM (n=10 in each group). **P<0.01 vs corresponding the approach duration for one object used in the training trial (Student's t-test), #P<0.05 vs the approach duration for novel object in the (PGBR + Dizocilpine 0 µg/kg)-treated group in the retention trial (Tukey's multiple comparison test).

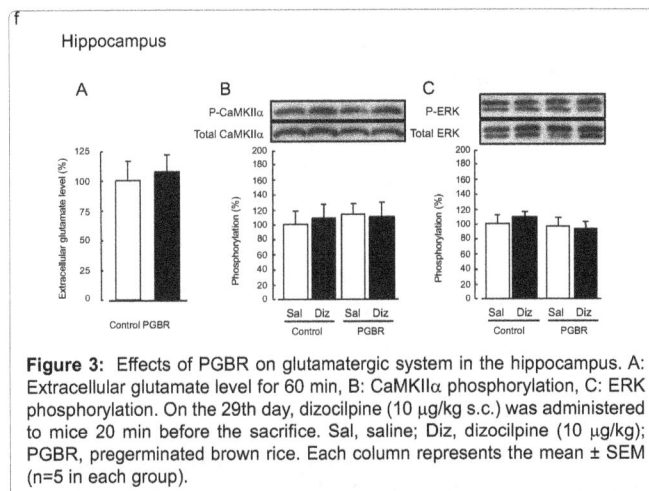

Figure 3: Effects of PGBR on glutamatergic system in the hippocampus. A: Extracellular glutamate level for 60 min, B: CaMKIIα phosphorylation, C: ERK phosphorylation. On the 29th day, dizocilpine (10 µg/kg s.c.) was administered to mice 20 min before the sacrifice. Sal, saline; Diz, dizocilpine (10 µg/kg); PGBR, pregerminated brown rice. Each column represents the mean ± SEM (n=5 in each group).

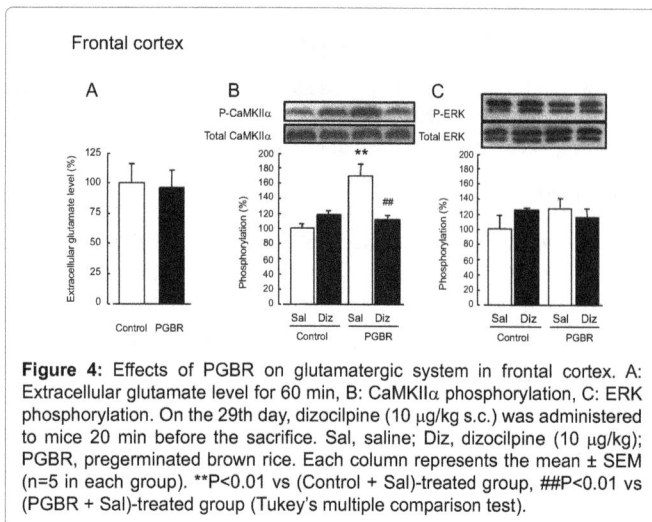

Figure 4: Effects of PGBR on glutamatergic system in frontal cortex. A: Extracellular glutamate level for 60 min, B: CaMKIIα phosphorylation, C: ERK phosphorylation. On the 29th day, dizocilpine (10 μg/kg s.c.) was administered to mice 20 min before the sacrifice. Sal, saline; Diz, dizocilpine (10 μg/kg); PGBR, pregerminated brown rice. Each column represents the mean ± SEM (n=5 in each group). **P<0.01 vs (Control + Sal)-treated group, ##P<0.01 vs (PGBR + Sal)-treated group (Tukey's multiple comparison test).

dizocilpine, an NMDA receptor antagonist, to the mice and subjected them to the Y-maze and novel object recognition tests. Dizocilpine (10 μg/kg) antagonized these enhancements in the PGBR-pellet-fed mice without exerting any abnormal behaviors by itself in the control. Taken together, it is possible that the PGBR-enhanced learning and memory in both tests may be accompanied by an increase in the activity of the endogenous glutamatergic system. Therefore, we assessed the glutamate release by *in vivo* brain microdialysis and the glutamate content in the frontal cortex and hippocampus. In both regions, there was no significant difference between the control- and PGBR-fed groups, suggesting that the synthesized and/or stored glutamate in presynapses and presynaptic releasing function may be similar in both groups and that postsynaptic changes may occur in the PGBR-fed group.

Accordingly, we examined the effects of PGBR on NMDA receptor function and signal transduction through the NMDA receptor. The two potential targets for immediate activation mediated by the NMDA receptor are CaMKII [15,16] and ERK [7,17]. Therefore, we measured CaMKIIα and ERK phosphorylation to assess the postsynaptic NMDA receptor function. CaMKIIα phosphorylation in the frontal cortex was markedly potentiated in the PGBR-fed mice compared with the control, but not ERK phosphorylation. Dizocilpine inhibited this increased phosphorylation in the PGBR-fed mice. In the hippocampus, we could not find any significant difference in the phosphorylation of CaMKIIα or ERK between the two groups. On the basis of the behavioral and biochemical results, it is likely that the NMDA receptor/CaMKII cascade, but not the NMDA receptor/ERK cascade, may be enhanced functionally in the frontal cortex of PGBR-fed mice.

Previously, we reported that continuous intake of PGBR also increases the serotonin content in the learned helplessness paradigm [18]. It has been reviewed that one of main energy sources in PGBR, glucose, modifies the brain function [19]. In addition to the function of PGBR in the brain, it decreases blood cholesterol level in rats [20], and it has been clarified that acylated steryl glucosides function as bioactive constituents in PGBR [21]. As our next step, we will try to examine the roles of PGBR constituents in the brain.

We conclude that PGBR has the potential to enhance the learning and memory capabilities; in particular, it activates the glutamatergic

system in the frontal cortex. Taken together, these results suggest that PGBR activates the NMDA receptor/CaMKIIα cascade in the brain, which leads to the enhancement of learning and memory ability in mice.

Acknowledgement

This work was supported by JSPS KAKENHI (Grant Numbers 24590304 and 22790233); the Nakatomi Foundation; the Takeda Science Foundation; the Sasakawa Scientific Research Grant from the Japan Science Society; and the Aichi Health Promotion Foundation. We are thankful to the late Professor Makoto Ukai, and Messrs. Keita Onogi, Yuya Hasegawa, Yoshiaki Kawai and Takamasa Asanuma for help in the experiment.

Conflict of Interest

The authors declare no conflict of interest.

References

1. Mamiya T, Asanuma T, Kise M, Ito Y, Mizukuchi A, et al. (2004) Effects of pre-germinated brown rice on beta-amyloid protein-induced learning and memory deficits in mice. Biol Pharm Bull 27: 1041-1045.

2. Riedel G, Platt B, Micheau J (2003) Glutamate receptor function in learning and memory. Behav Brain Res 140: 1-47.

3. Ahlander M, Misane I, Schött PA, Ogren SO (1999) A behavioral analysis of the spatial learning deficit induced by the NMDA receptor antagonist MK-801 (dizocilpine) in the rat. Neuropsychopharmacology 21: 414-426.

4. Sakimura K, Kutsuwada T, Ito I, Manabe T, Takayama C, et al. (1995) Reduced hippocampal LTP and spatial learning in mice lacking NMDA receptor epsilon 1 subunit. Nature 373: 151-155.

5. Tang YP, Shimizu E, Dube GR, Rampon C, Kerchner GA, et al. (1999) Genetic enhancement of learning and memory in mice. Nature 401: 63-69.

6. Atkins CM, Selcher JC, Petraitis JJ, Trzaskos JM, Sweatt JD (1998) The MAPK cascade is required for mammalian associative learning. Nat Neurosci 1: 602-609.

7. Enomoto T, Noda Y, Mouri A, Shin EJ, Wang D, et al. (2005) Long-lasting impairment of associative learning is correlated with a dysfunction of N-methyl-D-aspartate-extracellular signaling-regulated kinase signaling in mice after withdrawal from repeated administration of phencyclidine. Mol Pharmacol 68: 1765-1774.

8. Kornhauser JM, Greenberg ME (1997) A kinase to remember: dual roles for MAP kinase in long-term memory. Neuron 18: 839-842.

9. Mamiya T, Ukai M, Morikawa K, Kise M (2013) Intake of food pellets containing pre-germinated brown rice alleviates cognitive deficits caused by ß-amyloid peptide25-35 in mice: Implication of lipid peroxidation. J Rice Res 1: 116.

10. Maurice T, Hiramatsu M, Itoh J, Kameyama T, Hasegawa T, et al. (1994) Behavioral evidence for a modulating role of sigma ligands in memory processes. I. Attenuation of dizocilpine (MK-801)-induced amnesia. Brain Res 647: 44-56.

11. Mamiya T, Kise M, Morikawa K (2008) Ferulic acid attenuated cognitive deficits and increase in carbonyl proteins induced by buthionine-sulfoximine in mice. Neurosci Lett 430: 115-118.

12. VanderHorst VG, Ulfhake B (2006) The organization of the brainstem and spinal cord of the mouse: relationships between monoaminergic, cholinergic, and spinal projection systems. J Chem Neuroanat 31: 2-36.

13. Mamiya T, Noda Y, Nishi M, Takeshima H, Nabeshima T (1998) Enhancement of spatial attention in nociceptin/orphanin FQ receptor-knockout mice. Brain Res 783: 236-240.

14. Mamiya T, Yamada K, Miyamoto Y, König N, Watanabe Y, et al. (2003) Neuronal mechanism of nociceptin-induced modulation of learning and memory: involvement of N-methyl-D-aspartate receptors. Mol Psychiatry 8: 752-765.

15. Ehlers MD, Zhang S, Bernhadt JP, Huganir RL (1996) Inactivation of NMDA receptors by direct interaction of calmodulin with the NR1 subunit. Cell 84: 745-755.

16. Giese KP, Fedorov NB, Filipkowski RK, Silva AJ (1998) Autophosphorylation at Thr286 of the alpha calcium-calmodulin kinase II in LTP and learning. Science 279: 870-873.

17. Kennedy MB (1997) The postsynaptic density at glutamatergic synapses. Trends Neurosci 20: 264-268.

18. Mamiya T, Kise M, Morikawa K, Aoto H, Ukai M, et al. (2007) Effects of pre-germinated brown rice on depression-like behavior in mice. Pharmacol Biochem Behav 86: 62-67.

19. Messier C (2004) Glucose improvement of memory: a review. Eur J Pharmacol 490: 33-57.

20. Roohinejad S, Omidizadeh A, Mirhosseini H, Saari N, Mustafa S, et al. (2010) Effect of pre-germination time of brown rice on serum cholesterol levels of hypercholesterolaemic rats. J Sci Food Agric 90: 245-251.

21. Usuki S, Ariga T, Dasgupta S, Kasama T, Morikawa K, et al. (2008) Structural analysis of novel bioactive acylated steryl glucosides in pre-germinated brown rice bran. J Lipid Res 49: 2188-2196.

24

Price Variability, Co-integration and Exogeniety in the Market for Locally Produced Rice: A Case Study of Southwest Zone of Nigeria

Mafimisebi TE[1]*, Agunbiade BO[1] and Mafimisebi OE[2]

[1]*Department of Agricultural & Resource Economics, School of Agriculture & Agricultural Technology, Nigeria*
[2]*Department of Agricultural Technology, Rufus Giwa Polytechnic, Nigeria*

Abstract

Most studies on local rice in Nigeria were centred on increasing production, consumption or competitiveness with very few addressing the issue of marketing efficiency. Filling this gap requires a study on extent of pricing contacts in the market for local rice. In this study, secondary data consisting of urban monthly retail price series in the six southwest states of Nigeria were collected and analyzed. Analytical techniques used included Augmented Dickey Fuller (ADF), Johansen Co-integration and Granger Causality models. Empirical results indicated that growth in retail prices was highest in 2004 in Ogun Market (48.7%) and Ondo Market (45.4%) implying that local rice was more costly in these states. Retail prices were more volatile in Lagos Market (37.3%) and least volatile in Ogun Market (30.4%). The ADF test showed all price series were non-stationary at their levels but were stationary after first-difference. Pair-wise market integration model indicated that prices were co-integrated connoting high degree of marketing efficiency. The Multiple Co-integration model also indicated five co-integrating equations in six, validating the result of pair-wise market integration test. Granger causality model revealed that the supply-deficient markets in Lagos and Osun States were driving prices elsewhere. These results may have arisen from the storability of rice and closeness of the market locations examined. Despite high level of linkage, there is need for all stakeholders in the market to continue to effectively perform their roles so that economic benefits derivable from this scenario of strong pricing contacts can be fully realized and sustained.

Keywords: Rice; Local production; Price movements; Market linkage; Price leadership; Nigeria

Introduction

The status of rice in the average Nigerian diet has been transformed from being a luxury food item to that of a staple which is gradually taking part of the share formerly accounted for cassava and yam [1-3]. According to Akanji [4] and Akpokodge et al. [5], a combination of factors has triggered the structural increase in rice consumption. These include: rapid urbanization and ease of preparation that fits easily to the lifestyle of urban workers. Besides household's demand which keeps rising, there is an increase in fast food joints as a result of increasing urbanization. It is expected that the demand for rice will continue to increase [1,6-8]. Furthermore, as more women enter the workforce both in the formal and informal sectors, the opportunity cost of their time increases and convenience food such as rice; which can be prepared quickly, becomes the preferred choice [3,7].

Nigeria's annual rice demand is estimated at 5 million tonnes out of which only about 2.2 million tonnes is produced locally. The annual rice supply gap of about 2.8 million tonnes (or 56% of demand) is bridged by importation [2,8-10].

Over the years, most research efforts have been geared toward increasing local rice production to meet the quest for self-sufficiency in its production, make local rice compete favorably with imported rice, and also halt the excessive outflow of foreign exchange for importing rice by raising local rice consumption in Nigeria [2,7]. Sadly, little importance has been given to research and development of the country's rice marketing and distribution system to the extent it deserves. Only a few studies have been devoted to examining the competitiveness and efficiency of the local rice market in Nigeria. However, unless agricultural markets are integrated, producers and consumers will not realize the gains from trade liberalization, since the correct price signals will not be transmitted between and among contiguous market locations. The consequence of this is that, farmers will not be able to

specialize according to long-term comparative advantage [11,12]. Making the market and the distribution systems work better for farmers, processors and consumers is a continuing challenge [13] that should be adequately met through an expanded research programme.

The major objective of this paper therefore, is to examine price co-integration in the market for locally produced rice in South west, Nigeria. The specific objectives are to (i) determine the extent of variability in retail prices; (ii) determine the degree and extent of market integration among different spatial markets and (iii) identify markets exhibiting leadership positions in price formation and transmission.

Theoretical Framework

Co-integration is a statistical property possessed by some time series data that is defined by the concept of stationarity and order of integration of the series. It deals with relationship among a group of variables where (unconditionally) each has a unit root. It means that despite being individually non-stationary, a linear combination of two or more time series can be stationary.

A stationary series is one with a mean value which is time invariant. In contrast, a non-stationary series will exhibit a time varying mean. The order of integration of a series is given by the number of time the

*Corresponding author: Mafimisebi TE, Department of Agricultural & Resource Economics, School of Agriculture & Agricultural Technology, Nigeria
E-mail: temafis@yahoo.com

series must be differenced in order to produce a stationary series. A series generated by the first difference is integrated of order I denoted as I(1). Thus, if a time series is I(0), it is stationary; if it is I(1), then its change is stationary and its level is non-stationary.

The concept of co-integration and the method for estimating a co-integrated relation or system [14-17] provide a framework for estimating and testing for long run equilibrium relationships between or among non-stationary integrated variables. If two prices in spatially separated markets (or different levels of the supply chain) p_{1t} and p_{2t}, contain stochastic trends and are integrated of the same order, say I(d), the prices are said to be co-integrated if:

$$p_{1t} - \beta p_{2t} = \mu \text{ is } I(0)$$

β is referred to as the co-integrating vector (in the case of two variables, a scalar), whilst the equation p1t –βp2t is said to be the co-integrating regression. Co-integration implies that these prices move closely together in the long run, although in the short run they may drift apart, and this is consistent with the concept of market integration [18]. Co-integration analysis thus provides a powerful discriminating test for spurious correlation: conducting co-integration analysis between apparently correlated I (1) series and finding co-integration confirms the regression.

Several methods have been used to measure market integration. Advocated by Granger and Elliot (1967), simple bivariate correlation coefficient, also called the Law of One Price (LOP), has long been the most common measure used. Later, this method was strongly criticized, most notably by Ravallion [19]. Advances in time series econometrics led to the development of models that address some of the perceived weaknesses in the correlation coefficient approach. In this respect, Ravallion [19] proposed a dynamic model of spatial price differentials incorporating time lags.

One major drawback however remained. Both the LOP and Ravallion models test whether price changes in one market will be translated on a one-for-one basis to the other market, either instantaneously (LOP) or with lags (Ravallion model). But prices in different markets will only move on a one-for-one basis if the inter-market price differential is equal to transfer costs. Thus, price movements inside the band-with set by the transfer costs do not harm the hypothesis of market integration, whereas these models possibly reject this hypothesis.

Palakas and Harris-White [20], Alexander and Wyeth [21] extended Ravallion's model using co-integration and Granger causality ordinary least squares (OLS) techniques. This allowed testing for more general notions between markets and measures whether prices in two markets wander within a fixed band [11]. A limitation of these models, however, is that all models are in fact "static". Markets are either integrated or not. This requires the assumption of a constant market structure throughout the sample period. It implies that when observations for different sub-periods are limited, then doing integration analysis is not feasible.

Presently, the most common approach to test for market integration is the Johansen co-integration technique and Vector Error Correction Model applied among others by Mafimisebi [18], Rufino [22], Mohammad and Wim [23]. This paper used this approach.

Methodology

Sources of data and scope of study

The secondary data used for this study were sourced from National Bureau of Statistics (NBS), Nigeria. These are monthly retail prices of urban local rice markets in the six south western states of Nigeria which comprises Lagos, Oyo, Osun, Ogun, Ondo and Ekiti. The data covered from January 2001 to December 2010, giving a total of 120 data points per state.

Analytical procedure

The data analytical techniques that were used in this study comprised of descriptive statistical techniques and co-integration technique (Johansen co-integration test). The descriptive statistics that were used included frequency counts, means and co-efficient of variation. Augmented Dickey Fuller Tests (ADF) and Philip Perron (PP) tests were used for the stationarity tests. Johansen co-integration test was used to test for long run market integration between spatial markets that are stationary of the same order.

Mean spatial prices and variability index: Average monthly growth rate of prices for the whole period considered were computed as well as coefficients of variation (CV).

Test for order of econometric integration (unit root test): A stationary series is one with a mean value which will not vary with the sampling period. In contrast, a non-stationary series will exhibit a time varying mean [17]. Before examining integration relationships between or among variables, it is essential to test for unit root and identify the order of stationarity, denoted as I(0) or I(1). This is necessary to avoid spurious and misleading regression estimates. The framework of ADF methods is based on analysis of the following model

$$\Delta pt = \alpha + \beta p_{t-1} + yT + \sum_{k=1}^{n} \delta_k \Delta p_{t-k} + \mu_t \qquad (1)$$

Here, p_t is the rice price series being investigated for stationarity, Δ is first difference operator, T is time trend variable, μ_t represents zero-mean, serially uncorrelated, random disturbances, k is the lag length; α, β, γ and δ_k are the coefficient vectors. Unit root tests were conducted on the β parameters to determine whether or not each of the series is more closely identified as being I(1) or I(0) process. Test statistics is the t statistics for β. The test of the null hypothesis of equation (1) shows the existence of a unit root when $\beta=1$ against alternative hypothesis of no unit root when $\beta \neq 1$. The null hypothesis of non-stationarity is rejected when the absolute value of the test statistics is greater than the critical value. When p_t is non-stationary, it is then examined whether or not the first difference of p_t is stationary (i.e. to test $\Delta p_t . \Delta p_{t-1} \approx (1)$ by repeating the above procedure until the data were transformed to induce stationarity.

The Philips-Perron (PP) test is similar to the ADF test. PP test was conducted because the ADF test loses its power for sufficiently large values of "k", the number of lags [24]. It includes an automatic correction to the Dickey-Fuller process for auto-correlated residuals [25]. The regression is as follows:

$$\Delta p_t . \Delta p_{t-1} \Delta \ y_t = b_o + b_1 y_{t-1} + \mu_t \qquad (2)$$

Where y_t is the rice price series being investigated for stationarity, b_0 and b_1 are the coefficient vectors and μ_t is serially correlated.

5.2.3. Testing for Johansen co-integration (trace and eigenvalue tests): If two series are individually stationary at same order, the [16,17] model can be used to estimate the long run co-integrating vector using a Vector Auto regression (VAR) model of the form:

$$\Delta_{Pt=\alpha} \sum_{i=1}^{k-1} \Gamma i \Delta p_{t-1} + \Pi p_{t-1} + \mu_t \qquad (3)$$

Where p_t is a nx1vector containing the series of interest (rice price series) at time (t), Δ is the first difference operator.

Γi and Π are nxn matrix of parameters on the ith and kth lag of $p_{t,}$ $\Gamma i = \left(\Sigma_{i=1}^{k} A_i\right) - I_{g,}$ $\Pi = \left(\Sigma_{i=1}^{k} A_i\right) - I_g$, Ig is the identity matrix of dimension g, α is constant term, μ_t is nx1 white noise error vector. Throughout, p is restricted to be (at most) integrated of order one, denoted I(1), where I(j) variable requires jth differencing to make it stationary. Equation (2) tests the co-integrating relationship between stationary series. Johansen and Juselius [16] and Juselius [17] derived two maximum likelihood statistics for testing the rank of Π, and for identifying possible co-integration as the following equations show:

$$\lambda_{trace}(r) = -T\sum\nolimits_{i=r+1}^{m} In(1-\lambda) \tag{4}$$

$$\lambda_{trace}(r, r+1) = -T\mathrm{ln}\sum\nolimits_{i=r+1}^{m} In(1-\lambda) \tag{5}$$

Where r is the co-integration number of pair-wise vector, λ_i is ith eigenvalue of matrix Π. T is the number of observations. The λ_{trace} is not a dependent test, but a series of tests corresponding to different r -value. The λ_{max} tests each eigenvalue separately. The null hypothesis of the two statistical tests is that there is existence of r co-integration relations while the alternative hypothesis is that there is existence of more than r co-integration relations. This model was used to test for; (1) integration between pair-wise price series of local rice in the six markets, and (2) integration among the six rice price series in local rice market taken together as a unit.

Test for Granger-causality: After undertaking co-integration analysis of the long run linkages of the various market pairs, and having identified the market pair that are linked, an analysis of statistical causation was conducted. The causality test uses an error correction model (ECM) of the following form;

$$\Delta p_t^i = \beta_o + \beta_1 p^i(t-1) + \beta_2 p^j(t-1) +$$
$$\sum\nolimits_{k=1}^{m} \delta_k \Delta p^i(t-k) + \sum\nolimits_{h=1}^{n} \Delta \sigma \Delta_h \Delta p^j(t-h) + \mu_t$$

Where m and n are number of lags determined by Akaike Information Criterion (AIC).

If the null hypothesis that prices in market j do not Granger cause prices in market i is rejected (by a suitable F-test) that $\sigma_h = 0$ for h = 1, 2 ....n and β=0, this indicates that prices in market j Granger-cause prices in market i. If prices in i also Granger cause prices in j, then prices are said to be determined by a simultaneous feedback mechanism (SFM). This case is then referred to as bi-directional Granger causality. If the Granger-causality is in only one direction, it is called uni-directional Granger-causality and the market which exhibited sufficient strength to have Granger-caused the other is referred to as the exogenous market [12].

Results and Discussion

Descriptive analysis

Average growth rate in retail prices of local rice: Analysis of prices of local rice indicates that the growth in retail prices of local rice was highest in 2004 in most of the states in the region. Table 1 showed that growth was highest in Ogun (48.69%), followed by Ondo (45.36%) and Lagos (36.01%). This implied that local rice costs more in these states than in other states during that year. In 2010, growth in retail prices was negative in most states with the least growth occurring in Lagos (-11.91%), followed by Osun (-5.56%) and Ondo (-5.51%). This interpretation is that there was a fall in the price of the commodity in these states during that year. The average growth rate of retail prices for the entire period of observation was highest in Lagos (15.72%), followed by Ekiti (11.80%) and Osun (11.70%) but least in Ogun (10.49%).

The negative average growth rates in the retail prices of rice in most

Period	Lagos	Ogun	Ondo	Osun	Oyo	Ekiti
2002	34.90	7.95	4.84	7.41	-6.10	2.60
2003	9.70	14.78	7.02	23.27	21.36	19.05
2004	36.01	48.69	45.36	25.20	19.16	18.64
2005	26.47	-3.48	14.12	18.03	22.01	20.55
2006	-4.77	19.98	0.95	-5.16	0.70	11.70
2007	-5.00	-16.45	1.85	2.88	-5.25	-2.31
2008	39.74	16.82	6.12	33.88	39.15	12.06
2009	16.37	11.61	19.72	5.32	10.37	19.76
2010	-11.91	-5.51	-0.25	-5.56	0.09	4.15
Average	15.72	10.49	11.08	11.70	11.28	11.80

Source: Computed from National Bureau of Statistics (NBS) data

Table 1: Growth Rates in Retail Prices of Local Rice.

states in the year 2009 and 2010 possibly resulted from government policy on liberalization of rice imports to provide a short term solution to increasing rice prices in these years. If this negative growth rate in local rice prices continues, then the welfare of rice consumers may be secured at the expense of the self-sufficiency drive in rice production which can be achieved on the long run. This will result because local rice producers will experience relatively low price increases which may not yield sufficiently remunerative prices and profit incentive for continued production. This will serve as a disincentive to further investment in rice cultivation.

Another implication of negative growth rate is a further decline in the level of local rice output and its attendant increase in the prices of the product which may engender further increases in foreign rice importation and depletion of the foreign reserves of the country. The negative growth in rice prices observed in recent years could be a reflection of deliberate government policies toward securing cheap food items for its citizens. These findings concur with that of Akande and Akpokodje [1].

Variability in average retail price: Variability is one of the major attributes that explain the characteristics of most price data. This attribute has important implications for policy and the welfare of food consumers and a nation's economy. The degree of variability in the prices of rice is reflected in the coefficient of variation computed for local rice in the region (Table 2). The retail prices of local rice were more volatile in Lagos (37.29%) while the least price volatility was recorded in Ogun (30.41%). In general, the relatively low price variability in local rice implied that, all things being equal, consumers can effectively plan their expenditure with a fairly high degree of certainty that prices are not likely to substantially deviate from their prevailing levels. On the part of policy, this makes for effective planning of both production and consumption.

Order of econometric integration of local rice price series: The augmented Dickey-Fuller (ADF) test showed that all price series in the model were non-stationary at their level; this means that they all contained a unit root since the absolute values of their test statistics was less than their critical values at both 1% and 5% levels of significance. However, stationarity was reached after the first difference as shown in Table 3. As discussed in the methodology section, this means that all the price series were integrated of order one I(1), a requirement for Johansen's co-integration analysis [16,17].

To bolster our findings concerning the I (1) and I (0) nature of the price series at their level and their first difference, respectively, the Phillip-Perron (PP) test was also conducted. The PP test, like the ADF test, indicated significance for all variables, rejecting the null hypothesis of stationarity at the 1% and 5% levels of significance. The findings here

State	Coefficient of Variation (%)
Ekiti	33.81
Lagos	37.29
Ogun	30.41
Ondo	30.47
Osun	32.68
Oyo	33.64

Source: Computed from National Bureau of Statistics (NBS) Data

Table 2: Coefficient of Variation in Retail Prices of Local Rice.

Variables (Market Price Series)	Price Level 1(0)		First Difference 1(1)	
	ADF Statistics	PP Statistics	ADF Statistics	PP Statistics
Ekiti	-1.4409	-1.6042	-11.5337	-19.2139
Lagos	-1.0567	-1.7997	-8.8527	-41.2822
Ogun	-1.3519	-1.5975	-18.9991	-30.5026
Ondo	-1.5214	-1.6406	-11.6777	-18.4598
Osun	-1.2456	-2.2268	-8.9768	-43.6053
Oyo	-0.9827	-0.9979	-14.7398	-14.7208

Source: Compiled from results of stationarity test
Notes: Critical values are -3.4870 and -3.4861 at the 99% and -2.8859 and -2.8861 at the 95% Confidence levels for price level and first difference series, respectively.
If the absolute value of the ADF or PP is lower than their critical statistics, we fail to reject the null hypothesis of non-stationarity

Table 3: Results of Econometric Test of Price Series.

concur with earlier findings and conclusion by previous authors that food commodity price series are mostly stationary of order one i.e. I (1) [12,18,26-28]. According to Mafimisebi [18], the result is probably explained by the fact that most food price series contain trends arising from inflation, thus causing them to exhibit mean non-stationarity.

Long-run integration test results: The co-integration test result is presented in Table 4. The results indicate price co-integration in all the market pairs at both 1% and 5% levels of significance. Since the test statistics was greater than the critical value for all the market pairs, we reject the null hypothesis in favour of the alternative for both the maximal Eigen value and trace tests. Thus, it can be said that 100% of the markets for local rice in the Southwest Nigeria were strongly linked together in the long-run despite a potential short-run divergence among them. The implication of this is that there is high degree of marketing efficiency in local rice marketing in the region since market integration is a proxy for marketing efficiency [12,27,29].

Multiple co-integrations in local rice market: The result of Johansen's multiple co-integration model for local rice price series is displayed in Table 5. Both the Trace tests and Maximum eigenvalue statistics indicated five (5) co-integrating equations at the 5% level, meaning that there are five co-integrating relationships (out of six) existing in the local rice markets in the region. This, according to Johansen procedure means that there are five linear combinations that exist among the variables over the entire period of study. This result validates and strengthens the findings of pair-wise markets co-integration tests earlier reported. The overall economic implication of the result is that, local rice markets in Southwest, Nigeria, were strongly linked together thus suggesting a stable long-run equilibrium.

Exogeneity in local rice market price series: The result of pair-wise Granger causality test is shown in Table 6. Twenty (20) market pairs out of the 30 tested rejected the null hypothesis of no causality. Ten (10) market links of the 20 displayed bi-directional (two-way) Granger causality. The remaining 10 exhibited uni-directional (one-

way) causality. In the fifth (5th) market link, Lagos was stronger as it Granger-caused Osun prices at 1% level while the latter Granger-caused the former at 5% level. Ekiti was also stronger in the twelfth market link as it Granger caused Ogun prices at 1% level of significance while Ogun prices Granger-caused Ekiti prices at 5% level. In the 2nd, 3rd, 8th, 9th, 16th and 17th market pairs, the markets shown in these links demonstrated equal strength as they Granger-caused themselves at 1% level. This means that they influenced one another in terms of price formation and transmission probably stemming from the proximity between these states which facilitated free flow of price information. It is also interesting to note that, apart from Lagos Market which exhibited bi-directional causality with Ondo and Osun Markets, it also displayed uni-directional causality with Ogun, Oyo and Ekiti Markets.

Worthy of note is the case of Osun that displayed bi-directional causality with Lagos and also exhibited uni-directional causality with Ondo. Based on these results, Lagos and Osun Markets were identified as occupying leadership positions in the local rice price formation and transmission processes in Southwest Nigeria. While Lagos leads price formation process in Ekiti, Ogun and Oyo Markets, Osun Market however leads prices in Ondo Market. The implication of these findings is that the local rice deficit markets of Lagos and Osun drive the market for local rice in the region. Since rice production statistics for Southwest Nigeria revealed that Lagos and Osun States have the lowest local rice production output in the region, this may mean that the forces of demand are stronger than that of supply in local rice price formation.

Market Pairs Pi-Pj	Trace Test Statistics	Maximal Eigenvalue Test Statistics
Lagos/Ogun	26.671**	25.056**
Lagos/Ondo	28.903**	27.540**
Lagos/Osun	29.791**	28.077**
Lagos/Oyo	24.077**	23.452**
Lagos/Ekiti	24.984**	23.594**
Ogun/Ondo	29.815**	27.545**
Ogun/Osun	29.576**	27.021**
Ogun/Oyo	23.925**	23.635**
Ogun/Ekiti	32.098**	29.579**
Ondo/Osun	21.876**	19.300**
Ondo/Oyo	20.716**	20.362**
Ondo/Ekiti	27.349**	25.562**
Osun/Oyo	41.043**	39.479**
Osun/Ekiti	22.765**	20.721**
Ekiti/Oyo	26.396**	26.175**

Source: Compiled from the result of Co-integration Test
* (**) means significant at 5% (1%) level of significance
The critical values for trace test and maximal eigenvalue test are 19.937 and 18.520 at 99%, and 15.495 and 14.265 at 95% level of significance, respectively

Table 4: Pair-wise Co-integration Test Result.

Null hypothesis	Trace Statistics	95% critical value	Maximum eigenvalue	95% critical value
r=0	932.71*	95.75	619.56*	40.08
r=1	313.15*	69.82	146.08*	33.88
r=2	167.06*	47.86	83.69*	27.58
r=3	83.37*	29.80	53.26*	21.13
r=4	30.11*	15.49	28.78*	14.26
r=5	1.33	3.84	1.33	3.84

Source: Compiled from the result of Co-integration Test
Both Trace and Maximum eigenvalue tests indicate 5 co-integrating equations at the 0.05 level of significance
* denotes rejection of the null hypothesis at the 0.05 level of significance

Table 5: Multiple Co-integration Results.

Null hypothesis	F-Statistics	Probability
Lagos→Ogun	9.039**	0.0002
Ondo→Lagos	4.706**	0.0109
Lagos→Ondo	8.643**	0.0003
Osun→Lagos	3.997*	0.0210
Lagos→Osun	6.569*	0.0020
Lagos→Oyo	9.708**	0.0001
Lagos→Ekiti	7.638**	0.0008
Ondo→Ogun	6.701**	0.0018
Ogun→Ondo	6.353**	0.0024
Osun→Ogun	10.867**	5.E-05
Oyo→Ogun	7.717**	0.0007
Ekiti→Ogun	6.243**	0.0027
Ogun→Ekiti	3.668*	0.0286
Osun→Ondo	5.763**	0.0041
Oyo→Ondo	4.647**	0.0115
Ekiti→Ondo	5.276**	0.0064
Ondo→Ekiti	4.898**	0.0091
Osun→Oyo	13.457**	6.E-06
Osun→Ekiti	5.941**	0.0035
Oyo→Ekiti	10.285**	8.E-05

Source: Compiled from the result of Granger-Causality Test
* (**) means significant at 5% (1%) level of significance
→ indicates direction of causality

Table 6: Pair-wise Granger-causality Test for Local Rice Markets.

Summary, Recommendations and Conclusion

Summary and recommendations

This study examined spatial price linkage in local rice markets in Southwest Nigeria. The trend analysis in retail prices of local rice showed that there was less fluctuations in the retail prices of the commodity over the period studied. The smaller values of the coefficient of variation provided more evidence to support this position. The negative growth rate in the retail prices of local rice observed in some years could be a reflection of deliberate government policies toward securing cheap food items for its citizens. The economic implication of this is that if growth in price maintains this trend, then the welfare of rice consumers in the study area may be secured. This however, will be at the expense of local rice producers, who will be experiencing relatively small increases in the prices of their product. This could bring about disincentive to further investment in rice farming activities, increase in price occasioned by reduction in local rice output and loss of foreign exchange to other countries from which the country shall be forced to import rice to meet the increasing shortfall.

The result of the stationary tests indicated that the price series for local rice exhibited stationary after first differencing. From the result of the study was discovered existence of a high level of spatial pair-wise integration in local rice markets across the six states. All market pairs for local rice had long-run price linkages. The multiple (Johansen) co-integration tests also largely confirmed these results. This implied that short-run deviations from equilibrium will be readily corrected through efficient price setting and transmission of price signals.

The Granger causality tests conducted on all inter-state market pairs showed that Lagos and Osun Markets, which are situated in states with low volume of local rice production, lead price formation and transmission processes in local rice marketing in the region.

Based on the results of the study, some important policy implications and recommendations for the rice industry emerge for the various

Governments in Southwest Nigeria. It is recommended that the problem of highly inefficient and fragmented distribution and transportation systems be addressed for the rice traders to take advantage of the high level of spatial market integration. Also, development of infrastructures in inter-state rice markets, government price support and other market-oriented policies should be pursued as they will achieve intended goals.

Conclusion

The existence of high level of integration in the markets for locally produced rice in the Southwest, Nigeria, has been discovered in the study. The results of the study did not support findings by past researchers of low agricultural commodity market integration in Nigeria attributed generally to fragmented distribution system and oftentimes inefficient transportation system. This may be as a result of the fact that rice is not a highly perishable agricultural commodity that stores up to one year before spoilage, if well dried. The closeness of the markets examined may also have been a factor in the strong linkage detected among these markets. It should be noted that the result may have been widely different if samples of rice markets have been taken across all the six zones in Nigeria. There is, however, the need for all the stakeholders in the Southwest rice market to continue to effectively perform their roles so that the economic benefits derivable from this very high level of integration can be fully realized and sustained.

References

1. Akande SO, Akpokodge G (2003) Rice Prices and Market Integration in Selected Areas in Nigeria. A Study Report, WARDA-NISER Collaborative Study.

2. Daramola AG (2005) Government Policies and Competitiveness of Nigerian Rice Economy. A Paper Presented at the Workshop on Rice Policy and Food Security in Sub-Saharan Africa Organized by WARDA, Republic of Benin.

3. Odusina OA (2008) Urban Rice Demand Analysis: A Case Study of Ijebu-Ode Township. Middle-East Journal of Scientific Research 3: 62-66

4. Akanji BO (1995) Hedonic-Price Analysis of the Demand for Grain Crops in Nigeria: The Case of Rice and Cowpea. Nigeria.

5. Akpokodge G, Lancon F, Erenstein O (2001) Nigeria's Rice Economy: State of the Art. A Paper Presented at the WARDA-NISER Nigerian Rice Economy Workshop, Ibadan, Pp 4-9.

6. FAO (2005) Marketing Integration, Price Transmission and Import Surges. FAO Import Surge Working Paper 6: 1-9.

7. Bamidele FS, Abayomi OO, Esther OA (2010) Economic Analysis of Rice Consumption Patterns in Nigeria. Journal of Agricultural Science and Technology. 12: 1-11

8. Odularu GO (2010) Rice Trade Policy Options in an Open Developing Economy: The Nigerian Case Study. Journal of Development and Agricultural Economics. 2: 166-177.

9. Rahji MA, Adewumi OM (2008) Market Supply Response and Demand for Local Rice in Nigeria: Implications for Self-Sufficiency Policy". Journal of Central European Agriculture. 9: 567-574.

10. Kassali R, Kareem RO, Oluwasola O, Ohaegbulam OM (2010) Analysis of Demand for Rice in Ile-Ife, Osun State, Nigeria. Journal of Sustainable Development in Africa. 12: 63-78.

11. Baulch B (1997) Transfer Costs, Spatial Arbitrage and Testing Food Market Integration. American Journal of Agricultural Economics. 79: 477-487.

12. Mafimisebi TE (2012) Spatial Equilibrium, Market Integration and Price Exogeniety in Dry Fish Marketing in Nigeria. Journal of Economics, Finance and Administrative Sciences. 17: 31-37.

13. Intal, Jr P, Rani LO (2001) Literature Review of the Agricultural Distribution Services Sector: Performance, Efficiency and Research Issues. PIDS Discussion Paper Series.

14. Engle RF, Granger CWJ (1987) Co-integration and Error Correction: Representation, Estimation and Testing. Econometrica. 55: 251-276

15. Johansen S (1988) A Statistical Analysis of Co-integration Vectors. Journal of Econometric Dynamics and Control, 12: 231-254.

16. Johansen S, Juselius K (1990) Maximum Likelihood Estimation and Inference on Co -integration with Application to the Demand for Money. Oxford Bulletin of Economic Statistics, 52: 231-254.

17. Juselius K (2006) The Co-integrated VAR Model: Methodology and Applications. Oxford University Press

18. Mafimisebi TE (2008) Long-run Price Integration in the Nigerian Fresh Fish Market: Implication for Marketing and Development. Delhi Business Review, 9: 55-67

19. Ravallion M (1986) Testing Market Integration. American Journal of Agricultural Economics, 68: 102-109

20. Palakas T, Harris White B (1993) Testing Market Integration: New Approaches with case Materials from the West Bengal Economy. J of Development Studies, 30: 1-57.

21. Alexander C, Wyeth J (1994) Co-integration: An Application to the Indonesian Rice Market. Journal of Development Studies, 30: 303-328

22. Rufino CO (2008) Inter-regional Integration of the Philippine Rice Market. DLSU-AKL Working Paper Series 2008-06.

23. Mohammad IH, Wim V (2010) Evaluation of Rice Markets Integration in Bangladesh. The Lahore Journal, 15: 77-96

24. Ghosh A, Saidi R, Johnson K (1999) Who Moves the Asia-Pacific Stock Markets, US or Japan? Empirical Evidence Based on Theory of Co-integration. The Financial Review, 34: 159-170

25. Brooks C (2008) Introductory Econometrics for Finance 2nd Edition. Cambridge, University Press.

26. Mafimisebi TE (20t01) Spatial Price Equilibrium and Fish Market Integration in Nigeria. University of Ibadan, Nigeria Pp201.

27. Okon RN, Egbon PC (2003) The Integration of Nigeria's Rural and Urban Food Stuff Markets. A Report of African Economic Research Consortium (AERC) Sponsored Research. Pp. 1-17.

28. Oladapo OO (2003) Market Integration for Pineapples in Nigeria. An Unpublished Ph.D Thesis, University of Agriculture, Nigeria.

29. Hopcraft P (1987) Grain Marketing Policies and Institutions in Africa. Finance and Development, 24: 37

25

High Frequency Embryogenic Callus Induction and Whole Plant Regeneration in *Japonica* Rice Cv. Kitaake

Saroj Kumar Sah*, Ajinder Kaur and Jagdeep Singh Sandhu

School of Agricultural Biotechnology, Punjab Agricultural University, India

Abstract

A new and rapid protocol for production of embryogenic callus and plant regeneration has been reported in *Oryza sativa cv.* Kitaake. Callus cultures were established from seeds of Kitaake on MS medium supplemented with 2,4-D (3.0 mg/l), BAP (0.25 mg/l) and proline (0.6 g/l) gelled with phytagel (3.0 g/l), after 9 days of seed culturing. The embryogenic calli were regenerated on regeneration medium, i.e. MS medium supplemented with BAP (3.0 mg/l) and NAA (0.2 mg/l) and gelled with agar (8.0 g/l) in combination with phytagel (2.0 g/l) and regeneration occurred in 18 days. The regeneration percentage achieved was very high i.e 82.66% on the above medium. Rooting was achieved on half strength MS medium. The plantlets were hardened and transferred to soil in earthen pots. The regeneration protocol so developed through callus formation was highly reproducible. The plants showed normal growth and flowering under glass house conditions as well as in field conditions.

Keywords: MS medium; Kitaake; Shoot regeneration

Introduction

World population is increasing day by day and by the year 2050, it is expected to reach 9.1 billion. In order to feed this population, food production needs to be increased by 70% of the present production and global agricultural production must be increased by 60-110% [1-3]. Current rate of grain production are not sufficient to meet the increasing food demand. Nearly 2.4% increase in grain yield per year is needed to double the crop production by 2050. But at current rates, approximately 60%, 42%, 38% and 55% increase in maize, rice, wheat and soybean production, respectively is possible by that time [4]. Increase in rice yield by 1.0% per year, may not result in any change in the per capita rice harvest by 2050. Rice, wheat, maize and soybean provide nearly 43% of global dietary energy and 40% of daily protein supply [4]. More than two billion people world-wide totally depend upon rice and in Asia rice is the staple food, where it provides 40-70% of the total food calories consumed [5]. In Afghanistan, India, Bangladesh, Laos, Vietnam and Cambodia, rice production is doubling, but significant reduction of yield is found in some local parts of India like Uttar Pradesh, Maharashtra and Tamilnadu and there is no significant change in per capita rice harvest in Pakistan, Nepal, Malaysia and South Korea. Rice provides approximately 30% and 27% of the dietary energy in India and China, respectively [4]. Due to abiotic stresses like drought, submergence tolerance and salinity, production of rice is decreased, therefore, the production of abiotic stress-tolerant rice cultivars is the main priority [6]. Since conventional breeders have been trying to improve the quality and quantity of rice for many years, but they have not been able to solve these issues. With the help of biotechnological techniques the quality and quantity of rice can be improved. Many techniques of biotechnology are directed towards the improvement of conventional plant breeding processes, such as introduction of novel genes by genetic transformation, protoplast fusion for the production of male-sterile lines, production of haploids for attaining rapid homozygosity and somaclonal variation for introducing trait variability, plastid engineering [7-9]. For successful carrying out genetic transformation in rice, establishment of efficient plant regeneration *in vitro* is a pre-requisite [10-12]. Till now, different protocols have been developed to initiate callus from explants, such as mature embryos [13-18], immature embryos [19,20-22], mature seeds [22-25], root segments [15,26], coleoptile [27,28]

and leaf bases [9]. Genetic engineering is strongly dependent on genotype and availability of an efficient *in vitro* plant regeneration method. In general, because of poor regeneration abilities, cultivars are recalcitrant to various biotechnological advances [29]. Identification and screening of useful cultivars for embryogenic callus formation and subsequent plant regeneration *in vitro* are the key steps in rice genetic improvement program by using biotechnological application [30]. Genotype and nutrient media are two of the most important factors which affect callus induction and subsequent plant regeneration. Keeping in view the above facts, an attempt was made to develop an easy, rapid and highly efficient regeneration protocol through high frequency embryogenic callus induction using mature seeds as explants and found its successful applicability to japonica rice (cv. Kitaake). The cultivar Kitaake was selected for standardization of various factors critical for enhanced transformation and regeneration as it has a very short life cycle (9 weeks) and is widely used in genetic transformation [31]. Kitaake can be efficiently transformed by *A. tumefaciens-mediated* T-DNA approach. Jung et al. [32] have generated several thousand T-DNA insertional mutants and overexpressed or silenced several hundred genes in Kitaake.

In the present study, different components of the regeneration medium viz., growth hormones and gelling agents were standardized. The modified regeneration medium triggered production of a large number of shoots from a small number of calli and also promoted their fast growth, and hence has an edge over the existing protocols where the regeneration step requires maximum time. Using this protocol, significantly high regeneration frequency (up to 82%) was achieved in the tested cultivar Kitaake.

*Corresponding author: Saroj Kumar Sah, School of Agricultural Biotechnology, Punjab Agricultural University, India, E-mail: saroj-biotec@ pau.edu, saroj1021@gmail.com

This speedy, yet less labour intensive protocol overcomes major limitations associated with genetic manipulation in rice. Moreover, our protocol uses mature seeds as explants, which can easily be obtained in large quantity throughout the year and kept viable for a long time. Such an easy, efficient and generalized large protocol has the potential to be used for crop improvement and gene function studies on the model monocot plant rice.

Materials and Methods

Plant material

The experiment was conducted on *japonica* rice cv. Kitaake, the seeds of which were procured from the Department of Biotechnology, National Institute of Agrobiological Resources, Japan. The research work was carried out at the Plant Tissue Culture and Genetic Transformation Laboratories, G.S. Khush Labs, School of Agricultural Biotechnology, Punjab Agricultural University, Ludhiana, India during 2010-2013.

Methodology

Mature dry seeds of Kitaake (Figure 1A) were used and *in vitro* regeneration protocol was standardized following somatic embryogenesis mode. For this purpose, healthy dehusked seeds of Kitaake (Figure 1B) were surface sterilized with 70% ethanol (v/v) for 1 min, followed by 10-15 mins sterilization in mercuric chloride (0.1%). Seeds were then washed 5-6 times with sterile distilled water and dried on autoclaved Whatman paper (3mm) for 3-5 min. For callus induction, eight to ten seeds were inoculated per petriplate on callus induction medium (CIM) and incubated at 25 + 1°C in dark. .

Tissue culture media

Different media components viz., carbohydrate source, agar concentration and combination of phytagel, agar and hormones were optimized for rice callus induction and regeneration. Sucrose or maltose at a concentration of 3%, 4% or 5% was used as carbohydrate source in the callus induction and regeneration media. The media were solidified with 0.8%, 1% or 1.2% (w/v) agar. The callus induction and regeneration media used were supplemented with proline (600 mg/l). The callus induction media was also supplemented with 2,4-D (3.0 mg/l) and BAP(0.5 mg/l). The regeneration medium comprised of NAA (0.2 mg/l) and BAP (3.0 mg/l). After evaluating each of the agents mentioned above singly, the combined effect of these ingredients was also evaluated in a series of experiments and data were taken on days to callus formation, callus induction frequency (%), efficiency

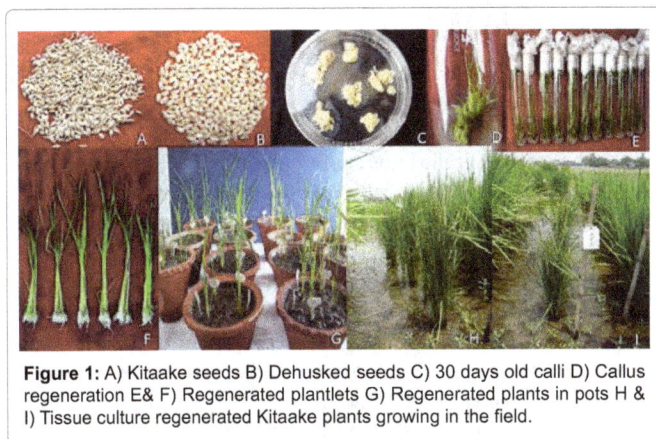

Figure 1: A) Kitaake seeds B) Dehusked seeds C) 30 days old calli D) Callus regeneration E& F) Regenerated plantlets G) Regenerated plants in pots H & I) Tissue culture regenerated Kitaake plants growing in the field.

of embryogenic calli formation efficiency (%), days to green bud formation and regeneration frequency (%).

Non embryogenic calli were discarded and only embryogenic calli were selected. The embryogenic calli were cut into approximately 3 equal pieces and subcultured again onto fresh CIM and kept in dark. Calls induction frequency (%) was calculated as follows:

$$\frac{\text{Number of seeds exhibiting callus formation}}{\text{Number of seeds cultured}} \times 100$$

The frequency of developing embryogenic calli was calculated as follows:

$$\frac{\text{Number of embryogenic calli}}{\text{Total number of calli}} \times 100\%$$

After a week of subculture, pieces of callus (2-4 mm in diameter) were placed in a test tube containing regeneration medium. Regeneration frequency (%) was calculated as follows:

$$\frac{\text{Number of regenerated calli}}{\text{Number of calli incubated}} \times 100$$

For the development of roots, the regenerated shoots were shifted to jam jars containing rooting medium (comprising half strength of MS salts (½ MS), 20 g/l sucrose and 5 g/l agar, pH 5.8) and maintained at 25 + 10°C in light for 10 days.

Statistical analysis

Statistical analysis was performed by Completely Randomized Design (CRD) with the help of CPCS1 (a computer program package for the analysis of commonly used experimental design) [33]. All the data were subjected to analysis of variance (ANOVA). CD values at 5% level of significance were calculated to compare mean values and interpretations were done accordingly.

Results and Discussion

For conducting genetic transformation of any crop, a rapid and robust tissue culture system is a prerequisite. More than 5 month old calli start losing their regeneration capacity, so calli age is an important factor that influence callus regeneration. Thus, selecting the most suitable medium to improve the quality of calli is an important step towards callus regeneration. Regeneration frequency also depends on genotype and its interaction with culture conditions [34]. Influence of genotype on plant regeneration has been observed in several rice cultivars belonging to *japonica, indica and japonica-indica* hybrids [15]. Some of these showed high capacity for plant regeneration while others did not. Abe and Futsuhara [16] reported that callus formation from mature seeds, and subsequent regeneration in 60 different *japonica, indica and javanica* cultivars indicated both intra and inter-varietal differences. Genetic markers have been used to find genes involved in the induction of embryogenic calli [35-37], but these genes have not yet been isolated [34]. In addition to genotype, various factors including physiological and developmental status of the explant, composition and concentration of the basal salts, organic components and plant growth regulators in the culture medium are known to influence callus induction and plant regeneration ability in rice [38]. The first success in regeneration of plants from callus, derived from mature seeds of rice, was obtained by Nishi *et al* [39]. Numerous explants have since been used by different workers for callus induction and plant regeneration like leaf sheath [40], anthers or pollen [41,42], immature embryos [43], immature panicles [44,45], roots [46], young inflorescences [47], leaf bases [48] and mature embryos [34]. Among these explants, immature embryos and mature seeds are most commonly used for callus induction and subsequent studies.

The frequency of plant regeneration depends upon the type of callus and various reports have described the nature and properties of highly regenerable calli, also termed as embryogenic calli. Some of the factors that affect plant regeneration efficiency from callus are the concentrations of gelling agents, osmoticum and different combinations of plant growth regulators [49]. Kavi-Kishor [50] reported that osmolarity of both growth and regeneration media was important for obtaining highly regenerable rice callus and retaining its regeneration potential. It has been observed that low water content of callus cultured on a medium containing mannitol and a high concentration of agar were key factors for efficient regeneration of plants from callus [51].

Therefore, an effort was made to obtain high frequency of regeneration using mature seed derived embryogenic callus in case of *japonica* rice cv Kitaake. The results reported here suggested that callus induction and regeneration in Kitaake were influenced by the type and concentration of carbon source.

Effect of different concentrations of sucrose and maltose was studied on different events like days to callus formation, callus induction, embryogenic calli formation, days to green bud formation and regeneration frequency in cv. Kitaake, and significant differences among different events were found. (Table 1) Supplementation of culture medium with 4% maltose showed best results for days to callus formation (~9 days), per cent callus induction (90.33%), embryogenic callus formation (57.66%) and shoot regeneration (82.66 %). For days to green bud formation, 5% maltose showed the best result however, the regeneration percentage was lesser than 4% maltose. Upon comparing 4% maltose with the other sugar concentrations it was found to produce more significant results in respect to all the events mentioned in the Table 1 except days to green bud formation. The highest amount of callus induction (90.33 %) and regeneration (82.66%) was achieved with 4% maltose. Media containing 3% maltose yielded 61.66 % callus induction and 31.66% regeneration. The presence of maltose (4%) instead of sucrose in the media also lowered the number of days to callus formation in Kitaake.

Previous reports have also found maltose to be a better carbon source for regeneration [52-54]. According to Lentini et al. [55], maltose has been described as an agent that may regulate osmotic potential of cellular environment of callus. Since sucrose promotes *in vitro* production of ethylene in excised tissues causing the browning of callus, the substitution of maltose for sucrose may help to protect the calli by reducing the production of ethylene. Similar results that maltose is better than sucrose as the carbohydrate source in both the

Event	Treatment means					CD (5%)
	Sucrose (%)			Maltose (%)		
	3	4	5	3	4	5	
Days to callus formation	10.66	9.66	12.33	11.66	8.66	11.33	1.027
Callus induction (%)	54.00	68.66	70.33	61.66	90.33	90.33	2.78
Embryogenic calli formation (%)	21.00	30.00	33.33	52.00	57.66	46.00	3.30
Days to green bud formation	17.33	15.66	15.33	17.66	17.66	14.66	1.03
Regeneration frequency (%)	20.33	29.66	29.00	31.66	82.66	50.33	4.39

Callus induction medium: [MS medium supplemented with 2,4-D (3.0 mg/l) + BAP (0.25 mg/l) + proline (0.6 g/l) + sucrose or maltose (i.e. 3 %, 4%, 5%)]+ phytagel 3 (g/l)

Regeneration medium: [MS medium + BAP (3.0 mg/l)+ NAA (0.2 mg/l)+ phytagel (2g/l)+ agar (8g/l)]

Table 1: Effect of different carbohydrates (sucrose and maltose) on callus induction and regeneration in japonica rice cv. Kitaake.

Event	Treatment means of agar concentration (%)			CD (5%)
	0.8	1.0	1.2	
Days to callus formation	11.65	8.67	6.67	1.15
Callus induction (%)	46.51	55.94	57.18	0.69
Embryogenic calli (%)	35.85	47.00	42.11	1.28
Days to green bud formation	19.67	17.67	17.67	1.15
Regeneration frequency (%)	32.14	45.55	38.04	2.72

Callus induction medium: [MS medium +2,4-D (3.0 mg/l) + BAP (0.25mg/l) + proline (0.6g/l) + agar (0.8%, 1, 1.2%)]

Regeneration medium: [MS medium + BAP (3.0 mg/l)+ NAA (0.2 mg/l)+ agar (0.8, 1.0 and 1.2)

Table 2: Effect of agar concentrations (0.8%, 1%, 1.2%) on callus induction and regeneration in japonica rice cv. Kitaake.

Gelling agent (%)	Treatment means	
	Callus induction	Regeneration
0.8 % agar	52.67 (46.51)	28.33 (32.14)
1.0 % agar	68.67 (55.94)	51.00 (45.56)
1.2 % agar	70.67 (57.12)	38.00 (38.04)
0.3% phytagel	92.00 (72.73)	60.54 (58.85)
0.8 % agar+ 0.2% Phytagel	87.00 (68.85)	90.00(71.83)
CD (p=0.05)	2.51	4.67

*The values in parentheses are arc sine transformed values

Callus induction medium: [MS +2,4-D (3.0 mg/l) + BAP (0.25 mg/l)+ proline (0.6 g/l)+ maltose (40 g/l) + agar (0.8, 1.0 , 1.2 %) , phytagel (0.3%) or agar in combination with phytagel

Regeneration medium: [MS+BAP (3.0mg/l) + NAA (0.2 mg/l)+ phytagel (2 g/l)+agar (8g/l)

Table 3: Effect of different concentrations of agar and phytagel on callus induction and regeneration in japonica rice cv. Kitaake.

subculture and differentiation media were obtained by Lin et al. [56]. It is likely that the two sugars exhibit different bioavailability to rice under culture condition.

Another factor enhancing the callus induction and regeneration in Kitaake is the gelling agent. Agar concentration in the callus induction and regeneration media was also important in enhancing the frequencies of callus induction, embryogenic calli and regeneration. Of different levels of agar (0.8%, 1% and 1.2%) evaluated, the medium containing 1% agar was the most favourable and 0.8% proved to be least favourable as shown in Table 2. While increasing the concentration of agar it increased the callus induction and also affected the regeneration as well. Use of 1% agar led to formation of 68.66% callus induction and subsequent regeneration of 46.77%. The use of 1.2% agar led to 57.18% callus induction and 38.04% regeneration.

Use of phytagel in combination with agar leads to increase in embryogenic calli frequency and regeneration frequency (71.83%). Effect of different concentrations of agar (0.8, 1.0 and 1.2%), phytagel (0.3%) and combination of agar and phytagel (0.8% agar + 0.2% phytagel) on percent callus was studied. Among these phytagel (0.3%) showed statistically higher callus induction frequency (72.73%) as compared to other gelling agents. The callus induction frequency also increased upon increasing the concentration of agar from 0.8 to 1.2 %, but to a lesser extent as compared to phytagel (0.3%; Table 3). For regeneration, 1.0% agar showed statistically better shoot induction (45.56%) as compared to other agar concentrations. Upon increasing the concentration of agar from 0.8 to 1.0 %, the regeneration percentage increased, but decreased at 1.2%. In combination agar and phytagel (0.8% agar + 0.2% phytagel) showed significantly better regeneration

(71.83 %) in Kitaake amongst all treatments (Table 3). Similar results that agar concentration influences callus induction and regeneration has been reported by Zaidi et al. [52]. According to Lee et al. [57], agar influences shoot regeneration by regulating the humidity of *in vitro* culture conditions. Agar contains agropectins and some other organic impurities that might have inhibitory effect on explant growth and callus proliferation. Phytagel has been reported to be free from such impurities [58]. Gelrite was found to enhance callus induction and *in vitro* regeneration in Swarna (89.9%) and Mahsuri (93.4%) rice varieties [52]. Although use of agar as gelling agent to promote *in vitro* regeneration of rice has been reported [59-61,57], but Saharan et al. [62] and Mensens et al. [63] have reported higher regeneration frequency in *indica* rice varieties on media solidified with gelrite. Lin and Zhang [56] also have reported that 0.3% phytagel give higher callus induction. The addition of cytokinin (BAP) and auxin (NAA) to regeneration medium enhanced the frequency of shoot regeneration from callus cultures [24,25]. Use of casein hydrolysate was found to be beneficial for generation of embryogenic calli in *japonica* rice [62,64,65]. Use of proline in the medium has also been reported to be effective for the initiation and maintenance of embryogenic calli [61,66].

The combined action of all the supplements led to higher efficiency of callus induction, formation of embryogenic calli and regeneration. For the callus induction, it is proposed that the best media among all the tested combinations is MS media supplemented with 40 g/l maltose, 0.3 g/l casein hydrolysate, 0.6 g/l proline, 3.0 mg/l 2,4-D, 0.25 mg/l BAP gelled with 3.0 g/l phytagel (Figure 1C). MS media supplemented with 30 g/l sucrose, 3 mg/l BAP, 0.2 mg/l NAA gelled with 8 g agar/l and 2 g/l phytagel (Figure 1D) is proposed and for rooting, half strength MS supplemented with 20 g/l sucrose and 5g/l agar was found to be highly responsive (Figures 1E and F). The plants were transferred to the pots in the green house (Figure 1G) after hardening and were transplanted in the field after 10 days (Figure 1H and I).

Thus in the present study, it was observed that the callus induction and regeneration was greatly affected by the carbon source and gelling agent. By using this protocol, a total of 49 transgenic lines of Kitaake using biolistic approach was developed. Thus, protocol reported here might help in crop improvement strategies.

Acknowledgement

The authors are thankful to Dr. Kuldeep Singh, Director, School of Agricultural Biotechnology, Punjab Agricultural University (PAU) for providing the necessary research facilities, and Dr. Darshan Singh Brar, Honorary Adjunct Professor, School of Agricultural Biotechnology, PAU for valuable suggestions during the entire work. We are also thankful to Department of Biotechnology, National Institute of Agrobiological Resources, Japan for providing rice seeds of Kitaake. Saroj Kumar Sah acknowledges Indian Council for Cultural Relations and Ministry of Foreign Affairs, New Delhi, India for providing fellowship during M.Sc. programme.

References

1. Tilman D, Balzer C, Hill J, Befort BL (2011) Global food demand and the sustainable intensification of agriculture. Proc Natl Acad Sci U S A 108: 20260-20264.

2. FAO (2009) Global agriculture towards 2050. Rome, FAO.

3. OECD/FAO (2012) OECD-FAO Agriculture outlook 2012-2021, OECD Publishing and FAO.

4. Ray DK, Mueller ND, West PC, Foley JA (2013) Yield Trends Are Insufficient to Double Global Crop Production by 2050. PLoS One 8: e66428.

5. Datta SK (2004) Rice biotechnology: A need for developing countries. AgBioForum 7: 31-35.

6. Grover A, Minhas D (2000) Towards production of abiotic stress tolerant transgenic rice plants: Issues, progress and future research needs. Proc Indian Natn Sci Acad (PINSA) B66: 13-32.

7. Lee SM, Kang K, Chung H, Yoo SH, Xu XM, et al. (2006) Plastid transformation in the monocotyledonous cereal crop, rice (Oryza sativa) and transmission of transgenes to their progeny. Mol Cells 21: 401-410.

8. Hoque HE, Mansfield JW (2004) Effect of genotype and explant age on callus induction and subsequent plant regeneration from root-derived callus of indica rice genotypes. Plant Cell Tiss Org Cult 78: 217-223.

9. Ramesh M, Murugiah V, Gupta AK (2009) Efficient in vitro plant regeneration via leaf base segments of indica rice (Oryza sativa L.). Indian J Exp Biol 47: 68-74.

10. Raemakers CJJM, Sofiari E, Jacobsen E, Visser RGF (1997) Regeneration and transformation of cassava. Euphytica 96: 153-161.

11. Sanyal I, Singh AK, Kaushik M, Amla DV (2005) Agrobacterium-mediated transformation of chickpea (Cicer arietinum L.) with Bacillus thuringiensis cry1Ac gene for resistance against pod borer insect Helicoverpa armigera. Plant Sci 168: 1135-1146.

12. Dabul ANG (2009) Screening of a broad range of rice (Oryza sativa L.) germplasm for in vitro rapid plant regeneration and development of an early prediction system. In vitro Cell Dev Biol-Plant 45: 414-420.

13. Seraj ZI, Islam Z, Faruque MO, Devi T, Ahmed S (1997) Identification of the regeneration potential of embryo derived calluses from various indica rice varieties. Plant Cell Tiss Organ Cult 48: 9-13.

14. Ramesh M, Gupta A K (2005) Transient expression of ß-glucuronidase gene in indica and japonica rice (Oryza sativa L.) callus cultures after different stages of co-ombardment. Afr J Biotechnol 4: 596-600.

15. Abe T, Futsuhara Y (1984) Varietal differences in plant regeneration from root callus tissues of rice. Jp J Breed 34: 147-155.

16. Abe T, Futsuhatra Y (1986) Genotypic variability for callus formation and plant regeneration in rice (Oryza sativa L.). Theor Appl Genet 72: 3-10.

17. Wang MS, Zapata FJ, De Castro DC (1987) Plant regeneration through somatic embryogenesis from mature seed and young inflorescence of wild rice (Oryza perennis Moench). Plant Cell Rep 6: 294-296.

18. Azria D, Bhalla PL (2000) Plant regeneration from mature embryo-derived callus of Australian rice (Oryza sativa L.) varieties. Aus J Agri Res 51: 305-312.

19. Koetje DS, Grimes HD, Wang YC, Hodges TK (1989) Regeneration of indica rice (Oryza sativa L.) from primary callus derived from immature embryos. J Plant Physiol 135: 184-190.

20. Nouri-Delawar M Z, Arzani A (2001) Study of callus induction and plant regeneration from immature embryo culture in rice cultivars JWSS - Isfahan University of Technology 4: 57-72.

21. Li ZN, Fang F, Liu GF, Bao MZ (2007) Stable Agrobacterium-mediated genetic transformation of London plane tree (Platanus acerifolia Willd.). Plant Cell Rep 26: 641-650.

22. Hiei Y, Komari T (2008) Agrobacterium-mediated transformation of rice using immature embryos or calli induced from mature seed. Nat Protoc 3: 824-834.

23. Aananthi N, Anandakumar CR, Ushakumari R, Shanthi P (2010) Regeneration study of some indica rice cultivars followed by Agrobacterium- mediated transformation of highly regenerable cultivar, Pusa Basmati 1. J Plant Breed 1: 1249-1256.

24. Wani SH, Sanghera GS, Gosal SS (2011) An efficient and reproducible method for regeneration of whole plants from mature seeds of a high yielding Indica rice (Oryza sativa L.) variety PAU 201. N Biotechnol 28: 418-422.

25. Sah SK, Kaur A (2013) Genotype independent tissue culture base line for high regeneration of japonica and indica rice. Res J. Biotechnol 8: 96-101.

26. Mandal AB, Maiti A, Biswas A (2003) Somatic embryogenesis in root derived callus of indica rice. Plant Tissue Cult 13: 125-133.

27. Oinam GS, Kothari SL (1995) Totipotency of coleoptile tissue in indica rice (Oryza sativa L. cv. ch 1039). Plant Cell Rep 14: 245-248.

28. Chang S, Sahrawat AK (1997) Somatic embryogenesis and plant regeneration from coleoptile tissue of indica rice. Rice Biotech Quarterly 32:12-16.

29. Kumar V, Shriram V, Nikam TD, Kavi Kishor PB, Jawali N, et al. (2008) Assessment of tissue culture and antibiotic selection parameters useful for transformation of an important Indica rice genotype Karjat-3. Asian Australas. J Plant Sci Biotechnol 2: 84-87.

30. Hoque ME, Ali MS, Karim NH (2007) Embryogenic callus induction and regeneration of elite Bangladeshi indica rice cultivars. Plant Tissue Cult Biotech 17: 65–70.

31. Washida H (2007) Agrobacterium-mediated genetic transformation for rice.

32. Jung KH, An G, Ronald PC (2008) Towards a better bowl of rice: assigning function to tens of thousands of rice genes. Nat Rev Genet 9: 91-101.

33. Cheema HS, Singh B (1993) CPCS1: A programme package for the Analysis of Commonly Used Experimental Designs. Punjab Agricultural University.

34. Ozawa K, Kawahigashi H, Kayano T, Ohkawa Y (2003) Enhancement of regeneration of rice (Oryza sativa L.) calli by integration of the gene involved in regeneration ability of the callus. Plant Sci 165: 395-402.

35. Armstrong CL, Severson JR, Hodges TK (1992) Improved tissue culture response of an elite maize inbred through backcross breeding, and identification of chromosome regions important for regeneration by RFLP analysis. Theor Appl Genet 84: 755-762.

36. Taguchi-Shiobara F, Lin SY, Tanno K, Komatsuda T, Yano M, et al. (1997) Mapping quantitative trait loci associated with regeneration ability of seed callus in rice (Oryza sativa L). Theor Appl Genet 95: 828-833.

37. Takeuchi Y, Abe T and Sasahara T (2000) RFLP mapping of QTLs influencing shoot regeneration from mature seed-derived calli in rice. Crop Sci 40: 245-247.

38. Ge X, Chu Z, Lin Y, Wang S (2006) A tissue culture system for different germplasms of indica rice. Plant Cell Rep 25: 392-402.

39. Nishi T, Yamada Y, Takahashi E (1968) Organ redifferentiation and plant restoration in rice callus. Nature 219: 508-509.

40. Bhattacharya P, Sen SK (1980) Potentiality of leaf sheath cells for regeneration of rice (Oryza sativa L.) plants. Theor Appl Genet 57: 87-90.

41. Chen Y, Wang R, Tian W, Zuo O, Zeng S, et al. (1980) Studies on pollen culture in vitro and induction of plantlets in Oryza sativa subsp. keng. Acta Genetica Sinica, 7: 46-54.

42. Toriyama K, Hinata K, Sasaki T (1986) Haploid and diploid plant regeneration from protoplasts of anther callus in rice. Theor Appl Genet 73: 16-19.

43. Lai K L and Liu L F (1982) Induction and plant regeneration of callus from immature embryos of rice plants (Oryza sativa L.). Jpn J Crop Sci 51: 70-74.

44. Ling DH, Chen WY, Chen MF, Ma ZR (1983) Direct development of plantlets from immature panicles of rice in vitro. Plant Cell Rep 2: 172-174.

45. Li Z, Xie Q, Rush MC, Murai N (1992) Fertile transgenic rice plants generated via protoplasts from US cultivar Labelle. Crop Sci 32: 810-814.

46. Abe T, Futsuhara Y (1985) Efficient plant regeneration by somatic embryogenesis from root callus tissue of rice (Oryza sativa L.) Plant Physiol 121: 111-118.

47. Chen TH, Lam L, Chen SC (1985) Somatic embryogenesis and plant regeneration from cultured young inflorescences of (Oryza sativa L) rice. Plant Cell Tiss Org Cult 4: 51-54.

48. Abdullah R, Cocking EC, Thompson JA (1986) Efficient plant regeneration from rice protoplasts through somatic embryogenesis. Biotechnol 4: 1087-1090.

49. Tsukahara M, Hirosawa T (1992) Simple dehydration treatment promotes plantlet regeneration of rice (Oryza sativa L.) callus. Plant Cell Rep 11: 550-553.

50. Kavi Kishor PB (1987) Energy and osmotic requirement for high frequency regeneration of rice plants from long-term cultures. Plant Sci 48: 189-194.

51. Lai KL, Liu LF (1988) Increased plant regeneration frequency in water stressed rice tissue culture. Jpn J Crop Sci 57: 553-557.

52. Zaidi MA, Narayanan M, Sardana R, Taga I, Postel S, et al. (2006) Optimizing tissue culture media for efficient transformation of different indica rice genotypes. Agro Research 4: 563-575.

53. Kumaria R, Waie B, Rajam MV (2001) Plant regeneration from transformed embryogenic callus of an elite indica rice via Agrobacterium. Plant Cell Tiss Org Cult 67: 63-71.

54. Kumaria R, Rajam MV (2002) Alteration in polyamine titres during Agrobacterium-mediated transformation of indica rice with ornithine decarboxylase gene affects plant regeneration potential. Plant Sci 162: 769-777.

55. Lentini Z, Reyes P, Marinez CP, Roca WM (1995) Androgenesis of highly recalcitrant rice genotypes with maltose and silver nitrate. Plant Sci 110: 127-138.

56. Lin YJ, Zhang Q (2005) Optimising the tissue culture conditions for high efficiency transformation of indica rice. Plant Cell Rep 23: 540-547.

57. Lee K, Jeon H, Kim M (2002) Optimization of a mature embryo-based in vitro culture system for high-frequency somatic embryogenic callus induction and plant regeneration from japonica rice cultivars. Plant Cell Tiss Org Cult 71: 237-244.

58. Bhojwani SS, Razdan MK (1996) Plant Tissue Culture, Theory and Practices. A revised edition. Pp 50-51. Elsevier Publishers, Amsterdam.

59. Wenzhong T, Rance I, Sivamani E, Fauquet C, Beachy RN (1994) Improvement of plant regeneration frequency in vitro in indica rice. China J Genet 21: 1-9.

60. Patpanukul T, Bunnag S, Theerakulpisut P, Kosittrakul M (2004) Transformation of indica rice (Oryza sativa L.) cv. RD6 meidated by Agrobacterium tumefaciens. Songklanakarin J Sci Technol 26: 1.

61. Datta SK, Datta K, Soltanifar N, Donn G, Potrykus I (1992) Herbicide-resistant Indica rice plants from IRRI breeding line IR72 after PEG-mediated transformation of protoplasts. Plant Mol Biol 20: 619-629.

62. Saharan V, Yadav R C, Yadav N R and Ram K (2004) Studies on improved Agrobacterium-mediated transformation in two indica rice (Oryza sativa L.). Afr J Biotechnol 3: 572-575.

63. Meneses A, Flores D, Muñoz M, Arrieta G, Espinoza AM (2005) Effect of 2,4-d, hydric stress and light on indica rice (Oryza sativa) somatic embryogenesis. Rev Biol Trop 53: 361-368.

64. Hiei Y, Ohta S, Komari T, Kumashiro T (1994) Efficient transformation of rice (Oryza sativa L.) mediated by Agrobacterium and sequence analysis of the boundaries of the T-DNA. Plant J 6: 271-282.

65. Toki S, Takamatsu S, Nojiri C, Ooba S, Anzai H, et al. (1992) Expression of a Maize Ubiquitin Gene Promoter-bar Chimeric Gene in Transgenic Rice Plants. Plant Physiol 100: 1503-1507.

66. Kishore PBK, Sangan S, Naidu KP (1999) Sodium, potassium, sugar, alcohol and proline mediated somatic embryogenesis and plant regeneration in recalcitrant rice callus. Proc Plant Tissue Culture and Biotechnology: Emerging Trends Symp. pp 78-85.

26

Genome-Wide Comparative Transcriptional Analysis of Developing Seeds among Seven *Oryza sativa* L. Subsp. *Japonica* Cultivars Grown near the Northern Limit of Rice Cultivation

Sho Takano[1], Shuichi Matsuda[1], Yuji Hirayama[2], Takashi Sato[2], Itsuro Takamure[3] and Kiyoaki Kato[1*]

[1]Department of Agro-Environmental Science, Obihiro University of Agriculture and Veterinary Medicine, Nishi 2-11 Inada, Obihiro, Hokkaido, 080-8555, Japan
[2]Rice Breeding Group, Kamikawa Agricultural Experiment Station, Local Independent Administrative Agency Hokkaido Research Organization, Minami 1-5, Pippu, Hokkaido 078-0397, Japan
[3]Graduate School of Agriculture, Hokkaido University, Kita 9 Nishi 9, Kita-ku, Sapporo, Hokkaido 060-811, Japan

Abstract

Improved grain quality is a major breeding target in rice (*Oryza sativa* L. subsp. *japonica*), owing to market demand. Rice cultivars grown in Hokkaido (42-45°N), the northernmost region of rice paddy cultivation in Japan and near the northern limit of rice cultivation, have been bred for over 100 years for adaptation to low temperature together with high yield and grain quality. In this study, for seven closely related rice cultivars released in Hokkaido in the last 70 years, we investigated the transcriptome profiles of developing seeds 8 days after flowering (DAF, middle stage) and 15 DAF (late stage) under natural conditions in Hokkaido (42.52°N) using a whole-genome oligonucleotide microarray. The transcriptome profiles were divided into two groups depending on stage and were more variable at the late than that at the middle stage. The genome-wide transcriptome data revealed the features of differentially expressed genes in two first-generation cultivars compared to that in the parental cultivars. Starch properties were varied and correlated with the expression of starch biosynthesis genes among Hokkaido cultivars.

Apparent amylose content was positively correlated with Waxy gene expression at the late stage. The expression of starch biosynthesis genes and starch properties were varied among Hokkaido cultivars. The transcriptomes of the most recent cultivars reveal the expression of ideal genes for the development of cultivars with high grain yield, grain quality, and grain traits essential for rice production near the northern and southern limits of rice cultivation.

Keywords: Grain quality; Rice; Transcriptome; Cultivar; Grain filling

Introduction

Rice (*Oryza sativa* L.) is a staple food for nearly one-half of the world's population. Rice productivity has been greatly improved during the Green Revolution since the 1960s, and scientists and breeders now focus on improving grain quality for different purposes and markets. Rice grain quality is determined mainly by four constituents: grain appearance, nutritional content, and the cooking and eating quality [1]. Grain quality involves many characteristics ranging from physical to biochemical and physiological properties [2]. Starch and protein are two main components of rice endosperm and, consequently, key components of grain quality. Rice grain is also an important source of micronutrients and vitamins, whose storage plays crucial roles in grain quality and nutritional value. Consumer preferences for grain quality vary worldwide based on food habits. Therefore, the improvement of rice grain cooking quality has become the most important research component of almost all rice improvement programs worldwide. For instance, people in the Far East, including Japan, prefer sticky and soft rice, whereas those in India prefer non-sticky rice. Consumers in developed countries demand mainly for grain with good cooking quality and eating characteristics, but in many developing regions, nutritional value is crucial.

Starch comprises two major components: amylose and amylopectin. Amylose is a linear molecule containing $\alpha((1,4)$-linked glucose units with few branches, whereas amylopectin is a highly branched biopolymer containing linear chains of $\alpha(1,4)$-linked glucose residues joined by $\alpha(1,6)$-linkages. Within the granule, the chains are considered to be arranged in clusters at intervals of 9 nm within which chains are associated to form double helices. These helices are packed in ordered arrays to form concentric crystalline lamellae, and the distribution of branch lengths in amylopectin is important

in determining the crystalline nature of the starch granule and the physical properties of the starch [3-6]. The percentage of amylose in total starch, measured as Apparent Amylose Content (AAC), is the key determinant of rice cooking properties. The suggested classification of Amylose Content (AC) identifies classes as waxy (0-5%), very low (5-12%), low (12-20%), intermediate (20-25%), or high (25-33%), whereas rice is commercially classified by amylose content as low (<20% amylose), medium (21-25%), or high (26-33%) [7,8]. Rice grains with high AC (≥26%), like risotto varieties, cook dry and become less tender and hard upon cooling, whereas rice grains with low AC (<20%) cook moist and become sticky, glossy, and cohesive after cooking. In most rice-growing or consuming regions of the world, rice with intermediate AC is preferred. In contrast, low-AC *japonicas* are preferred in Japan.

In addition to AAC (AC), alteration of amylopectin content and structure affects cooking quality and consumption traits [9,10]. The high degree of organization within amylopectin, a non-random distribution of linear chains, the clustered positioning of branch linkages, and the

---

*Corresponding author: Kiyoaki Kato, Department of Agro-Environmental Science, Obihiro University of Agriculture and Veterinary Medicine, Nishi 2-11 Inada, Obihiro, Hokkaido, 080-8555, Japan, E-mail: kiyoaki@obihiro.ac.jp

double helical conformation between adjacent parallel side chains contribute to the crystalline features of amylopectin molecules. In the current model of the multiple cluster structure of amylopectin, A chains, which lack other chains, are linked to other chains at their reducing ends, whereas B chains comprise one or more chains belonging to a cluster. Amylopectins exhibit polymodal chain-length distributions with periodic waves of varying Degrees of Polymerization (DP). The chains are grouped into four different fractions with DP ranging in the intervals 6-12 (A), 13-24 (B1), 25-36 (B2), and >37 (B3) [11]. Based on the ratio of the number of short chains of DP ≤ 10 to the short and intermediate chains of DP ≥ 24 (short chain ratio, SCR), amylopectin with lower SCR is called L-type and that with a higher SCR is called S-type [12]. Cooked rice with the S-type amylopectin of *japonica* cultivars is much softer after cooling than rice with the L-type amylopectin of *indica* cultivars [13]. Among *japonica* rice cultivars, a higher content of short chain amylopectin is also positively correlated with softness and stickiness of cooked rice [14].

The variations in starch structures arise from a differential expression of various isoforms of starch biosynthesis enzymes. Starch biosynthesis in higher plants, including rice, is catalyzed by four classes of enzymes: ADP-Glc pyrophosphorylase (AGPase), starch synthase (SS), starch branching enzymes (SBE), and starch debranching enzymes (DBE). Several studies have led to the identification of various SS isoforms via gene sequencing in several plant genomes. In rice, there are 10 SS isoforms, which can be grouped into five types: GBSS (I, II); SSI; SSII (SSIIa, SSIIb, and SSIIc); SSIII (SSIIIa and SSIIIb); and SSIV (SSIVa and SSIVb) [15-17]. All of these isoforms of starch-synthesizing enzymes coordinate and form a regulatory network to control starch synthesis in rice endosperm, affecting grain cooking and eating quality [18-20]. Amylose content is regulated primarily by the *Waxy* (*Wx*) gene, which encodes the granule-bound starch synthase I (GBSSI) enzyme, and the level of grain amylose is directly associated with the amount of GBSSI in the endosperm [21-23]. Amylopectin biosynthesis is controlled by SS, SBE, and DBE [9,17, 23-26] The gene identified as being responsible for amylopectin structure and processing suitability is the *starch synthase IIa* gene (*SSIIa*), known as the *Alk* (*alkali disintegration*) gene [27]. SSIIa is responsible for the elongation of short chains with DP 5-10 to form longer A and B1 chains with DP 11-24 within the cluster of amylopectin. Therefore, the difference in allele of the *SSIIa* is responsible for the structural alteration of amylopectin between *indica-* and *japonica-*type rice varieties having active SSIIa and inactive SSIIa, respectively [9, 28]. SBEIIb plays a specific role in the formation of A chains of the amylopectin cluster [29], and the levels of SBEIIb activity affect the extent of structural changes in the cluster [25]. Isoamylase (ISA1), one of the DBE, is essential for the formation of the highly organized cluster structure of amylopectin [30], and the level of ISA1 expression influences both the fine structure of amylopectin and the thermal properties of starch [31,32]. In addition, quantitative trait locus (QTL) analysis has revealed that many grain quality traits are controlled by multiple regions in the rice genome [33-37].

Rice seed development can be divided into four stages: the initiation stage [1-3 days after flowering (DAF)], during which starch is synthesized exclusively in the pericarp; the early developmental stage (3-5 DAF), indicated by endosperm starch accumulation with a marked increase in seed weight; the middle stage (5-10 DAF), with a rapid increase in starch deposition and grain weight; and the late stage (10 DAF and beyond), in which seed maturation occurs [38]. The expressions of genes involved in starch biosynthesis are differentially regulated depending on the developmental stage of the seed. [15,39,40]. The challenge lies in understanding the genetic basis and identification

of genomic regions controlling these isoforms in order to develop cultivars with desirable combinations of alleles that control these isoforms depending on the developmental stage of the seed.

Grain quality is determined by a variety of developmental processes, which are affected not only by the genetic constitution but also by environmental conditions such as temperature, water availability, fertilizer application, and drought and salinity stresses [2]. Strong genotype × environment effects have been found for several quality traits in rice cultivars grown at different locations and during different seasons [13,41,42]. *Wx* gene expression and Wx protein were increased when rice plants were exposed to low temperature (18°C) [21,43]. The G-T polymorphism in the first intron of the *Wx* gene (*Wx^b*) is associated with a different efficiency of RNA splicing and processing under low and high temperature during grain development [44]. In addition, lower temperatures increased the molar and weight ratios (A + B1)/(B2 + B3) of amylopectin unit chains in *japonica* cultivars [45], and thus in Hokkaido (*japonica*) cultivars leading, under cool conditions during grain filling, to low eating quality [46]. The molecular basis of the increased number of short chains at low temperature remains to be elucidated. Nonetheless, for the past 30 years, rice breeders have selected genotypes optimizing the activity of *Wx* and other genes, affecting amylose content and amylopectin chain length, to adapt to environmental conditions during grain filling in Hokkaido.

The rice cultivars growing in Hokkaido (42-45°N), the northernmost region of rice paddy cultivation in Japan and one of the northern limits of rice cultivation in the world, have been improved to be adapted to low temperatures during the grain-filling stage along with high yield and good eating quality. There is little information about transcriptional profiles in developing seeds of Hokkaido cultivars under natural conditions. In the present study, we investigated the gene expression in developing seeds of seven very closely related cultivars released in Hokkaido during the last 70 years using Agilent 44K rice oligoarrays. Using genome-wide transcriptome profiling, we addressed the breeding history of these seven cultivars in the transcriptome, transmission of transcriptome profiles to two first-generation cultivars, and the transcriptional profile unique to the most recent Hokkaido cultivar. These genome-wide transcriptome data provide a foundation for deep exploration of gene-trait relationships, and for rice improvement near the northern and southern limits of world rice cultivation.

Materials and Methods

Plant materials and growth conditions

Rice (*Oryza sativa* L. ssp. *japonica*) plants of seven Hokkaido cultivars, Tomoemasari, Kitahikari, Kuuiku99, Tomohikari, Yukihikari, Kirara397, and Jouiku462 (Figure S1), that carry *Wx^b* were grown in 3-L pots filled with 2 kg of soil including 1.2 g each of N, P_2O_5 and K_2O per pot under natural conditions at Obihiro, Hokkaido (42.52°N) in 2013. Immature seeds of each cultivar were collected at 8 DAF (middle stage), and 15 DAF (late stage). Depending on flowering time, immature seeds of Kuuiku99, Tomohikari, Yukihikari, Kirara397, and Jouiku462 were collected on August 19 (8 DAF) and August 26 (15 DAF), immature seeds of Kitahikari were collected on August 23 (8 DAF) and August 30 (15 DAF), and Tomoemasari seeds were collected on August 29 (8 DAF) and September 5 (15 DAF). In all, 50 immature seeds of each cultivar were collected from 10-12 plants of each cultivar and bulked for comprehensive gene expression analysis (Figure S2). For amylose and amylopectin analysis, >300 plants of each cultivar were grown on the experimental farm of the Hokkaido University

Field Science Centre for Northern Bio-Research (Sapporo, Hokkaido, 43.074°N) in 2013.

RNA isolation and microarray analysis

Agilent 44 K rice oligoarrays (Agilent Technologies), which contain 44,000 features, were used as one-color oligoarrays. Each feature comprises a 60-mer oligonucleotide corresponding to a full-length cDNA of rice. Total RNA was extracted from the endosperm of all cultivars at 8 and 15 DAF using the RNeasy plant mini kit (QIAGEN). Microarray experiments were performed according to the manufacturer's manual. Feature Extraction software 10.7.3.1 (Agilent Technologies) was used to delineate and measure the signal intensity of each spot on the array. The resulting data were normalized by a 75 percentile shift [47]. Poor quality features that were either nonuniform or saturated were flagged and not used for further analysis.

Cluster analysis of SNPs/InDels between seven cultivars

Whole-genome sequencing and detection of SNP/InDel positional data between Nipponbare (Pseudomolecules build 5.0, http://rapdblegacy.dna.affrc.go.jp/) and seven Hokkaido cultivars were performed using our previous data [48], yielding 83,738 unique SNP/InDel sites. A dendrogram was created using complete linkage in Cluster 3.0 [49] (http://bonsai.ims.u-tokyo.ac.jp/~mdehoon/software/cluster/), and the result was visualized with Java TreeView software 1.1.6r4 [50] (http://jtreeview.sourceforge.net/). The colors used in the clustergram representation for each of the seven cultivars were based on a comparison with the Nipponbare genome. If the SNP/InDel type matched that of Nipponbare, it was colored blue; otherwise, it was colored yellow.

Apparent amylose content analysis

According to the protocol described by Williams et al. [51] with modifications by Inatsu [52], the AAC of milled grain was measured with an auto-analyzer (AA-III, Bran Luebbe), which is based on a flow injection of a 0.09% NaOH solution into the sample, addition of an iodine solution, and spectrometric determination of the absorbance at 620 nm was determined by a spectrometer. Calibration was performed using the absorbance of standard rice samples carrying 21.1% AAC. ACC analysis was performed in triplicates for each genotype.

Branch chain-length distribution of α-glucan

Isolation of starch and analysis of the branch chain-length distribution of α-glucan were performed as described by [53] Wong et al. Mature kernels from each cultivar were dehulled and the endosperm was separated from the germ and milled in a coffee mill. The ground endosperm powder (7 g) was suspended in 100 mL of distilled water and stirred for 1 h at 4°C. The suspension containing starch was centrifuged at 700 ×g for 10 min at 4°C. The precipitate was washed five times with 100 mL of distilled water, followed by centrifugation at 700 ×g for 10 min at 4°C and a final centrifugation at 10,000 ×g. The resulting starch pellet was suspended by stirring in 100 mL of 2% SDS for 2 h at room temperature, and was then centrifuged at 700 ×g for 10 min at room temperature to remove protein, and the supernatant was discarded. The procedure for removing protein was repeated three times. The precipitate was then washed six times with 100 mL of distilled water, followed by centrifugation at 700 ×g for 10 min at room temperature and final centrifugation at 10,000 ×g. Starch samples were dried *in vacuo*.

The starch was suspended in 5 mL of methanol and heated in a boiling water bath for 10 min. The suspension was centrifuged at 1,000 ×g for 10 min, and the precipitate was collected and washed twice with 5 mL of 90% (v/v) methanol and then re-suspended in 300 mL of 0.25 M NaOH. The suspension was heated in a boiling water bath for 5 min to gelatinize the starch. The pH and concentration of α-glucan sample solutions were adjusted by the addition of 9.6 μL of 100% acetic acid, 100 μL of 600 mM Na-acetate buffer (pH 4.4), 15 μL of 2% NaN$_3$, and 1.09 mL of distilled water. The α-glucan sample was debranched with *Pseudomonas amyloderamosa* isoamylase (354 units, Seikagaku-Kogyo, Tokyo) at 37°C for 24 h. The debranched sample was heated in a boiling water bath for 20 min to stop enzyme activity, and was then centrifuged. The supernatant was deionized with ion exchange resin [Bio-Rad AG 501-X8(D)] in a microtube. An aliquot containing the equivalent of 5 nmol of reducing sugar, determined by the modified Park Johnson method [54], was evaporated to dryness in a centrifugal vacuum evaporator (Taitec, Tokyo). Fluorescence labeling of α-1,4-glucan at the reducing end with 8-amino-1,3,6-pyrenetrisulfonic acid (APTS) was performed according to the method described by O'Shea et al. [54] The APTS-labeled α-1,4-glucan was analyzed using an eCAP N-linked oligosaccharide profiling kit and a P/ACE System 5000 high-resolution capillary electrophoresis system equipped with a laser-induced fluorescence detector (Beckman Instruments, Coulter, CA, U.S.A.).

Results

Gene expression profiles of seven Hokkaido rice cultivars

Genes involved in adaptation to cool summers in Hokkaido during grain filling, or possibly unknown genetic factors that are essential for grain filling in Hokkaido, may be positively selected in flanking chromosomal regions. To assess the correlation between expression profile and genomic structure in seven Hokkaido cultivars, we compared the dendrograms based on hierarchical clustering of transcriptome profiles of seven cultivars at 8 and 15 DAF and of the previous comprehensive SNPs/InDel data set [48] (Figure 1). The dendrogram from hierarchical clustering revealed that the transcriptome profiling was divided into two large clusters depending on the seed development stage (Figure 1A). The branch length of the dendrogram of 15 DAF was longer than that of 8 DAF, suggesting that the transcriptome of each cultivar at 15 DAF varied as compared to those at 8 DAF. At both stages, six cultivars formed similar dendrograms with respect to four features; (i) Tomoemasari was the farthest from the other six cultivars, (ii) Tomohikari, Yukihikari, and Kuuiku99 formed the primary group, (iii) Yukihikari and Tomohikari formed a sister pair, and (iv) Kirara397 and Jouiku462 formed a sister pair. Kitahikari was positioned differently depending on the stage; Kitahikari clustered in the primary group with Kirara397 and Jouiku462 at 15 DAF, but was separate from these cultivars at 8 DAF.

Based on genome-wide SNPs/InDels, the genome structures of the seven cultivars were divided into two groups, the first containing Tomoemasari, Kuuiku99, Tomohikari, and Yukihikari and the second containing Kitahikari, Kirara397, and Jouiku462 (Figure 1B). In the first group, Yukihikari and Tomohikari formed a sister pair. In the second group, Kirara397 and Jouiku462 formed a sister pair. Thus, the shape of the dendrogram based on comprehensive SNPs/InDels was similar to that of the dendrogram based on the transcriptome profiles at 15 DAF. These findings, taken together, suggest that breeders have positively selected genotypes associated with the transcriptome profiles at 15 DAF to improve grain-filling traits in Hokkaido in the past 70 years.

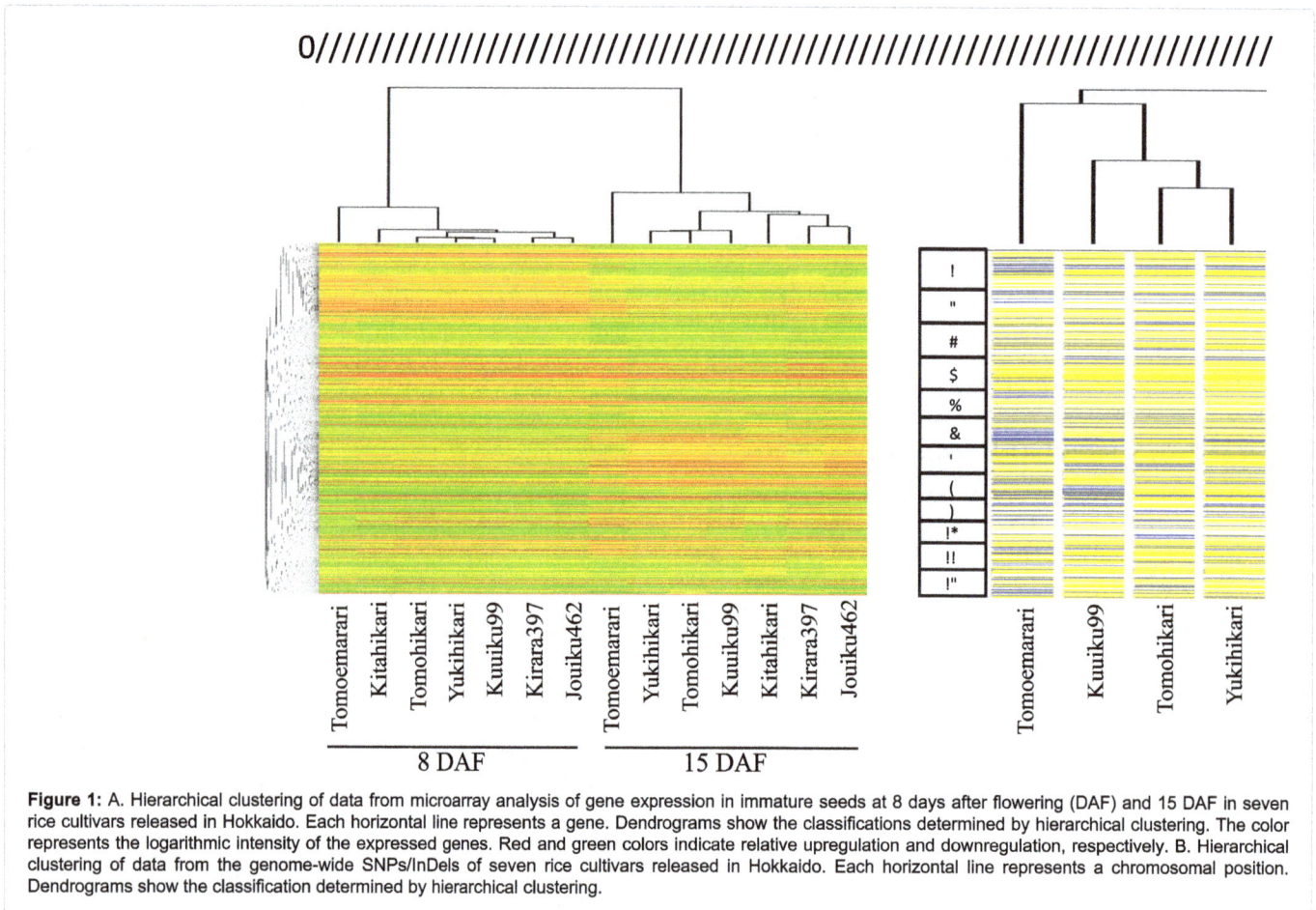

Figure 1: A. Hierarchical clustering of data from microarray analysis of gene expression in immature seeds at 8 days after flowering (DAF) and 15 DAF in seven rice cultivars released in Hokkaido. Each horizontal line represents a gene. Dendrograms show the classifications determined by hierarchical clustering. The color represents the logarithmic intensity of the expressed genes. Red and green colors indicate relative upregulation and downregulation, respectively. B. Hierarchical clustering of data from the genome-wide SNPs/InDels of seven rice cultivars released in Hokkaido. Each horizontal line represents a chromosomal position. Dendrograms show the classification determined by hierarchical clustering.

Transmission of gene expression in the first generation cultivars

Yukihikari and Tomohikari were first-generation cultivars and were developed from (Kitahikari/Tomoemasari)/Kuuiku99 (Figure S1). We assessed the transmission of gene expression from each parental cultivar to the first generation cultivars. As described above, the dendrogram of hierarchical clustering based on the transcriptome showed that Yukihikari and Tomohikari formed a sister pair and clustered with Kuuiku99 at both stages, and that Tomoemasari was the farthest from Yukihikari and Tomohikari at both stages. These features were very similar to those of clustering based on genome-wide SNPs/InDels (Figure 1B).

In Yukihikari, 4,714 genes (8 DAF) and 2,348 genes (15 DAF) were similar to at least one parental line (showing between 1/1.2 and 1.2 times the Yukihikari reference expression). Among them, 3,651 (8 DAF) and 1,341 (15 DAF) were conserved between Yukihikari and all three parental cultivars. Thirty (8 DAF) and 42 (15 DAF) genes were conserved with Kitahikari, 30 (8 DAF) and 75 (15 DAF) with Tomoemasari, and 157 (8 DAF) and 285 (15 DAF) with Kuuiku99. Like Yukihikari, 4,310 (8 DAF) and 2,893 (15 DAF) genes in Tomohikari were similar in at least one parental line (between 1/1.2 and 1.2 times Tomohikari reference expression). Among these, 3,367 (8 DAF) and 1,231 (15 DAF) genes were conserved between Tomohikari and the three parental cultivars. Thirty-seven (8 DAF) and 57 (15 DAF) genes were conserved with Kitahikari, 25 (8 DAF) and 40 (15 DAF) with Tomoemasari, and 173 (8 DAF) and 328 (15 DAF) with Kuuiku99.

Some genes were found to be differentially expressed in Yukihikari and/or Tomohikari compared to either parental cultivar, and were therefore considered "transgressive." In Yukihikari, 63 (8 DAF) and 45 (15 DAF) genes were upregulated at least two-fold compared to either parental cultivar, while 50 (8 DAF) and 41 (15 DAF) were downregulated at least two-fold compared to either parental cultivar. In Tomohikari, 61 (8 DAF) and 344 (15 DAF) genes were upregulated at least two-fold compared to either parental cultivar, while 95 (8 DAF) and 80 (15 DAF) were downregulated to less than 1/2 times compared to either parental cultivar. Among these differentially expressed genes, 6 (8 DAF) and 16 (15 DAF) genes were consistently upregulated in Yukihikari and Tomohikari compared to either parental cultivar (Table S1). In contrast, 10 (8 DAF) and 18 (15 DAF) genes were consistently downregulated in Yukihikari and Tomohikari compared to either parental cultivar (Table S1).

Transgressive gene expression and flanking genomic regions

Our previous study of genome-wide SNPs/InDels revealed that Yukihikari and Tomoemasari have clusters of unique SNPs/InDels (haplotype blocks) compared to the parental cultivars (Takano et al. 2014, Figure 2). Each unique haplotype block was thought to be transmitted from the parental accession used originally for the breeding of the cultivars but not the parental accession used in the genomic sequencing study [48]and the present transcriptome study. To assess the possibility that differentially expressed genes are present in these unique genomic regions, we compared the SNPs/InDel sites unique to Yukihikari and/or Tomohikari with the positions of differentially

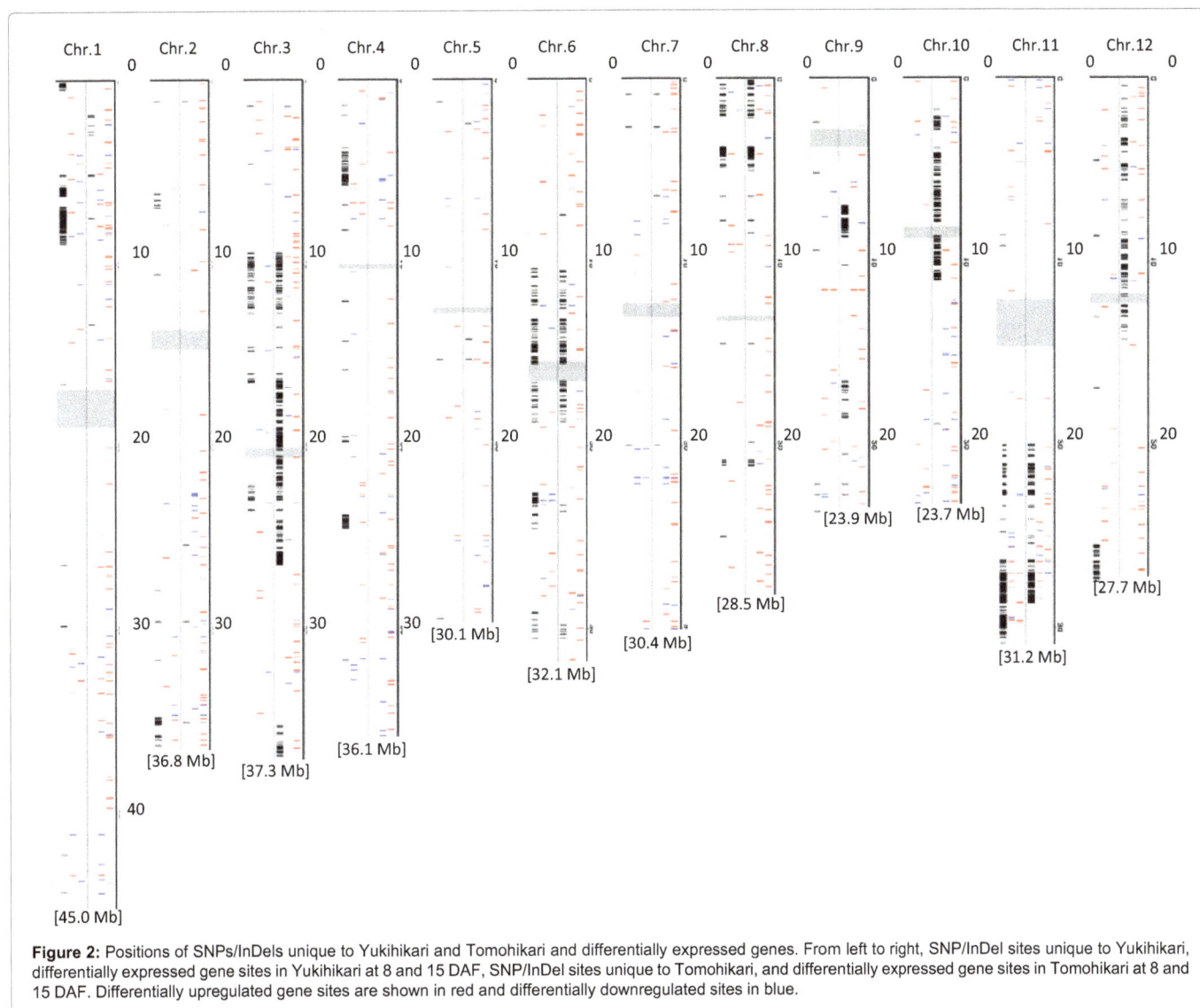

Figure 2: Positions of SNPs/InDels unique to Yukihikari and Tomohikari and differentially expressed genes. From left to right, SNP/InDel sites unique to Yukihikari, differentially expressed gene sites in Yukihikari at 8 and 15 DAF, SNP/InDel sites unique to Tomohikari, and differentially expressed gene sites in Tomohikari at 8 and 15 DAF. Differentially upregulated gene sites are shown in red and differentially downregulated sites in blue.

expressed genes in Yukihikari and/or Tomohikari (Figure 2). The within-chromosome distributions of differentially expressed genes in Yukihikari and Tomoemasari were nonuniform (Figure 2). Among 113 (8 DAF) and 86 (15 DAF) differentially expressed genes in Yukihikari, 17 (15.0% of 113 genes, 8 DAF) and 21 (24.4% of 86 genes, 15 DAF) were located in a unique genomic region. As with Yukihikari, 22 (14.1% of 156 genes, 8 DAF) and 56 (13.2% of 424 genes, 15 DAF) differentially expressed genes were located in a unique genomic region in Tomohikari. Differentially expressed genes located in unique genomic regions in the first generation cultivars may be controlled by cis-regulatory elements from the parental accession used originally for breeding the cultivars. In contrast, 96 (85.0% of 113 genes, 8 DAF) and 65 genes (75.6% of 86 genes, 15 DAF) in Yukihikari and 134 (85.9% of 156 genes, 8 DAF) and 368 (86.8% of 424 genes, 15 DAF) in Tomohikari may be controlled by trans-regulatory elements and/or novel combinations of cis-regulatory elements.

For example, Os11g0453900 (*OsRAB16D*) expression was increased in Yukihikari and Tomoemasari compared to three parental cultivars (Table S1) and the gene is located between 17,466,009 and 17,467,247 bp on chromosome 11 (IRGSP/RAP build 5). On chromosome 11,

Yukihikari and Tomoemasari shared unique SNPs/InDels from 20.0 to 22.7 Mb and from 26.3 to 28.6 Mb [48] (Figure 3). Thus, *OsRAB16D* was outside the unique genomic region, suggesting that differential expression of *OsRAB16D* in Yukihikari and Tomoemasari is a result of selection for trans-regulatory element(s) and/or novel combinations of cis-regulatory elements.

Transcripts unique to the most recent cultivar, Jouiku462

To characterize changes in gene expression profiles of seed development during >70 years of breeding history in Hokkaido, we compared the gene expression between the most recent cultivar, Jouiku462, and the six older cultivars. A total of 2,594 (8 DAF) and 489 (15 DAF) genes were expressed at similar levels in the seven cultivars (between 1/1.2 times and 1.2 times those of Jouiku462). Compared to Jouiku462, numbers of similarly expressed genes varied among cultivars and fell between 16,214 (Kirara397) and 9,054 (Tomohikari) at 8 DAF and between 9,721 (Yukihikari) and 4,922 (Tomohikari) at 15 DAF (Figure 3). In contrast, numbers of differentially expressed genes fell between 507 (Kirara397) and 2,439 (Tomohikari) at 8 DAF and between 894 (Yukihikari) and 4,474 (Tomohikari) at 15 DAF

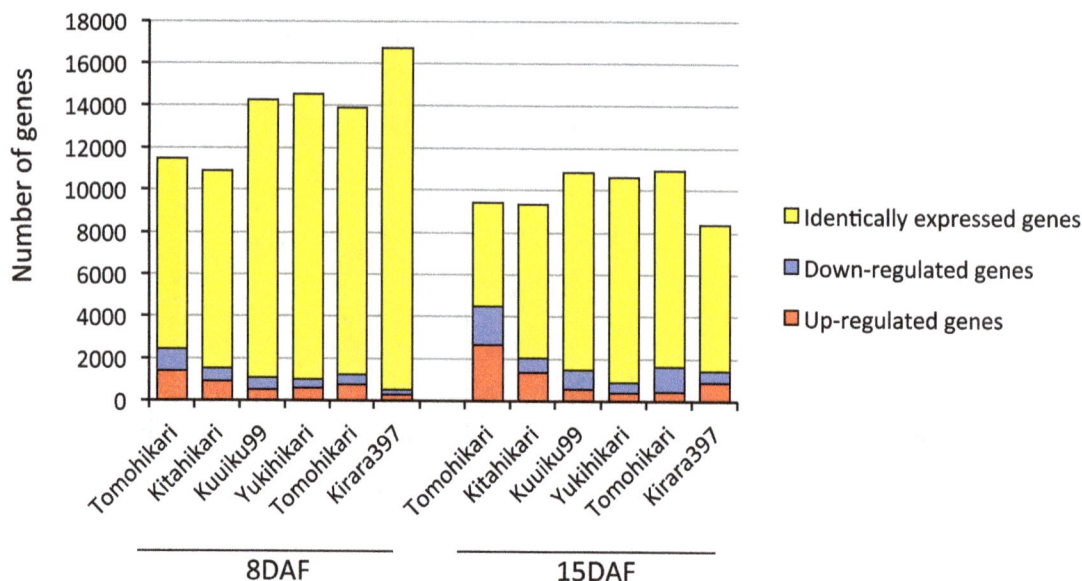

Figure 3: Numbers of genes identically expressed and differentially expressed in Jouiku462 compared to six older cultivars, Tomohikari, Kitahikari, Kuuiku99, Yukihikari, Tomohikari, and Kirara397.

compared to Jouiku462 (Fig. 3). We list the genes unique to Jouiku462 in Tables S2 (8 DAF) and S3 (15 DAF). In Jouiku462, 25 (8 DAF) and 17 (15 DAF) were consistently upregulated compared to the six older cultivars, while 45 (8 DAF) and 31 (15 DAF) were consistently downregulated. As at 8 DAF, 75 genes were differently expressed from all of the other six cultivars by more than 2-fold. Among them, 17 genes were upregulated in Jouiku462 and 31 were downregulated in Jouiku462 compared to all of the other six cultivars (Table S3).

Expression profiling of genes involved in starch synthesis

AACs of the seven cultivars were varied, ranging from 16.6% (Jouiku462) to 20.2% (Tomoemasari) (Table S4). AACs of the seven cultivars were divided into four groups by statistical analysis. Jouiku462 was classified as the lowest AAC group. Kitahikari (20.1%) and Tomoemasari (20.2%) were classified as the highest-AAC group. Kuuiku99 (18.2%) and Yukihikari (18.4%) were classified as a lower AAC group, and Kirara397 (18.8%) and Tomohikari (19.1%) were classified as a higher AAC group, respectively. Thus, breeders have succeeded in reducing the amylose content in rice grain for improvement of the sticky and soft eating quality of cooked rice.

Properties of amylopectin chain length also varied in the seven cultivars (Table S5). Two older cultivars Tomoemasari (released in 1945) and Kuuiku99 (released in 1972) showed opposite properties of amylopectin chain length. Tomoemasari contained more short chains (Fa; $5 \leq DP \leq 12$), fewer intermediate-size (Fb$_2$; $25 \leq DP \leq 36$), and long chains (Fb$_3$; $DP \geq 37$). Kuuiku99 contained fewer short chains and more intermediate size and long chains. Thus, Tomoemasari had longer chains (Fb$_2$ + Fb$_3$), and fewer short chains (Fa + Fb$_1$), whereas Kuuiku99 had shorter chains and fewer long chains. The other five cultivars had intermediate properties of amylopectin chain length between Kuuiku99 and Tomoemasari. The most recent cultivar, Jouiku462, was at the midpoint among the seven cultivars. These results suggest that breeders have selected the intermediate type of amylopectin chain length in the current cultivars.

To assess the relationship between starch properties and gene expression, we evaluated the correlation between starch properties

and the expression of genes involved in starch synthesis in the seven cultivars (Figure 4, Table 1). AAC was positively correlated with GBSSI (*Wx*) expression at 15 DAF. The marked reduction of *Wx* expression (15 DAF) was evident in the recent cultivar Jouiku462 compared to the six older cultivars, and its signal intensity was between 70.2% (Kitahikari) and 50.1% (Tomoemasari) (Figure 4). AAC was negatively correlated with *OsAPL3* expression at 15 DAF. The marked reduction of *OsAPL3* expression was evident in Tomoemasari and Kitahikari. The signal intensities in these two cultivars were approximately half those of the other five. The reduced expression of *Wx* and increased expression of *OsAPL3* might thus contribute, at least in part, to the reduction of amylose content in the most recent cultivar, Jouiku462.

We evaluated the expression of genes leading to synthesis of amylopectin. *OsSSIIa* expression (both stages) was positively correlated with Fb$_3$, but negatively correlated with Fa and the (Fa + FB$_1$)/(Fb$_2$ + Fb$_3$) ratio (Table 1). The reduction of *OsSSIIa* expression was evident in Tomoemasari. The signal intensity of *OsSSIIa* was reduced in Tomoemasari to 60.1% of Tomohikari and 80.1% (Jouiku462) at 8 DAF, and 68.5% (Tomohikari) and 93.7% (Kitahikari) at 15 DAF. In addition, the (Fa + FB$_1$)/(Fb$_2$ + Fb$_3$) ratio was positively correlated with the expression of *OsAGPS1* (15 DAF), *OsAPL4* (8 DAF), and *OsSSIVb* (8 DAF), but negatively correlated with expression of *OsAPL3* (15 DAF), *SBE-I* (8 DAF), *ISA* (8 DAF), and *PUL* (8 DAF).

Discussion

Possible positive selection for gene expression in hokkaido rice cultivars

We have shown several lines of evidence that support the preferential selection by breeders, over the past 70 years, of the genotype associated with gene expression at 15 DAF rather than that at 8 DAF. First, the expression profiles in developing seeds of seven rice cultivars were distinct in the two development stages of 8 DAF and 15 DAF (Figure 1A). This difference in temporal expression pattern across cultivars between middle (8 DAF) and late (15 DAF) stages seem to be related to the distinct spatial pattern of starch deposition within a

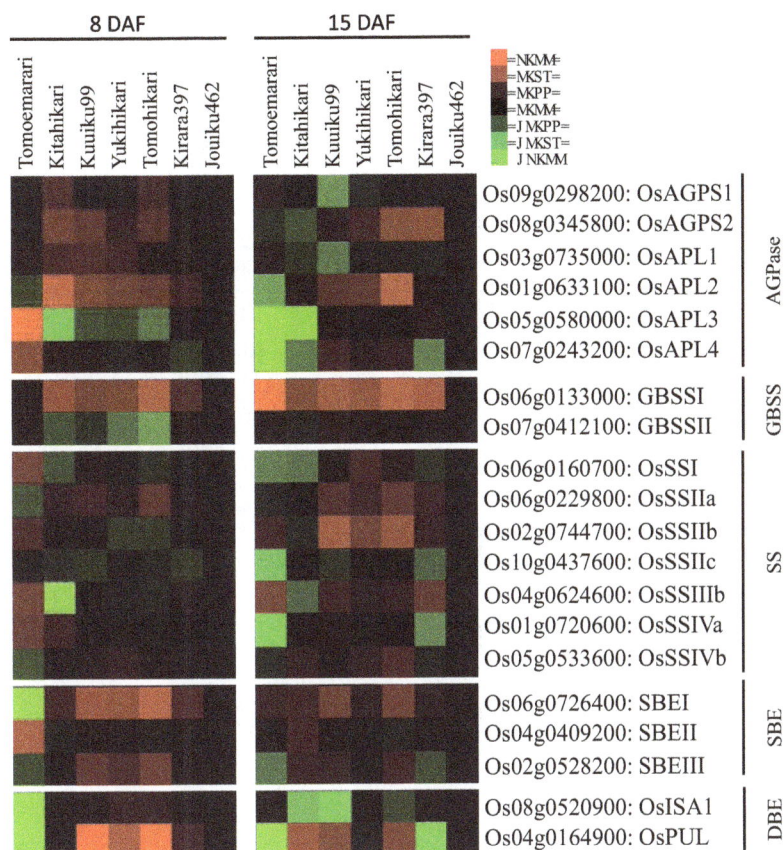

Figure 4: Expression patterns of genes involved in starch synthesis during rice seed development in seven Hokkaido cultivars at 8 days after flowering (DAF) and 15 DAF. Heat map shows the relative expression levels normalized against the expression level in Joiku462 and indicated by a graded color scale, representing relative expression levels of −1, −0.67, −0.33, 0, 0.33, 0.67, and 1 (log2 fold changes). Red, green, and black represent positive, negative, and zero, respectively.

caryopsis and embryo development. Second, the expression profile of seven cultivars at 15 DAF was more variable than that at 8 DAF, as shown by the longer branch length at 15 DAF than at 8 DAF (Figure 1A). Third, the number of differentially expressed genes between the most recent cultivar, Jouiku462, and the other six cultivars was much greater at 15 than at 8 DAF (Figure 3). In contrast, the number of genes expressed at the same level in the most recent cultivar Jouiku462 and the other six cultivars was lower at 15 than at 8 DAF. Fourth, the clustering dendrogram of genome-wide SNPs/InDels showed high similarity to that from clustering by comprehensive expression profiling at 15 but not at 8 DAF (Figure 1B) suggesting that the genome-wide genotype was closely correlated with the transcriptome at 15 DAF.

Rice seed development is divided into four stages: the initiation stage (1-3 DAF), during which starch is synthesized exclusively in the pericarp; the early developmental stage (3-5 DAF), indicated by endosperm starch accumulation with a pronounced increase in seed weight; the middle stage (5-10 DAF), with a rapid increase in starch deposition and grain weight; and the late stage (10 DAF and beyond), in which seed maturation occurs [15,38]. Thus, the gene expression profiles at the middle stage were relatively conserved among the seven cultivars studied suggesting that grain traits associated with the increase in starch deposition and grain weight were selected more than 70 years ago and are critical for rice production in Hokkaido. Grain traits associated with seed maturation may have been improved by the change in gene expression profiles during the last 70 years, as shown in the most recent cultivar, Jouiku462 (Figure 3, Table S3). Further

research is needed to clarify the possible association between grain traits and gene expression profiles.

The expression profile of Tomoemasari, the oldest cultivar (released in 1945), was far from that of the other six cultivars at both stages. Because the present study aimed to compare the profiles of gene expression under natural condition, immature seeds of each cultivar were collected at different days (three groups) depending on its flowering time. The flowering date of Tomoemasari was 10 days later than those of Kuuiku99, Tomohikari, Yukihikari, Kirara397, and Jouiku462 and was 6 days later than that of Kitahikari. The distinct expression profile of Tomoemasari might thus represent both the genetic difference and the effects of different temperatures in late summer between August 25 (flowering date) and September 5 (15 DAF) in Hokkaido.

Transgressive gene expression in first-generation cultivars

We showed that some of the gene expression changes in the first generation cultivars are far (by ≥ 2 times) outside the range of gene expression of either parent, and can accordingly be considered transgressive. These differentially expressed genes were classified into three groups: 97 (8 DAF) and 52 (15 DAF) genes were differentially expressed only in Yukihikari (first group), 140 (8 DAF) and 390 (15 DAF) were differentially expressed only in Tomohikari (second group), and 16 (8 DAF) and 34 (15 DAF) were differentially expressed in both cultivars (third group) (Table S1). Among differentially expressed genes, 12 (8 DAF) and 13 (15 DAF) of the first group, 16 (8 DAF) and 50 (15 DAF) of the second group, and 5 (8 DAF) and 6 (15 DAF) of

Gene	RAP ID / Stage	AAC	Fa	Fb_1	Fb_2	Fb_3	(Fa+Fb1)/(Fb2+Fb3)
OsAGPS1	Os09g0298200 8DAF 15DAF		0.71*		-0.75*	-0.71*	0.73*
OsAGPS2b/a						0.72*	
OsAPL2	Os01g0633100 8DAF 15DAF						
OsAPL3	Os05g0580000 8DAF 15DAF	-0.73*	-0.85**		0.80*	-0.86**	-0.86**
OsAPL4	Os07g0243200 8DAF 15DAF		0.73*	-0.85**		-0.76	O.76*
OsAPL1	Os03g0735000 8DAF 15DAF				-0.73*		
GBSSI(Wx)	Os06g0133000 8DAF 15DAF	0.78*					
GBSSII	Os07g0412100 8DAF 15DAF						
SSI	Os06g0160700 8DAF 15DAF						
OsSSIIa	Os06g0229800 8DAF 15DAF		-0.71* -0.83**		0.72*	0.73* 0.83**	-0.72* -0.80*
OsSSIIc	Os10g0437600 8DAF 15DAF						
OsSSIIb	Os02g0744700 8DAF 15DAF						
OsSSIIIb	Os04g0624600 8DAF 15DAF						
OsSSIVa	Os01g0720600 8DAF 15DAF		0.83**		-0.74*	-0.84**	0.83**
OsSSIVb	Os05g0533600 8DAF 15DAF						
SBE-I	Os06g0726400 8DAF 15DAF		-0.74*			0.75*	-0.74*
SBE-III	Os02g0528200 8DAF 15DAF						
SBE-II	Os04g0409200 8DAF 15DAF			-0.82*			
Isoamylase (ISA)	Os08g0520900 8DAF 15DAF		-0.71*			0.74*	-0.74*
Pullulanase (PUL)	Os04g0164900 8DAF 15DAF		-0.82**		0.76*	0.83**	-0.82**

Table 1: [a]Tarit: AAC, apparent amylose content; Fa, chain length of amylopectins with degrees of polymerization (DP) ranging in the intervals 6–12; Fb_1, DP=13–24; Fb_2, DP=25–36; Fb_3, DP≥ 37. **, *; correlation is significant at 0.01 and 0.05 level, respectively.

the third group were located in genomic regions unique to each of the first-generation cultivars, which are expected to have been transmitted from the parental accession used originally for breeding the cultivars (Figure 2, [48]. In contrast, 85 (8 DAF) and 39 (15 DAF) genes in the first group, 124 (8 DAF) and 340 (15 DAF) in the second group, and 11 (8 DAF) and 28 (15 DAF) in the third group were located outside of unique genomic regions, suggesting that regulation of their expression originated from trans-regulatory factors and/or new combinations of cis-regulatory factors. Recently, expression quantitative trait loci (eQTLs) have been identified as genomic regions that regulate gene expression variation in a genetically segregating population [56, 58]. eQTL analysis can detect genetic elements that regulate the expression variation of the e-trait acting in cis if the QTL is located in the vicinity of an e-trait, or in trans if it is located in a distant position. eQTL analysis could suggest the existence of a potential regulator of a given gene [56,57]. Further study is needed to clarify, by eQTL analysis, the genetic basis of the differentially expressed genes identified in the first-generation cultivars. In addition, there is some possibility that breeders have selected the genotype associated with the differentially expressed gene to improve grain traits during grain filling. The contributions of the genomic regions harboring regulatory factors for the differentially expressed genes to grain traits during grain filling also remain to be clarified.

Expressed genes unique to the most recent Hokkaido cultivar Jouiku462

The expression profile of the most recent Hokkaido cultivar, Jouiku462, represents an ideal gene expression profile essential for high grain yield, high grain quality, and grain traits for grain filling under low-temperature conditions in Hokkaido, the northern limit of rice cultivation. In the present study, we identified 39 genes that were differentially upregulated in Jouiku462 and 64 genes that were differentially downregulated in Jouiku462 compared to the six older cultivars (Tables S2, S3). For example, Os07g0689600 (*Nicotianamine synthase 3*; *OsNAS3*) was upregulated in Jouiku462 at 15 DAF (Table S3). Nicotianamine (NA), a chelator of metals, is ubiquitous in higher plants and a key component of metal homeostasis [59]. Thus, manipulation of cellular NA concentrations would seem to be another approach for improving Fe concentrations *in planta* [59]. NA is synthesized from three molecules of S-adenosyl methionine by Nicotianamine Synthase (NAS). Transgenic rice seeds expressing *HvNAS1* driven by the seed-specific *pGluB-1* promoter also show greater amounts of NA (4-fold) and Fe (1.5-fold) [60] .Among the three *OsNAS* genes present in rice, *OsNAS1* and *OsNAS2* transcripts are markedly elevated in both roots and leaves in response to Fe deficiency, whereas *OsNAS3* expression is induced in roots but suppressed in leaves when Fe supply is inadequate [61]. Seeds from *OsNAS3* activation-tagged plants (*OsNAS3-D1*) in which *OsNAS3* expression was increased in seedling leaves, flag leaves, flowers, and immature seeds, grown in a paddy field, contained elevated amounts of Fe (2.9-fold), Zn (2.2-fold), and Cu (1.7-fold) accompanied with a 9.6-fold increase in the NA level [62]. Cu content of grains was significantly correlated with gel consistency and AC. In addition, significant associations were found between protein content, and Zn and Cu content. Jouiku462 was bred for high grain quality of cooked rice but was not directly selected for mineral content. In Jouiku462, improvement of mineral content to accompany its nutritional properties and eating quality by enhancing *NAS3* expression remains to be investigated.

Os03g0315400 (*OsMYB2*) was also upregulated by ≥ 2-fold in Jouiku462 at 15 DAF compared to the six older cultivars (Table S3). Transgenic rice upregulating *OsMYB2* showed enhanced seedling-stage salinity-stress tolerance along with drought and cold-stress tolerance without reduced growth rate, compared with a control (Yang et al. 2012). *OsLEA3*, *OsRab16A*, and *OsDREB2A*, were upregulated, suggesting that OsMYB2 encodes a stress-responsive MYB transcription factor that may act as a master switch in stress tolerance

[63]. Seed development is initiated by a stage of morphogenesis in which the mature embryo develops from a single fertilized cell. Following morphogenesis, developing seeds enter a maturation stage, during which storage compounds are synthesized [64]. The early and middle phases of maturation are dominated by the action of the Phytohormone, Abscisic Acid (ABA) [65]. Subsequently, ABA levels decline and the seed enters the desiccation phase. This phase is associated with a major loss of water, and the water content of mature dry seeds is generally less than 20% on a fresh weight basis. Such severe desiccation would kill cells in vegetative parts of the plant [66]. After 10 DAF, the water content of the developing seed drops markedly, indicating that this period is the beginning stage of dehydration [67]. However, these severely desiccated seeds germinate and develop into normal seedlings after imbibition, indicating that the developing seeds must acquire sufficient desiccation tolerance before the end of the desiccation phase, when their water content falls below the threshold of tolerance for vegetative cells. The contribution of the upregulation of *OsMYB2* to dehydration stress tolerance during grain filling awaits further study.

GBSSII (*Wx*) expression was reduced in Jouiku462 (Figure 4), although all seven cultivars carry the *Wx^b* allele. Jouiku462 was bred as a low-amylose cultivar by introgression of *qAC9.3*, which is a low-amylose-content QTL [68], using marker-assisted selection (Sato et al. unpublished data). The molecular mechanism underlying the reduction of AAC by *qAC9.3* is still unknown. The possibility that *qAC9.3* reduces *Wx* expression at transcriptional and post-transcriptional levels awaits further study.

Conclusion

Our study provides genome-wide transcriptome data of developing seeds at 8 DAF (middle stage) and 15 DAF (late stage) under natural conditions in Hokkaido (42.52°N) for very closely related Hokkaido rice cultivars released in the past 70 years. These genome-wide transcriptome data reveal the genetic control of differentially expressed genes in first-generation cultivars. The transcriptome of a current cultivar shows possible ideal expressed genes associated with high grain yield and quality and grain traits essential for rice production near the northern and southern limits of rice cultivation.

Acknowledgements

This study was supported in part by Grants-in-Aid for Regional R&D Proposal-Based Program from Northern Advancement Centre for Science & Technology of Hokkaido Japan and the Tojuro Iijima Foundation for Food Science and Technology. Seeds of Hokkaido rice cultivars were provided by the Central Agricultural Experiment Station, Local Independent Administrative Agency Hokkaido Research Organization. We thank Reina Asa, Shoko Yoshikawa, Naomi Shimoda, Hideyuki Kitazawa, Hidetaka Nagasawa, Noriko Kinoshita, and Hiroki Yamamoto for technical support.

References

1. Juliano BO, Perez CM, Kaosa M (1990) Grain quality characteristics of export rice in selected markers. Cereal Chem 67: 192-197.

2. Chen Y, Wang M, Ouwerkerk PBF (2012) Molecular and environmental factors determining grain quality in rice. Food Energy Sec 1: 111-132.

3. French D (1984) Organization of starch granules. In: Whistler RL, BeMiller JN, Paschall EF (eds) Starch: Chemistry and technology, Academic Press, Orlando, FL: 183-247.

4. Hizukuri S (1986) Polymodal distribution of the chain lengths of amylopectins, and its significance. Carbohyd Res 147: 342-347.

5. Gidley M (1992) Structural order in starch granules and its loss during gelatinisation. In: Phillips GO, Williams PA, Wedlock DJ (eds) Gums and stabilisers for the food industry 6, Oxford, UK, IRL: 87-92.

6. Jenkins PJ, Cameron RE, Donald AM (1993) A universal feature in the structure of starch granules from different botanical sources. Starch 45: 405-409.

7. Juliano BO (1992) Structure and function of the rice grain and its fractions. Cereal Foods World 37: 772-774.

8. Suwannaporn P, Pitiphunpong S, Champangern S (2007) Classification of rice amylose content by discriminant analysis of physicochemical properties. Starch 59: 171-177.

9. Nakamura Y, Francisco PB Jr, Hosaka Y, Sato A, Sawada T, et al. (2005) Essential amino acids of starch synthase IIa differentiate amylopectin structure and starch quality between japonica and indica rice varieties. Plant Mol Biol 58: 213-227.

10. Waters DLE, Henry RJ, Reinke RF, Fitzgerald MA (2006) Gelatinization temperature of rice explained by polymorphisms in starch synthase. Plant Biotechnol J 4: 115-122.

11. Hanashiro I, Abe J, Hizukuri S (1996) A periodic distribution of the chain length of amylopectin as revealed by high-performance anion-exchange chromatography. Carbohyd Res 283: 151-159.

12. Nakamura Y, Sakurai A, Inaba Y, Kimura K, Iwasawa N, et al. (2002) The fine structure of amylopectin in endosperm from Asian cultivated rice can be largely classified into two classes. Starch 54: 117-131.

13. Cameron DK, Wang YJ, Moldenhauer KA (2008) Comparison of physical and chemical properties of medium-grain rice cultivars grown in California and Arkansas. J Food Sci 73: C72-C78.

14. Takahashi S, Sugiura T, Naito H, Shibuya N, Kainuma K (1998) Correlation between taste of cooked rice and structural characteristics of rice starch. J Appl Glycosic 45: 99-106.

15. Hirose T, Terao T (2004) A comprehensive expression analysis of the starch synthase gene family in rice (Oryza sativa L.). Planta 220: 9-16.

16. Tetlow IJ, Morell MK, Emes MJ (2004) Recent developments in understanding the regulation of starch metabolism in higher plants. J Exp Bot 55: 2131-2145.

17. Zhang X, Cheng, Z, Guo X, Su N, Jiang, L, et al. (2011) Double repression of soluble starch synthase genes SSIIa and SSIIIa in rice (Oryza sativa L.) uncovers interactive effects on the physicochemical properties of starch. Genome 54:448-459

18. James MG, Denyer K, Myers AM (2003) Starch synthesis in the cereal endosperm. Curr Opin Plant Biol 6: 215-222.

19. Jeon JS, Ryoo N, Hahn TR, Walia H, Nakamura Y (2010) Starch biosynthesis in cereal endosperm. Plant Physiol Biochem 48: 383-392.

20. Tian Z, Qian Q, Liu Q, Yan M, Liu X, et al. (2010) Allelic diversities in rice starch biosynthesis lead to a diverse array of rice eating and cooking qualities. Proc Natl Acad Sci USA 106: 21760-21765.

21. Sano Y, Maekawa M, Kikuchi H (1985) Temperature effects on the Wx protein level and amylose content in the endosperm of rice. J Hered 76: 221-222.

22. Preiss J (1991) Biology and molecular biology of starch synthesis and its regulation. Oxf Surv Plant Mol Cell Biol 7: 59-114.

23. Smith AM, Denyer K, Martin C (1997) The synthesis of the starch granule. Annu Rev Plant Physiol Plant Mol Biol 48: 67-87.

24. Myers AM, Morell MK, James MG, Ball SG (2000) Recent progress toward understanding biosynthesis of the amylopectin crystal. Plant Physiol 122: 989-998.

25. Tanaka N, Fujita N, Nishi A, Satoh H, Hosaka Y, et al. (2004) The structure of starch can be manipulated by changing the expression levels of starch branching enzyme IIb in rice endosperm. Plant Biotechnol J 2: 507-516.

26. Gao ZY, Zheng DL, Cui X, Zhou YH, Yan MX, et al. (2003) Map-based cloning of the ALK gene, which controls the gelatinization temperature of rice. Sci China (Ser. C) 46: 661-668.

27. Umemoto T, Yano M, Shomura A, Satoh H, Nakamura Y (2002) Mapping of a gene responsible for the difference in amylopectin structure between japonica-type and indica-type rice varieties. Theor Appl Genet 104: 1-8.

28. Nishi A, Nakamura Y, Tanaka N, Satoh H (2001) Biochemical and genetic analysis of the effects of amylose-extender mutation in rice endosperm. Plant Physiol 127: 459-472.

29. Kubo A, Fujita N, Harada K, Matsuda T, Satoh H, et al. (1999) The starch-debranching enzyme isoamylase and pullulanase are both involved in amylopectin biosynthesis in rice endosperm. Plant Physiol 121: 399-409

30. Fujita N, Kubo A, Suh DS, Wong KS, Jane JL, et al. (2003) Antisense inhibition of isoamylase alters the structure of amylopectin and the physicochemical properties of starch in rice endosperm. Plant Cell Physiol 44: 607-618.

31. Kubo A, Rahman S, Utsumi Y, Li Z, Mukai Y, et al. (2005) Complementation of sugary-1 phenotype in rice endosperm with the wheat Isoamylase1 gene supports a direct role for isoamylase1 in amylopectin biosynthesis. Plant Physiol 137: 43-56.

32. Tanaka I, Kobayashi A, Tomita K, Takeuchi Y, Yamagishi M, et al. (2006) Detection of quantitative trait loci for stickiness and appearance based on eating quality test in japonica rice cultivar. Breed Res 8: 39-47.

33. Takeuchi Y, Nonoue Y, Ebitani T, Suzuki K, Aoki N, et al. (2007) QTL detection for eating quality including glossiness, stickiness, taste and hardness of cooked rice. Breed Sci 57: 231-242.

34. Kobayashi A, Tomita K, Yu F, Takeuchi Y, Yano M (2008) Verification of quantitative trait locus for stickiness of cooked rice and amylose content by developing near-isogenic lines. Breed Sci 58: 235-242.

35. Wada T, Ogata T, Tsubone M, Uchimura Y, Matsue Y (2008) Mapping of QTLs for eating quality and physicochemical properties of the japonica rice 'Koshihikari'. Breed Sci 58: 427-435.

36. Wada T, Yasui H, Inoue T, Tsubone M, Ogata T, et al. (2013) Validation of QTLs for eating quality of japonica rice 'Koshihikari' using backcross inbred lines. Plant Prod Sci 16: 131-140.

37. Counce PA, Keisling TC, Mitchell AJ (2000) A uniform, objective, and adaptive system for expressing rice development. Crop Sci 40: 436-443.

38. Dian W, Jiang H, Wu P (2005) Evolution and expression analysis of starch synthase III and IV in rice. J Exp Bot 56: 623-632.

39. Ohdan T, Francisco PB, Jr, Sawada T, Hirose T, Terao T, et al. (2005) Expression profiling of genes involved in starch synthesis in sink and source organs of rice. J Exp Bot 56: 3229-3244.

40. Bao J, Kong X, Xie J, Xu L (2004) Analysis of genotypic and environmental effects on rice starch. 1. Apparent amylose content, pasting viscosity, and gel texture. J Agric Food Chem 52: 6010-6016

41. Sharifi P, Dehghani H, Mumeni A, Moghaddam M (2009) Genetic and genotype 9 environment interaction effects for appearance quality of rice. Agric Sci China 8: 891-901.

42. Hirano H-Y, Sano Y (1998) Enhancement of Wx gene expression and the accumulation of amylose in response to cool temperatures during seed development in rice. Plant Cell Physiol 39: 807-812.

43. Larkin PD, Park WD (1999) Transcript accumulation and utilization of alternate and non-consensus splice sites in rice granule-bound starch synthase are temperature-sensitive and controlled by a single-nucleotide polymorphism. Plant Mol Biol 40: 719-727.

44. Umemoto T, Terashima K, Nakamura Y, Satoh H (1999) Differences in amylopectin structure between two rice varieties in relation to the effects of temperature during grain-filling. Starch 51: 58-62.

45. Igarashi T (2010) Molecular structure of amylopectin from Hokkaido rice starch and its effect on eating quality. J Appl Glycosci 57: 25-32.

46. Bolstad BM, Irizarry RA, Astrand M, Speed TP (2003) A comparison of normalization methods for high density oligonucleotide array data based on variance and bias. Bioinformatics 19: 185-193.

47. Takano S, Matsuda S, Kinoshita N, Shimoda N, Sato T, et al. (2014) Genome-wide single nucleotide polymorphisms and insertion-deletions of Oryza sativa L. subsp. japonica cultivars grown near the northern limit of rice cultivation. Mol Breed 34: 1007-1021.

48. de Hoon MJ, Imoto S, Nolan J, Miyano S (2004) Open source clustering software. Bioinformatics 20: 1453-1454.

49. Saldanha AJ (2004) Java Treeview--extensible visualization of microarray data. Bioinformatics 20: 3246-3248.

50. Williams VR, Wu WT, Tsai HY, Bates HG (1958) Varietal differences in amylose content of rice starch. J Agric Food Chem 6: 47-48.

51. Inatsu O (1988) Studies on improving the eating quality of Hokkaido Rice. Report of Hokkaido Prefectural Agric Exp Stn 66: 3-7.

52. Wong KS, Kubo A, Jane JI, Harada K, Satoh H, et al. (2003) Structures and properties of Amylopectin and phytoglycogen in the endosperm of sugary-1 mutants of rice. J Cereal Sci 37: 139-149.

53. Hizukuri S, Takeda Y, Yasuda M, Suzuki, A (1981) Multibranched nature of amylose and the action of debranching enzymes. Carbohyd Res 94: 205-213.

54. O'Shea MG, Samuel MS, Konik CM, Morell MK (1998) Fluorophore-assisted carbohydrate electrophoresis (FACE) of oligosaccharides: efficiency of labeling and high-resolution separation. Carbohyd Res 105: 1-12.

55. Gilad Y, Rifkin SA, Pritchard JK (2008) Revealing the architecture of gene regulation: the promise of eQTL studies. Trends Genet 24: 408-415.

56. Hansen BG, Halkier BA, Kliebenstein DJ (2008) Identifying the molecular basis of QTLs: eQTLs add a new dimension. Trends Plant Sci 13: 72-77.

57. Kliebenstein D (2009) Quantitative genomics: analyzing intraspecific variation using global gene expression polymorphisms or eQTLs. Ann Rev Plant Biol 60: 93-114.

58. Douchkov D, Gryczka C, Stephan UW, Hell R, Baumlein H (2005) Ectopic expression of nicotianamine synthase genes results in improved iron accumulation and increased nickel tolerance in transgenic tobacco. Plant Cell Environ 28: 365-374.

59. Usuda K, Wada Y, Ishimaru Y, Kobayashi T, Takahashi M, et al. (2009) Genetically engineered rice containing larger amounts of nicotianamine to enhance the antihypertensive effect. Plant Biotechnol J 71: 87-95.

60. Inoue H, Higuchi K, Takahashi M, Nakanishi H, Mori S, et al. (2003) Three rice nicotianamine synthase genes, OsNAS1, OsNAS2, and OsNAS3 are expressed in cells involved in long-distance transport of iron and differentially regulated by iron. Plant J 36: 366-381.

61. Lee S, Jeon US, Lee SJ, Kim Y-K, Persson DP, et al. (2009) Iron fortification of rice seeds through activation of the nicotianamine synthase gene. Proc Natl Acad Sci USA 106: 22014-22019.

62. Yang A, Dai X, Zhang WH (2012) A R2R3-type MYB gene, OsMYB2, is involved in salt, cold, and dehydration tolerance in rice. J Exp Bot 63: 2541-2556.

63. Angelovici R, Galili G, Fernie AR, Fait A (2010) Seed desiccation: a bridge between maturation and germination. Trends Plant Sci 15: 211-218.

64. Nambara E, Marion-Poll A (2003) ABA action and interactions in seeds. Trends Plant Sci 8: 213-217.

65. Jensen AB, Busk PK, Figueras M, Albà MM, Peracchia G, et al. (1996) Drought signal transduction in plants. Plant Growth Regul 20: 105-110

66. Sano N, Masaki S, Tanabata T, Yamada T, Hirasawa T, et al. (2013) Proteomic analysis of stress-related proteins in rice seeds during the desiccation phase of grain filling. Plant Biotech 30: 147-156.

67. Ando I, Sato H, Aoki N, Suzuki Y, Hirabayashi H, et al. (2010) Genetic analysis of the low-amylose characteristics of rice cultivars Oborozuki and Hokkai-PL9. Breed Sci 60: 187-194.

27

Comparative Study of Population Dynamics and Breeding Patterns of *Mastomys natalensis* in System Rice Intensification (SRI) and Conventional Rice Production in Irrigated Rice Ecosystems in Tanzania

Loth Mulungu S[1]*, Happy Lopa[2] and Mashaka Mdangi E[3,4]

[1]Pest Management Centre, Sokoine University of Agriculture, PO Box 3110 Morogoro, Tanzania
[2]Rodent Control Centre, Ministry of Agriculture, Food Security and Cooperatives, P.O. Box 3047 Morogoro, Tanzania
[3]MATI-Ilonga, P.O. Box 66, Kilosa, Tanzania
[4]Crop Science and Production, Sokoine University of Agriculture, P.O. Box 3005, Morogoro, Tanzania

Abstract

Mastomys natalensis is among the most important rodent pests in Sub-Saharan Africa. This study investigated the population dynamics and breeding patterns of this mouse in system rice intensification (SRI) and in conventional cropping systems in irrigated rice ecosystem in eastern Tanzania. The *Mastomys natalensis* population varied with years and season, but not with either SRI or conventional cropping system which would be expected as the all fields are in the same area. The highest population peak was observed during the dry season i.e., August to September. Breeding patterns of this rodent pest was not influenced by the cropping system or season, indicating that *M. natalensis* is sexually active throughout the year and does not be affected by the rice production systems. Regular control and sustainable operations, such as the use trap barrier system (TBS), are therefore essential if the populations are to be kept within tolerable limits.

Keywords: *Mastomys natalensis*; Population dynamics; Breeding patterns; SRI; Conventional cropping system

Introduction

Tanzania ranks, second to Madagascar, as a major rice producer in Africa [1]. However, in several areas of Tanzania, rice has been grown in the conventional way, where farmers flood the field with water, followed by planting several seedlings per hole [2]. Farmers in some of the irrigated scheme in Tanzania, such as Mkindo, have started practicing System Rice Intensification (SRI) for rice crop production [3]. The SRI is an agro-ecological method of growing rice that enhances crop yields which is resilient to the adverse effects of climate change. It is being recognized for its impact on the availability, affordability, accessibility and adequacy of food. The main factors that explain the impacts of SRI management are: high vigor of plant, more effective root systems and the promotion of greater abundance and diversity of beneficial soil organisms, which are factors outside the Green Revolution paradigm [4,5]. In areas where farmers are not growing the rice crop using SRI, they can get paddy rice yields of an average of 2 tons ha^{-1} from their very poor soils with conventional methods as compared to an average of 8 tons ha^{-2} in SRI [3]. Therefore, SRI methods have been shown to increase crop yields by 20 to 50% with significant reductions in water and seed requirements [6].

SRI is a package of practices especially developed to improve the productivity of rice grown in paddies. Unlike the conventional method of continuous flooding of paddy fields, SRI involves intermittent wetting and drying of paddies as well as specific soil and agronomic management practices. Therefore, the advantages of SRI are that it gives higher rice yields compared to continuously flooded paddies; requires less water (e.g. it can save 25-50% water used in irrigation); uses less seeds. Moreover, the SRI, rice has a harder grain, thus less breakage during milling, fetches higher price. Nearly all rice varieties give higher yield with SRI, but some high-yielding varieties respond better than others [7,8].

Despite these advantages of rice production using SRI, rodents could be the major threat to crop. Mulungu et al. [9] reported that crop losses caused by rodents in rice crop are largely attributed to *Mastomys natalensis*, which is the most widespread rodent pest across sub-Saharan [10]). However, it is not known whether the SRI has any effect on rodent population dynamics due to availability of more food in the field for *M. natalensis*.

Mulungu et al. [9] reported that the population of *M. natalensis* in rice fields produced under the conventional cropping systems varied seasonally with the population being higher during the dry season. The rodents are sexually active throughout the year, although the population reaches the highest level when rice is at the maturity stage. This suggests that breeding is highly influenced by the presence of a rice crop in the field. Furthermore, Mulungu et al. [11] reported that continuously rodent breeding in rice grown under conventional cropping systems is attributed to the availability of vegetative plant materials and seeds largely consumed but also by the fact that the rodent are highly specialized herbivorous/grainivorous in nature.

This study aimed at investigating the population dynamics and breeding patterns of the rodent pest species in system rice intensification (SRI) in irrigated rice cropping systems in eastern Tanzania.

*Corresponding author: Loth Mulungu S, Pest Management Centre, Sokoine University of Agriculture, PO Box 3110 Morogoro, Tanzania
E-mail: lothmulungu@yahoo.co.uk

Materials and Methods

Study area

This study was conducted at Mkindo village (6° 16' S, 37° 36'E; altitude 365 m a.s.l.) in Mvomero District, Morogoro, Tanzania. The study area has a bimodal rainfall pattern in which short rains fall from October to December while long rains fall from March to June. The soil is clayey loam with infiltration rate of 12 cm/day. Soils have a pH of 6.2, K 12.75 mg/kg, P 0.532 mg/kg and N (%) 1.00. Farmers in the study area produce the rice crop twice per year. The first cropping calendar is in the wet season from February to July, while the second is in the dry season, from September to January. The latter purely depends on irrigation. Land preparation and rice transplanting are done in February and September during the wet and dry seasons, respectively. The rice crops reaches physiological maturity in July and January, and farmers harvest in July and January, for the wet and the dry cropping seasons, respectively. For the remaining months, the crop is at a vegetative stage.

Trapping rodents

A capture–mark–recapture study was carried out from August 2012 to July 2014. Three 70 × 70 m permanent trapping grid in each three cropping systems (SRI, Farmer Field School with SRI, and conversional rice cropping systems) of the rice crop were established. Each cropping system consisted of seven parallel lines, 10 m apart, and seven trapping stations per line, also 10 m apart (a total of 49 trapping stations field^{-1}). One Sherman LFA live trap (8 × 9 × 23 cm; H.B Sherman Traps Inc., Tallahassee, FL) was placed at each trapping station, for three consecutive nights at intervals of 4 weeks. The traps were baited with peanut butter, mixed with maize bran/maize flour and were placed in the afternoon and inspected in the morning of the following day. During flooding, the traps were placed above dried grasses within the rice field.

Captured animals

All the captured animals were taken to the field laboratory and identified to species level [12]. On the first day of capture, the animals were marked by toe clipping using specific number coding. Their weights, trapping stations, sex and reproductive status were recorded. The sex and reproductive conditions considered either a perforated or closed vagina, in females, and scrotal or non-scrotal testes, in males. The animals were then released at the same station of capture. The recorded data were then entered into a CMR data input program for analysis. New animals captured on subsequent days and during subsequent rounds of trapping were similarly marked, recorded and released.

Data collection

Species composition: The percentages of each species relative to others were calculated by dividing the number of captured individuals of each species by the total number of captured animals in a particular habitat, and multiplied by 100.

Population dynamics: The population size was estimated monthly for each three day trapping session using the M(h) estimator of the program CAPTURE for a closed population, which allows for individual heterogeneity in a trapping probability [13].

Breeding patterns: Sexual activity and breeding patterns were determined by establishing the proportion of active and non-active individuals of female mice in both habitats and month. In females, the definition of sexual activity followed that of [14], who defined sexual

activity in females as a physiological condition and not as a typical behavior. Thus, females were considered to be sexually active when the vagina was perforated, when their nipples were swollen on account of lactation, and when they were pregnant.

Data analysis

To analyze factors that influence abundance and percentage of breeding animals, we used a generalized linear model with logit-link function. A correction for over dispersion was done as the residual deviation was always higher than the degrees of freedom of the models. We used the simultaneous test for general linear hypothesis with Tukey contrast (multcomp package in R) to investigate which years differed significantly from other years.

Results

Species composition

A total of 1882 individual animals belonging to three species of rodents and one species of shrew (*Crocidura* sp.), were captured from a total of 11025 trap nights (17.1% trap success).The rodent species comprised *Mastomys natalensis, Rattus rattus,* and *Mus species. Mastomys natalensis* comprised more than 96% of the total captures in the study area. The SRI had more individuals trapped than other rice cropping systems (Table 1).

Population dynamics

Mastomys natalensis was the predominant species in the study site. This observation justified analysis of population dynamics and breeding patterns for this species. Generally, the trend of this species was similar in both cropping systems. The abundances in the study area were significantly influenced by year (GLM, log link, df=2, p<0.001), with significant differences between the years 2013 - 2014 (p=0.004) and 2012 - 2013 (p=0.003). Similarly, the abundances were significantly influenced by season (GLM, log link, df=1, p<0.001), with a higher overall abundance during the dry season and a significant interaction effect between season and year (GLM, log link, df=2, p=0.002). However, no significant effect of cropping system on abundance (GLM, log link, df=1, p=0.054) was observed in the current study (Figure 1).

Breeding patterns

Results show that the percentage of breeding females was only significantly influenced by year (GLM, logit link, df=1, p<0.001), with significant differences between the years 2014-2013 (p=0.016), 2012-2013 (p<0.001) and 2014-2012 (p<0.001). However, no significant effect of season (GLM, logit link, df=1, p=0.618) and field rice cropping system (GLM, logit link, df=2, p=0.414) interactions, between the tested factors on the percentage of breeding females was observed. This indicated that the proportion of active and non-active female individuals were the same throughout the year and cropping systems. Generally, sexually active females were observed in all months during the entire study period in both rice cropping systems (Figure 2).

	Species	SRI	Farmer Field School	Conventional
1	*Mastomys natalensis*	746 (96.76)	567 (97.09)	511 (96.96)
2	*Crocidura* spp	25 (3.24)	15 (2.57)	15 (2.85)
3	*Rattus rattus*	-	2 (0.34)	-
4	*Mus* spp	-	-	1 (0.19)
	Total	771 (100)	584 (100)	527 (100)

Table 1: Species composition of different rodent species and shrew in SRI, Farmers Field School and Conversional rice production in Irrigated Rice production scheme.

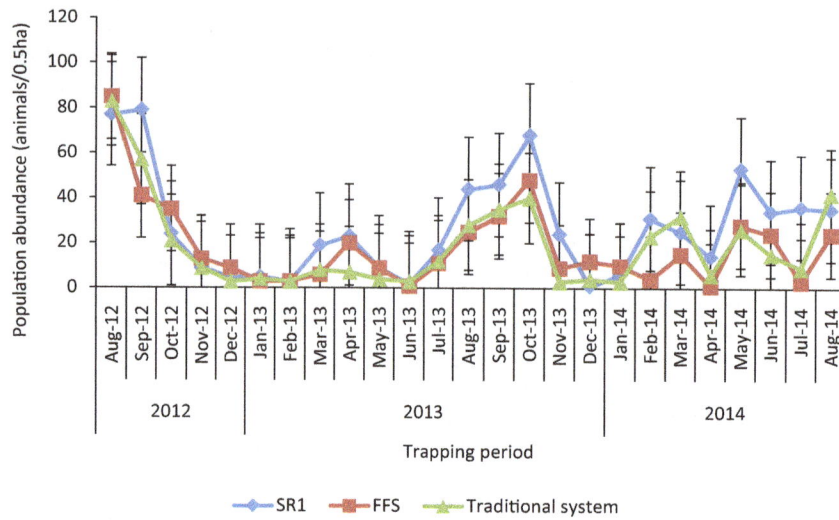

Figure 1: Population dynamics of *M. natalensis* in different cropping systems in Mkindo village.

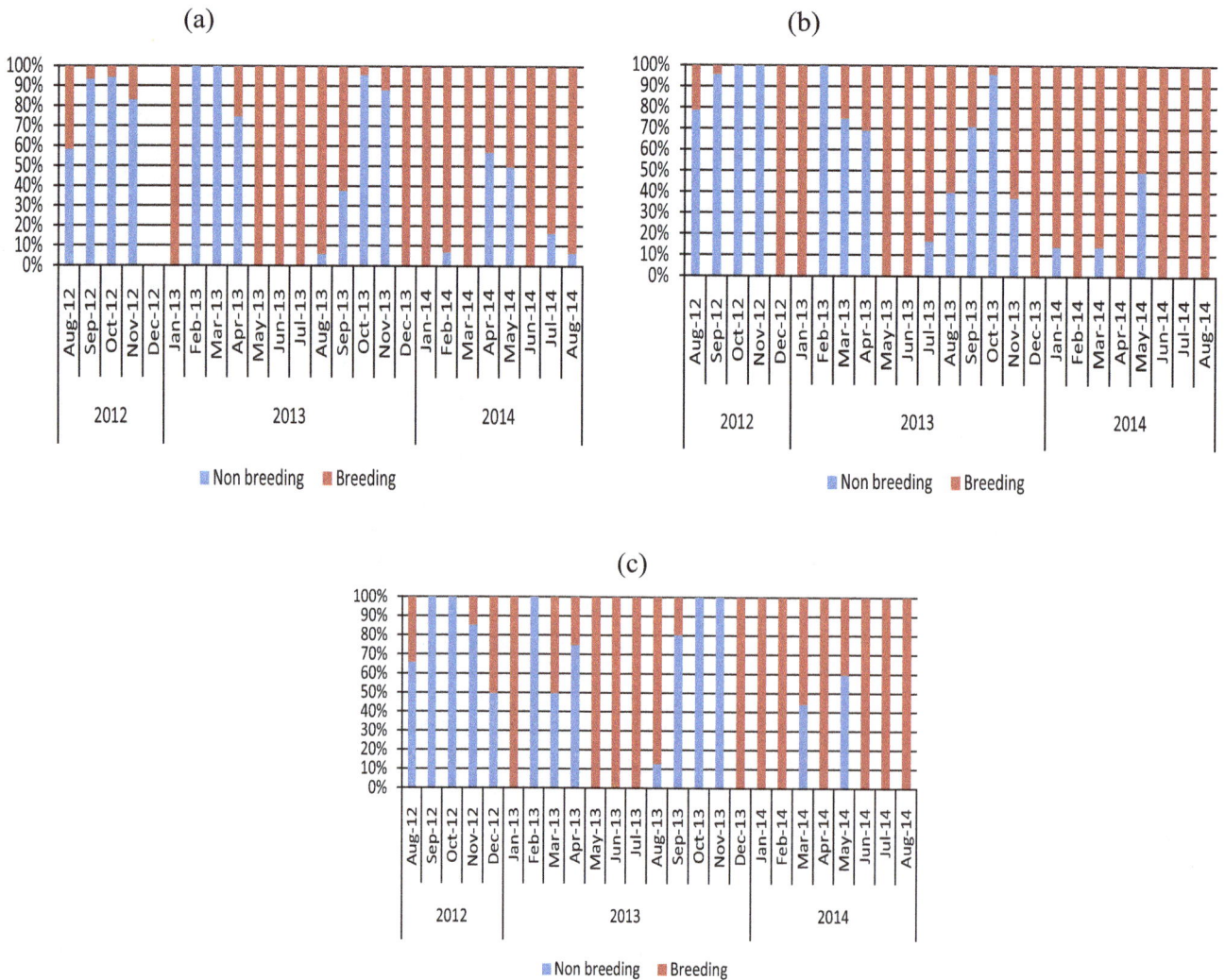

Figure 2: Breeding patterns of female *M. natalensis* in (a) SRI Fields, (b) Farmer field school, and (c) conventional rice production system.

Discussion

Rodent and shrew species were captured in the study area across the study period. *Mastomys natalensis* was the most abundant rodent species captured in the study area. This agrees with the findings reported by Mulungu et al. [9] who observed that >95% of individuals captured in irrigated rice were *M. natalensis*. These observations are consistent with those reported for areas where *M. natalensis* was most common in sites with many bushes or tall grasses [15,16].

Many small mammal species including rodents, show large spatial and temporal fluctuations in numbers. The abundances of *M. natalensis* in this study fluctuated over years and season in rice agro-ecosystem. This could be attributed to food availability in a particular year and season. Habitat quality for rodent pest species will likely vary according to food availability changes between years and seasons. Therefore, it is expected that the population dynamics of resident animals will exhibit abundance differences [17]. It was observed that in both years rodent abundance was higher in August and September, which corresponds with the dry season. The current findings are consistent with other findings [9,14,18,19].

It has been reported that in mosaic habitats, *M. natalensis* breeds during the long rains, and normally starts one month after the usual peak rainfall (March and April), lasting until the dry season [20,21]. This kind of breeding is related to variations in rainfall, peaking towards the end of the rainy season when food resources are plenty [18,22]. Their breeding therefore decreases during the drier months [23] where the population is higher [14,24]. However, in the current study, breeding of *M. natalensis* was observed to have no differences between season and rice cropping systems, indicating that both rice cropping systems had the equally adequate resources to support individuals which would be expected as the three cropping fields are in the same area, and hence breeding was continuous. Similar findings were reported by Mulungu et al. [9] in a conventional rice production in irrigated agro-system. Mulungu et al. [11] pointed out that continuous breeding in a conventional rice cropping system is attributed by vegetative plant material and seeds consumed. Vegetative productivity in the study area was influenced by continuous irrigation in rice fields which trigger sprouting of new green grasses. It has been reported that large amount of the chemical, 6-methoxy-2-benzoxazolinone, which is found to be present in large quantities in sprouting grass and it acts as a trigger for reproduction and influences growth in several rodents [25,26]. Similarly, Leirs and Verheyen [20] and Field [27] reported that seeds are an important food item during the breeding season and are required to meet the high energy needs of a reproducing organism.

Conclusion

Mastomys natalensis is the most abundant pest species in the study area and its population varies with years and season but not with the cropping systems. The highest population peak was observed during the dry season from August to September. The breeding patterns of this rodent pest species was also influenced by years and not by cropping systems and cropping seasons, indicating that *M. natalensis* is sexually active throughout the year in all cropping systems. Therefore, regular and sustainable control operations such as general field hygiene are necessary in order to keep the rodent pest population within tolerable limits in the study area.

Acknowledgement

This work was supported by Commission for Science and Technology (COSTECH). The authors appreciate the field assistance from Khalid S Kibwana, Omary Kibwana, and Ramadhani Kigunguli of the Pest Management Centre, Sokoine University of Agriculture, Morogoro, Tanzania.

Reference

1. Oteng JW, Sant'Anna R (2015) Rice production in Africa: Current situation and issues.

2. MAFSC (2009) National Rice Development Strategy, Ministry of Agriculture and Food Security.

3. Mdemu MV, Magayane MD, Lankford B, Hatibu N, Kadigi RMJ (2004) Co joining rainfall and irrigation seasonality to enhance productivity of water in rice irrigated farms in the Upper Ruaha River Basin, Tanzania Phys Chem Earth 29: 1119-1124.

4. Lu XM, Huang Q, Liu HZ (2006) Research of some physiological characteristics under the System of Rice Intensification, Journal of South China Agricultural University 27: 5-7.

5. Horie T, Shiraiwa T, Homma K, Katsura K, Maeda S (2005) Can yields of lowland rice resumes the increases that they showed in the 1980s? Plant Production Science 8: 257-272.

6. McDonald AJ, Hobbs PR, Riha SJ (2006) Does the system of rice intensification outperform conventional best management? A synopsis of the empirical record, Field Crops Research 96: 31-36.

7. Dobermann A (2004) A critical assessment of the system of rice intensification (SRI). Agricultural Systems 79: 261-281.

8. Sinha SK, Talati J (2005) Impact of System of Rice Intensification (SRI) on Rice Yields: Results of a New Sample Study in Purulia District, India.

9. Mulungu LS, Ngowo V, Mdangi M, Katakweba AS, Tesha P, et al. (2013) Population dynamics and breeding patterns of Multi-mammate mouse, Mastomys natalensis (Smith 1834) in irrigated rice field in Eastern Tanzania. Pest Management Science 69: 371-377.

10. Fiedler LA (1994) Rodent Pest Management in East Africa. FAO (Food and Agriculture Organization of the United Nations) Plant Production and Protection, Italy.

11. Mulungu LS, Mlyashimbi ECM, Ngowo V, Mdangi M, Katakweba AS, et al. (2014) Food preferences of the Multi-mammate mouse, Mastomys natalensis (Smith 1834) in irrigated rice field in eastern Tanzania. International Journal of Pest Management 60: 1-8.

12. Kingdon J (1997) The Kingdon Field Guide to African Mammals. Academic Press, London, UK, pp. 464.

13. White GC, Anderson DR, Burnham KP, Otis DL (1982) Capture-Recapture and Removal Methods for Sampling Closed Populations. Los Alamos National Laboratory, Los Alamos, New Mexico, pp. 235.

14. Leirs H (1995) Population ecology of Mastomys natalensis (Smith 1834): implication for rodent control in East Africa. Agricultural Edition No. 35, Belgian Administration for Development Cooperation, Brussels, Belgium 1-268.

15. Dieterlen F (1967) A study of the population of rodents in Kivu (Congo) (in German). Teil I Zool Jb Syst 94: 369-426.

16. Martin GHG, Dickinson NM (1985) Small mammal abundance in relation to microhabitat in dry sub-humid grassland in Kenya. Afr J Ecol 23: 223-234.

17. Mulungu LS, Sixbert V, Ngowo V, Mdangi M, Katakweba AS, et al. (2015) Spatio-temporal patterns in the distribution of the multimammate mouse, Mastomys natalensis, in rice crop and fallow land habitats in Tanzania. Mammalia 79: 177-184.

18. Leirs H, Verhagen R, Verheyen W, Mwanjabe P, Mbise T (1996) Forecasting rodent outbreaks in Africa: an ecological basis for Mastomys control in Tanzania. J Appl Ecol 33: 937-943.

19. Monadjem A (1998) Reproduction biology age structure and diet of Mastomys natalensis (Muridae: Rodentia) in a Swaziland grassland Z Säugetierkunde 63: 347-356.

20. Leirs H, Verhagen W (1995) Population ecology of Mastomys natalensis. Implications for rodent control in Africa Agricultural Editions No 35. Brussels: Belgium Administration for Development Cooperation pp. 268.

21. Massawe AW, Mulungu LS, Makundi RH, Dlamini N, Eiseb SJ, et al. (2011) Spatial and temporal population dynamics of rodents in three geographically

different regions in Africa: implication for ecologically-based rodent management. Afr Zool 46: 393-405.

22. Leirs H (1992) Population ecology of Mastomys natalensis (Smith, 1834) implication for rodent control in Africa. PhD Thesis, University of Antwerp, Antwerp, Belgium, pp. 268.

23. Wube T (2005) Reproductive rhythm of the grass rat, Arvicanthis abyssinicus at the Entoto Mountain, Ethiopia. Belg J Zool 135: 53-56.

24. Massawe AW, Makundi RH, Mulungu LS, Katakweba A, Shayo TN (2012) Breeding dynamics of rodent species inhabiting farm-fallow mosaic fields in Central Tanzania, African Zoology 47: 128-137.

25. Negus NC, Berger PT (1987) Mammalian reproductive physiology: adaptive responses to changing environments Curr Mammal 1: 149-173.

26. Korn H (1989) A feeding experiment with 6-methoxy benzoxalinone and a wild population of the deer mouse (Peromyscus maniculatus). Can J Zool 9: 2220-2224.

27. Field AC (1975) Seasonal changes in reproduction, diet and body composition of two equatorial rodents. East African Wildlife Journal 13: 221-235.

28

AMMI Biplot Analysis for Stability of Grain Yield in Hybrid Rice (*Oryza sativa L.*)

Anowara Akter[1*], Jamil Hassan M[1], Umma Kulsum M[1], Islam MR[1], Kamal Hossain[1] and Mamunur Rahman M[2*]

[1]Plant Breeding Division, Bangladesh Rice Research Institute, Bangladesh
[2]Senior Scientific Officer, Farm Management Division, Bangladesh

Abstract

Genotype x environment interaction and stability performance were investigated on grain yield with 12 rice genotypes in five environments. The ANOVA for grain yield revealed highly significant (P<0.01) for genotypes, environments and their interactions. The significant interaction indicated that the genotypes respond differently across the different environments. The mean grain yield value of genotypes averaged over environments indicated that BRRI 10A/ BRRI 10R (G3) had the highest (5.99 tha^{-1}) and BRRI dhan39 (G12) the lowest yield (3.19 tha^{-1}), respectively. In AMMI analysis, AMMI 1 biplot showed the hybrids BRRI 1A/ BRRI 827R (G1), IR58025A/ BRRI 10R(G2), BRRI 10A/BRRI 10R(G3) and BRRI hybrid dhan1(G4) have higher average mean yields with high main (additive) effects with positive IPCA1 score, but the hybrid BRRI 10A/BRRI 10R(G3) being the overall best. Hence, the genotype G3 would be considered more adapted to a wide range of environments than the rest of genotypes. Environments, such as Gazipur (E1) and Jessore (E5) could be regarded as a more stable site for high yielding hybrid rice improvement than other location for grain yield due to IPCA score near zero which had no interaction effect. In AMMI 2 biplot, Comilla (E2) and Rangpur (E4) are the most discriminating environments, while BRRI 1A/ BRRI 827R (G1) and Heera 99-5 (G9) are the most responsive genotypes.

Keywords: AMMI analysis; Stability; GEI structure; Hybrid rice

Introduction

Rice has a special significant position as a source of food providing over 75% of Asian's staple food and more than three billion of world population's meal which represents 50 to 80% of their daily calorie intake [1,2]. This population will increase to over 4.6 billion by 2050 [3] which will demand more than 50% of rice needs to be produced what is produced present to cope with the growing population [4,5]. Yields of improved inbreed rice varieties in favorable conditions have reached to a plateau or even subsequently declined in many countries including Bangladesh. It is recommended that a large number of high yielding stable hybrids with high adaption capability to diverse environments are required to accomplish specific socio-economic and agricultural needs. Hence, we need new hybrid rice because it gives 15-30% yield advantage over inbred rice. Moreover, hybrid rice has also shown better performance under adverse conditions like drought and saline conditions. If we can develop high yielding stable hybrid rice adopted on diverse environments, we can find most diverse stable heterotic hybrid combinations to increase food production for increasing world population.

Yield is a complex character which is dependent on a number of other characters and is highly influenced by many genetic factors as well as environmental fluctuations. On the other hand, the G x E interaction is an important aspect of both plant breeding program and the introduction of new crop cultivars [6-8]. The AMMI model is a hybrid model involving both additive and multiplicative components of two way data structure which enabled a breeder to get precise prediction on genotypic potentiality and environmental influences on it. AMMI uses ordinary ANOVA to analyze the main effects (additive part) and principal component analysis (PCA) to analyze the non-additive residual left over by the ANOVA [9]. The effectiveness of AMMI procedure has been clearly demonstrated by various authors using multilocation data in soybean [10], maize [11], Wheat [12-14], pearl millet [15], Okra [16], Field pea [17] and rice [18,19]. The main objectives of the present study are to identify more high yielding stable promising hybrids and to determine the areas where rice hybrids would be adapted by AMMI model. Therefore, using the AMMI analysis with biplot facility, yield data were analyzed to determine the nature and magnitude of G x E interaction effects on grain yield in diverse environments.

Materials and Methods

The experiments were conducted at five districts namely Gazipur (E1), Comilla (E2), Barisal (E3), Rangpur (E4) and Jessore (E5) representing five different agro-ecological zones (AEZ) of Bangladesh. Twelve genotypes consisting of 3 advanced lines (BRRI 1A/ BRRI 827R (G1), IR58025A/ BRRI 10R (G2) and BRRI 10A/ BRRI 10R (G3)), 6 released hybrids (BRRI hybrid dhan1(G4), Tea (G5), Mayna (G6), Richer (G7), Heera-2 (G8) and Heeta 99-5 (G9)), and 3 inbreed check varieties (BRRI dhan31 (G10), BRRI dhan33 (G11) and BRRI dhan39 (G12)) were used as experimental materials. The experiments were carried out in a randomized complete block design (RCBD), with three replications. Twenty-one-days old seedlings were transplanted in 20 square meter plot using single seedling per hill at a spacing of 20 cm×15 cm. Fertilizers were applied @ 150:100:70:60:10 kg/ha Urea, TSP, MP, gypsum and ZnSO$_4$, respectively. Standard agronomic practices were followed and plant protection measures were taken as required

*Corresponding authors: Mamunur Rahman M, Senior Scientific Officer, Farm Management Division, Bangladesh, E-mail: rahmanmmamunur@gmail.com
Anowara Akter, Plant Breeding Division, Bangladesh Rice Research Institute, Bangladesh, E-mail: anowaraa@yahoo.com

following the recommendation of BRRI [20]. Two border rows were used to minimize the border effects. The grain yield (tha^{-1}) data were collected at 14% moisture level. Data were collected followed by standard method as described by [21]. The grain yield data for twelve (12) genotypes in five (5) environments were subjected to AMMI analysis of variance using statistical analysis package software Cropstat version 6.1 (Cropstat, Tutorial Mannual Part 2, Revised April, 2008).

Results and Discussion

AMMI analysis of variance

The AMMI analysis of variance for grain yield (tha^{-1}) of 12 genotypes tested in five environments showed that the main effects of G and E accounted for 67.11% and 18.46% variation, respectively, and G x E interaction effects represent 13.11% of the total variation for grain yield (Table 1). The analysis revealed that variances due to environments, and G x E interactions are significant (P<0.01). The large sum of squares for genotypes indicated that the genotypes were diverse, with large differences among genotypic means causing most of the variation in grain yield, which is in harmony with the findings of [22,23]. The presence of genotype-environment interaction (GEI) was clearly demonstrated by the AMMI model, when the interaction was partitioned among the first three interaction principal component axis (IPCA) as they were significant in postdictive assessment. The

IPCA1 explained 9.68% of the interaction sum of square in 14% of the interaction degree of freedom (DF). Similarly, the second and third principal component axis (IPCA 2-3) explained a further 2.02% and 1.23% of the GEI sum of squares, respectively (Table 1). This implied that the interaction of the rice genotypes with five environments was predicted by the first three components of genotypes and environments, which is in agreement with the recommendation of Sivapalan et al. [24]. However, this contradicted the findings of Gauch and Zobel [25] which recommended that the most accurate model for AMMI can be predicted using the first two IPCAs. These results indicate that the number of terms to be included in an AMMI model cannot be specified a prior without first trying AMMI predictive assessment [26]. In general, factors like type of crop, diversity of the germplasm and range of environmental conditions will affect the degree of complexity of the best predictive model [11].

Stability analysis by AMMI model

Biplot analysis is possibly the most powerful interpretive tool for AMMI models. There are two basic AMMI biplots, the AMMI 1 biplot where the main effects (genotype mean and environment mean) and IPCA1 scores for both genotypes and environments are plotted against each other. On the other hand, the second biplot is AMMI 2 biplot where scores for IPCA1 and IPCA2 are plotted (Table 2). The mean grain yield value of genotypes averaged over environments indicated

Source of Variation	d.f	SS	MS	Explained SS (%)
Genotypes (G)	11	40.498	3.682**	67.11
Environments (E)	4	11.142	2.785**	18.46
G x E Interaction (GEI)	44	7.908	0.179**	13.11
IPCA1	14	5.842	0.417**	9.68
IPCA2	12	1.220	0.102**	2.02
IPCA3	10	0.741	0.074**	1.23
Eroor	120	0.803	0.007	
Total	179	60.353		

** Significant at P<0.01

Table 1: Additive main effects and multiplicative interaction (AMMI) analysis of variance for grain yield (tha^{-1}) of 12 rice genotypes across 5 environments.

Genotypes/Environments	Gazipur (E1)	Comilla (E2)	Barisal (E3)	Rangpur (E4)	Jessore (E5)	Mean	Index	IPCA1	IPCA2
BRRI 1A/ BRRI 827R (G1)	5.10	5.04	4.41	3.37	5.13	4.61	0.57	0.852	0.421
IR58025A/ BRRI 10R (G2)	5.78	5.11	4.52	4.31	5.12	4.97	0.92	0.502	0.147
BRRI 10A/ BRRI 10R (G3)	6.47	6.05	5.86	5.67	5.41	5.99	1.95	0.293	0.254
BRRI hybrid dhan1 (G4)	5.83	4.66	4.26	4.19	5.19	4.83	0.78	0.356	0.001
Tea (G5)	4.40	3.19	2.99	3.58	3.22	3.48	-0.56	-0.008	0.670
Mayna (G6)	4.23	2.37	3.11	3.67	3.56	3.39	-0.66	-0.461	0.155
Richer (G7)	4.43	2.45	3.42	3.67	4.19	3.63	-0.41	-0.449	-0.294
Heera-2 (G8)	4.33	3.03	3.66	3.88	3.86	3.75	-0.29	-0.273	0.103
Heeta 99-5 (G9)	4.17	2.47	4.00	4.03	4.32	3.80	-0.25	-0.647	-0.460
BRRI dhan31 (G10)	4.34	3.35	3.11	2.76	4.02	3.52	-0.53	0.391	-0.239
BRRI dhan33 (G11)	4.24	2.62	3.03	3.34	3.65	3.38	-0.66	-0.193	0.049
BRRI dhan39 (G12)	3.80	2.32	3.08	3.40	3.37	3.19	-0.86	-0.358	0.033
Mean	4.76	3.56	3.79	3.82	4.30				
Index	0.72	-0.49	-0.26	-0.22	0.25				
IPCA 1	0.040	1.179	-0.234	-0.985	-0.001				
IPCA 2	0.278	0.328	-0.313	0.478	-0.771	**GM=4.05**			
SE	0.08	0.09	0.06	0.10	0.07				
CV(%)	3	4	3	5	3				
5% LSD	0.24	0.25	0.18	0.30	0.21				

Table 2: Stability parameters for grain yield (tha^{-1}) of 12 rice genotypes in 5 environments.

that the genotypes G3 and G12 had the highest (5.99 tha^{-1}) and the lowest (3.19 tha^{-1}) yield, respectively. Different genotypes showed inconsistent performance across all environments. The environments mean grain yield ranged from 4.76 tha^{-1} for E1 to 3.56 tha^{-1} for E2 and averaged grain yield over environments and genotypes is 4.05 tha^{-1}. On the basis of environmental index value in terms of negative and positive, E2, E3 and E4 are poor, and E1 and E5 are rich environment. Within the genotypes G1, G2, G3 and G4 have higher average yields and these genotypes adapted to favorable environments, while genotypes G5 to G12 adapted to poor environments.

AMMI 1 biplot display

Biplots are graphs where aspects of both genotypes and environments are plotted on the same axes so that inter relationships can be visualized. In the AMMI 1 biplot, the usual interpretation of biplot is that the displacements along the abscissa indicate differences in main (additive) effects, whereas displacements along the ordinate indicate differences in interaction effects. Genotypes that group together have similar adaptation while environments which group together influences the genotypes in the same way [27]. The best adapted genotype can plot far from the environment. If a genotype or an environment has a IPCA1 score of nearly zero, it has small interaction effects and considered as stable. When a genotype and environment have the same sign on the PCA axis, their interaction is positive and if different, their interaction is negative. The AMMI 1 biplot expected yield clearly indicated for any genotype and environment combination can be calculated from Figure 1 following standard procedures suggested by Zobel et al. [10].

The AMMI 1 biplot gave a model fit 96.5%. This result is in agreement with the findings of Naveed et al. [28] and Gauch and Zobel [25]. Genotypes and environments on the same parallel line, relative or ordinate have similar yields and a genotype or environment on the right side of the mid point of this axis has higher yields than those of left hand side. Consequently, among the hybrids, (G1), (G2), (G3) and (G4) were generally exhibited high yield with high main (additive) effects showing positive IPCA1 score, but the hybrid (G3) being the over all best. Hence, the hybrid (G3) was identified as specially adapted

culture to the mentioned environments and these environments were considered as the wide range suitable environments for this genotype. Similar outcomes have reported by Das et al. [29], and Kulsum et al. [30]. Since, the environments E1 and E5 had positive IPCA1 score near zero and hence had small interaction effects indicating that all the genotypes performed well in these locations. Adugna et al. [31] and Anandan et al. [32] reported similar pattern of interactions. Thus these two locations were considered as the favorable environments for the genotypes G1, G2, G3 and G4. The genotype G5 showed IPCA1 score close to zero, indicating that the variety was stable and less influenced by the environments [33]. Other genotypes showed below average yield. Similarly, the genotype G10 was moderately stable across environments (low positive IPCA1 score) and below average yield. On the other hand, G8, G11 and environment, E3 had below average yield with negative IPCA1 score near zero indicating that these varieties were less influenced by the environments. Likewise, the environment E3, were found favorable environment for the genotype G11 and G8. Finally, The AMMI 1 biplot statistical model has been used to diagnose the G x E interaction pattern of grain yield of hybrid rice. The hybrids (G1), (G2), (G3) and (G4) were hardly affected by the G x E interaction and thus will perform well across a wide range of environments. Locations, such as E1 and E5 could be regarded as a good selection site for rice hybrid improvement due to stable yields.

AMMI 2 biplot display

In AMMI 2 biplot, (Figure 2) the environmental scores are joined to the origin by side lines. Sites with short spokes do not exert strong interactive forces. Those with long spokes exert strong interaction. An example of this is shown in Figure 2 where the points representing the environments E1, E2, E3, E4 and E5 are connected to the origin. The environments E1 and E3 had short spokes and they do not exert strong interactive forces. The genotypes occurring close together on the plot will tend to have similar yields in all environments, while genotypes far apart may either differ in mean yield or show a different pattern of response over the environments. Hence, the genotypes near the origin are not sensitive to environmental interaction and those distant from the origins are sensitive and have large interaction. In the present study,

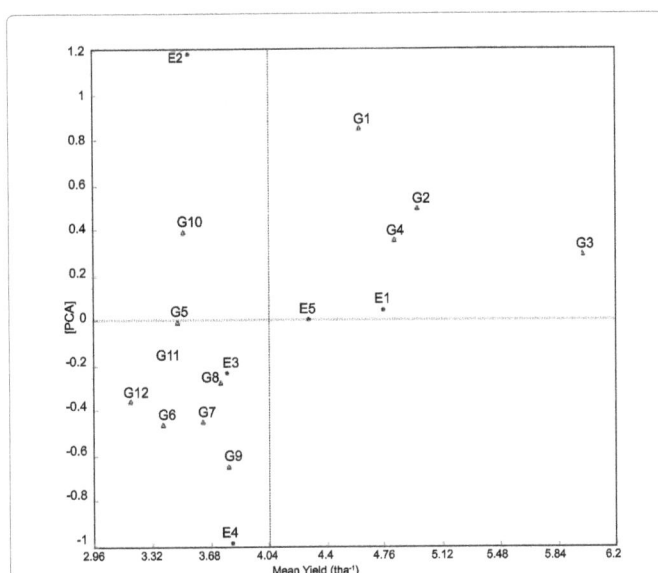

Figure 1: AMMI 1 Biplot for grain yield (tha^{-1}) of 12 rice genotypes (G) and five environments (E) using genotypic and environmental scores.

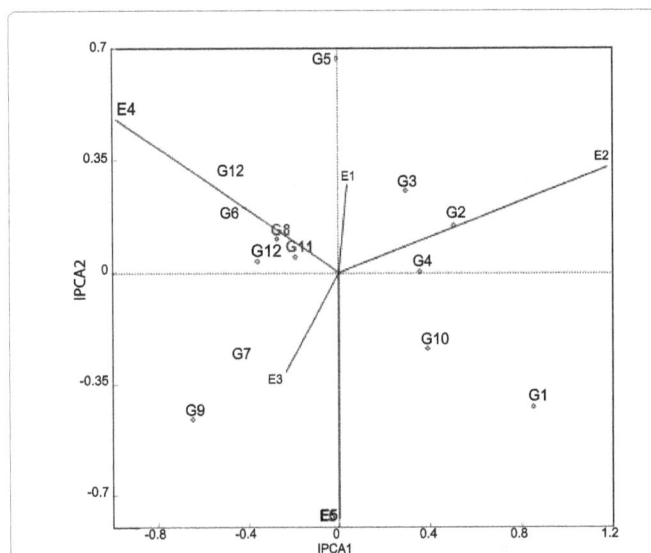

Figure 2: AMMI 2 Biplot for grain yield (tha^{-1}) showing the interaction of IPCA2 against IPCA1 scores of 12 rice genotypes (G) in five environments (E).

G1 and G9 had more responsive since they were away from the origin whereas the genotypes G8, G11, G12 and G4 were close to the origin and hence they were non sensitive to environmental interactive forces.

Conclusions

Crop yield is a complex trait that is influenced by a number of component characters along with the environment directly or indirectly. If we could develop high yielding stable hybrid rice for diverse environments, we could offer most diverse stable heterotic hybrids for the rice growers. AMMI statistical model could be a great tool to select the most suitable and stable high yielding hybrids for specific as well as for diverse environments. In the present study, AMMI model has shown that the largest proportion of the total variation in grain yield was attributed to environments. Here most of the genotypes showed environment specificity. The mean grain yield value of genotypes averaged over environments indicated that G3 had the highest (5.99 tha^{-1}) and G12 the lowest yield (3.19 tha^{-1}), respectively. It is noted that the variety G3 showed higher grain yield than all other varieties over all the environments. The genotypes (G1), (G2), (G3) and (G4) were hardly affected by the G x E interaction and thus would perform well across a wide range of environments.

References

1. Khush GS (2005) What it will take to feed 5.0 billion rice consumers in 2030. Plant Mol Biol 59: 1-6.

2. Amirjani MR (2011) Effect of salinity stress on growth, sugar content, pigments and enzyme activity of rice. International Journal of Botany 7: 73-81.

3. Honarnejad RS, Abdollahi MS, Mohammad-Salehi, Dorosti H (2000) Consideration of adaptability and stability of grain yield of progressive rice (Oryza sativa L.) lines. Research in Agriccultural Science 1: 1-9.

4. Ashikari M, Sakakibara H, Lin S, Yamamoto T, Takashi T, et al. (2005) Cytokinin oxidase regulates rice grain production. Science 309: 741-745.

5. Srividya A, Vemireddy LR, Hariprasad AS, Jayaprada M, Sridhar S (2010) Identification and mapping of landrace derived QTL associated with yield and its components in rice under different nitrogen levels and environments. International Journal of Plant Breeding and Genetics 4: 210-227.

6. McLaren CG, Chaudhary C, (1994) Use of additive main effects and multiplicative interaction models to analyse multilocation rice variety trials. Paper presented at the FCSSP Conference, Puerton Princesa, Palawan, Philippines.

7. Prasad KV, Singh RL (1990) Stability analysis of yield and yield components and construction of selection indices of direct seeded rice in frost season. Annual Review conference Proceeding. 20-23 October 1992. National Agril. Res. Inst. Caribbean Agricultural Research and development Institute, Guyana, pp. 63-71.

8. Freeman GH (1985) The analysis and interpretation of interaction. Journal of Applied Statistics. 12: 3-10.

9. Gauch HG (1993) Matmodel version 2.0. AMMI and related analysis for two-way data matrics. Micro computer power, Ithaca, New York, USA.

10. Zobel RW, Wright MJ, Gauch HG (1988) Statistical analysis of a yield trial. Agronomy Journal. 80: 388-393.

11. Crossa J, Gauch HGJ, Zobel RW (1990) Additive main effects and multiplicative interaction analysis of two international maize cultivar trials. Crop Science. 30: 493-500.

12. Crossa J, Fox PN, Pfeiffer WH, Rajaram S, Gauch HG Jr (1991) AMMI adjustment for statistical analysis of an international wheat yield trial. Theor Appl Genet 81: 27-37.

13. Yan W, Hunt LA (2001) Interpretation of genotype x environment interaction for winter wheat yield in Ontario. Crop Science 41: 19-25.

14. Tarakanovas T, Rugas V (2006) Additive main effect and multiplicative interaction analysis of grain yield of wheat varieties in Lithuania. Agronomy Research 4: 91-98.

15. Shinde GC, Bhingarde MT, Mehetre SS (2002) AMMI analysis for stability of grain yield of pearl millet (Pennisetum typhoides L.) hybrids. International Journal of Genetics 62: 215-217.

16. Ariyo OJ, Ayo-Vaughan MA (2000) Analysis of genotype x environment interaction of okra (Abelmoschus esculentus (L) Moench). Journal of Genetics and Breeding 54: 33-40.

17. Taye G, Getachew T, Bejiga G (2000) AMMI adjustment for yield estimate and classifications of genotypes and environments in field pea (Pisum sativum L.). Journal of Genetics and Breeding 54: 183-191.

18. Das S, Misra RC, Patnaik MC (2009) GxE interaction of mid-late rice genotypes in LR and AMMI model and evaluation of adaptability and yield stability. Environment and Ecology 27: 529-535.

19. Nassir AL (2013) Genotype x Environment analysis of some yield components of upland rice (Oryza sativa L.) under two ecologies in Nigeria. International Journal of Plant Breeding and Genetics 7: 105-114.

20. BRRI (2010) Adhunik dhaner chash. (15thedn), Bangladesh Rice Research Institute, Gazipur-1700, Bangladesh. p. 20-50.

21. Yoshida S, Forno DA, Cock JH, Gomez KA (1976) Laboratory manual for physiological studies of rice. (3rdedn), International Rice Research Institute, Los Banos, Philippines.

22. Misra RC, Das S, Patnaik MC (2009) AMMI Model Analysis of Stability and Adaptability of Late Duration Finger Millet (Eleusine coracana) Genotypes. World Applied Sciences Journal 6: 1650-1654.

23. Fentie M Assefa A, Belete K (2013) Ammi Analysis of Yield Performance and Stability of Finger Millet Genotypes Across Different Environments. World Journal of Agricultural Sciences 9: 231-237.

24. Sivapalan S, Brien LO, Ferrana GO, Hollamby GL, Barelay I, et al. (2000) An adaption analysis of Australian and CIMMYT/ ICARDA wheat germplasm in Austrlian production environments. Australian Journal of Agriculture Research 51: 903-915.

25. Gauch HG, Zobel RW (1996) AMMI analysis of yield trials. In M.S. Kang & H.G. Gauch. (eds.). Genotype-by-environment interaction, pp: 85-122.

26. Kaya Y, Palta C, Taner S (2002) Additive main effects and multiplicative interactions analysis of yield performance in bread wheat genotypes across environments. Turuk Journal of Agriculture 26: 275-279.

27. Kempton RA (1984) The use of biplots in interpreting variety by environment interactions. Journal of Agricultural Science 103: 123-135.

28. Nadeem NM, Islam N (2007) AMMI analysis of some upland cotton genotypes for yield stability in different milieus. World Journal of Agricultural Sciences 3: 39-44.

29. Das S, Misra RC, Patnaik MC, Das SR (2010) GxE interaction, adaptability and yield stability of mid-early rice genotypes. Indian Journal of Agricultural Research. 44: 104-111.

30. Kulsum MU, M Jamil Hasan, Anowara Akter, Hafizar Rahman, Priyalal Biswas (2013) Genotype-environment interaction and stability analysis in hybrid rice: an application of additive main effects and multiplicative interaction. Bangladesh J Bot 42: 73-81.

31. Adugna A (2007) Assessment of Yield Stability in Sorghum. African Crop Science Journal. 15: 83-92.

32. Anandan A, Eswaran R, Sabesan T, Prakash M (2009) Additive main effects and multiplicative interactions analysis of yield performances in rice genotypes under coastal saline environments. Advances in Biological Research 3: 43-48.

33. Yau SK (1995) Regression and AMMI analyses of genotype x environment interactions: An empirical comparison. Agronomy Journal 87: 121-126.

29

Ferric Pyrophosphate-fortified Rice Given Once Weekly Does not Increase Hemoglobin Levels in Preschoolers

Francisco Placido Nogueira Arcanjo[1]*, Paulo Roberto Santos[1] and Sergio Duarte Segall[2]

[1]*Federal University of Ceara, Brazil*
[2]*Federal University of Minas Gerais, Brazil*

Abstract

Background: The objective of this study was to evaluate the effect of rice fortified with iron (Ultrarice®), given once weekly, and on hemoglobin levels and anemia prevalence compared with standard rice.

Methods: In this prospective quasi-experimental study, we evaluated preschoolers aged 2 to 5 years (n=303) from 2 public schools in Sobral, Brazil. Intervention lasted 18 weeks. The once weekly 50 g individual portion (uncooked) of fortified rice provided 56.4 mg of elemental iron as ferric pyrophosphate. Capillary blood samples to test for hemoglobin levels were taken at baseline and after intervention. Student's *t*-test was used to assess the difference in hemoglobin within / between the groups.

Results: For fortified rice school: baseline mean hemoglobin was 12.06 ± 1.01 g/dL, and after intervention 12.14 ± 1.06 g/dL, p=0.52; anemia prevalence was 8.9% (11/120) at baseline, and 10.5% (13/120) at end of study, p=0.67. For the standard rice school: baseline mean hemoglobin was 12.40 ± 4.14 g/dL, and after intervention 12.29 ± 2.48, p=0.78; anemia prevalence was 20.8% (30/144) at baseline, and 37.5% (54/144) at the end of study, p=0.002. Considering only anemic participants, there was a significant increase in hemoglobin means before and after intervention, p=0.003 in the fortified rice school. Relative Risk was 0.29 and the Number Needed to Treat was 4.

Conclusions: This study shows that consumption of the iron-fortified rice compared to the control rice does not change hemoglobin or anemia prevalence in preschoolers.

Keywords: Anemia; Rice; Fortification; Iron; Preschoolers

Background

Iron deficiency (ID) has been identified as the most prevalent micronutrient deficiency in the world, afflicting over 50% of the world population. An estimated 47% of children under five years worldwide experience Iron Deficiency Anemia (IDA) [1-3] and poor and/or minority children are at increased risk [4]. Inadequate intake of iron and consumption of foods with low iron bioavailability are the major causes of this problem. Iron deficiency in the first years of life can cause serious consequences later in life including lack of concentration, behavioral and cognitive disorders, and growth impairment [5,6].

WHO regards food fortification as a safe and effective means to supplement diets with low-iron content [5], whereas international suggestions for food fortification often require changes in local diet habits and even the importation of foodstuffs not locally available. Interventions should focus on the fortification of locally consumed foodstuffs, as this type of fortification may be implemented sustainably on a large scale, allowing people to get more nutritional value from the food they already eat. This current study focuses on rice fortification, as in Brazilian households rice is probably the most widely consumed foodstuff. The impact of iron-fortified rice has already been the object of other studies and its effect on anemia and hemoglobin levels, but up-to-date no studies have investigated the use of ferric pyrophosphate in weekly dosages with preschoolers [7-13].

The objective of our study was to evaluate the impact of rice fortified with iron, given once weekly, in preschoolers aged 2-5 years compared with control (standard household rice) on hemoglobin values and anemia levels.

Materials and Methods

This prospective quasi-experimental study was conducted in the City of Sobral-Ceará, in the northeast of Brazil, between August and December 2010. The study population comprised of preschoolers (2 to 5 years) from two chosen public schools (n=303).

Prior to the study, each school was identified to receive one type of rice to avoid contamination between the different groups, as the meal was served at the same time in the same school refectory. The first school was designated to receive fortified rice once weekly in schools meals–School A, and the latter received standard rice-School B. The menus at the two schools were equal in content; the study rice was consumed with poultry, which was the customarily consumed meal for Tuesdays at the schools. Staff at the schools were not aware of the rice that was being served (fortified or standard) as the rice was provided by the study team in non-identifying packages. Data collection team was also unaware of intervention and control groups. Intervention period was adapted to the 20-week school semester, the first 2 weeks were used for intervention setup and training; the following 18 weeks constituted the study period, starting and ending on the same date.

All preschoolers from the two schools were invited to participate in our study. Exclusion criteria were parents' refusal to participate and

*Corresponding author: Francisco Placido Nogueira Arcanjo, Federal University of Ceara, Brazil, E-mail: placidoarcanjo@yahoo.com.br; franciscoplacidoarcanjo@gmail.com

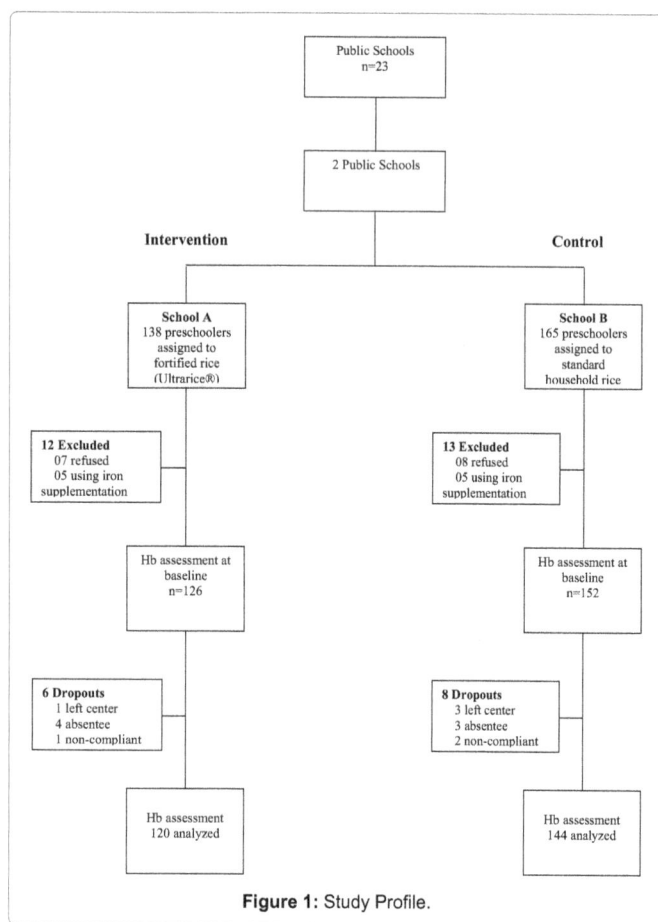

Figure 1: Study Profile.

preschoolers already using iron supplementation (Figure 1). Upon enrollment, baseline characteristics of participants were defined by means of questionnaire with the children's mothers. Characteristics analyzed for this study were age, gender, mother's schooling, and family income.

Ultra Rice grains (Adorella Alimentos LTDA, Indaiatuba, Sao Paulo, Brazil) are made with rice flour and selected micronutrients, which are combined and then extruded through a rice-shaped mold. Ultra Rice can be tailor made with different micronutrients, including iron, thiamin, zinc, and folic acid, while mimicking the look of standard rice.

In our nutritional intervention, Ultra Rice fortified with non-encapsulated ferric pyrophosphate with an average particle size of 3 microns in micronized form (donated to the Secretariat of Education–Sobral by the Program for Appropriate Technology in Health PATH) was used at 6%. This proportioned a weekly ingestion of 56.4 mg of elemental iron as ferric pyrophosphate per 50 g uncooked individual portion. The meals were prepared in the school kitchen and served at lunchtime (11 a.m.). This quantity of rice provided a satisfactory weekly amount of elemental iron considering dietary reference intakes (DRI) for this mixed age range (children aged 2 to 3 years=7 mg Fe/day and 4 to 5 years=10 mg) [14].

The main outcomes analyzed were hemoglobin values and anemia prevalence, by means of two biochemical evaluations, before and after intervention. Capillary blood samples were taken from a finger prick using aseptic techniques. Hemoglobin concentrations were promptly analyzed with a portable HemoCue B-hemoglobin photometer (Hb 301-HemoCue AB, Ängelholm, Sweden) by technician. Cutoff point to define anemia for children <5 years: Hb <11.0 g/dL [6].

Anemia prevalence in the study population was estimated at 20%, according to the global prevalence for anemia in this age range, in Brazil, is 20–25% (varies according to region) [15]. To achieve a reduction in global anemia prevalence from 20 to 10%, with 80% power, 2-sided, type I error of 1%, accounting for 10% losses to follow-up, each group required a minimum of 107 participants [16].

Data were entered into a database in double-entry EPI Info version 6.04 (Centers for Disease Control and Prevention–CDC, Atlanta, Georgia, USA). At baseline, the participants from the two schools were compared. To compare ratios and means we used, respectively, the χ^2-test and the paired student's t-test to assess the difference in hemoglobin within the groups, and unpaired student's t-test between the groups before and after intervention. Data had normal distribution. Processing and analysis of data was made using the statistics package PASW (Predictive Analytic Software, Windows Version 17.0, SPSS Inc., Chicago, IL). We considered $\alpha<0.05$. Analyses were by intention to treat.

Relative risk to anemia was calculated upon completion of intervention. The χ^2-test was used to compare the ratios between the study groups. The independent variable (intervention or control) was organized and examined in the form of a dichotomy: (a) fortified rice (intervention), (b) standard household rice (control). From this point, using 2x2 contingency tables, the following measures of association were calculated: reduction of absolute risk (RAR), relative risk (RR), reduction of relative risk (RRR) and number needed for treatment (NNT), which in this study is the number of preschoolers submitted to intervention for the prevention of unfavorable outcome (anemia).

Approval for this study was obtained from the Ethics Committee at the Federal University of Ceara, with necessary prior written consent from school directors and parents/guardians. Medical support was available upon request. After intervention, anemic children were referred for treatment.

Results

At baseline, the following parameters were analyzed: age, gender, mother's schooling, and family income. However, there were no statistically significant differences between the two schools (Table 1). The mean hemoglobin values for the groups were 12.06 ± 1.01 g/dL for school A and 12.40 ± 4.14 g/dL for school B, p=0.38; anemia prevalence was 8.9% and 20.8%, for schools A and B, respectively, p=0.009. During our study, there were six dropouts from school A (1 left center, 4 absentee, 1 non-compliant); in school B there were eight dropouts (3 left center, 3 absentee, 2 non-compliant) (Figure 1and Table 1).

	School A (n=138)	School B (n=165)	p-value
Age in months	41.3 (10.1)	42.5 (9.96)	0.29[a]
Gender M:F	65:73	79:86	0.89[b]
Mother with ≤8y schooling	103	135	0.13[b]
Family income ≤300USD	126	154	0.50[b]

M:F male:female
( ) standard deviation
[a] descriptive level of unpaired student t-test
[b] descriptive level of chi-square.

Table 1: Baseline characteristics of study participants, by intervention center.

Mean hemoglobin value at baseline was 12.06 ± 1.01 g/dL for the fortified rice group (school A), and after intervention 12.14 ± 1.06 g/dL, p=0.52. For the standard household rice/control group (school B) mean hemoglobin value at baseline was 12.40 ± 4.14 g/dL, and after intervention 12.29 ± 2.48, p=0.78. In the intervention group (school A), there was no significant change in anemia prevalence during the study, p=0.67; however, there was a statistically significant increase in anemia prevalence for the control group (school B), from 20.8 to 37.5%, p=0.002 (Table 2).

At baseline, the groups were similar for hemoglobin means, p=0.38; after intervention, there was no significant difference between the schools, p=0.56 (Table 2). However, for anemia prevalence, the groups were different at baseline: 8.9% (11/120 were anemic) in school A, and 20.8% (30/144 were anemic) in school B, p=0.009; at the end of the study, the groups remained different, 10.5% (13/120) at school A, and 37.5% (54/144) at school B, p<0.001 (Table 2).

Considering only anemic participants, in school A before intervention mean hemoglobin value was 10.12 ± 0.85 g/dL (n=11) and 11.56 ± 0.86 after intervention, p=0.0003; in school B (n=30), mean hemoglobin values went from 10.83 ± 1.11 g/dL before intervention to 10.94 ± 1.18 g/dL after intervention, p=0.18 (Table 3). For non-anemic participants, there was no significant alteration in the mean hemoglobin values before and after intervention for schools A (12.20 ± 0.76 before and 12.24 ± 1.04 after) and B (12.44 ± 3.27 before and 12.48 ± 3.01 after), p=0.72 and 0.74, respectively (Table 3).

In this study, the following indicators were compared: fortified rice versus standard household rice, for a favorable (absence of anemia) or adverse (anemia) outcome.

After intervention, adverse outcome was present in 37.5% of control subjects and 10.8% of experimental subjects. The difference, the Reduction of Absolute Risk (RAR), was 26.7%. The 95% confidence interval for this difference ranges from 17.00 to 36.33%. The Number Needed to Treat (NNT) was 4. This means that about

one in every 4 preschoolers will benefit from this anemia prevention intervention. The 95% confidence interval for the NNT ranges from 2.8 to 5.9. Measurements of efficacy were: Relative Risk (RR)=0.29; 95% Confidence Interval (CI)=0.166, 0.503; Relative Risk Reduction (RRR)=0.73, expressing that the preschoolers submitted to this intervention had 73% less likelihood of developing anemia.

Discussion

Main findings of this study

This 18-week study demonstrated that rice fortified with iron once weekly does not change hemoglobin and anemia in preschoolers; nevertheless, it was capable of preventing anemia in preschoolers. There was a statistically significant increase in anemia prevalence in the control group; however, the fortified rice group did not present a statistically significant alteration in the number of anemic children after intervention. Furthermore, it prevented anemia onset for 1 in every 4 preschoolers submitted to the intervention, preschoolers submitted to this intervention had 73% less likelihood of developing anemia (RRR), a significant result taking into consideration that the population submitted to the fortified rice intervention had low anemia prevalence, 8.9%. Additionally, when we analyzed only anemic participants there was a significant increase in hemoglobin means for the ´participants in the fortified rice school, compared to no change in the standard rice group, strengthening the results from the fortified rice.

In our study, low anemia prevalence and hemoglobin means above 12.0 g/dL probably represented a population with low iron deficit, situation that may have impeded a greater response from the proposed intervention. It is likely that the results from this kind of intervention would be more significant than in populations with more widespread ID, anemia prevalence and lower mean hemoglobin.

Our study was conducted in compliance with DRI recommendations for this age range [14] and also taking into consideration the low absorption of this iron composition [17], we used a once weekly

	School A – Fortified Rice (n=120)			School B – Standard Rice (n=144)				
	Before	After	p-value	Before	After	p-value	AxB[a]baseline p-value	AxB[a]after p-value
Hemoglobin (g/dL)	12.06	12.14	0.52[b]	12.40	12.29	0.78[b]	0.38[c]	0.56[c]
SD[d] CI[e]	1.01 11.87-12.25	1.06 11.96-12.33		4.14 11.84-12.96	2.48 11.73-12.84			
Anemia[f]	11 (9.2%)	13 (10.8%)	0.67[g]	30 (20.8%)	54 (37.5%)	0.002[g]	0.009[g]	<0.001[g]

[a]School A vs. School B
[b]Based on paired Student's *t*-tests
[c]Based on unpaired Student's *t*-tests
[d]SD = Standard deviation
[e] CI = Confidence interval
[f]Anemia defined as Hb concentration <11.0 g/dL
[g] Descriptive level of chi-square

Table 2: Effect of iron fortified rice, compared with standard rice before and after intervention and comparison of hemoglobin means and anemia prevalence between schools.

	School A – Fortified Rice (n=11)			School B – Standard Rice (n=30)		
	Before	After	p-value	Before	After	p-value
Hemoglobin (g/dL)	10.12	11.56	0.0003[a]	10.83	10.94	0.18[a]
SD	0.85	0.86		1.11	1.18	

[a]Based on paired Student's *t*-tests
SD = Standard deviation

Table 3: Effect of iron fortified rice, compared with standard rice before and after intervention on hemoglobin means for anemic participants.

56 mg elemental iron dosage; this was the amount offered to the preschooler and does not necessarily represent the amount consumed; individual ingestion was not measured in this study. Additionally, this intervention sought not to alter the established school menu where rice was consumed only once weekly.

What we already know

According to Hurrel and Egli [18], bioavailability of fortification iron varies widely with the iron compound used. Furthermore, iron status of the individual and other host factors largely influence iron bioavailability such as obesity, nutritional deficiencies, infection/inflammation, and genetic disorders. Additionally, iron status generally has a greater effect than the vehicle that is used in fortification.

Two recent studies conducted in infants with the similar intervention designs, have achieved statistically significant results increasing hemoglobin values and reducing anemia prevalence in the fortified rice groups when compared to the standard rice groups. In the first study [13] mean hemoglobin values increased from 11.37 to 11.95 g/dL, p<0.0001 and anemia prevalence reduced from 27.8 to 11.1%, p=0.012; in the second study [12] hemoglobin values rose from 11.44 to 11.67, p=0.029, and anemia prevalence decreased from 31.25 to 18.75%, p=0.045. In these studies, anemia prevalence at baseline in the intervention groups was higher than that of this intervention, which may have contributed to better results.

One 5-month home-based randomized controlled trial (RCT) conducted in Brazil with anemic infants used the same iron to fortify rice, comparing the daily consumption of iron-fortified rice with orally administered iron drops (10 mg iron/daily). The authors concluded that both interventions improved iron status, p<0.01 for hemoglobin and p<0.02 for serum ferritin, reducing anemia prevalence from 100 to 61.8% (fortified rice) and 85.6% (iron drops). It is likely that daily small doses will have a greater impact than the same quantity delivered once weekly [10]. However in a double-blind, 8-month, placebo-controlled trial using micronized ferric pyrophosphate (MFP) in extruded rice kernels mixed in a rice-based meal (19 mg Fe/day), this intervention did not increase hemoglobin levels nor reduce anemia in children aged 5 to 10 years, but iron stores were improved when compared to control [11].

Other studies using extruded rice grains fortified with MFP have achieved positive outcomes. Moreti et al. [7] obtained a 50% reduction in IDA (30% from 15%) in a 7-month RCT with preschoolers, which fortified a rice meal with 20 mg Fe daily. Angeles et al. [8] compared different rice fortifications (ferrous sulfate and MFP) compared to a control group with non-fortified rice, in a six-month intervention on schooldays with anemic schoolchildren. There was significant reduction in anemia prevalence, from 100 to 38 and 33% in the intervention groups compared to 63% in the control. However, in a study by Bagni et al. [19] there was apparently no impact on hemoglobin levels in preschoolers (when compared to control), in a 16-week intervention with rice fortified with iron chelate once weekly.

Food fortification, in recent years, has drawn the attention of care professionals and health authorities around the world; in the Global Report, published in 2009 [20], food fortification was defended as an approach that has shown great success, whether in packets for in-home use or delivered through mass fortification programs, this kind of intervention warrants urgent and wide expansion. Additionally, in May 2008 the Copenhagen Consensus Panel [21] in analyzing 30 options for the world's best investment for development, considered the provision of micronutrients as the best alternative.

Limitations of this study

Some limitations need to be acknowledged and addressed regarding the present study. The study was limited with respect to measurement of iron status; IDA was estimated and not measured. This study was conducted as a quasi-experimental study, not randomized; therefore, it was not possible to control anemia levels at baseline between the different schools. We acknowledge that with non-encapsulated ferric pyrophosphate is of low availability and that other food components consumed with the rice may have inhibited or enhanced iron bioavailability. However, these differences are inherent to studies that measure the effectiveness of interventions in real environments, without changing the daily routine of the schools. Nevertheless, at baseline, the variables assessed did not justify the difference in anemia levels. Additionally, individual consumption was not measured due to the nature of the study, conducted in the customary school ambit, with the intention to treat.

What this study adds

Although the reduction in hemoglobin was not statistically significant in the control group, there was a significant increase in anemia prevalence. This may have been caused by the fact that several children (12 children) in the study were on the borderline between anemia and non-anemia, (for example hemoglobin measurements approximately 11.0 or 11.1 g/dL before intervention, and 10.9 g/dL after intervention) thus, the non-anemic children in the control group were classified as anemic after the intervention.

Nevertheless, as there was no change in hemoglobin values or anemia rates in the intervention group we may assume this intervention prevented preschoolers from becoming anemic. Furthermore as this intervention was conducted once weekly, it may constitute a useful strategy to reduce operational costs and intervention follow-ups in poor communities. As far as our review shows this study is singular as it is the first to fortify rice with ferric pyrophosphate in preschoolers with weekly iron dosages. This intention-to-treat intervention was conducted in public schools with anemic and non-anemic participants. For traditionally highly anemic populations, which have rice as a staple in their diet, this could provide an alternative method to prevent anemia.

Acknowledgement

The author would like to thank the preschoolers and teachers at the schools for their participation and cooperation during this study; Secretariat of Education and Secretariat of Health at the Municipal City Hall-Sobral-CE for their support during the project; Program for Appropriate Technology in Health (PATH) for donating the rice to the Secretariat of Education at the Municipal City Hall-Sobral-CE.

References

1. Stoltzfus RJ, Mullany L, Black RE (2004) Iron deficiency anaemia. In: Ezzati M, Lopez D, Rodgers A, Murray CJL, eds. Comparative Quantification of Hkealth Risks: Global and Regional Burden of Disease Attributable to Selected Major Risk Factors. Geneva: WHO.

2. McLean E, Cogswell M, Egli I, Wojdyla D, de Benoist B (2009) Worldwide prevalence of anaemia, WHO Vitamin and Mineral Nutrition Information System, 1993-2005. Public Health Nutr 12: 444-454.

3. Benoist B, Mclean E, Cogswell M, Egli I, Wojdyla D (2008) Worldwide Prevalence of Anemia 1993–2005 WHO Global Database on Anemia, Geneva, WHO.

4. Brotanek JM, Halterman JS, Auinger P, Flores G, Weitzman M (2005) Iron deficiency, prolonged bottle-feeding, and racial/ethnic disparities in young children. Arch Pediatr Adolesc Med 159: 1038-1042.

5. Allen L, de Benoist B, Dary O, Hurrell R (2006) Guidelines on food fortification with micronutrients. Geneva, Switzerland/Rome, Italy: World Health Organization/Food and Agriculture Organization of the United Nations.

6. ficiency anaemia: assessment, prevention, and control (2001). A guide for programme managers. Geneva, World Health Organization.

7. Moretti D, Zimmermann MB, Muthayya S, Thankachan P, Lee TC, et al. (2006) Extruded rice fortified with micronized ground ferric pyrophosphate reduces iron deficiency in Indian schoolchildren: a double-blind randomized controlled trial. Am J Clin Nutr 84: 822-829.

8. Angeles-Agdeppa I, Capanzana MV, Barba CV, Florentino RF, Takanashi K (2008) Efficacy of iron-fortified rice in reducing anemia among schoolchildren in the Philippines. Int J Vitam Nutr Res 78: 74-86.

9. Hotz C, Porcayo M, Onofre G, García-Guerra A, Elliott T, et al. (2008) Efficacy of iron-fortified Ultra Rice in improving the iron status of women in Mexico. Food Nutr Bull 29: 140-149.

10. Beinner MA, Velasquez-Meléndez G, Pessoa MC, Greiner T (2010) Iron-fortified rice is as efficacious as supplemental iron drops in infants and young children. J Nutr 140: 49-53.

11. Radhika MS, Nair KM, Kumar RH, Rao MV, Ravinder P, et al. (2011) Micronized ferric pyrophosphate supplied through extruded rice kernels improves body iron stores in children: a double-blind, randomized, placebo-controlled midday meal feeding trial in Indian schoolchildren. Am J Clin Nutr. 94: 1202-10.

12. Nogueira Arcanjo FP, Santos PR, Arcanjo CP, Amancio OM, Braga JA (2012) Use of iron-fortified rice reduces anemia in infants. J Trop Pediatr 58: 475-480.

13. Arcanjo FPN, Santos PR, Leite AJM, Mota FSB, Segall SD. Rice fortified with iron given once weekly increases hemoglobin levels and reduces anemia in infants: a randomized, double-blind, placebo-controlled trial.

14. Food and Nutrition Board. Dietary Reference Intakes for Vitamin A, Vitamin K, Arsenic, Boron, Chromium, Copper, Iodine, Iron, Manganese, Molybdenum, Nickel, Silicon, Vanadium, and Zinc: A Report of the Panel on Micronutrients Subcommittees on Upper Reference Levels of Nutrients and of Interpretation and Uses of Dietary Reference Intakes, and the Standing Committee on the Scientific Evaluation of Dietary Reference Intakes. Washington, DC: National Academic Press, 2001.

15. Brasil. Ministerio da Saude. Centro Brasileiro de Analise e Planejamento. Pesquisa Nacional de Demografia e Saude da Crianca e da Mulher PNDS, 2006. Serie G. Estatistica e Informacao em Saude. Brasilia. DF 2009.

16. Lwanga SK, Lemeshow S (1991) Sample Size Determination in Health Studies: A Practical Manual. Geneva: WHO.

17. Moretti D, Zimmermann MB, Wegmüller R, Walczyk T, Zeder C, et al. (2006) Iron status and food matrix strongly affect the relative bioavailability of ferric pyrophosphate in humans. Am J Clin Nutr 83: 632-638.

18. Hurrell R, Egli I (2010) Iron bioavailability and dietary reference values. Am J Clin Nutr 91: 1461S-1467S.

19. Bagni UV, Baiao MR, Santos MM, Luiz RR, Veiga GV (2009) Effect of weekly rice fortification with iron on anemia prevalence and hemoglobin concentration among children attending public daycare centers in Rio de Janeiro, Brazil. Cad Saude Publica. 25: 291-302.

20. Micronutrient Initiative Investing in the future, a united call to action on vitamin and mineral deficiencies, Global Report 2009. Ontario: Micronutrient Initiative (2009).

21. Copenhagen's Consensus (2008) The world's best investment: Vitamins for undernourished children, according to top economists, including 5 Nobel Laureates. Copenhagen Consensus Center. Copenhagen, Denmark.

30

The Effects of DNA Methylation on the Expression of Non-imprinted Genes in Rice

Hongyu Zhang#, Yutong Liu#, Mipeng Han, Limei Wu, Zhijian Liu, Xiaotong Chen, Peizhou Xu and Xianjun Wu*

Rice Research Institute, Sichuan Agricultural University, Wenjiang, Chengdu, P.R.China
#These authors are equal contribute to this work

Abstract

The expression of imprinted genes is regulated by well-known genetic mechanisms such as DNA and histone methylation. However the mechanism regulating the expression of non-imprinted genes that are specifically expressed in endosperm is currently unknown. To determine whether DNA methylation is involved in the regulation of non-imprinted gene expression in endosperm, we used rice seeds from a reciprocal cross between cultivars Nipponbare and 9311 treated with a methylation inhibitor 5-aza-2'-deoxycytidine to investigate the expression patterns of four non-imprinted genes in seedlings. We found these endosperm specific genes were activated in F1 with two types of expression patterns: 1) either both parental alleles were expressed in F1 leaves; 2) or only one parental allele was expressed in the leaves of the progeny. We speculate that the altered expression patterns of parental alleles may be associated with F1 heterosis. We also observed that the expression of non-imprinted genes underwent dynamic changes at different development stages with two showing imprinted expression patterns, suggesting that DNA methylation is involved in regulating the expression of some imprinted as well as non-imprinted genes. The results of this study provide a reference for further exploring epigenetic mechanisms underlying seed development and, potentially, the association of dynamic changes of DNA methylation with heterosis.

Keywords: Rice; Imprinted genes; Non-imprinted genes; Endosperm; 5-azadC; DNA methylation

Introduction

Cereal grain consists of two major components: embryo and endosperm that are the products of double fertilization specifically occurring in flowering plants. Endosperm occupies most of the volume of the grain and is a major source of human nutrition. Thus it is important to investigate grain development to improve crop and food production to fulfill the growth of world population.

Genomic imprinting almost exclusively occurs in plant endosperm and is modulated by the variation of DNA and histone methylation [1-7]. Gene imprinting refers to the phenomenon that one of the parental gene copies is exclusively or preferentially expressed depending on parent-of-origin [8-11]. Proper endosperm development requires a set of genes being imprinted and these genes may affect grain size [12]. The imbalance of parental genome dosage in the endosperm observed in interploidy crosses and DNA methylation mutants is the main reason for endosperm abortion [13-18]. Endosperm development is regulated by epigenetic mechanisms such as DNA and histone methylation that regulate the expression of imprinted genes [19-27]. Because endosperm is a terminally differentiated tissue, the genetic and epigenetic variations that occur in imprinted genes will not be passed to the next generation.

Changes of DAN methylation patterns and levels of spring are closely related to the formation of heterosis [28-33]. The patterns of plant DNA methylation are tissue- and organ-specific [34,35]. For example, elevated DNA methylation levels at some loci are considered to be positively associated with plant heterosis [32,36], while at other loci it is the decreased DNA methylation levels that correlates with the plant heterosis [31,37]. Specifically expressed genes in endosperm may influence embryonic development via influencing endosperm development [22]. Ectopic activation of these endosperm specific genes in seedlings or other organs may impact on plant growth and thereby heterosis. Zemach [38] identified 165 endosperm-specific genes using gene chip technology and found that not all were imprinted

genes. Luo et al. [23] identified 262 candidate imprinted loci in rice by Illumina high-throughput sequencing and verified 56 imprinted loci in the endosperm. These authors identified a few imprinted rice candidates that are highly homologous to genes in Arabidopsis thaliana associating with epigenetic regulation, including DNA methylation, histone methylation, and small RNA pathways. Wang et al. [39] found that the expression of genes silenced by DNA methylation could be activated through the addition of the methylation inhibitor 5-aza-2'-deoxycytidine (5-azadC) [40]. However, it is still unknown whether DNA methylation is involved in the regulation of non-imprinted genes during endosperm development.

In this study, we treated the seeds of a hybrid rice (*Oryza sativa*) from a reciprocal cross between japonica cultivar Nipponbare and indica cultivar 9311 with the DNA methylation inhibitor 5-azadC (thereby altering the expression of genes in the resulting seedlings). We then analyzed the regulatory effect of DNA methylation on non-imprinted genes. The results of the present study provide a reference for further elucidating the genetic mechanism underlying heterosis and the molecular mechanism of imprinting during endosperm development.

Materials and Methods

Plant materials

Rice (*Oryza sativa*) line Parents Nip (*Nipponbare*) line 9311, planted

*Corresponding author: Xianjun Wu, Rice Research Institute, Sichuan Agricultural University, Wenjiang, Chengdu, P.R.China
E-mail: wuxjsau@126.com

in Sichuan Wenjiang test fields with reciprocal cross in the flowering phase, taking the seed endosperm with the development of 3-5D and the parental roots, stems, endosperm, leaves, freezing in liquid nitrogen and preserving at -80°C to prepare RNA extraction.

The mature hybrid seeds were collected and preserved at -4°C for one month, to break seed dormancy for the preparation of the subsequent pharmaceutical treatment and germination. RNA isolation, reverse transcription and sequencing.

Total RNA was isolated using the RNeasy Plant Mini kit (Qiagen, Hilden, Germany) according manufacturers' instruction. The genomic DNA contamination was removed by using RQ1 RNase-Free DNase (Promega, Madison, WI). cDNA was synthesized from the 100 ng of total RNA samples by using the PrimeScript First Strand cDNA Synthesis kit (Takara, Dalian, Chian). After synthesizing cDNA, PCR amplification was performed using the TaKaRa PCR Amplification kit (Takara) using the primers listed in Table 1. The PCR condition was 95°C for 3 min for pre-denaturing, and then 34 cycles for denaturing at 72°C for 45S, annealing at 56°C and polymerization at 72°C for 7

min. PCR products were examined on a 1% agarose gel followed by sequencing in BGI (Shenzhen, China).

Gene annotation and primer sequences

According to the gene annotation provided by http://bioinformatics.cau.edu.cn/ neweasygo/ to confirm the genes encoding genes encoding transcription factors in Assaf According to http://blast.ncbi.nlm.nih.gov/ to detect the differences of polymorphic loci between japonica rice and indica rice, using Primer Premier 5 to design the primers. A total of 18 genes were identified.

Seed treatment by methylation inhibitor 5-azadC

The concentration of 5-azadC was diluted to 30, 50 μmol/L dissolving with DMSO, diluting with distilled water to the corresponding concentration; the control group was a mixed solution of DMSO and distilled water. The seeds were steeped in the condition of light incubator temperature (28°C/25°C) and photoperiod(16h D/8h N) for 3-5d,observing the changes of plant properties after treatment according to the germination, The drug concentrations used finally was confirmed to be 3 μmol/L.

Gene Name	Gene Annotation	Primer sequences(5'-3')
LOC_Os01g01290	histone-like transcription factor and archaeal histone, putative, expressed (CCAAT Family)	CACCAAAGGCTCAACAACAA
		ACGGTTATGGGATTGAGCAG
LOC_Os01g01470	no apical meristem protein, putative, expressed (NAC Family)	TGGGTCATGCACGAGTACAG
		AGTGAGAGTGAAGCGGTGGT
LOC_Os01g24460	histone-like transcription factor and archaeal histone, putative, expressed (CCAATFamily)	GGCCAAGAAGAACAACATGAG
		GCCATTACTGGTGCTTGGAT
LOC_Os01g29840	no apical meristem protein, putative, expressed (NAC Family)	CAGATGCCCTCCATGTCTG
		GACACCACCAGCGACGAC
LOC_Os01g33350	Expressed protein	GAATTTGTGCCTCCATGGTT
		CGTGGTATGATGATCGCCTA
LOC_Os01g39850	histone-like transcription factor and archaeal histone, putative, expressed (CCAAT Family)	CCATTGCCTCCACAGAGTC
		GACCCCTTGCTATGTTGTGAA
LOC_Os02g12310	no apical meristem protein, putative, expressed (NAC Family)	TCTTCGGTGAATCGTCTTCC
		CCACCATGGTTTCTTTGCAT
LOC_Os02g15350	Transcriptional activator can collaborate LOC_Os07g08420 regulate seed storage protein (SSP)gene expression (C2C2-DOF Family) in Rice	ATTATCCCCGGTGGAGGAG
		CAGGAGCAGGAGGAGACG
LOC_Os04g10260	bZIP transcription factor basic region leucine zipper domain containing protein	TCAGTTAAGCCGGAGGTCAC
		TGAATTTCACATTCGCAAGC
LOC_Os04g35010	helix-loop-helix DNA-binding domain containing protein, expressed (BHLH Family)	TCTGGTAAGGTCGATTAAAGCA
		CATCTTCTTCCTCCGCTCTC
LOC_Os05g34310	no apical meristem protein, putative, expressed (NAC Family)	CTTCAACCCGTGGGAGCTT
		CGAGCACTGTAACCGTGAGA
LOC_Os07g08420	bZIP transcription factor,Transcriptional activator can collaborate LOC_Os02g15350regulate seed storage protein (SSP)gene expression (C2C2-DOF Family) in Rice	ACAACTTCACCAGGCCATTC
		GCTCCATGTTGACAAGCTCA
LOC_Os09g34880	basic region leucine zipper domain containing protein, expressed (bZIP Transcription Factor)	CTCCCTTCCTCGGTCCTCT
		TTTGGCTGTGGAAACCCTAC
LOC_Os10g25850	Nuclear transcription factor Y subunit A-7	AAACCTGAGTGCAACCAACC
		ATGCCTCAATTTTGCTTGCT
LOC_Os11g31330	no apical meristem protein	TCGGAGGTGCCCATCTATTA
		AGGGTGGCTCTGAACCATT
LOC_Os11g31340	no apical meristem protein, putative, expressed (NAC Family)	CTGCTGGTGATGGGTTCTG
		GCGATGGTCGTTCCTGTG
LOC_Os11g31360	no apical meristem protein, putative, expressed conserved seed development associated transcription factors, arabidopsis its analogues function is to regulate embryo (NAC Family)	CAAGGAGAACAGCCACCCTA
		CATGAGTATGGGCAGCAGAC
LOC_Os11g31380	no apical meristem protein	CTCAAAACCACCCTGCAACT
		CTCCCTTGCATTGCCATT

Table 1: Preliminary screening, gene annotation, and primer sequences.

Results

Identification of the genes specifically expressed in endosperm

Based on the endosperm-specific genes identified by Zemach [38], 26 genes encoding transcription factors were chosen in our analysis. We investigated whether there were SNPs in these genes in *Japonica* rice Nip and non-glutinous rice 9311 based on information obtained from the PubMed website. The gene regions that are different compared with parents were used to design PCR primers, and endosperm cDNA (shown in experimental materials) was used as a template for PCR. A total of 18 genes could be amplified by PCR and RT-PCR results of 8 genes are shown in Figure 1.

To confirm if these 18 genes are specifically expressed in endosperm, we synthesized cDNA using PCR amplification from RNA that were isolated from leaves, endosperm, roots, and stems of Nip. All 18 genes could only be amplified from endosperm tissue but not from leaf, root, or stem tissue, demonstrating that these 18 genes were specifically expressed in endosperm (Figure 2).

Identification of non-imprinted genes

We used Sanger sequencing to test the imprinting status of selected endosperm specific genes. The amplified bands were sequenced to examine the expression of SNPs loci containing the parental differences in the progeny plants. If the differences in the expression of SNPs of the parents simultaneously expressed in the progeny, indicating that the gene was non-imprinted gene; conversely, if the progeny combinations only expressed one of the parental different bases, then the gene was imprinted gene. The sequencing results found that among 18 genes, there were 4 non-imprinted genes and 14 imprinted genes. The Sanger sequencing results of 6 genes (4 non-imprinted genes, 2 imprinted genes) were shown in in Table 2.

The effect of 5-azadC-mediated demethylation on the expression of endosperm specific genes in hybrid seedlings

DNA methylation regulates the expression of imprinted genes in the endosperm, but it is not yet known whether methylation is also involved in regulating the expression of non-imprinted genes. Therefore, we treated rice seeds with the methylation inhibitor 5-azadC (30 μmol/L) for 3-5 d, followed by germination, to examine the effects of methylation on the regulation of non-imprinted genes.

We first examined the true and false hybrid plants and to ensure the authenticity of the hybrids, we used 38 pairs of microsatellite primers to differentiate between genes from Nipponbare and 9311 and obtained 11 different markers (RM1, RM4, RM9, RM18, RM47, RM52,

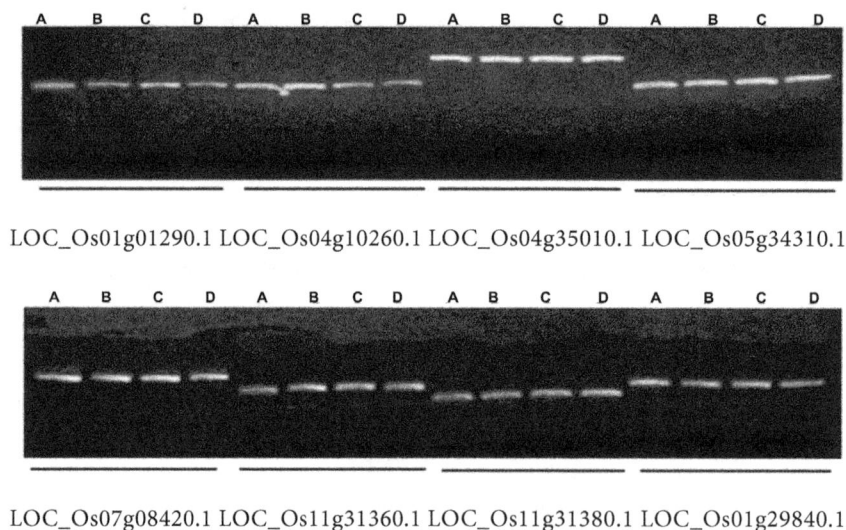

LOC_Os01g01290.1 LOC_Os04g10260.1 LOC_Os04g35010.1 LOC_Os05g34310.1

LOC_Os07g08420.1 LOC_Os11g31360.1 LOC_Os11g31380.1 LOC_Os01g29840.1

A: Nipponbare (Nip), B: 9311, C: Nip × 9311, D: 9311 × NIP.

Figure 1: RT-PCR analysis of the expression of 8 endosperm-specific transcription factor genes in the endosperm of four plant lines.

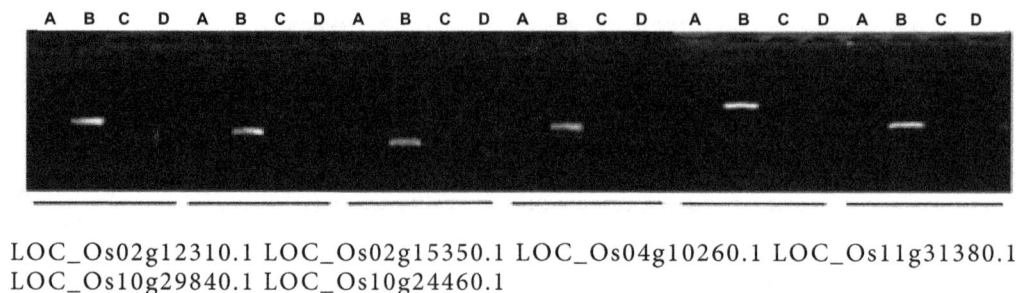

LOC_Os02g12310.1 LOC_Os02g15350.1 LOC_Os04g10260.1 LOC_Os11g31380.1
LOC_Os10g29840.1 LOC_Os10g24460.1

Note: A: Leaves, B: Endosperm, C: Roots, D: Stems.

Figure 2: Expression detection of 6 selected genes in four organs of Nipponbare.

Combination	Nip	9311	Nip × 9311	9311 × Nip
SNPs	CACTG	CATTG	CAC (T) TG	CAT (C) TG
LOC_Os04g35010.1 (non-imprinted)	CACTG	CATTG	CACTG	CATTG
SNPs	GACAT	GATAT	GAC (T) AT	GAT (C) AT
LOC_Os07g08420 (non-imprinted)	GACAT	GATAT	GACAT	GACAT
SNPs	TGCCA	TGTCA	TG (T) CA	TGT (C) CA
LOC_Os02g12310.1 (non-imprinted)	TGCCA	TGTCA	TGCCA	TGCCA
SNPs	TCTTA	TCCTA	TCT (C) TA	TCC (T) TA
LOC_Os01g24460.1 (non-imprinted)	TCTTA	TCCTA	TCTTA	TCCTA
	TCATA	TCGTA	TCGTA	TCATA
LOC_Os01g33350.1 (imprinted)	TCATA	TCGTA	TCGTA	TCATA
SNPs	CAGAG	CAAAG	CAGAG	CAAAG
LOC_Os10g25850.1 (imprinted)	CAGAG	CAAAG	CAGAG	CAAAG

Table 2: Comparison of SNP expression of endosperm-specific genes in parents and hybrid seedlings.

RM72, RM335, RM337, RM339, and RM341). Markers RM1 and RM72 were used to confirm the identity of the hybrids by examining the leaves of the F1 hybrid plants, as both markers should be amplified simultaneously in true hybrids. The results confirm that the hybrids utilized in this study were true hybrids (Figure 3).

After inhibitor treatment, the 5-azadC-treated seeds exhibited a delayed germination and the plants became stunted compared with the control (Figure 4). The latter phenomenon was changed in most plants after 5 days of growth except some plants wilted and gradually died. These results are consistent with the findings of Sano et al. [41].

The expression of the 18 endosperm-specific genes in the leaves of plants produced from 5-azadC-treated seeds were investigated using RT-PCR. It turned out that only four genes could be amplified from these leaves (Figure 5), indicating that the expression of these four genes is regulated by DNA methylation, i.e., de-methylation enabled these genes to be expressed in leaves.

Activation of various parental alleles by demethylation in hybrid seedlings

RNA was extracted from the leaves of seedlings derived from 5-azadC-treated seeds and subjected to RT-PCR to investigate the expression of parental alleles at the four non-imprinted loci aforementioned and to see whether the expression patterns of SNP loci in leaves were consistent with those in endosperm. In the endosperm of hybrids, gene LOC_Os07g08420 expressed the SNPs C and T simultaneously that come from mother and father, respectively. In the seeds treated with 5-azadC, the new leaves (demethylated leaves) still expressed the bases C and T, suggesting that 5-azadC treatment did not influence the expression of parental alleles of this gene in the treated hybrid seedlings. As LOC_Os 07g08420 is only expressed in endosperm and silenced in embryos and other tissues. We speculate that DNA methylation plays an essential role in silencing the activity of this gene in non-endosperm tissues. On the same line, we observed that seeds treated with 5-azadC showed the expression of both parental alleles in the seedlings at this locus (Table 3).

Interestingly, gene LOC_Os04g35010.1 was similarly activated like LOC_Os 07g08420 (Table 4) thus suggesting that DNA methylation is involved in regulating the expression of endosperm specific non-imprinted genes.

Furthermore, gene LOC_Os02g12310 expressed the polymorphism loci C and T in the 3d and 5d endosperm simultaneously as we expected. However, we observed that only the Nip allele (C) was activated after 5-azadC treatment while the 9311 remains silenced. Obviously there are differential effects of DNA methylation on the Nip and 9311 alleles. But it is not clear why 9311 is not activated (Table 5).

Similarly, LOC_Os01g24460 expressed the SNPs C and T in the 5d endosperm simultaneously (Table 6). Interestingly at 3d only the maternal allele is detected in the endosperm, suggesting that the

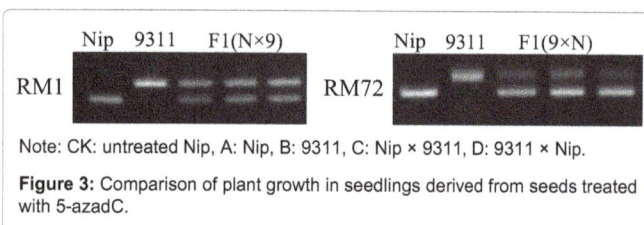

Note: CK1: no-treatment Nip, CK2: no-treatment 9311, CK3: no-treatment Nip × 9311, CK4: no-treatment 9311 × Nip.
A: treatment Nip, B: treatment 9311, C: treatment Nip × 9311, D: treatment 9311 × Nip.

Figure 4: Comparison of plant growth when the seeds were treated by 5-azadC.

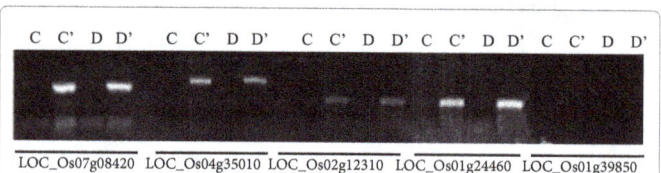

LOC_Os07g08420 LOC_Os04g35010 LOC_Os02g12310 LOC_Os01g24460 LOC_Os01g39850

Note: C: Nip × 9311, C': 5-azadC-treated Nip × 9311, D: 9311 × Nip, D': 5-azadC-treated 9311 × Nip.

Figure 5: Gene expression in leaves of seedlings derived from seeds treated with 5-azadC.

imprinting process could be dynamic (Table 6). Strikingly the hybrid seedlings with reduced DNA methylation after inhibitor treatment showed the same maternal expression at this locus as shown in the endosperm of 3d. But the underlying mechanisms driving the imprinted expression in early endosperm and treated seedlings remain to be determined.

Together, expression patterns at LOC_Os01g24460 are related to development stages, namely imprinted expression occurs in the 3d endosperm, but not in the 5d endosperm. This dynamic process may require the conversion of imprinted locus into non-imprint one at later development stage and DNA de-methylation may be involved in this process. Our results also showed that DNA methylation is also involved in the regulation of the selective demethylation of the maternal alleles of the seedlings being treated by 5-azadC at this locus.

Discussion

DNA methylation regulates on endosperm specific genes and may repress the activity of deleterious gene in seedlings

The expression levels of many genes are dynamic in different organs and at different developmental stages. The mechanisms regulating the spatial and temporal expression of genes include both epigenetic modifications and the role of regulatory factors. It is known that methylation inhibitor 5-azadC can reduce DNA methylation levels, thereby leading to the change of gene expression [39-42]. For endosperm-specific genes, we have tested whether the silencing in non-endosperm tissues are due to DNA methylation. We detected the expression of endosperm-specific genes in the seedlings derived from rice seeds treated with 5-azadC. Further sequencing analysis of the seedling-expressed genes showed that there were two types of expression patterns: either both parental alleles as represented by

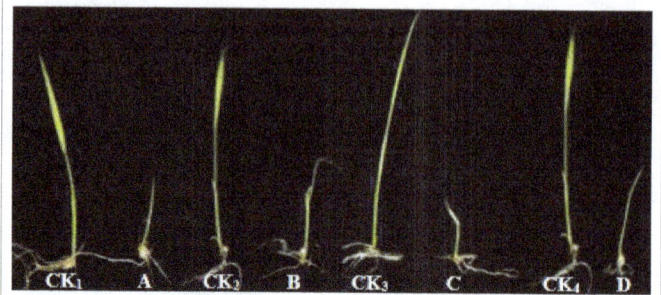

Note: CK: untreated Nip, A: Nip, B: 9311, C: Nip × 9311, D: 9311 × Nip.

Figure 3: Comparison of plant growth in seedlings derived from seeds treated with 5-azadC.

Endosperm-3d		Endosperm-5d		5-azadC-leaf	
Nip × 9311	9311 × Nip	Nip × 9311	9311 × Nip	Nip × 9311	9311 × Nip
GAC(T)AT	GA T(C)AT	GAC(T)AT	GAT(C)AT	GAC(T)AT	GAC(T)AT

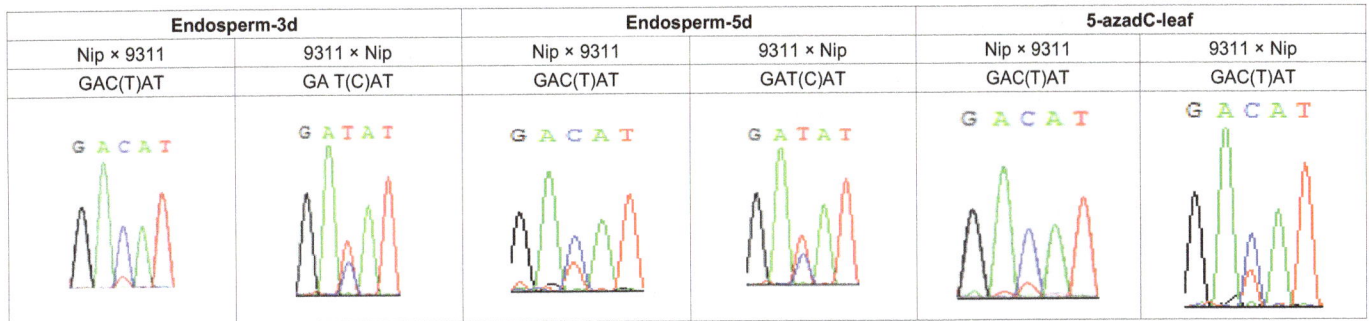

Table 3: The expression of C and T SNPs at LOC_Os07g08420 in two developmental stages of endosperm and the demethylated leaves after 5-azadC treatment.

Endosperm-3d		Endosperm-5d		AzadC-leaf	
Nip × 9311	9311 × Nip	Nip × 9311	9311 × Nip	Nip × 9311	9311 × Nip
CAC(T)TG	CAT(C)TG	CAC(T)TG	CAT(C)TG	CAC(T)TG	CAC(T)TG

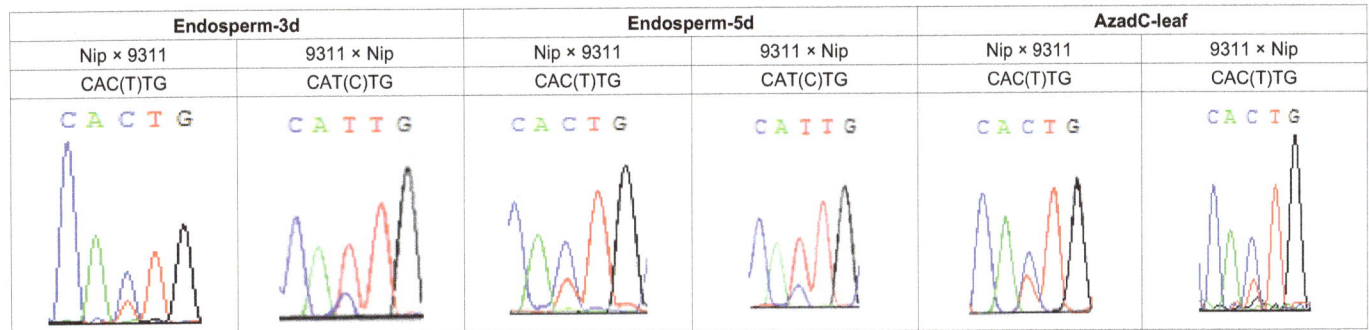

Table 4: The expression of C and T SNPs at LOC_Os04g35010.1 in two development stageal s of endosperm and the demethylationed leaves after 5-azadC treatment.

Endosperm-3d		Endosperm-5d		AzadC-leaf	
Nip × 9311	9311 × Nip	Nip × 9311	9311 × Nip	Nip × 9311	9311 × Nip
TGC(T)CA	TGT(C)CA	TGC(T)CA	TGT(C)CA	TGCCA	TGCCA

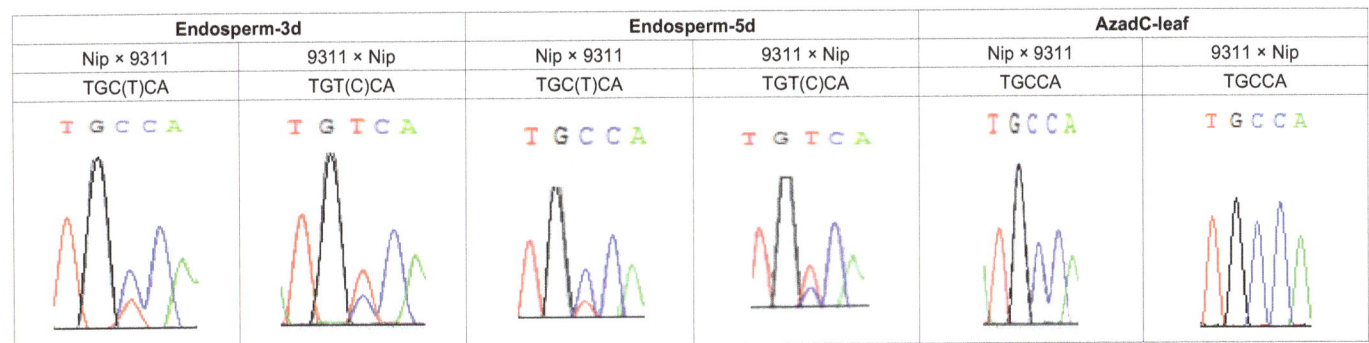

Table 5: The expression of C and T SNPs at LOC_Os02g12310 in two developmental stages of endosperm and the seedlings after 5-azadC treatment.

Endosperm-3d		Endosperm-5d		AzadC-leaf	
Nip × 9311	9311 × Nip	Nip × 9311	9311 × Nip	Nip × 9311	9311 × Nip
TCTTA	TCCTA	TCT(C)TA	TCC(T)TA	TCTTA	TCCTA

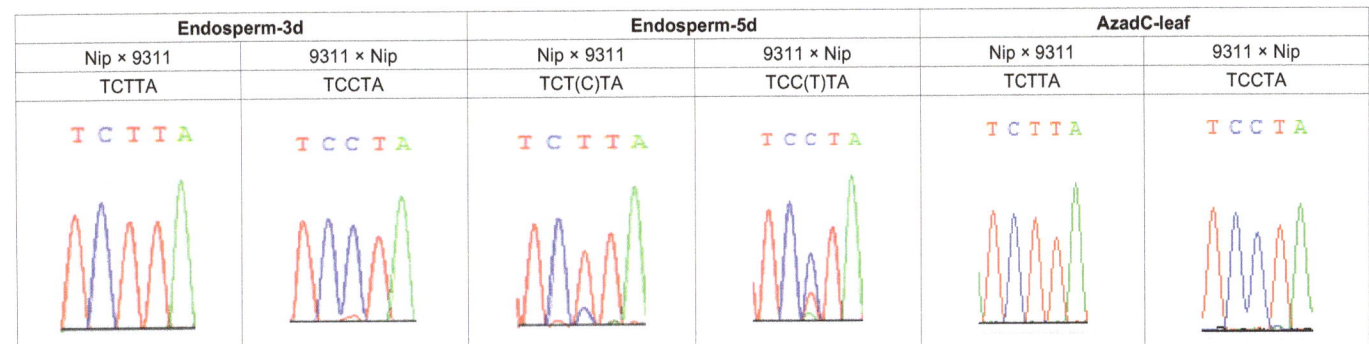

Table 6: The expression of C and T bases SNPs at LOC_Os01g24460.1 in Endosperm two developmental stages of e and the leaves seedlings after 5-azadC treatment.

polymorphic loci (SNPs) or only one of the polymorphic loci (SNPs) were expressed in the leaves derived from inhibitor treated seeds.

According to functional annotation [38], LOC_Os07g08420 plays an important role in grain filling of rice, LOC_Os04g35010.1 encodes DNA helix domain protein, LOC_Os02g12310 encodes no apical meristem protein (NAC family), and LOC_Os01g24460.1 encodes histone-like transcription factors. Seeds treated with 5-azadC showed delayed germination, dwarfing of plant height, death at late development stage and other detrimental phenotypes. It remains to be determined whether activation of the endosperm-specific genes of diverse functions contributes to the detrimental phenotypes after seed treatment. It is possible that demethylation mediated by 5-azadC would activate many more genes that would be silenced in seedlings and, most likely these ectopic activations will cause abnormal plant development. Consistently, it is hypothesized that evolutionary selection of imprinted genes (which usually are endosperm specific and silenced by DNA methylation) is driven by silencing deleterious gene activity in somatic tissues [3]. Interestingly, LOC_Os01g24460 showed imprinted expression in early endosperm development.

It has been shown DNA methylation patterns and the changes in its expression levels were closely related to the formation of heterosis [28-33]. Chodavarapu et al. [30] found that the difference of cytosine methylation level between rice parents (Nip and 9311) and F1 was 0.79%, thus they hypothesized that the heterosis of F1 was related to the reduction of DNA methylation levels. Zhiben et al. [28-29,31-32] also found that DNA methylation had a close relationship to heterosis in rice, maize and sorghum, respectively. At the LOC_Os01g24460 locus only the Nip allele was detected in the hybrid seedlings after removal of DNA methylation, suggesting that the Nip locus had a heterosis in the progeny. The other three genes also showed activation of parental alleles in F1 hybrid derived from treated seeds.

LOC_Os01g24460 in the 3d endosperm displayed maternal expression, but the expression at 5d is bi-allelic, indicating that imprinting is a dynamic process. Strikingly, after removing DNA methylation, LOC_Os01g24460 also expressed as a maternal imprinting gene in the hybrid leaves, suggesting that the maternal alleles of this gene in both endosperm and 5-azadC treated seedlings are prone to demethylation process that drives imprinted expression. What causes the differential responses to 5-azadC mediated demethylation process in hybrid seedlings is unknown. In contrast, DEMETER mediated demethylation process at the maternal alleles in endosperm has been clearly demonstrated in various plants.

Du [42] also found that an imprinted gene yellow2-like in the seedling derived from rice seeds treated with DNA methylation inhibitor 5-azadC or zebularine could produce novel transcripts, indicating that DNA methylation can inhibit the transcription initiation of abnormal promoters in plants, thereby resulting in specific expression. This finding together with our study provide evidence that DNA methylation at least is one of the main mechanisms that drive tissue-specific gene expression and DNA methylation is important for plant development.

DNA methylation is associated with Heterosis and imprinting

In angiosperms, seed development and germination are related processes because embryo development and germination require nutriment flow from endosperm. Most imprinted expression is observed in the endosperm of developing seed. Previous studies suggest that the transposon insertion that affect the expression of neighboring genes is the main driving force of the evolution of genomic imprinting [25-27,43]. DNA methylation is considered to silence the deleterious insertion of transposons. Therefore, large numbers of imprinted genes are endosperm specific because these genes are silenced in non-endosperm tissues due to DNA methylation and activated by DEMETER that remove DNA methylation. For non-imprinted endosperm-specific genes, their role and the regulation of their expression are largely unknown. Current study analyzed the role of DNA methylation modification in the regulation of non-imprinted genes and detected expression of the parental SNPs of four non-imprinted genes in demethylated hybrid seedlings. Our results support the idea that there is a close relationship between the dynamic changes in DNA methylation and the establishment of imprinting as well as, potentially, heterosis.

In this study we have only selected 4 endosperm-specific transcription factors to analyze the role of DNA methylation in the regulation of gene expression, it is still required to perform s more in-depth analysis on the regulation of other transcription factors and non-transcription factors that are expressed specifically in endosperm. It is known that DNA methylation, histone H3K27 modification, POIIV-dependent siRNAs (p4-siRNAs) and other epigenetic mechanisms are involved in the expression of imprinted genes and the formation of heterosis [25,33].

Financial Support

Support by the National Natural Science Fund (31301049).

References

1. Jullien PE, Katz A, Oliva M, Ohad N, Berger F (2006) Polycomb group complexes self-regulate imprinting of the polycomb group gene MEDEA in Arabidopsis. Curr Biol 16: 486-492.

2. Huh JH, Bauer MJ, Hsieh TF, Fischer R (2007) Endosperm gene imprinting and seed development. Curr Opin Genet Dev 17: 480-485.

3. Berger F, Chaudhury A (2009) Parental memones shape seeds. Trends Plant Sci 14: 550-556.

4. Bauer MJ, Fischer RL (2011) Genome demethylation and imprinting in the endosperm. Curr Opin Plant Biol 14: 162-167.

5. Hsieh TF, Ibarra CA, Silva P, Zemach A, Eshed-Williams L, (2009) Genome-wide demethylation of Arabidopsis endosperm. Science 324: 1451-1454.

6. Haun WJ, Springer NM (2008) Maternal and paternal alleles exhibit differential histone methylation and acetylation at maize imprinted genes. Plant J 56: 903-912.

7. Gutierrez-Marcos JF, Costa LM, Pra DM, Scholten S, Kranz E, et al. (2006) Epigenetic asymmetry of imprinted genes in plant gametes. Nat Genet 38: 876-878.

8. Haig D, Wilczek A (2006) Sexual conflict and the alternation of haploid and diploid generations. Philos Trans R Soc Lond B Biol Sci 361: 335-343.

9. Dilkes BP, Comai L (2004) A differential dosage hypothesis for parental effects in seed development. Plant Cell 16: 3174-3180.

10. Babak T, Deveale B, Armour C, Raymond C, Cleary MA, et al. (2008) Global survey of genomic imprinting by transcriptome sequencing. Curr Biol 18: 1735-1741.

11. Hagan JP, O'Neill BL, Stewart CL, Kozlov SV, Croce CM (2009) At least ten genes define the imprinted Dlk1-Dio3 cluster on mouse chromosome 12qF1. PLoS One 4: e4352.

12. Arnaud P, Feil R (2006) MEDEA takes control of its own imprinting. Cell 124: 468-470.

13. Stoute AI, Varenko V, King CJ, Scott RJ, Kurup S (2012) Parental genome imbalance in brassica oleracea causes asymmetric triploid block. Plant J 71: 503-516.

14. Scott RJ, Spielman M, Bailey J, Dickinson HG (1998) Parent-of-origin effects on seed development in Arabidopsis thaliana. Development 125: 3329-334.

15. Rapp RA, Udall JA, Wendel JF (2009) Genomic expression dominance in allopolyploids. BMC Biology 7: 18.

16. Sekine D, Ohnishi T, Furuumi H, Ono A, Yamada T, et al. (2013) Dissection of two major components of the post-zygotic hybridization barrier in rice endosperm. Plant J 76: 792002D799.

17. Kirkbride RC, Yu HH, Nah G, Zhang C, Shi X, et al. (2015) An epigenetic role for disrupted paternal gene expression in postzygotic seed abortion in Arabidopsis interspecific hybrids. Mol Plant 8: 1766-1775.

18. Jullien PE, Berger F (2010) Parental genome dosage imbalance deregulates imprinting in Arabidopsis. PLoS Genet 6: p.e1000885.

19. Choi Y, Gehring M, Johnson L, Hannon M, Harada JJ, et al. (2002) DEMETER, a DNA glycosylase domain protein, is required for endosperm gene imprinting and seed viability in Arabidopsis. Cell 110: 33-42.

20. Xiao W, Gehring M, Choi Y, Margossian L, Pu H, et al. (2003) Imprinting

of the MEA polycomb gene is controlled by antagonism between METI methyltransferase and DME glycosylase. Dev Cell 5: 891-901.

21. Kinoshita T, Miura A, Choi Y, Klnoshlta Y Cao X, et al. (2004) One-way control of FWA imprinting in Arabidopsis endosperm by DNA methylation. Science 303: 521-523.

22. Zhang WW, Cao SX, Jiang L, Zhu SS, Wan JM (2005) Genomic imprinting and seed development. Genetics 27: 665-670.

23. Luo M, Taylor JM, Spriggs A, Zhang HY, Wu XJ, et al. (2011) A genome-wide survey of impnnted genes in rice seeds reveals imprinting pdmadly occurs in the endosperm. PLoS Genet7.e1002125

24. Hsieh TF, Shin J, Uzawa R, Silva P, Cohen S, et al. (2011) Regulation of imprinted gene expression in Arabidopsis endosperm. Proc Natl Acad Sci U S A 108: 1755-1762.

25. Zhang M, Liu B (2012) Epigenetic regulation of endosperm development in plant. Acta BOtanica Sinica 47: 101-110.

26. Zhang H, Chaudhury A, Wu X (2013) Imprinting in Plants and Its Underlying Mechanisms. J Genet Genomics 40: 239-247.

27. Hongyu Z, Peizhou X, Hua Y (2010) Emprinted genes in Arabidopsis and expression regulation of imprinting. Genetics 32: 670-676.

28. Zhiben Y, Yi S, Tiantang N (2005) Study on difference of cytosine methylation of sorghum genome DNA between hybrids and parents. Crop Sinica 31: 1138-1143.

29. Hepburn PA, Margison GP, Tisdale MJ (1991) Enzymatic methylation of cytosine in DNA is prevented by adjacent O6-methylguanine residues. J Biol Chem 266: 7985-7987.

30. Chodavarapu RK, Feng S, Ding B, Simon SA, Lopez D, et al. (2012) Transcriptome and methylome interactions in rice hybrids. Proc Natl Acad Sci USA 109: 12040-12045.

31. Tsaftaris AS, Kafka M (1998) Mechanism of heterosis in crop plants. J Crop Prod 1: 95-111.

32. Xiong LZ, Xu CG, Maroof MAS, Zhang Q (1999) Patterns of cytosine methylation in an elite rice hybrid and its parental lines, detected by a methylation-sensitive amplification polymorphism technique. Mol Gene Genet 261: 439-446.

33. Huihui C, Chao X, Yingyao S, Wensheng W, Yongming G, et al. (2015) Research progress in epigenetics of heterosis. Journal of Plant Genetic Resources 16: 933-939.

34. Na L, Yang Z, Linan X (2012) Research progress in plant DNA methylation. Plant Physiology Journal 48: 1027-1036.

35. Guangping G, Jinling Y, Xiaoli W (2011) Application status and prospects of DNA methylation in plant research. Journal of Plant Genetic Resources 12: 425-430.

36. Jin H, Hu W, Wei Z, Wan L, Li G, et al. (2008) Alterations in cytosine methylation and species-specific transcription induced by interspecific hybridization hybridization between Oryza sativa and O. Officinalis. Theor Appl Genet 117: 1271-1279.

37. Zhao Y, Yu S, Xing C, Fan S, Song M (2008) Analysis of DNA methylation in cotton hybrids and their parents. Mol Biol (Mosk) 42: 195-205.

38. Zemach A, Kim MY, Silva P, Rodrigues JA, Doston P, et al. (2010) Local DNA hypomethylation activates genes in rice endosperm. Proc Natl Acad Sci USA 107: 18729-18734.

39. Wang J, Tian L, Madlung A, Lee HS, Chen M, et al. (2004) Stochastic and epigenetic changes of gene expression in Arabidopsis polyploids. Genetics 167: 1961-1973.

40. Lee HS, Chen ZJ (2001) Protein-coding genes are epigenetically regulated in Arabidopsis polyploids. Proc Natl Acad Sci USA 98: 6753-6758.

41. Sano H, Kamada I, Youssefian S, Katsumi M, Wabiko H (1990) A single treatment of rice seedlings with 5-azacytidine induces heritable dwarfism and undermethylation of genomic DNA. Molecular and General Genetics MGG 220: 441-447.

42. Du M, Luo M, Zhang R, Finnegan EJ, Koltunow AM (2014) Imprinting in rice: the role of DNA and histone methylation in modulating parent-of-origin specific expression and determining transcript start sites. Plant J 79: 232-242.

43. Ishikawa R, Kinoshita T (2009) Epigenetic programming: The Challenge to Species Hybridization. Mol Plant 2: 589-599.

31

Water Response of Upland Rice Varieties Adopted in Sub-Saharan Africa: A Water Application Experiment

Shunsuke Matsumoto[1], Tatsushi Tsuboi[2], Godfrey Asea[3], Atsushi Maruyama[1], Masao Kikuchi[1] and Michiko Takagaki[1*]

[1]*Graduate School of Horticulture, Chiba University, Chiba, Japan*
[2]*JICA Expert, JICA Uganda Office, Kampala, Uganda*
[3]*National Crops Resources Research Institute, Kampala, Uganda*

Abstract

Whether a rice Green Revolution in sub-Saharan Africa becomes a reality critically hinges on how far productive upland rice cultivation diffuses in the region. In order to quantify the drought tolerance, the rate of water response and the contribution of yield components to changes in yield due to water availability of upland rice varieties used in sub-Saharan Africa, we conducted water application experiments in Namulonge, Uganda, using NERICA 4, NERICA 10, NARIC 2 and Yumenohatamochi, with five different levels of water application. We found that the NERICA varieties were most drought tolerant, followed by NARIC 2. Yumenohatamochi did not withstand the lowest amount of water application of 378 mm. The results suggested that the minimum water requirement was around 311-400 mm per season for the three varieties used widely in East Africa, and around 420-600 mm for Yumenohatamochi, an upland variety in Japan famous in its drought tolerance. It was estimated that an additional water application of 1 mm increased rice yield by 11-12 kg /ha for the upland varieties tested. The high water response of upland rice was brought about by high water response of four yield components, among which the rate of grain filling contributed most to the increase in yield, followed by number of panicles/m², number of grains per panicle and 1000 - grain weight, in the order of the degree of contribution, for all the varieties tested.

Keywords: Drought tolerance; NERICA; Minimum water requirement; Uganda; Yield component

Introduction

Since a series of NERICA varieties (interspecific *Oryza sativa x O. glaberrima* progenie) were developed in the late 1990s by the Africa Rice Center (then WARDA), upland rice has been expected to catalyze a rice green revolution in sub-Saharan Africa where nearly 50% of land area planted to rice is upland, with the other half under rainfed lowland [1,2]. Such expectation has met mixed results due partly to the vulnerability of upland rice cultivation to drought [3,4].

There have been studies on the impacts of water deficit, water stress and drought on the growth and yield of upland rice [5-13], but all of them are conducted in Asia and Australia using *Oryza sativa* cultivars. Studies on the relationship between water and the yield of NERICA varieties in sub-Saharan Africa have been burgeoning, but reliable studies on drought tolerance and water response are still scarce in spite of its importance in promoting upland rice cultivation in the region. The yield response of NERICA varieties under different levels of water availability is reported [14,15], but the water application levels in these studies are 700-1200 mm per crop in the former and 1700-3000 mm per crop in the latter, both without controlling rainfalls. The relationship between yield and soil moisture using NERICA varieties is reported [16], but the soil moisture contents tested are 50-70%. Farmers in sub-Saharan Africa are attempting to plant upland rice in many upland areas that are prone to drought with rainfall less than 600 mm per crop season [2], and critical soil moisture contents for the growth of upland rice are 40% and below [8,13]. More studies that focus on lower levels of water availability are needed. A rice production manual for East African countries recommends upland rice cultivation using NERICA varieties in areas where 5-day rainfalls of 20 mm for about 90 days (360 mm /crop) are available, based on the results of experiments that test the effects of water application on the yield of NERICA 4, without presenting any detail of the experiments [17].

Information on the drought tolerance of NERICA and other upland rice varieties grown in sub-Saharan Africa and their response to water is critically important not only for rice research in the region, in particular for breeding drought-tolerant upland rice varieties [18], but also for developing adequate cultural practices suited to the environmental conditions in the region and disseminating upland rice cultivation among upland farmers there. With a basic purpose of reinforcing related information for researchers, extension workers and policy makers in sub-Saharan Africa, this paper reports the results of water-application experiments conducted in Uganda, aiming at clarifying the drought tolerance and water response of upland rice varieties adopted in sub-Saharan Africa. More specifically, this paper intends to (1) elucidate drought tolerance in terms of the minimum water requirement of upland rice varieties adopted widely in sub-Saharan Africa, (2) quantify how the upland rice varieties respond to changes in water availability, and (3) examine how yield components contribute to the increase in yield as water availability increases.

Materials and Methods

Experiments

Experiments were conducted from March to July 2012 (126 days) in the National Crops Resources Research Institute (NaCRRI) in Namulonge, Uganda (latitude 00°30'46.4N, longitude 32°38'03.6E, altitude 1120 m), using wooden boxes placed in a glass-roofed, screen-

*Corresponding author: Michiko Takagaki, Chiba University, Chiba, Japan
E-mail: mtgaki@faculty.chiba-u.jp

walled greenhouse that shut out rainfall perfectly but kept sunlight and atmosphere the same as outside.

The planting boxes, measuring 1 m × 1 m × 0.7 m (width × length × depth), were made in such a way that the four walls were shielded by plastic sheet and the bottom was with holes the total area of which took one tenth of the bottom area, and filled up with soil taken from the top soil (0-50 cm deep from the soil surface) of an upland field in the NaCRRI. The soil composed of 55% of sand, 35% of clay and 10% silt, with pH of 5.6.

Upland rice varieties used in the experiments were NERICA 4 (henceforth denoted N4), NERICA 10 (N10), NARIC 2 (Naric), and Yumenohatamoti (Yume).The characteristics of the varieties used are found in previous studies[19-23]. N4, N10 and Naric are upland rice varieties formally released in Uganda [24], and hereafter grouped as 'African varieties' for the convenience of comparison with Yume, an upland variety in Japan famous in its drought tolerance.

Seeds were sown by dibbling 6 hills at a space of 30 cm ×15 cm (22.2 hills /m²) and 4 varieties were planted in a box. Fertilization was conducted following the recommended rates: N-P-K = 60-30-30 kg / ha, 30 kg N, 30 kg P and 30 kg K applied as basal, 30 kg N applied at 30 days after sowing.Weeds were pulled out whenever emerged and no pest control was needed.

Five treatments for water application were adopted: T-1 = 21 mm /week (3 mm/day) x 18 (total water application = 378 mm), T-2 = 28 mm /week (4 mm /day) x 18 (504 mm), T-3 =35 mm /week (5 mm/day) x 18 (630 mm), T-4 = 42 mm/week (6 mm /day)x 18 (756 mm), T-5 = 49 mm /week (7 mm/day) x 18 (882 mm).The lowest water application treatment was set to be lower than the level of 4 mm /day, which is stated as the minimum rainfall for NERICA cultivation [17]. Actual water applications were made every Monday and Friday at the ratio of 4:3, for 18 weeks. Each water application treatment was replicated four times.

Data were collected on the following parameters: (1) total above-ground dry matter weight at harvest, (2) yield, (3) yield components; number of panicles /hill, number of grains /panicle, rate of grain filling and 1,000-grain weight, and (4) soil moisture content measured every day until 17th week after sowing by a soil moisture meter (DIK-312A; Daiki Rika Kogyo Co. Ltd.). Yield componentswere fully enumerated for all the six hills, and yield and above- ground dry matter weight were measured by uprooting the entire plants.

Methods

First, we examined the drought tolerance of the rice varieties by means of a simple correlation graph between yield and the total quantity of water applied, with fitted regression lines, distinguishing between 'African varieties' and Yume. Rough, but practically useful estimates of the minimum level of water required for the upland varieties were obtained by reading the coordinate in the graph of the intercepts made by the regression lines on the axis of the total quantity of water applied.

Second, in order to quantify the response of yield, total dry matter and the four yield components to the total quantity of water applied, we estimated linear water-response functions for these factors. Taking yield as an example, the water-response function we estimate is expressed as follows:

$$Y_i = \alpha + \beta W_i + \left(\sum_{K=1}^{K} \gamma_k V_{k,i}\right) + \left(\sum_{K=1}^{K} \delta_k V_{k,i} W_i\right) + \epsilon_i \quad i = 1, 2, \ldots, N \quad (1)$$

where Y = yield (kg/ha), W = total water applied (mm), V_k = dummy

variable for variety ($V_k = 1$ if k-th variety, = 0 otherwise), ε = random error, N = number of observations, and $\alpha, \beta, \gamma, \delta$ are regression coefficients to be estimated. Note that Equation (1) has the same structure as the two-factors-with-cross-effect ANOVA model which can be expressed as $Y_{ikw} = \mu + \rho_k + \varphi_w + \omega_{kw} + \epsilon_{ikw}$, where μ = overall mean, ρ_k =effect of variety, φ_w =effect of water application, ω_{kw} =interaction effect between variety and water application, ε_{ikw} = random error, k = four varieties, and w = five water treatments. A difference in Equation (1) from the ANOVA is that water is treated as a continuous variable, not categorical as in the ANOVA. Indeed, the estimation of Equation (1) performs what the ANOVA model is designed to perform, and in addition, quantifies the effects of water, variety and their cross-term on yield [25]. A simple rearrangement of Equation (1) gives:

$$Y_i = \alpha + \left(\sum_{K=1}^{K} \gamma_k V_{k,i}\right) + \left(\beta + \sum_{K=1}^{K} \delta_k V_{k,i}\right) W_i + \epsilon_i \quad (2)$$

The second term in the right hand side of Equation (2) shows that the intercept of this water- response function could be different by variety (intercept dummies), and the third term shows that the effects of water on yield (the slope of the response function) could be different by variety (slope dummies). In the estimation, all the explanatory variables were 'centered' by converting all the observations to mean deviations in order to avoid multicolinearlity [25]. Note that the overall intercept term (α) in a 'centered' regression is the mean of dependent variable, yield in this example. The estimation of Equation (1) was made for all the varieties and for 'African varieties' separately. In the estimation for all the varieties, dummy variables were set for N4, N10 and Naric, using Yume as the base of comparison.In the estimation for 'African varieties', dummy variables were set for N10 and Naric, using N4 as the base of comparison.

Third, the contribution of changes in the yield components due to the changes in water availability to the change in yield is assessed by means of the additive decomposition of changes in a variable which is a product of other variables. Define yield as Y = P·S·R·G, where Y = yield, P = number of panicle/m², S = number of grains /panicle, R = rate of grain filling and G = 1000-grain weight. Differentiating Y with respect to the total quantity of water applied (W) and dividing through the differentiated equation by Y, we obtain $\left(\frac{dY}{dW}\right) / Y \left(\frac{dP}{dW}\right) / P \left(\frac{dS}{dW}\right) / S + \left(\frac{dR}{dW}\right) / R + \left(\frac{dG}{dW}\right) / G$, that is, the rate of change in Y is decomposed into the rates changes in P, S, R and G. Taking the derivative of Equation (2) with respect to W for P, S, R and G, and computing the rate of change for respective components for respective varieties, we can compute the relative contribution of these components to the changes in yield by variety. Since the decomposition equation is an approximation when differences (Δx) are used instead of differentials (dx), the left-hand side of the equation does not necessarily exactly tally with the right-hand side. For the computation of the percentage contribution of the components, we used the summation of the rates of changes in the right-hand side as the rate of change in yield. The decomposition was made for lower and higher water application levels separately, and for the 'mean' using the derivatives obtained from the estimated water response functions.

As will be presented in the next section, Yume did not withstand low levels of water application, resulting in no yield in five replications. Since the inclusion of zero-yield observations results in overestimations of water response, we excluded these five observations, which made the total number of observations used in the analyses 75. Throughout the paper, we adopted three levels of significance levels for hypothesis testing, p<0.001, p<0.01 and p<0.05, with the symbols of ‡, † and *, respectively.

Results and Discussion

The mean minimum daily temperature in the greenhouse during the period of the experiment was 15.7°C (standard deviation = 1.1°C), the mean maximum daily temperature was 28.4°C (1.3°C) and the average daily temperature was 22.1°C (0.9°C).

Weekly average soil moisture contents were shown in Figure 1 for T-1 (the lowest water application) and T-5 (the highest water application), which ranged 6.9 – 13.8% with the mean of 9.9% for T-1 and 10.1 – 20.0% with the mean of 14.5% for T-5. The trends of the moisture contents at the later stages of growth indicated that the rice plants reached the harvesting stage by 14th week (98 days) after sowing. These levels of soil moisture contents are comparable to those of a study that uses upland rice varieties obtained from IRRI grown in chambers [8] and those of a study that uses NERICA varieties grown in open upland fields at IITA, Nigeria [15], but lower than the levels of soil moisture contents reported by previous studies conducted in Australia and Asia [7,13].

The results of the experiment were shown in Table 1. At the lowest level of water application, N4 and N10 yielded 0.50 t /ha and 0.68 t /ha, respectively. Though very low, Naric also yielded 0.10 t /ha. However, Yume failed to reach the flowering stage. The yield of N4 and N10 jumped up to more than 3 t /ha at the second water application and that of Naric to 1.8 t /ha. In contrast, Yume gave no yield in one of replications at the second water application level, with the average yield in the rest of three replications of 0.35 t/ha, and it was at the fourth level of water application for the yield to exceed the 3 t /ha line. At higher levels of water application, yield reached around 6 t /ha for all the varieties tested. Total dry matter weight and all the four yield components also increased as the water application level went up.

Drought tolerance

Seventy five non-zero yield observations were plotted in Figure 2 against the total quantity of water applied, distinguishing 'African' varieties from Yume. Higher drought tolerance of 'African varieties' over Yume was apparent. The regression line for 'African varieties' intersected with the horizontal axis at 310 mm of water applied, and gave the yield of around 1 t /ha at the water availability of 400 mm per crop season. Taking into account the result that the rice plants reached maturity by 98 days after sowing (Figure 1), 311 mm (400 mm /126 days x 98 days) of water brought about the yield of around 1 t /ha for 'African varieties', which may be called the safe minimum water requirement. For Yume, the corresponding safe minimum water requirement was computed as 467 mm. These results suggested that 'African varieties' were more tolerant to drought than Yume by 33% in terms of the safe minimum water requirement. The results for 'African varieties' confirm that the recommended level of rainfall for NERICA cultivation of 360 mm per crop [17] is beyond the safe minimum level of 311 mm.

We failed to find out any literature on the minimum water requirement of upland varieties used in sub-Saharan Africa to compare with our results. For outside Africa, it is reported that the threshold rainfall for upland rice cultivation in Asia and Latin America is 200 mm/month, or 600 mm for a crop season of three months [26,27]. Comparing the results of water-application experiments for upland rice conducted in Australia [28,29], USA [30] and Asia [9,13,31], it is shown that the minimum water supply level tested is 419 mm /crop [12]. Though for rainfed lowland ecosystem, it is reported that rice

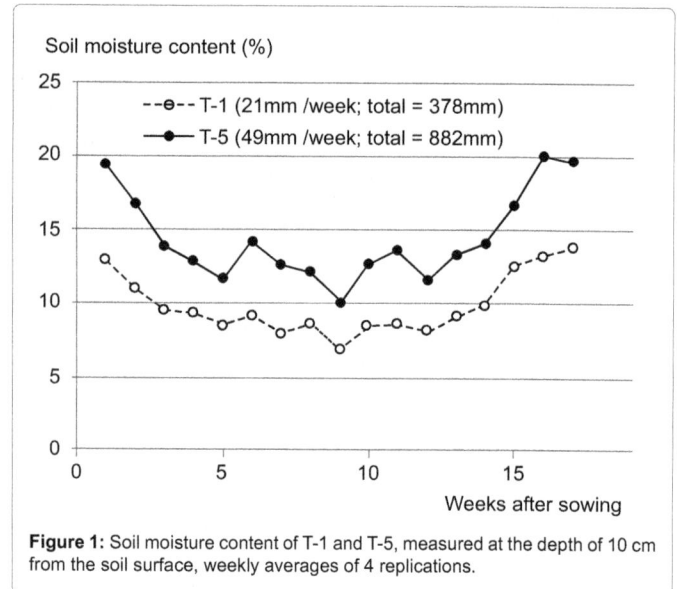

Figure 1: Soil moisture content of T-1 and T-5, measured at the depth of 10 cm from the soil surface, weekly averages of 4 replications.

production is virtually impossible with rainfall below 450 mm/crop [32]. These observations point to the high drought tolerance of 'African varieties', of N4 and N10 in particular.

It should be reminded that in our experiments water was applied regularly twice a week for 18 weeks. The distribution of rainfall during a crop season is not even as such. Since the water deficit at the reproductive stage affects yield more seriously than at the vegetative stage [5-7,12], a slight bias in rainfall distribution towards earlier or later stages could easily make the safe minimum water requirement of 311-400 mm for 'African varieties' out of range.

Water response

The results of the estimation of water response functions were presented in Table 2. Water was a positive, significant factor for all the 12 regression equations estimated. Remarkable were high significance levels at which the coefficients of water were estimated: Except for Regression VIII for which the significance level was 5%, it far exceeded 0.1% for all the rest.

First, let us look at the water response of yield for all the varieties (Regression I). The intercept of 3.46 t/ha showed the mean yield for the observations as a whole. The coefficient of water, 0.0123, the slope of the water response regression line for the base variety of Yume, indicated that a 1 mm increase in water availability increased yield by 12.3 kg /ha. The coefficients of three variety dummies for intercept were all positive and significant, indicating that the water response regression lines for these African varieties were located significantly above that of Yume, as shown in Figure 2 for 'African varieties' and Yume. The coefficients of two cross terms, Water x N4 and Water x N10, were negative and significant, indicating that the slopes of the water response regression lines for these NERICA varieties were less steep than that for Yume. These less-steeper slopes for N4 and N10 were resulted from their higher drought tolerance that kept their yields high relative to Yume at lower levels of water application. Regression VII, the water response function for yield estimated by using only the observations for 'African varieties', showed few differences in their water response regression lines, except for the cross-term between Water and Naric. The slope of the regression line was steeper for Naric for the same reason mentioned for Yume.

The estimated 11-12 kg/ha/mm of water response of yield is high compared to earlier studies. Linear water response functions estimated for upland rice grown in North China show 5.2 - 5.7 kg /mm of water response of yield [10]. The comparison of six water application studies mentioned above gives the rate of yield increase of 2.5 kg/ha/mm [12]. A study using rainfed rice farmers' field data reports that the rate is 1.2 kg /ha per 1 mm of total rainfall per season [33].

Regressions II and VIII show that the rate of water response of total dry matter did not differ significantly among the varieties tested, though N4 and Naric produced significantly more total dry matter than N10 and Yume. 'African varieties' varieties N10 produced total dry matter significantly less than others. Although total dry matter increased significantly as water availability increased (Table 2), its rate of increase was far less than that of yield (Table 1), implying that the harvest index increased at the rate closer to that of yield.

Number of panicle/m² and number of grains per panicle were significantly larger for 'African varieties' than Yume (Regressions III and IV), while opposite was the case for 1000 -grain weight (Regression VI). In contrast, there was no difference between 'African varieties' and Yume for the rate of grain filling, except for N10 that showed a significantly higher rate (Regression V). Also except for N10 for number of panicle /m² and number of grains per panicle, the rate of water response (the slope) was not significantly different between 'African varieties' and Yume. For 'African varieties', the water response functions of four yield components differed little among the varieties. There were two exceptions for this; the rate of grain filling was significantly higher for N10 (Regression XI) and 1000-grain weight was significantly heavier for Naric (Regression XII).

Contribution of yield components to yield increase

Yield increased significantly as water availability increased (Table 1). For 'African' varieties, the rate of yield increase was very high at lower levels of water application from T-1 to T-2 for N4 and N10 and from T-2 to T-3 for Naric, whereas for Yume, high rates of yield increase occurred at higher levels of water application from T-3 to T-5. The results of the decomposition of the yield increase were presented in Table 3. The decomposition based on the estimated water response functions in Table 2 (Mean) showed that the component that contributed most to the increase in yield was the rate of grain filling for all the varieties tested, followed by number of panicle/m², number of grains per panicle, and 1000-grain weight, in the order of the importance. The importance of the rate of grain filling was particularly large in the lower levels of water application. At higher levels of water availability, the contribution of number of panicles/m² remained to be an important contributing component and number of grains per panicle increased its importance.

These results are consistent with findings of earlier studies in Australia and Asia. It is found in Australia that water stress during panicle development reduces the rate of grain filling to zero [7]. It is found in Asia that the rate of grain filling is the key factor contributing to high harvest index of upland rice varieties under upland conditions [10], that decreasing water supply during 20-40 days before heading reduces the number of grains per unit area and harvest index [12],

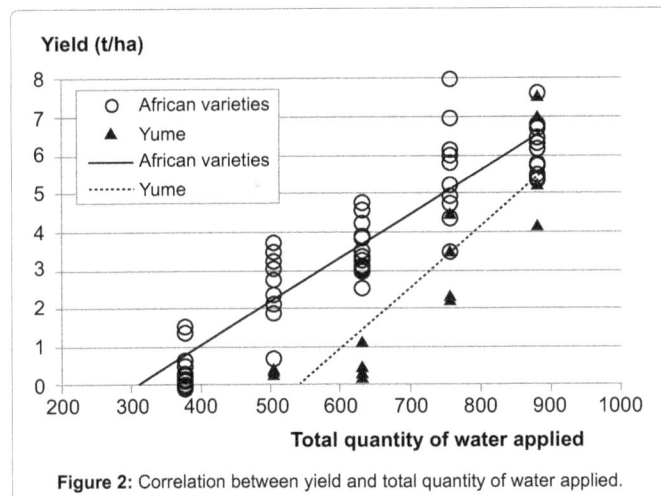

Figure 2: Correlation between yield and total quantity of water applied.

	Water treatment[a]	Yield (t/ha)	Total dry matter weight (t/ha)	No. of panicle /m²	No. of grains /panicle	% grain filling	1000 grain weight (g)
N4	T-1	0.50	9.1	157	75.8	14.1	20.5
	T-2	3.04	11.2	317	102.2	37.1	25.4
	T-3	3.29	11.3	318	99.5	42.9	25.6
	T-4	4.47	13.4	349	115.2	43.0	26.2
	T-5	5.98	13.8	373	114.3	53.5	26.5
N10	T-1	0.68	4.3	177	84.5	19.5	22.8
	T-2	3.12	8.2	291	85.1	51.9	24.1
	T-3	3.71	9.0	307	101.2	49.9	24.1
	T-4	5.72	11.5	355	102.5	59.6	26.5
	T-5	5.77	12.1	354	103.2	62.7	25.5
Naric	T-1	0.10	7.0	128	62.9	5.2	22.1
	T-2	1.77	9.8	278	95.5	23.7	26.6
	T-3	3.75	12.1	334	95.5	44.2	27.3
	T-4	6.40	17.4	408	101.8	56.4	27.2
	T-5	6.69	16.3	376	113.6	55.5	28.8
Yume	T-1	0	0	0	0	0	0
	T-2	0.35	6.7	101	54.8	24.5	29.1
	T-3	0.55	8.6	141	38.9	31.6	31.1
	T-4	3.14	12.7	296	58.0	54.8	32.6
	T-5	5.97	13.1	299	101.6	58.7	33.4

Table 1: Yield, total dry matter weight, harvest index and four yield components of upland rice varieties by level of water applied.

a) The total quantity of water applied: T-1 (378 mm), T-2 (504 mm), t-3 (630 mm), T-4 (756 mm), T-5 (882 mm).

	Yield (t/ha)	Total dry matter (t/ha)	No. of panicle /m²	No. of grains /panicle	% grain filling	1000 grain weight (g)
All varieties (base: Yume, n=75):	I	II	III	IV	V	VI
Intercept	3.46 ‡	11.0 ‡	284 ‡	90.3 ‡	41.7 ‡	26.6 ‡
Water (mm)	0.0123 ‡	0.0157 ‡	0.438 ‡	0.0810 ‡	0.0862 ‡	0.00956 ‡
N4	1.93 ‡	2.51 †	128 ‡	46.9 ‡	1.60	-6.04 ‡
N10	2.27 ‡	-0.141	122 ‡	40.4 ‡	12.3 *	-6.33 ‡
Naric	2.27 ‡	3.45 †	132 ‡	39.6 ‡	1.04	-4.45 ‡
Water x N4	-0.00613 †	-0.00902	-0.230	-0.067	-0.0332	-0.00121
Water x N10	-0.00584 †	-0.00332	-0.264 *	-0.094 *	-0.0258	-0.00507
Water x Naric	-0.00184	0.00241	-0.100	-0.0524	0.0052	-0.000121
R²	0.861	0.607	0.741	0.595	0.696	0.703
African varieties (base:N4,n=60)	VII	VIII	IX	X	XI	XII
Intercept	3.67 ‡	11.1 ‡	301 ‡	96.9 ‡	41.3 ‡	25.3 ‡
Water (mm)	0.0114 ‡	0.0150 *	0.399 ‡	0.0668 ‡	0.0826 ‡	0.00913 ‡
N10	0.343	-2.73 ‡	-5.92	-6.08	10.60 †	-0.233
Naric	0.284	0.768	1.66	-7.55	-1.14	1.57 *
Water x N10	0.000285	0.00570	-0.0345	-0.0278	0.00740	-0.00386
Water x Naric	0.00429 †	0.0114 *	0.130	0.0141	0.0384	0.00109
R²	0.876	0.632	0.689	0.417	0.695	0.456

Table 2: Water response functions of upland rice varieties: yield, total dry matter and four yield components.
a) Explanatory variables: Water = total quantity of water applied. N4, N10 and Naric are dummy variables that take 1 if the variety is N4, N10 or Naric and 0 otherwise, respectively. Water*N4, Water*N10 and Water*Naric are the products of Water and respective variety dummy variable. The symbols ‡, † and * show that the estimated regression coefficients are statistically different from 0 at the significance level of $p<0.001$, $P<0.01$ and $p<0.05$, respectively.

		Rate of change in yield (%)	Contribution to the change in yield (%)				
			Yield /ha	No. of panicle /m²	No. of grains /panicle	% grain filling	1000 grain weight
N4	T-3 / T-1	363	100	28	9	56	7
	T-5 / T-3	60	100	29	25	41	5
	Mean	44	100	30	16	46	8
N10	T-3 / T-1	255	100	29	8	61	2
	T-5 / T-3	49	100	31	4	53	12
	Mean	40	100	30	10	55	5
Naric	T-3 / T-1	991	100	16	5	76	2
	T-5 / T-3	62	100	20	30	41	9
	Mean	60	100	29	15	49	7
Yume	T-3 / T-2	46	100	85	-63	63	15
	T-5 / T-3	367	100	31	44	23	2
	Mean	49	100	32	18	42	7

Table 3: Rate of change in yield and percentage contributions of yield components to yield increase due to changes in water applied
a) Rates of change between T-1 and T-3 and between T-3 and T-5 are computed from Table 1. For Yume, since T-1 gives no yield, the growth rate for the lower water application levels is taken between T-2 and T-3. Mean rates of change are obtained from regression equations in Table 2 by taking derivatives with respect to water applied. Regression equations used are Regression VII - XII for 'African' varieties and Regression I - VI for Yume. The rate of change in yield is calculated as the summation of the rates of change in four yield components.

and that the number of grains per panicle is the most important factor responsible for yield gap between aerobic and flooded rice [34].

It is stated that a high potential yield and harvest index, as well as yield stability under different water regimes, are important putative plant characters for developing new elite upland rice varieties [12]. In addition to this, our findings suggest it would be necessary that two different, yet closely related, breeding strategies have to be sought in developing new upland rice varieties suited to sub-Saharan Africa. To enhance drought tolerance for wider dissemination of upland rice cultivation in the region, it is critical to build in a higher ability for grains to reach maturity. To enhance the yield potential of upland rice varieties to be planted in areas with relatively favorable rainfall conditions, it is effective to build in a higher ability for forming larger sink size (number of panicles/m² and number of grains per panicle).

Conclusions

In view of the importance that upland rice varieties play in the rice Green Revolution in sub-Saharan Africa where many farmers are trying to plant rice in upland areas prone to drought, we conducted experiments in Namulonge, Uganda, to clarify water response of upland rice varieties using NERICA 4, NERICA 10, NARIC-2 and Yumenohatamochi with five different levels of water application. We found that the NERICA varieties were most drought tolerant, followed by NARIC-2. Yumenohatamochi did not withstand the lowest amount of water application of 378 mm that was applied evenly throughout the growing stages until harvesting. The results suggest that the minimum water requirement was 311-400 mm per season for three varieties used widely in East Africa, and about 420-600 mm for Yumenohatamochi. It was estimated that an additional water application of 1 mm increased rice yield by 11-12 kg /ha for the upland varieties tested. The high water response of upland rice was brought about by high water response of four yield components. Among the components, the contribution by the rate of grain filling was highest, followed by number of panicles /m², number of grains per panicle and 1000-grain weight, in the order of the degree of contribution, for all the varieties tested.

Acknowledgement

The authors are grateful to Dr. Kunihiro Tokida and the two anonymous reviewers for their helpful comments made on our earlier manuscripts.

References

1. Otsuka K, Kalirajan K (2006) Rice green revolution in Asia and its transferability to Africa. Dev Econ 44: 107-122.

2. Balasubramanian V, Sie M, Hijmans RJ, Otsuka K (2007) Increasing rice production in sub-Saharan Africa: Challenges and opportunities. AdvAgron 94: 55-133.

3. Fujiie H, Maruyama A, Fujiie M, Kurauchi N, Takagaki M, et al. (2010) Potential of NERICA production in Uganda: based on the simulation results of cropland optimization. Trop AgricDev 54: 44-50.

4. Kijima Y, Otsuka K, Sserunkuuma D (2011) An inquiry into constraints on a green revolution in sub-Saharan Africa: the case of NERICA rice in Uganda. World Dev 39: 77-86.

5. O'Toole JC, Moya TB (1981) Water deficits and yield in upland rice. Field Crop Res 4: 247-259.

6. Lilley JM, Fukai S (1994) Effect of timing and severity of water deficit on four diverse cultivars: III. Phenological development, crop growth and grain yield. Field Crop Res 37: 225-234.

7. Boonjung H, Fukai S (1996) Effects of soil water deficit at different growth stages on rice growth and yield under upland conditions: 2. Phenology, biomass production and yield. Field Crop Res 48: 47-55.

8. Price AH, Steele KA, Gorham J, Bridges JN, Moore BJ, et al. (2002) Upland rice growth in soil-filled chambers and exposed to contrasting water deficit regimes: I. Root distribution, water use and plant water status. Field Crop Res 76: 11-24.

9. Bouman BAM, Peng S, Castaneda AR, Visperas RM (2005) Yield and water use of irrigated tropical aerobic rice systems.Agr Water Manage 74: 87-105.

10. Bouman BAM, Xiaoguang Y, Huaqi W, Zhimin W, Junfang Z, et al. (2006) Performance of aerobic rice varieties under irrigated conditions in North China. Field Crop Res 97: 53-63.

11. Kato Y, Kamoshita A, Yamagishi J, Abe J (2006) Growth of three rice (Oryza sativa L.) cultivars under upland conditions with different levels of water supply: 1. Nitrogen content and dry matter production. Plant Prod Sci 9: 422-434.

12. Kato Y, Kamoshita A, Yamagishi J, Abe J (2006) Growth of three rice (Oryza sativa L.) cultivars under upland conditions with different levels of water supply: 2. Grain yield. Plant Prod Sci 9: 435-445.

13. Kato Y, Kamoshita A, Yamagishi J, Abe J (2007) Growth of Rice (Oryza sativa L.) Cultivars under Upland Conditions with Different Levels of Water Supply 3. Root System Development, Soil Moisture Change and Plant Water Status. Plant Prod Sci 10: 3-13.

14. Akinbile CO, Sangodoyin AY, Nwilene FE, Futakuchi K (2007) Growth and yield response of upland rice (NERICA 2) under different water regimes in Ibadan, Nigeria. Res J Agron 1: 71-75.

15. Akinbile CO (2010) Crop water use response of upland rice to different water distribution under sprinkler irrigation system. AdvApplSci Res 1: 133-144.

16. Odunze AC, Kudi TM, Daudu C, Adeosun J, Ayoola G, et al. (2010) Soil moisture stress mitigation for sustainable upland rice production in the Northern Guinea Savanna of Nigeria. AgricBiol J N Am 1: 1193-1198.

17. NaCRRI (National Crops Resources Research Institute) (2010). Rice cultivation handbook.NaCRRI and JICA, Kampala, Uganda.

18. Bernier J, Atlin GN, Serraj R, Kumar A, Spaner D (2008) Breeding upland rice for drought tolerance. J Sci Food Agric 88: 927-939.

19. Africa Rice Center (WARDA) (2008) NERICA: the new rice for Africa- a compendium. FAO, Tokyo.

20. Kaneda C (2007) Breeding and dissemination efforts of "NERICA": (1) Breeding of upland rice. Jpn J Trop Agric 51: 1-4.

21. Kaneda C (2007) Breeding and dissemination efforts of "NERICA": (2) Evaluation of important characteristics. Jpn J Trop Agric 51: 41-45.

22. Onaga G, Asea G, Lamo J, Kitafunda J, Bigirwa G (2012) Comparison of response to nitrogen between upland NERICAs and ITA (Oryza sativa) rice varieties. J AgricSci 4: 197-205.

23. Hirasawa H, Nemoto H, Suga R, Ishihara M, Hirayama M, et al (1998) Breeding of a new upland rice variety "Yumenohatamochi" with high drought resistance and good eating quality. Breed Sci 48: 415-419.

24. Ministry of Agriculture, Animal Industry and Fisheries (2009) Uganda National Rice Development Strategy (2nd draft). Ministry of Agriculture, Animal Industry and Fisheries. Entebbe, Uganda.

25. Aiken LS, West SG (1991) Multiple regression: Testing and interpreting interactions. Sage Publication, London.

26. De Datta SK (1981) Principles and practices of rice production. John Wiley & Sons, New York.

27. Brown FB (1969) Upland rice in Latin America. Int Rice CommNewslett 18: 1-5.

28. Blackwell J, Meyer WS, Smith RCG (1985) Growth and yield of rice under sprinkler irrigation on a free-draining soil. Aust J ExpAgric 25: 636-641.

29. Fukai S, Inthapan P (1988) Growth and yield of rice cultivars under sprinkler irrigation in south-eastern Queensland, I: Effects of sowing time. Aust J ExpAgric 28: 237-242.

30. McCauley GN (1990) Sprinkler vs flood irrigation in traditional rice production regions of Southeast Texas. Aglon J 82: 677-683.

31. Yang X, Bouman BAM, Wang H, Wang Z, Zhao J, et al. (2005) Performance of temperate aerobic rice under different water regimes in North China. Agric Water Manage 74: 107-122.

32. Hijiman RJ, Serraj R (2008) Modeling spatial and temporal variation of drought in rice production. In Serraj J, et al. (eds.) Drought frontiers in rice: crop improvement for increased rainfed production. World Scientific Publishing and International Rice research Institute, Singapore and Los Banos.

33. Haneishi Y, Maruyama A, Asea G, Okello SE, Tsuboi T, et al. (2013) Exploration of rainfed rice farming in Uganda based on a nationwide survey: Regionality, varieties and yield. Afr J Agric Res 8: 4038-4048.

34. Patel DP, Das A, Mundal GC, Ghosh PK, Bordoloi JS, et al. (2010) Evaluation of yield and physiological attributes of high-yielding rice varieties under aerobic and flood-irrigated management practices in mid-hills ecosystem. Agric Water Manage 97: 1269-1276.

32

Molecular and Morphological Characterisation of Back Cross Generations for Yield and Blast Resistance in Rice

Divya Balakrishnan[1]*, Robin S[2], Rabindran R[2] and John Joel A[2]

[1]*National Professor Project, Crop Improvement Section, Directorate of Rice Research-ICAR, Hyderabad, India*
[2]*Department of Rice, Centre for Plant Breeding and Genetics, Tamil Nadu Agricultural University, India*

Abstract

Blast caused by *Pyricularia oryzae* is one of the major constraints limiting rice production globally. Study was undertaken at TNAU, India using a set of selected genotypes including cultivated varieties and (Near Isogenic Lines) NILs and they were screened for blast resistance both genotypically and phenotypically. Genetic diversity was studied among the genotypes and those which are diverse in blast resistance but had similar morphological and quality traits were selected as parents for breeding programme. Selected genotypes were crossed and advanced to further generations up to BC_4F_1 by marker assisted back crossing. Genetic variability and heritability parameters were assessed among these segregating generations for yield and resistance traits and their trend in each generation was analysed. The disease resistance traits showed high variability in segregating generations as the parents were selected based on the diversity in disease reaction and therefore gives scope for improvement through selection. High heritability coupled with low genetic advance was found in all characters in the segregating generations except single plant yield and panicle length.

Keywords: Genetic advance; Heritability; Molecular markers; Yield; Blast resistance

Introduction

Rice (*Oryza sativa L.*) being one of the prime staple crops of the world has a crucial role in the global food security and contributes for the livelihood of majority of the Asian population. Rice production and productivity are constrained by biotic and abiotic stresses in the growing areas. Rice blast caused by the fungus *Pyricularia oryzae* has been recognized as one of the most serious diseases and is distributed across 85 countries globally. Host plant resistance has been recognized as an important strategy to manage blast disease compared to chemical control measures. More than 100 blast resistance genes have been identified in rice [1,2] but effective and durable use of these reported genes has been limited because of evolution of new virulent races of the fungus [3]. Enhancing host plant resistance through pyramiding of multiple resistance genes is one of the useful strategies to avoid frequent breakdown of resistance. In addition to overlapping resistance, it could provide cross protection by minimizing the race evolution in the fungus. Combining many resistant traits or resistant genes without appropriate study of their effect on the genetic background may derive undesirable genotypes.

The application of molecular tools in identification of resistance sources and selection of desirable genotypes in segregating generations can accelerate the development of blast-resistant cultivars in an adapted genetic background. Simple Sequence Repeats (SSRs) linked to report resistance have been used for screening the parents and to identify the polymorphism between the genotypes because of their quantity and high polymorphism among rice varieties [4-8]. Rice genetic diversity on the basis of morphological characterization has been studied previously and used in selection of parents in breeding programs [9-13].

Along with this, genetics of resistance traits and their effect on yield is to be assessed to obtain high yielding resistant lines. Study on inheritance and genetic basis of the resistance traits is of high importance, in initial segregating generations or prior to crop improvement programs. It will be helpful in implementing appropriate breeding strategy, and precise screening of advanced generations. Variability parameters for different morpho-agronomic as well as resistance traits were estimated in the segregating generations. It is evident that assessing the variability in the segregating lines under varying environmental conditions is essential for the improvement of popular varieties with stable yield potential. Study was conducted to improve blast resistance in popular Indian rice varieties through marker assisted backcross breeding approach [14]. The parental genotypes were screened for identifying suitable genotypes to develop a high yielding variety with blast resistance. Each segregating populations generated from selfing and back crossing of the selected cross were phenotyped for yield and resistance traits. Information on the phenotypic data of different generations was subjected to analysis of variability and heritability parameters and the trend in each generation was assessed.

Materials and Methods

In this study, a group of blast susceptible and resistant genotypes (Table 1) were subjected to diversity analysis and they were crossed to obtain F_1s. The parental genotypes and advanced generations obtained from a selected cross (ADT43 × CT13432-3R) were screened for blast resistance and yield parameters using both genotypic and phenotypic measures. Popular Indian cultivar, ADT 43 was selected as recurrent parent for improvement because of its wide adaptability and acceptability in South India owing to its high yield, short duration and grain quality, but it is highly susceptible to blast disease. CO 39 near isogenic line (NIL), CT13432-3R pyramided with four blast resistance genes [15] developed from CIRAD was selected as resistant donor in

*Corresponding author: Dr. Divya Balakrishnan, Scientist (Plant Breeding and Genetics), National Professor Project, Crop Improvement Section, Directorate of Rice Research-ICAR, Rajendranagar-30, Hyderabad, India
E-mail: divyabalakrishnan05@gmail.com or divyab0005@gmail.com

Sl No.	Parents	Blast Response	Pedigree	Varietal group	Origin	Features
P1	ADT 43	Susceptible	IR 50×Improved White Ponni	Indica	TRRI, Aduthurai, Tamil Nadu	High yielding, semi dwarf popular rice variety in Tamil Nadu (110 days)
P2	Improved White Ponni	Susceptible	Selection from Ponni	Indica	TNAU, Coimbatore, Tamil Nadu	High yielding, tall, medium slender rice variety (125-130 days)
P3	BPT 5204	Susceptible	GEB 24/ TN1/ Mahsuri	Indica	ANGRAU, Bapatla, Andhra Pradesh	High yielding, semi dwarf, fine medium slender grain variety grown across India for its high yield and quality. (140-145 days)
P4	CO 39	Susceptible	Cul. 340×Kannagi	Indica	TNAU, Coimbatore, Tamil Nadu	High yielding short duration variety Blast susceptible
P5	CT 13432-3R	Resistant	CO 39 NIL	Indica	CIRAD	Blast resistant NIL with genes Pi1, Pi2, Pi33
P6	C101 A51	Resistant	CO 39×A5173 NIL	Indica	CIRAD	Blast resistant NIL with genes Pi2/Piz5
P7	CI01 PKT	Resistant	CO 39×Pai-kan-tao NIL	Indica	CIRAD	Blast resistant NIL with genes Pi 4a, Pi-3
P8	CI01 LAC	Resistant	CO 39×Lac23 NIL	Indica	CIRAD	Blast resistant NIL with genes Pi1, Pi33
P9	TORIREI	Resistant	CO 39 NIL	Indica	CIRAD	Blast resistant NIL with genes Piz^t
P10	CO 39 (CIRAD)	Resistant	CO 39 NIL	Indica	CIRAD	
P11	TETEP	Resistant		Indica	Vietnam	Semi dwarf, Blast resistant variety with genes Pikh/ Pi54, Pi ta (130-135 days)
P12	MOROBOREKAN	Resistant		Japonica	Guinea (West Africa)	Blast resistant variety with genes Pi 5(t), Pi7(t)

Table 1: Description of the genotypes used in the study.

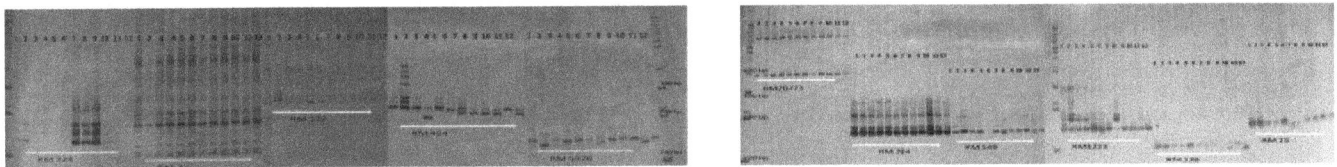

Figure 1: Parental polymorphism of blast resistant genes linked markers along with CO 39 NILs.

this study. Parental genotypes and advanced generations were raised during 2009 to 2012 in the experimental plots of the Paddy Breeding Station (PBS), Coimbatore and at Hybrid Rice Evaluation Centre (HREC), Gudalur which is an endemic location for blast disease. Advanced generations from the cross of ADT 43 × CT13432-3R viz., F_2, F_3, BC_1, BC_2 and BC_3 were raised and screened in normal as well as epiphytotic conditions. The data obtained from the segregating generations were assessed statistically for various yield and disease related traits viz., plant height, number of tillers, productive tillers, leaf length, leaf width, panicle length, days to first flowering, days to maturity, filled grains per panicle, total grains per panicle, spikelet fertility, 100 grain weight, single plant yield, lesion number, infected leaf area, potential disease incidence percentage, leaf blast and lesion type (Supplementary Table 1). The experimental details are given in Supplementary Figure 1.

Diversity analysis

The parental genotypes were screened for reported molecular markers linked to blast resistance (Supplementary Table 2). For the microsatellite assay, twenty markers linked to four blast resistance genes Pi1, Pi2, Pi33 and Pi54 were used viz., RM 224, RM 1223, RM 5926, RM 1233, PR 10, RM 527, RM 136, RM 549, RM6836, AP5659-3, RM 72, RM 331,RM 404, RM 483, RM 3374, RM 284 and RM 25 (Figure 1). The allelic variation of these markers were surveyed at the parental level was and the polymorphic SSR markers were selected for the further screening of the crosses and segregating populations. Diversity of the parental genotypes were studied using phenotypical data and molecular marker information using NTSYs-pc UPGMA cluster analysis and Jaccard's coefficient [16].

Estimation of genetic parameters

The various genetic parameters like variability, genotypic co-

efficient of variation (GCV), phenotypic co-efficient of variation (PCV), heritability and genetic advance were calculated by adopting the given formulae [17]. The average variance observed in the genotypes was considered as environmental variance. The genotypic variance of each progeny was estimated by subtracting the estimated environmental variance from the phenotypic variance.

Environmental variance (VE) = Average phenotypic variance of both the parents

Phenotypic variance (VP) = VG + VE

Genotypic variance (VG) = VP - VE

Data on quantitative characters were analysed for variances and significance of treatments. The genotypic co-efficient of variation (GCV) and phenotypic co-efficient of variation (PCV) were estimated.

$$\text{Phenotypic Coefficient of variation} = \frac{\sqrt{\text{Phenotypic variance}}}{\text{Mean of the character under study}} \times 100$$

The PCV and GCV values were categorized as follows [18].

PCV and GCV	Category
<10%	Low
10-20%	Moderate
>20%	High

Heritability

Heritability in the broad sense was calculated according to the formula [17] suggested as follows

$$\text{Heritability} = \frac{\text{Genotypic variance}}{\text{Phenotypic variance}} \times 100$$

The heritability was categorized as follows.

GENOTYPES	PH	NT	PRT	LL	LW	PL	DFF	DM	FGP	TGP	SPF	SPS	100GW	SPY	GL	GB	LBR	LN	ILA	PDI%
ADT 43	62.30	28.00	25.33	27.17	0.97	23.13	86.00	111.00	186.00	191.00	97.38	2.69	1.52	25.72	8.00	2.00	4.00	35.00	80.00	100.00
CT 13432-3R	87.33	23.00	21.00	28.00	1.30	20.67	82.67	107.67	157.00	184.00	85.33	17.20	2.31	24.58	6.83	3.17	2.16	5.00	3.00	11.11
IMP. W. PONNI	113.67	19.33	19.00	30.67	1.03	24.70	119.33	144.33	121.00	175.00	69.14	44.63	2.43	50.50	7.00	1.67	4.20	22.00	60.00	77.78
BPT 5204	65.00	22.00	21.00	26.00	1.20	21.00	125.33	150.33	113.00	171.00	66.08	51.33	2.13	35.87	7.17	1.83	3.91	25.00	75.00	55.56
TETEP	133.33	23.67	23.67	30.00	0.83	25.53	114.00	139.00	125.00	158.00	79.11	26.40	2.06	29.68	6.33	2.33	2.71	5.00	5.00	11.11
MOROBOREKAN	117.33	9.33	9.33	33.00	1.50	26.67	109.00	134.00	143.00	148.00	96.62	3.50	2.84	32.56	8.17	3.17	2.58	15.00	10.00	33.33
TORIREI	61.00	14.33	10.33	17.00	0.63	16.40	75.00	100.00	50.00	69.00	72.46	38.00	2.02	35.17	6.57	3.00	2.19	25.00	10.00	22.22
CIOI A51	62.67	12.00	7.33	23.53	0.67	17.50	77.00	102.00	80.00	94.00	85.11	17.50	2.71	31.09	7.33	3.17	2.32	3.00	5.00	11.11
CIO1 PKT	58.33	14.00	9.33	14.00	0.57	15.63	69.67	94.67	66.00	79.00	83.54	19.70	2.34	35.14	7.00	2.50	2.80	10.00	10.00	33.33
CIOI LAC	55.00	7.33	5.33	18.77	0.57	15.33	70.00	95.00	71.00	91.00	78.02	28.17	2.35	35.53	7.00	3.00	2.33	15.00	50.00	55.56
CO 39 CIRAD	56.67	10.67	9.00	17.50	0.53	15.70	69.67	94.67	70.00	73.00	95.89	4.29	2.11	39.99	6.00	3.00	2.00	25.00	10.00	44.44
CO 39	68.33	10.00	6.67	21.00	0.97	20.17	69.67	94.67	49.00	53.00	92.45	8.16	2.16	13.51	7.00	3.00	2.33	45.00	85.00	77.78
MEAN	78.41	16.14	13.94	23.89	0.90	20.20	88.95	113.95	102.58	123.83	83.43	21.80	2.25	32.45	7.03	2.65	2.79	19.17	33.58	44.44
MAX	55.00	7.33	5.33	14.00	0.53	15.33	69.67	94.67	49.00	53.00	66.08	2.69	1.52	13.51	6.00	1.67	2.00	3.00	3.00	11.11
MIN	133.33	28.00	25.33	33.00	1.50	26.67	125.33	150.33	186.00	191.00	97.38	51.33	2.84	50.50	8.17	3.17	4.20	45.00	85.00	100.00
SD	27.58	6.78	7.39	6.17	0.32	4.12	21.61	21.61	44.77	51.62	10.81	16.39	0.34	9.01	0.61	0.56	0.79	12.83	33.41	29.59
VAR	760.86	45.96	54.57	38.11	0.10	16.96	466.89	466.89	2004.27	2664.70	116.83	268.56	0.12	81.27	0.38	0.32	0.62	164.52	1115.90	875.51
SKEWNESS	1.00	0.36	0.37	-0.12	0.48	0.22	0.63	0.63	0.42	0.01	-0.15	0.45	-0.23	-0.05	0.02	-0.18	0.11	0.50	0.51	0.52
KURTOSIS	-0.62	-1.25	-1.56	-1.29	-0.94	-1.38	-1.29	-1.29	-1.02	-1.69	-1.21	-0.97	0.35	-0.65	-1.10	-0.79	-1.07	-0.55	-1.52	0.64
CV%	33.68	40.22	50.73	24.74	34.07	19.52	23.26	18.16	41.78	39.91	12.40	71.98	14.51	31.82	37.46	30.75	16.06	64.07	95.23	-0.74
S.E	7.62	1.87	2.04	1.71	0.09	1.14	5.97	5.97	12.37	14.27	2.99	4.53	0.09	3.27	2.15	1.71	0.05	3.55	9.23	1.23

Table 2: Mean performance of the parental genotypes in field conditions. Plant height (PH) in cm, number of tillers (NT), number of productive tillers (PRT), leaf length (LL) in cm, leaf width (LW) in mm, panicle length (PL) in cm, days to first flowering (DFF), days to maturity(DM), filled grains per panicle (FGP), total grains per panicle (TGP), spikelet fertility (SPF), spikelet sterility (SPS), 100 grain weight (100GW), single plant yield (SPY), lesion number(LN), infested leaf area (ILA), potential disease incidence percentage (PDI%), grain length (GL), grain breadth (GB),length-breadth ratio (LBR).

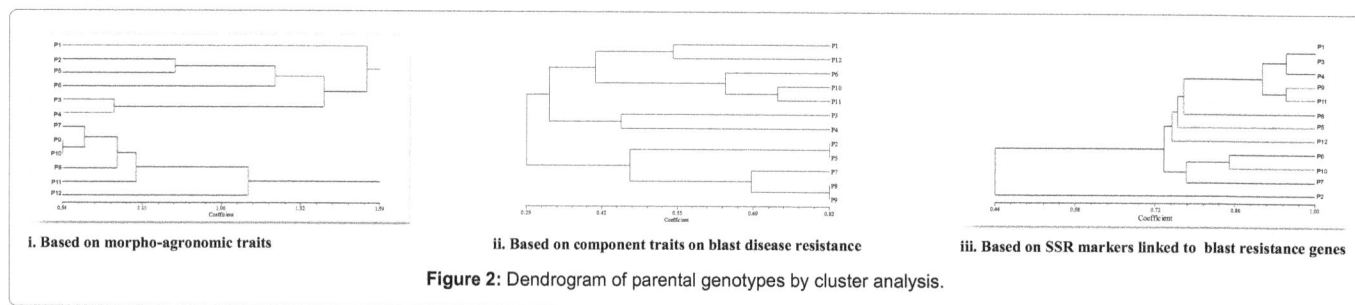

Figure 2: Dendrogram of parental genotypes by cluster analysis.

i. Based on morpho-agronomic traits   ii. Based on component traits on blast disease resistance   iii. Based on SSR markers linked to blast resistance genes

Heritability in per cent	Category
<30	Low
31 - 60	Medium
>60	High

GA per cent value	Category
<10%	Low
10-20%	Moderate
>20%	High

Genetic advance

Genetic advance was estimated by following formula [17] given as

Genetic advance = $k \times h^2 \times \sigma_p$

where,

h^2 = Heritability in broad sense

σ_p = Phenotypic standard deviation

k = Selection differential (2.06 at 5% selection intensity)

Genetic gain

Expected genetic gain under selection was calculated by the following formula [17] suggested as

$$\text{Genetic gain} = \frac{\text{Genetic advance}}{\text{Grand mean}} \times 100$$

Genetic gain was categorized as

Results and Discussion

Estimation of diversity among parental genotypes

The parents ADT 43 and CT13432-3R were grouped into different clusters based on blast resistance parameters, but came under similar cluster on morpho-agronomic trait (Table 2) information (Figure 2). The F_1s of these parents were selected for further advancement, so that recovery of the parental genotype was faster with not much compromise on local adaptability and acceptability. Among the parents, resistant donor CT13432-3R was having alleles of all the *Pi* genes and the susceptible lines were devoid of those alleles which confirmed the linkage of markers to the resistance genes. Many workers employed the technique of molecular analysis along with morpho-agronomic studies and their association in rice [19-21] for the simultaneous improvement of various traits.

Cluster analysis based on SSR marker data on parental genotypes revealed that four susceptible genotype (ADT 43, Improved White Ponni, BPT 5204 and CO39) and moderately resistant genotypes

(C101PKT, CO39 CIRAD, C101A51 and Tetep) were grouped into cluster I. Three other blast-resistant genotypes (Moroborekan, C101 LAC, and TORIDE1) were grouped into cluster II, and Cluster III had only one genotype, CT 13432-3R which is highly resistant with four pyramided *Pi* genes. The susceptible and resistant genotypes exhibited polymorphic alleles for *Pi* gene linked SSR markers. Efficiency of detecting blast-resistance gene depended on genotypes and gene-linked markers. These results indicated scope for utilization of blast-resistant genotypes with detected gene-linked markers in marker-assisted selection. CO 39 NILs were screened for blast resistance using SSR markers linked to 14 *Pi* genes and obtained similar results [15]. Advanced backcross population consisting of 80 BC₃F₃ lines derived from rice vars. Vandana/Moroborekan were genotyped for blast resistance with 50 candidate genes and 23 SSR markers and cluster analysis was carried out [22]. Similarly eight genotypes of rice were studied for genetic analysis with 85 gene linked SSR markers for rice blast resistance and relatedness among eight genotypes were examined through clustering analysis [23].

Estimation of variability parameters

The values of variability parameters like genotypic variance (GV), phenotypic variance (PV), genotypic coefficient of variation (GCV), phenotypic coefficient of variation (PCV), heritability (h^2) and genetic advance as per cent of mean for plant height, number of tillers, productive tillers, leaf length, leaf width, panicle length, days to first flowering, days to maturity, filled grains per panicle, total grains per panicle, spikelet fertility, 100 grain weight, single plant yield, lesion number, infected leaf area, potential disease incidence percentage, leaf blast and lesion type pertaining to different segregating populations of the cross ADT 43 × CT13432-3R are presented in Tables 3 and 4. The co-efficient of variations at genotypic (GCV per cent) and phenotypic level (PCV per cent) calculated for the morphological characters

observed in various segregating generations are presented in Table 4. In general, the PCV was higher than the corresponding GCV for all the traits under study. The percentage of GCV was less than 20 per cent (low variability) for plant height, leaf length, leaf width, panicle length, days to 50 per cent flowering days to maturity, spikelet fertility, 100 grain weight, single plant yield; 20 to 25 per cent (medium variability) for number of tillers and productive number of tillers; above 25 per cent (high variability) for grains per panicle, lesion number, infected leaf area, potential disease percentage, leaf blast and lesion type.

Selection of parents is one of the most important steps in any breeding program and will helps to extract good cultivars if the parents used in the program were suitable. Therefore, emphasis was given to choose appropriate parents in order to obtain useful segregants. Inclusion of at least one agronomically superior, locally adapted, popular cultivar as parent (ADT 43) in the breeding programme will largely help to ensure the recovery of a high proportion of progenies with adaptation and quality that were acceptable to farmers. The variability in parental genotypes was studied for different morpho-agronomic as well as resistance traits. Mean, standard deviation and variance values for all the traits under study were exhibited less difference in the segregating generations of the cross ADT 43 × CT13432-3R (F₂, F₃, BC₁F₁ and BC₂F₁). But in BC₃F₁ a reduction in mean plant height, leaf length, panicle length, total grain per panicle and 100 grain weight was observed. Similarly, there was an increase in single plant yield, number of tillers and productive tillers which indicate the maximum restoration of the desirable parent in the third backcross generation as the population mean was more towards the recurrent parent ADT 43 compared to previous generations.

In the segregating generations of the cross ADT 43/ CT13432-3R, co-efficient of variations, PCV and GCV were classified following the low, moderate and high scale [18]. In the segregating populations

		PH	NT	PRT	LL	LW	PL	DFF	DM	FGP	TGP	SPF	SPS	100GW	SPY	LN	ILA	PDI%	LB	LT
Mean	F₂	78.96	24.89	21.78	27.67	-	20.49	-	-	119.64	143.91	83.26	16.74	-	-	-	-	-	-	-
	F₃	79.94	15.31	11.29	26.06	0.89	22.28	89.71	114.71	128.64	143.16	90.72	10.82	2.10	25.00	31.60	45.82	78.52	7.07	6.51
	B₁	78.99	25.56	22.89	26.87	-	20.53	-	-	190.13	207.56	91.70	9.62	1.82	22.06	-	-	-	-	
	B₂	78.36	24.96	22.02	27.73	-	20.67	-	-	190.13	207.56	91.70	9.62	1.82	-	13.13	23.27	29.14	2.62	3.13
	B₃	63.02	27.78	27.67	25.78	0.90	18.09	95.18	120.18	120.84	136.64	88.22	11.78	1.57	46.25	-	-	-	-	-
SD	F₂	8.23	5.78	5.61	3.83	-	2.40	-	-	27.43	30.36	8.54	8.54	-	-	-	-	-	-	-
	F₃	8.74	3.97	3.27	4.26	0.19	2.46	1.89	1.89	42.73	51.57	6.36	8.60	0.29	10.51	11.70	31.48	24.89	2.24	2.30
	B₁	8.38	6.33	6.10	3.75	-	2.34	-	-	54.72	58.58	6.22	8.67	0.25	10.22	-	-	-	-	-
	B₂	8.33	5.80	5.50	3.80	-	2.41	-	-	54.72	58.58	6.22	8.67	0.25	-	10.05	21.40	19.29	1.74	2.11
	B₃	7.93	11.06	10.96	4.21	0.13	2.26	1.86	1.86	34.70	35.66	7.27	7.27	0.18	21.13	-	-	-	-	-
VAR	F₂	67.70	33.46	31.45	14.68	-	5.76	-	-	752.64	921.76	72.94	72.94	-	-	-	-	-	-	-
	F₃	76.40	15.76	10.66	18.17	0.03	6.07	3.57	3.57	1825.60	2659.54	40.45	74.00	0.08	110.43	136.79	991.10	619.53	5.02	5.30
	B₁	70.24	40.07	37.24	14.07	-	5.48	-	-	2993.98	3431.34	38.69	75.11	0.06	104.55	-	-	-	-	-
	B₂	69.42	33.68	30.25	14.43	-	5.82	-	-	2993.98	3431.34	38.69	75.11	0.06	-	101.07	458.11	371.99	3.01	4.44
	B₃	62.96	122.36	120.14	17.72	0.02	5.13	3.47	3.47	1204.18	1271.73	52.88	52.88	0.03	446.66	-	-	-	-	-
%CV	F₂	10.42	23.22	25.76	13.84		11.71	-	-	22.93	21.10	10.26	51.02	-	-	-	-	-	-	-
	F₃	10.93	25.93	28.96	16.35	21.35	11.04	2.11	1.65	33.22	36.02	7.01	79.48	13.81	42.04	37.03	68.70	31.70	31.68	35.33
	B₁	10.61	24.77	26.65	13.96		11.40	-	-	28.78	28.22	6.78	90.12	13.74	46.33	-	-	-	-	-
	B₂	10.63	23.24	24.98	13.70		11.66	-	-	28.78	28.22	6.78	90.12	13.74	-	76.54	91.96	66.20	66.41	67.41
	B₃	12.58	39.81	39.61	16.33	14.44	12.49	1.95	1.55	28.72	26.10	8.24	61.71	11.46	45.69	-	-	-	-	-

Table 3: Variability parameters in Segregating generations for different quantitative traits. Plant height (PH) in cm, number of tillers (NT), number of productive tillers (PRT), leaf length (LL) in cm, leaf width (LW) in mm, panicle length (PL) in cm, days to first flowering (DFF), days to maturity(DM), filled grains per panicle (FGP), total grains per panicle (TGP), spikelet fertility (SPF), spikelet sterility (SPS), 100 grain weight (100GW), single plant yield (SPY), lesion number(LN), infested leaf area (ILA), potential disease incidence percentage (PDI%), leaf blast (LB), lesion type (LT).

		PH	NT	PRT	LL	LW	PL	DFF	DM	FGP	TGP	SPF	SPS	100GW	SPY	LN	ILA	PDI%	LB	LT
PCV	F₂	10.42	23.24	25.75	13.85	-	11.71	-	-	22.93	21.10	10.26	51.03	-	-	-	-	-	-	-
	F₃	10.93	25.93	28.93	16.35	20.86	11.06	2.11	1.65	33.21	36.02	7.01	79.54	13.74	42.03	37.01	68.70	31.70	31.70	35.36
	B₁	10.61	24.77	26.66	13.96	-	11.40	-	-	28.78	28.22	6.78	90.05	13.95	46.35	-	-	-	-	-
	B₂	10.63	23.26	24.97	13.70	-	11.67	-	-	28.78	28.22	6.78	90.05	13.95	-	76.55	91.99	66.20	66.20	67.22
	B₃	12.59	39.82	39.62	16.33	14.08	12.52	1.96	1.55	28.72	26.10	8.24	61.71	11.13	45.69	-	-	-	-	-
GCV	F₂	10.22	21.90	24.30	9.86	-	4.49	-	-	22.69	20.06	8.99	44.75	-	-	-	-	-	-	-
	F₃	10.75	22.63	23.81	12.68	19.10	4.84	1.94	1.59	33.07	35.42	5.36	69.90	12.49	15.51	32.08	68.22	29.61	29.61	32.00
	B₁	10.41	23.58	25.40	9.73	-	3.69	-	-	28.70	27.86	5.10	79.31	12.29	13.71	-	-	-	-	-
	B₂	10.43	21.92	23.51	9.67	-	4.62	-	-	28.70	27.86	5.10	79.31	12.29		62.36	90.58	58.75	58.75	59.51
	B₃	12.33	39.21	39.05	12.56	11.37	2.59	1.79	1.49	28.53	25.17	6.80	50.94	8.09	40.52	-	-	-	-	-
h²	F₂	96.22	88.76	89.06	50.69	-	14.73	-	-	97.91	90.38	76.91	76.91	-	-	-	-	-	-	-
	F₃	96.65	76.15	67.74	60.15	83.87	19.15	84.33	93.28	99.14	96.67	58.36	77.24	82.58	13.62	75.14	98.59	87.25	87.25	81.89
	B₁	96.36	90.62	90.76	48.55	-	10.47	-	-	99.47	97.42	56.46	77.58	77.59	8.75	-	-	-	-	-
	B₂	96.31	88.84	88.63	49.82	-	15.64	-	-	99.47	97.42	56.46	77.58	77.59		66.36	96.94	78.76	78.76	78.36
	B₃	95.93	96.93	97.14	59.15	65.28	4.30	83.85	93.08	98.69	93.03	68.15	68.15	52.87	78.64	-	-	-	-	-
GA	F₂	2.51	10.58	2.50	1.71	-	0.45	-	-	4.02	5.71	3.21	3.21	-	-	-	-	-	-	-
	F₃	2.52	6.23	1.90	2.03	0.47	0.59	1.50	1.35	4.07	6.11	2.44	3.22	0.59	0.88	3.74	3.93	5.36	1.61	1.67
	B₁	2.51	11.82	2.55	1.64	-	0.32	-	-	4.08	6.16	2.36	3.24	0.55	0.56	-	-	-	-	-
	B₂	2.51	10.62	2.49	1.68	-	0.48	-	-	4.08	6.16	2.36	3.24	0.55		3.30	3.86	4.84	1.45	1.60
	B₃	2.50	22.09	2.73	2.00	0.37	0.13	1.49	1.34	4.05	5.88	2.84	2.84	0.38	5.06	-	-	-	-	-
GA(%)	F₂	3.18	42.50	11.47	6.19	-	2.20	-	-	3.36	3.97	3.85	19.18	-	-	-	-	-	-	-
	F₃	3.15	40.68	16.84	7.80	52.91	2.64	1.68	1.17	3.16	4.27	2.68	29.80	28.13	3.51	11.83	8.57	6.82	22.75	25.65
	B₁	3.18	46.24	11.12	6.11	-	1.56	-	-	2.15	2.97	2.57	33.64	30.44	2.55	-	-	-	-	-
	B₂	3.20	42.56	11.29	6.07	-	2.32	-	-	2.15	2.97	2.57	33.64	30.44		25.13	16.60	16.60	55.34	50.99
	B₃	3.97	79.51	9.85	7.75	40.78	0.73	1.57	1.12	3.35	4.30	3.22	24.13	23.99	10.95	-	-	-	-	-

Table 4: Heritability and GA in segregating generation for different quantitative traits. Plant height (PH) in cm, number of tillers (NT), number of productive tillers (PRT), leaf length (LL) in cm, leaf width (LW) in mm, panicle length (PL) in cm, days to first flowering (DFF), days to maturity(DM), filled grains per panicle (FGP), total grains per panicle (TGP), spikelet fertility (SPF), spikelet sterility (SPS), 100 grain weight (100GW), single plant yield (SPY), lesion number(LN), infested leaf area (ILA), potential disease incidence percentage (PDI%), leaf blast (LB), lesion type (LT).

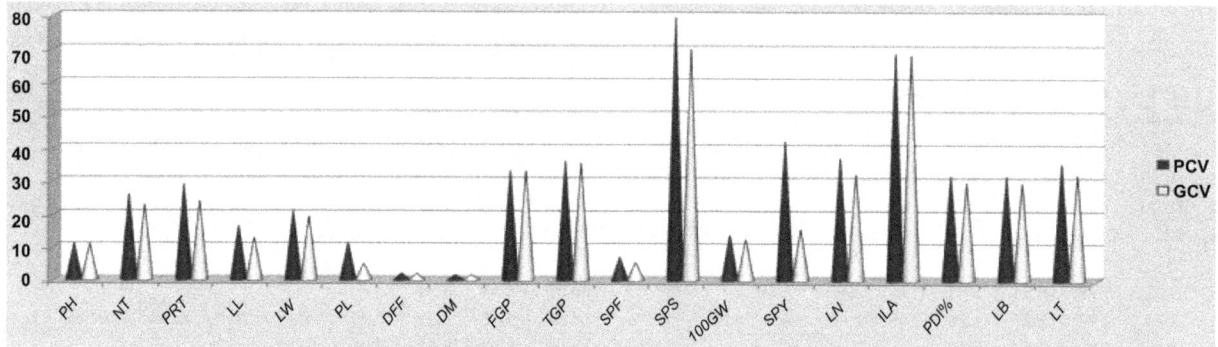

Figure 3: Genotypic and phenotypic components of variance for yield and resistance traits in BC3 generation.

variability for plant height, panicle length, spikelet fertility and 100 grain weight was low as it was evident from low GCV and high variability was exhibited by filled grains per panicle, spikelet sterility and disease parameters like lesion number infested leaf area, potential disease incidence per cent, leaf blast and lesion type. All other characters have shown an intermediate genotypic component of variation. The disease resistance traits showed high variability in segregating generations as the parents were selected based on the diversity in disease response. The high GCV gives an indication of justifiable variability among the genotypes with respect to these characters and therefore gives scope for improvement through selection. In general, the PCV was higher than the corresponding GCV for all the traits under study (Figure 3). The minor variation between values of GCV and PCV shows the limited role of environment in these characters and the heritability was very high for these traits. Selection for improvement of such characters will

be rewarding in this situation. Similar results were reported by Bisne et al. [24-27] in rice.

Estimation of heritability (h²) and genetic advance

Broad sense heritability was estimated for different traits and computed across five generations (Table 4). The trait such as plant height, number of tillers, days to 50 per cent flowering, number of grains per panicle, spikelet fertility per cent and 100 grain weight had high heritability (h²) values in all the segregating generations. The blast scoring traits viz., lesion number, infected leaf area, potential disease percentage, leaf blast and lesion type also recorded the high h² values of 75.14, 98.59, 87.25, 87.25 and 81.89 respectively, under stress condition. Other traits such as leaf length and leaf width have medium heritability and panicle length was having low heritability (14.73) values. Among those traits single plant yield recorded low h² values of 13.62 and

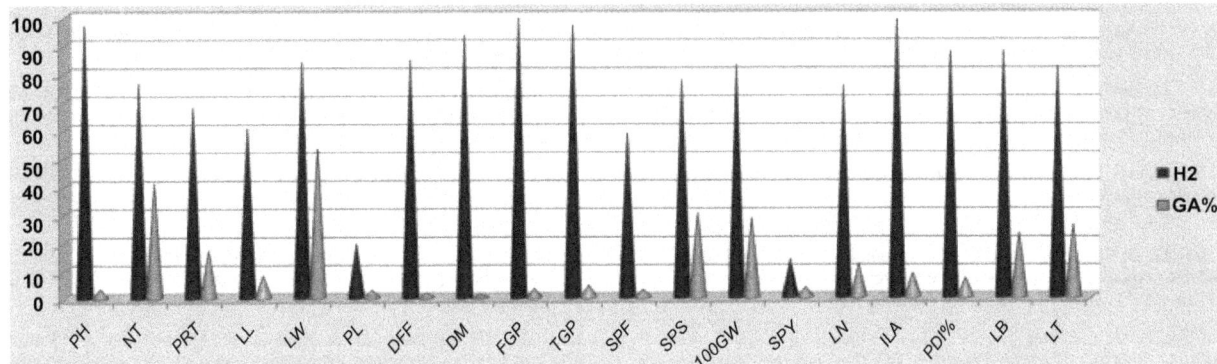

Figure 4: Heritability and genetic advance in percentage for yield and resistance traits in BC3 generation.

8.75 in F_3 and BC_1F_1 respectively but high heritability was recorded in (78.64) BC_3F_1 generation. Genetic advance was higher in case of number of tillers (10.58) and all other traits exhibited lower genetic advance value. Genetic gain or genetic advance per cent value was highest for number of tillers (42.50) and leaf width (52.91); medium value was exhibited by number of productive tillers (11.47), spikelet sterility (19.18), 100 grain weight (28.13), lesion number (11.83), leaf blast (22.75) and lesion type (25.65) and all other traits were recorded a lower genetic gain of less than 10 per cent. The heritability percentage was categorized [28]. The genetic advance as per cent of mean was categorized as low, moderate or high [17]. High heritability coupled with low genetic advance was found in all characters in the segregating generations except single plant yield and panicle length. This is the indication of predominance of epistasis and dominant gene action. The traits like number of tillers, spikelet sterility, 100 grain weight, and leaf blast and lesion type recorded simultaneous higher heritability and genetic gain compared to other traits and could be further improved through individual line selection (Figure 4). Frequency distribution curve of different agronomic yield and resistance traits were made in segregating generations with the normal distribution and given in the Figure 4. Similarly high heritability for most of the yield contributing traits was reported in rice [25,27,29,30].

Conclusion

Clustering of the parental genotypes on morphological traits on yield parameters grouped the major parents under study ADT 43 and CT13432-3R into same cluster but came under different clusters when grouped based on leaf blast score and molecular marker data generated using polymorphic SSR markers. In the segregating populations high variability was exhibited by filled grains per panicle, spikelet sterility and disease parameters like lesion number, infested leaf area, potential disease incidence percent, and leaf blast and lesion type. In general, the PCV was higher than the corresponding GCV for all the traits under study. Traits like number of tillers, spikelet sterility, 100 grain weight, leaf blast and lesion type recorded simultaneous higher heritability and genetic gain compared to other traits indicating the predominance of epistasis and dominant gene action.

Selection of genetically divergent parents and clear understanding of genetics of the selected traits have brought forth gene introgressed lines in adapted genetic background. Any marker assisted breeding programme has to be initiated with a detailed phenotypic and genotypic study of parental genotypes. It is essential to understand the behaviour of reported markers in the novel genetic background of the parents for a successful breeding programme for varietal development. It is also important to understand the genetics of the traits involved and their interactions in advanced segregating generations for the simultaneous improvement of various traits in an existing variety. In this study both the phenotypic and genotypic information was equally employed for the development of a high yielding blast resistant variety with better adaptability and acceptability.

Acknowledgments

We acknowledge Dr. D. Tharreau, CIRAD, France for providing pyramided donor line, CT13432-3R for this research work. First author thanks Indian council of agriculture research (ICAR) for Senior Research Fellowship awarded for her PhD work. We thank Centre for Plant Molecular Biology and Biotechnology, Centre for Plant Protection Studies (CPPS) and Hybrid rice Evaluation Centre, Gudalur, TNAU for providing facilities to support this work.

Conflict of Interest

Authors declare no conflict of Interest.

References

1. Koide Y, Kobayashi N, Xu D, Fukuta Y (2009) Resistance Genes and Selection DNA Markers for Blast Disease in Rice (*Oryza sativa* L.). Japan Agriculture Research 43: 255-280.

2. Koizumi S (2007) Durability of resistance to rice blast disease. JIRCAS Working Report 53: 1-10.

3. Sharma TR, Gupta SK, Vijayan J, Devanna BN, Ray S (2012) Review: Rice Blast Management Through Host-Plant Resistance: Retrospect and Prospects. Agriculture Research 1: 37-52.

4. Temnykh S, Park WD, Ayers N, Cartinhour S (2000) Mapping and genome organization of microsatellite sequences in rice (*Oryza sativa* L.). Theoretical Applied Genetics 100: 697-712.

5. Gao LZ, Zhang CH, Chang LP, Jia JZ, Qiu ZE, et al. (2005) Microsatellite diversity within *Oryza sativa* with emphasis on indica-japonica divergence. Genet Res 85: 1-14.

6. Zhang H, Sun J, Wang M, Liao D, Zeng Y, et al. (2007) Genetic structure and phylogeography of rice landraces in Yunnan, China, revealed by SSR. Genome 50: 72-83.

7. Thomson MJ, Polato NR, Prasetiyono J, Trijatmiko KR, Silitonga TS, et al. (2009) Genetic diversity of isolated populations of Indonesian landraces of rice (*Oryza sativa* L.) collected in East Kalimantan on the island of Borneo. Rice 2: 80-92.

8. Wang XQ, Kwon SW, Park YJ (2013) Evaluation of genetic diversity and linkage disequilibrium in Korean-bred rice varieties using SSR markers. Electronic Journal of Biotechnology 16: 1-20.

9. Brondani C, Caldeira KS, Borba TCO, Rangel PN, Morais OP, et al. (2006) Genetic variability analysis of elite upland rice genotypes with SSR markers. Crop Breeding and Applied Biotechnology 6: 9-17.

10. Moukoumbi YD, Sie M, Vodouhe R, Bonou W, Toulou B, et al. (2011) Screening of rice varieties for their weed competitiveness. African Journal of Agricultural Research 6: 5446-5456.

11. Bhadru D, Tirumala RV, Chandra MY, Bharathi D (2012) Genetic variability and diversity studies in yield and its component traits in rice (*Oryza Sativa L.*) Society for the Advancement of Breeding Researches in Asia and Oceania. Journal of Breeding and Genetics 44: 129-137.

12. Behera L, Mohanty S, Pradhan SK, Singh S, Singh ON, et al. (2013) Assessment of genetic diversity of rainfed lowland rice genotypes using microsatellite markers. Indian Journal of Genetics 73: 142-152.

13. Das B, Sengupta S, Parida SK, Roy B, Ghosh M, et al. (2013) Genetic diversity and population structure of rice landraces from Eastern and North Eastern States of India. BMC Genet 14: 71.

14. Divya B, Robin S, Rabindran R, Senthil S, Raveendran M, et al. (2014)Marker assisted backcross breeding approach to improve blast resistance in Indian rice (*Oryza sativa*) variety ADT43. Euphytica 200: 1-17.

15. Yanoria TMJ, Koide Y, Fukuta Y, Imbe T, Tsunematsu H, et al. (2011) A set of near-isogenic lines of Indica-type rice variety CO 39 as differential varieties for blast resistance. Molecular Breeding 27: 357-373.

16. Rohlf FJ (2002) NTSYS-pc: Numerical taxonomy and multivariate analysis system. Setauket, NY: Exeter Software.

17. Johanson HW, Robinson HF, Comstock RE (1955) Genotypic and genotypic correlations in soybean and their implications in selection. Agronomy Journal 47: 314-318.

18. Subramanian S, Menon MP (1973) Genotypic and phenotypic variability in rice. Madras Agricultural Journal 60: 1093-1096.

19. Kumar V, Sharma S, Sharma KA, Sharma S, Bhat KV (2009) Comparative analysis of diversity based on morphoagronomic traits and microsatellite markers in common bean. Euphytica 170: 249-262.

20. Selvaraj CI, Nagarajan P, Thiyagarajan K, Bharathi M, Rabindran R (2011) Genetic parameters of variability, correlation and path coefficient studies for grain yield and other yield Attributes among rice blast disease resistant genotypes of rice (*Oryza sativa L.*). African Journal of Biotechnology 10: 3322-3334.

21. Gowda SJM, Randhawa GJ, Bisht IS, Firke PK, Singh AK, et al. (2012) Morpho-agronomic and simple sequence repeat-based diversity in colored rice (*Oryza sativa L.*) germplasm from peninsular India. Genetics Resources and Crop Evolution 59: 179-189.

22. Wu JL, Sinha PK, Variar M, Zheng KL, Leach JE, et al. (2004) Association between molecular markers and blast resistance in an advanced backcross population of rice. Theor Appl Genet 108: 1024-1032.

23. GirijaRani MG, Adilakshmi D (2011) Genetic analysis of blast resistance in rice with simple sequence repeats (SSR). Journal of Crop Improvement 25: 232-238.

24. Bisne R, Sarawgi AK, Verulkar SB (2009) Study of heritability, genetic advance and variability for yield contributing characters in rice. Bangladesh Journal of Agriculture Research 34: 175-179.

25. Laxuman P, Salimath PM, Shashidhar HE, Mohankumar HD, Patil SS, et al. (2010) Analysis of genetics variability in interspecific backcross inbred lines in rice (*Oryza sativa L.*). Karnataka Journal of Agricultural Science 23: 563-565.

26. Anandrao SD, Singh CM, Suresh BG, Lavanya GR (2011) Evaluation of rice hybrids for yield and yield component characters under North East Plain Zone. The Allahabad Farmer 67: 63-68.

27. Prajapati M, Singh CM, Suresh BG, Lavanya GR, Jadhav P (2011) Genetic parameters for grain yield and its component characters in rice. Electronic Journal of Plant Breeding 2: 235-238.

28. Robinson HF, Comstock RE, Harvey PH (1949) Estimates of heritability and degree of dominance in Corn. Agronomy Journal 41: 353-359.

29. Seyoum M, Sentayehu A, Kassahun B (2012) Genetic variability, heritability, correlation coefficient and path analysis for yield and yield related traits in upland rice. Journal of plant science 7: 13-22.

30. Kumar A, Gupta S, Pandey A, Pattanayak A, Ngachan SV (2014) Studies on Aluminium Tolerance and Morphological Traits in Rice Lines from North Eastern India. Proc Natl Acad Sci India Sect B Biol Sci 86: 71-81.

PERMISSIONS

LIST OF CONTRIBUTORS

Hiroshi Ikehashi
Kyoto University, Fujisawa city, Japan

Grace Sharon Arul Selvi
Department of Genetics and Plant Breeding, University of Agricultural Sciences, Bangalore-65, India

Farhad Kahani
Marker Assisted Selection Laboratory, Genetics and Plant Breeding, University of Agricultural Sciences, GKVK, Bangalore 560065, India

Shailaja Hittalmani
University Head, Genetics and Plant Breeding, University of Agricultural Sciences, GKVK, Bangalore 560065, India

Maji AT, Odoba A, Gbanguba AU and Audu SD
National Cereals Research Institute Badeggi, Niger state. Nigeria

Bashir M
National Biotechnology Development Agency, Lugbe, Abuja, Nigeria

Ariful Islam MD
Dept of GPB, EXIM Bank Agricultural University, Bangladesh

Khaleque Mian MA and Golam Rasul
Department of GPB, Bangabandhu Sheikh Mujibur Rahman Agricultural University, Bangladesh

Khaliq QA
Department of Agronomy, Bangabandhu Sheikh Mujibur Rahman Agricultural University, Bangladesh

Mannan Akanda MA
Department of Plant Pathology, Bangabandhu Sheikh Mujibur Rahman Agricultural University, Bangladesh

Mawia Musyoki A, Wambua Kioko F, Agyirifo Daniel, Nyamai Wavinya D, Matheri Felix, Langat Chemtai R, Njagi Mwenda S, Arika Arika W, Gaichu Muthee D, Ngari Ngithi L and Ngugi Piero M
Department of Biochemistry and Biotechnology, School of Pure and Applied Sciences, Kenyatta University, P.O. Box 43844-00100, Nairobi, Kenya

Karau Muriira G
Molecular Biology Laboratory, Kenya Bureau of Standards, Nairobi, Kenya

Hedieh Jafari, Gadir Nouri-Ganbalani and Bahram Naseri
Department of Plant Protection, College of Agriculture science, University of Mohaghegh Ardebili, Ardebil, Iran

Arash Zibaee
Department of Plant Protection, Faculty of Agricultural Sciences, University of Guilan, Rasht, Iran

Wambua F Kioko, Musyoki A Mawia, Ngugi M Piero, Nyamai D Wavinya, Lagat R Chemutai, Matheri Felix, Arika W Makori and Njagi S Mwenda
Department of Biochemistry and Biotechnology, School of Pure and Applied Sciences, Kenyatta University, P.O. Box 43844-00100, Nairobi, Kenya

Karau G Muriira
Molecular Biology Laboratory, Kenya Bureau of standards, P.O. Box 54974-00200, Nairobi, Kenya

Fatema-Tuj-Johora
Department of Crop Botany, EXIM Bank Agricultural University, Bangladesh

Jalal Uddin Ahmed
Dept of Crop Botany, BSMRAU, Gazipur1706, Bangladesh

Morshed M
Senior Scientist, IRRI, Philippines

Mian MAK
Department of Genetics and Plant Breeding, BSMRAU, Gazipur1706, Bangladesh

Ariful Islam M
Department of GPB, EXIM Bank Agricultural University, Bangladesh

Grace Sharon Arul Selvi, Shailaja Hittalmani and Uday G
Department of Genetics and Plant Breeding, University of Agricultural Sciences, Bangalore-65, India

Ariful Islam
Department of GPB, EXIM Bank Agricultural University, Bangladesh

Mian MAK and Rasul G
Department of Genetics and Plant Breeding, BSMRAU, Gazipur1706, Bangladesh

Bashar K
International Potato Research Centre (CIP), Bangladesh

Fatema-Tuj-Johora
Department of Crop Botany, EXIM Bank Agricultural University, Bangladesh

Gbanguba AU
National Cereals Research Institute, Badeggi. P. M. B. 8. Bida, Niger State. Nigeria

Kolo MGM and Gana AS
Department of Crop Production, Federal University of Technology, Minna, Niger State, Nigeria

Odofin AJ
Department of Soil Science, Federal University of Technology, Minna, Niger State, Nigeria

Shajedur Hossain
Supreme Seed Company Limited, Mymenshing, Bangladesh

Maksudul Haque MD
Plant Breeding Division, Bangladesh Rice Research Institute, Gazipur 1701, Bangladesh

Jamilur Rahman
Department of Genetics & Plant Breeding, Sher-e-Bangla Agricultural University, Dhaka 1207, Bangladesh

Yang-Seok Lee and Gynheung An
Crop Biotech Institute & Graduate School of Biotechnology, Kyung Hee University, Yongin 446-701, Korea

Saroj Kumar Sah, Amandeep Kaur, Gurwinder Kaur and Gurvinder Singh Cheema
School of Agricultural Biotechnology, Punjab Agricultural University, Ludhiana 141 004, Punjab, India

Farhad Kahani and Shailaja Hittalmani
Marker Assisted Selection Laboratory, Genetics and Plant Breeding, University of Agricultural Sciences, GKVK, Bangalore 560065, India

Professor and University Head, Genetics and Plant Breeding, University of Agricultural Sciences, GKVK, Bangalore-560065, India

Adyati Putriekasari Handayani, Roselina Karim and Kharidah Muhammad
Universiti Putra Malaysia, Serdang 43400, Selangor Darul Ehsan, Malaysia

Komol Singha and Sneha Mishra
Department of Economics, Sikkim University, Gangtok, India

U Ismaila
National Cereals Research Institute, Badeggi, P.M.B. 08, Bida, Nigeria

MGM Kolo, JA Odofin and AS Gana
Federal University of Technology P.M.B. 65, Minna, Nigeria

Emily Gichuh
Graduate School of Environmental and Life Sciences, Okayama University, Japan

Eiko Himi and Masahiko Maekawa
Institute of Plant Science and Resources, Okayama University, Japan

Hidekazu Takahashi
Bioresource Sciences, Akita Prefectural University, Japan

NSBM Atapattu, KP Wickramasinghe and Thaksala serasinghe
Faculty of Agriculture, University of Ruhuna, Sri Lanka

SP Gunarathne
Faculty of Veterinary Science, University of Peradeniya, Sri Lanka

Arash Zibaee
Department of Plant Protection, Faculty of Agricultural Sciences, University of Guilan, Rasht, Iran

Harry Luiz Pilz Júnior, Neiva Knaak and Lidia Mariana Fiuza
Microbiology and Toxicology Laboratory; PPG Biology / University Valley River Unisinos Bells, São Leopoldo-RS, Brazil

Denize Righetto Ziegler and Renata Cristina Ramos
Technological Institute of Food for Health - ITT NUTRIFOR, São Leopoldo-RS, Brazil

Jeremias Pakulski Panizzon
Microbiology and Toxicology Laboratory; PPG Biology / University Valley River Unisinos Bells, São Leopoldo-RS, Brazil

Technological Institute of Food for Health - ITT NUTRIFOR, São Leopoldo-RS, Brazil

Takayoshi Mamiya
of Chemical Pharmacology, Faculty of Pharmacy, Meijo University, Japan

Keiko Morikawa and Mitsuo Kise
FANCL Institute, FANCL Corporation, Yokohama, Japan

Mafimisebi TE and Agunbiade BO
Department of Agricultural & Resource Economics, School of Agriculture & Agricultural Technology, Nigeria

Mafimisebi OE
Department of Agricultural Technology, Rufus Giwa Polytechnic, Nigeria

Saroj Kumar Sah, Ajinder Kaur and Jagdeep Singh Sandhu
School of Agricultural Biotechnology, Punjab Agricultural University, India

Sho Takano, Shuichi Matsuda, and Kiyoaki Kato
Department of Agro-Environmental Science, Obihiro University of Agriculture and Veterinary Medicine, Nishi 2-11 Inada, Obihiro, Hokkaido, 080-8555, Japan

Yuji Hirayama and Takashi Sato
Department of Agro-Environmental Science, Obihiro University of Agriculture and Veterinary Medicine, Nishi 2-11 Inada, Obihiro, Hokkaido, 080-8555, Japan

Itsuro Takamure
Graduate School of Agriculture, Hokkaido University, Kita 9 Nishi 9, Kita-ku, Sapporo, Hokkaido 060-811, Japan

Loth Mulungu S
Pest Management Centre, Sokoine University of Agriculture, PO Box 3110 Morogoro, Tanzania

Happy Lopa
Rodent Control Centre, Ministry of Agriculture, Food Security and Cooperatives, P.O. Box 3047 Morogoro, Tanzania

Mashaka Mdangi E
MATI-Ilonga, P.O. Box 66, Kilosa, Tanzania
Crop Science and Production, Sokoine University of Agriculture, P.O. Box 3005, Morogoro, Tanzania

Anowara Akter, Jamil Hassan M, Umma Kulsum M, Islam MR and Kamal Hossain
Plant Breeding Division, Bangladesh Rice Research Institute, Bangladesh

Mamunur Rahman M
Senior Scientific Officer, Farm Management Division, Bangladesh

Francisco Placido Nogueira Arcanjo and Paulo Roberto Santos
Federal University of Ceara, Brazil

Sergio Duarte Segall
Federal University of Minas Gerais, Brazil

Hongyu Zhang, Yutong Liu, Mipeng Han, Limei Wu, Zhijian Liu, Xiaotong Chen, Peizhou Xu and Xianjun Wu
Rice Research Institute, Sichuan Agricultural University, Wenjiang, Chengdu, P.R.China These authors are equal contribute to this work

Shunsuke Matsumoto, Atsushi Maruyama, Masao Kikuchi and Michiko Takagaki
Graduate School of Horticulture, Chiba University, Chiba, Japan

Tatsushi Tsuboi
JICA Expert, JICA Uganda Office, Kampala, Uganda

Godfrey Asea
National Crops Resources Research Institute, Kampala, Uganda

Divya Balakrishnan
National Professor Project, Crop Improvement Section, Directorate of Rice Research-ICAR, Hyderabad, India

Robin S, Rabindran R and John Joel A
Department of Rice, Centre for Plant Breeding and Genetics, Tamil Nadu Agricultural University, India

Index